Scorpion Biology and Research

SCORPION BIOLOGY AND RESEARCH

Edited by

PHILIP BROWNELL

GARY POLIS

OXFORD
UNIVERSITY PRESS

2001

OXFORD
UNIVERSITY PRESS

Oxford New York
Athens Auckland Bangkok Bogotá Buenos Aires Calcutta
Cape Town Chennai Dar es Salaam Delhi Florence Hong Kong Istanbul
Karachi Kuala Lumpur Madrid Melbourne Mexico City Mumbai
Nairobi Paris São Paulo Shanghai Singapore Taipei Tokyo Toronto Warsaw

and associated companies in
Berlin Ibadan

Published by Oxford University Press, Inc.
198 Madison Avenue, New York, New York 10016

Library of Congress Cataloging-in-Publication Data
Scorpion biology and research / edited by
Philip Brownell and Gary Polis.
p. cm.
Includes bibliographical references and index.
ISBN 0-19-508434-9
1. Scorpions. 2. Scorpions—Research.
I. Brownell, Philip. II. Polis, Gary A. 1946–2000.
QL458.7.S26 1999
595.4′6—dc21 98-37237

9 7 6 5 4 3 2 1

Printed in the United States of America
on acid free paper

We dedicate this book to Roger Farley, our graduate mentor and unfailing supporter through the years of discovering what this wonderful animal could offer to biology. It is a mark of Roger's patience and intellectual breadth that he so ably guided two students with such different ways of seeing the same thing.

Special recognition is due Victor Fet for his passionate dedication to all things scorpion, and for his unhesitating offers to stand in when editorial duties went beyond us.

Finally, we are forever indebted to our families, especially Kathy and Sharon, for allowing our inspirations to grow into this reality.

In Memoriam

In the final weeks of writing this book, Gary Polis died during a research expedition in the Sea of Cortez. Gary and the four people who perished with him—Michael Rose, Takuya Abe, Masahiko Higashi, and Shigeru Nakano— were doing what they loved best, but the loss to those of us who knew them as family, friends, students, and colleagues remains inexpressible. May these pages, each so much affected by their lives, inspire those who would continue their worthy adventure.

PREFACE

This volume grew from the common interest of its authors to display, by examples from their current research, the interesting biology of scorpions and their exceptional qualities for basic, interdisciplinary study. For those taking up the study of scorpions for the first time, it will seem paradoxical that these organisms have been so long neglected as subjects for research. In a phylum defined mostly through study of insects, scorpions remain something of an oddity even though they are among the most primitive, successful, and potentially dangerous of all arthropods. For these reasons, and a host of others contained within these chapters, it hardly seems fitting that organismal biology for the twentieth century should close its books before reconsidering one of its most curious and compelling subjects.

Our intent as authors and editors is nothing less than to propose a new animal model for organismal biology, a chelicerate arthropod to rival the best of the many mandibulate arthropods already under intensive investigation. Twenty biologists from varying disciplines have written the 18 chapters here. While each contribution stands alone as an interesting example of basic science, the combined works emphasize the scorpion's role as a bridge for movement of information between traditionally separated fields of study. Thus, knowledge of scorpion ecology and natural history is meaningful to those analyzing the group's biogeographic distribution, venom evolution, or unusual sensory capabilities; knowledge of the scorpion fossil record, especially during their transition from water to land, informs the analysis of body plan evolution and the peculiarities of in utero embryonic development. Additionally, because scorpions have strong affinities to other, possibly derived, orders of arachnida (e.g., the amblypygids, uropygids, and solifugae), a thorough study of the Scorpionida sheds light on the biology of this class at large. What we most hope to impress upon the reader is the central importance of scorpions in the study of other arthropods and their potential as models for broad-ranging and integrative studies in biology as a whole.

In the writing and editing of this work, two phantom readers have been nitpicking at our side. The first we imagined to be an interested layperson or biologist, someone seeking a deeper appreciation of animals and the processes that shape their physiology, behavior, and evolution. For these readers we made clear the rationale for each line of investigation, with explanations of why and how the work was conducted. It is our hope that beginning students of biology will discover the same excitement that we have for these animals and consider the opportunities they contain for future research. The second of our phantom readers was a practicing scientist or teacher of biology specializing in one of the several disciplines considered in this book. Our intent was to provide advanced readers with a complete and current text on the biology of scorpions, one usable in the classroom, laboratory, or field for reference to current information, techniques, and literature. For this reader, it was important to include some details of recent findings and to relate these to transcendent problems under investigation in biology. Finally, for all readers, the editors carefully reviewed the text and illustrations to ensure clarity, accuracy, and efficiency of communication with a minimum of technical jargon. We hope these efforts will make the book accessible to the broadest possible audience, all of whom we expect to fascinate.

Contents

Contributors

Tim Benton
Institute of Biological Sciences
University of Stirling
Stirling FK9 4LA, UK

Philip Brownell
Department of Zoology
Oregon State University
Corvallis, Oregon 97331 USA

Denise Due
Department of Biology
Vanderbilt University
Nashville, Tennessee 37235 USA

Roger Farley
Department of Biology
University of California
Riverside, California 92521 USA

Victor Fet
Department of Biological Sciences
Marshall University
Huntington, West Virginia 25755 USA

Gerta Fleissner
Zoologisches Institut
Universitat Frankfurt
Frankfurt/Main 60054 Germany

Günther Fleissner
Zoologisches Institut
Universitat Frankfurt
Frankfurt/Main 60054 Germany

Douglas Gaffin
Department of Zoology
University of Oklahoma
Norman, Oklahoma 73019 USA

Max Goyffon
L.E.R.A.I.
Museum National d'Histoire Naturelle
Paris 75005 France

Bruce K. Hammock
Department of Entomology
University of California
Davis, California 95616 USA

Andrew J. Jeram
Department of Geology
Ulster Museum
Belfast BT9 5AB
Northern Ireland, UK

N. Adam Locket
Department of Anatomy and Histology
University of Adelaide
South Australia 5005

Erwann Loret
Centre National de la Recherche Scientifique
Institut de Biologie Structurale et Microbiologie
13402 Marseille Cedex 20, France

Wilson Lourenço
Laboratoire de Zoologie (Arthropods)
Muséum d'Histoire Naturelle
61, rue de Buffon 75005
Paris, France

Dieter Mahsberg
Biozentrum der Universität,
Zoologie III
Universitat Würzburg Am Hubland
Würzburg D-97074 Germany

Christopher Myers
School of Interdisciplinary Studies
Miami University
Oxford, Ohio 45056 USA

Gary Polis
Department of Environmental Science and Policy
University of California
Davis, California 95616 USA

Vincent Roman
L.E.R.A.I.
Museum National d'Histoire Naturelle
Paris F-75005 France

Robert M.S. Schofield
Department of Physics
University of Oregon
Eugene, Oregon 97403 USA

Tsunemi Yamashita
Department of Biology
Northern Louisiana University
Monroe, Louisiana 71209 USA

Scorpion Biology and Research

1

Introduction

PHILIP BROWNELL
GARY POLIS

Scorpions are chelicerate arthropods, members of the class Arachnida and distant relatives to other arthropods such as the crustacea, insects, myriapods, and onychophora. Compared to spiders and mites, scorpions are a modest group containing 1259 described species in 16 living families and 155 genera (Fet et al., 2000). Scorpions are common animals in temperate, desert, and tropical habitats of the world, but they are generally nocturnal, obscure, and inactive. As arthropods go, scorpions rank among the most ancestral in origin and primitive in body form, having first appeared as aquatic organisms in the Silurian (450–425 MYA) and having changed little since. In those times large landmasses were emerging from the oceans, forming the shallow estuaries and terrestrial habitats that scorpions and insects would soon colonize. The dominant aquatic predators then were other chelicerates, the eurypterids ("giant water scorpions," now all extinct), and xiphosurans, surviving today as the horseshoe crabs. The first scorpions were aquatic too and were likely descendants of the eurypterids. The order Scorpiones quickly radiated into more than 50 families, an indication of their early and great success. But only one or a few of these lines survived and radiated into the present, and all of them now live on land. So it was into this niche—as terrestrial predators of other arthropods—that scorpions escaped from extinction in the sea and attained their current success. Currently some of them occupy the most extreme thermal and xeric environments on Earth, often holding a major portion of total animal biomass in these environments. From their beginnings, then, scorpions have had a significant history in the animal kingdom, one that remains both fascinating and full of questions to explore.

For most of this century the study of scorpion biology has been a phenomenological enterprise aimed primarily at descriptive anatomy, taxonomy, and biogeography, accompanied by detached medical research on venom biochemistry. In the late 1970s basic research on the group expanded sharply to encompass physiology, behavior, ecology, and evolutionary biology. It was these achievements that prompted the first integrative summary of scorpion biology (Polis, 1990). An important outcome of these studies was the recognition of the scorpion's utility as an arachnid model to investigate broader questions of organismal biology. Moreover, scorpions displayed several novel traits previously unstudied or absent in other animals. The current volume is both a compilation of this new information and an acknowledgment of the scorpion's emerging importance as a vehicle for basic research.

SCORPION BIOLOGY AND RESEARCH

Paleontology, Systematics, and Biogeography

Scorpions are quintessentially arthropod and thus carry special value for comparative studies of this tremendously diverse assemblage of organisms. The scorpion body is highly segmented with only partial condensation, or tagmazation, of serial segments into discrete body regions. The anterior tagma, or prosoma, has six pairs of postoral appendages—the chelicera, chelate pedipalps, and four pairs of walking legs covered dorsally by the cephalothoracic shield. The posterior opisthosoma is further subdivided into the mesosoma (or abdomen), with seven segments containing genital structures, pectinal appendages, and four pairs of book lungs, and the metasoma (or the postabdominal "tail"), with five segments lacking appendages but terminating in the anus and sting. The leg appendages are also minimally fused, with eight podomeres including two terminal tarsal segments. This basic body plan first evolved in the Silurian and remains unchanged today.

By comparison, other arachnids appear more derived in structure. By the end of the Devonian (~350 MYA), the arachnid lineage had differentiated broadly into at least 10 orders. Most of these animals displayed more condensed and derived features suited to their specialized ecologies and behaviors. Average body size and body segmentation was reduced in these groups (e.g., Palpigradi, Pseudoscorpiones, Ricinulei, Acari), and their appendages showed fewer segments and became specialized for particular functions. As measured by species diversity, the spiders (Araneae, >50,000 species described) and the mites and ticks (Acari, >40,000 species described; this number is estimated to be less than 10% of the total) are the most successful arachnids; their modern forms epitomize the evolutionary trend for arachnids. In many respects, arachnids exploited the new terrestrial environment in much the same way as the insects, converging on similar microhabitats through miniaturization, structural specialization, and physiological adaptation. How, then, did scorpions manage their success as land animals with such minimal change in external form?

Recent scorpions are all terrestrial and probably constitute a monophyletic taxon. Beginning about 400 MYA in the late Silurian, scorpions were among the first immigrants to estuarine, and then to land, environments, finally consolidating their dramatic shift in biology by the late Carboniferous or Permian. As reasoned by Jeram (see chapter 16), scorpions are especially valuable as models to study the water-to-land transformation because they are the only terrestrial arthropods with unequivocally identifiable aquatic ancestors in the same order. Moreover, the transformations of structure were well preserved in the fossil record. Evidently, terrestrialization required unexpectedly minor changes in key organ systems, most notably in the legs, which were thickened to accept greater weight, the abdominal gills, which were modified to become the book lungs, and the "mouth" (more accurately, a preoral chamber), which was formed by expansion of leg coxae to create a space for external digestion and liquid feeding. Fortunately, scorpion cuticle has an unusual propensity to retain detailed structure during fossilization, more so than any other terrestrial arthropod. Because the behavior and ecological milieu of arthropods are most strongly revealed in the structure of their exoskeletons, the scorpion fossil record now provides an unprecedented chronicle of the *biology* of terrestrialization.

The phylogenetic relationships of extant scorpions remain poorly resolved, even though studies of their taxonomy and systematics outnumber all other works published. Several plesiomorphic (ancestral) characteristics define the group as a whole, but scorpions have few apomorphic (derived) characters useful for taxonomic morphometry (in contrast, for example, to spiders and mites). Shape of the metasoma, location of bristles and trichobothria on appendages, and the number of pectinal teeth are among the few characters used to distinguish species. The near absence of apomorphic morphological characteristics—and therefore, of a confident taxonomy based on morphology—may relate to the scorpion's conservative ecological status as a generalized predator. In chapter 15, Yamashita and Fet describe how molecular techniques can be applied to resolve species and populational variation and thus to overcome these long-standing problems in scorpion taxonomy. With such new techniques, the great potential of scorpions for high-resolution biogeographical studies can be realized.

Scorpions are common animals of cosmopolitan distribution, having radiated broadly throughout Pangea some 300–250 MYA. The number of genera inhabiting desert environments is largest, with montane and tropical genera following in that order. Modern species rarely show relictal distributions, and in regions showing peak species diversity, like Baja California, it is possible to find from 5 to as many as 15 species living sympatrically within a 1-km^2 area. This provides a rich natural diversity of animals from temperate to tropical habits for comparative analysis of their ecology and biogeography.

Widespread as they are throughout the world, scorpions are not good dispersers on a geographical scale. This is attributable to their often strict dependence on particular microhabitats (e.g., sandy or rocky substrates), inability to swim, and poor phoretic ability. Their relative immobility simplifies the analysis of biogeographic distribution through time, as exemplified by the studies of Due (chapter 14). Likewise, Lourenço (chapter 18) shows how knowledge of scorpion biogeographical distribution in rainforests of South America can be of practical value to identify areas of high species diversity and endemism for preservation as national parks. Previous decisions were based exclusively on distributions of vertebrate and plant species. Scorpions now offer a valuable invertebrate alternative for comparison and confirmation.

Population Biology and Ecology

The same life-history characteristics that make scorpions interesting exceptions to the arthropod norm also favor their use as models for research conducted in the field. Most arthropods are opportunistic animals, *r* strategists that rely on fast reproduction and growth of numerous offspring to maximize resource assimilation in fluctuating environments. Scorpions, by contrast, have life-history characteristics more like those of vertebrates (i.e., they are *K* strategists). They tend to live in relatively stable, if harsh, environments where resources are predictable but limiting to population growth rate and carrying capacity. Their reproduction is slow and iteroparous, with internal development of embryos and relatively few but quite large offspring. Young scorpions pass through as many as six immature stages, reaching reproductive maturity only after one to several years. Mature scorpions can live 2 to 10 years, with some species living as many as 25 years, so their presence in the landscape is long and substantial.

A significant determinant of natural history for most scorpions is their great size. There are no minute forms less than 3–5 mm in length, and some species attain a 10–30 g weight and a 10–20 cm length. By these measures, scorpions are easily among the largest of land invertebrates, many even larger than one-quarter of terrestrial vertebrates. As described by Myers (chapter 13), body size has broad influences on an organism's biology—for example, its metabolism, fecundity, life span, and ecological status among competitors and predators—so natural selection should act strongly to fix size for particular environments. However, some scorpion species have adult body size that varies as a cline over their geographic range. These species should be good experimental systems to analyze the genetic and natural selective processes that maintain various body size and clinal variation within populations.

Ecologically, scorpions are hardly the passive and inconsequential participants in their communities they were once believed to be. In deserts, for example, population densities approach 5000–10,000 animals per hectare, a sizable fragment of total animal biomass exceeded only by ants and termites. At these densities one investigator can collect several hundred animals in an evening using portable black lights to locate them (see below). Such abundances facilitate efficient collection of large data sets in the field. It is possible, for example, to census the prey captured by hundreds of individual animals. Polis (chapter 12) has used these attributes to make accurate measures of feeding phenology, survivorship, age-specific ecology, population biomass, the effects of cannibalism on population regulation, and predator–prey interactions. Such detailed information applies broadly to problems in ecological research.

Scorpions are relatively sedentary animals, especially those inhabiting burrows and foraging in a small home range. As such, individuals within a population can be marked and observed repetitively for longer-term studies in the field. One disadvantage for observers of field behavior is that scorpions tend to be surface "time minimizers," spending most of their time in burrows or under rocks, away from predators and invisible to interested researchers. Imaginative approaches to observing their subsurface activities would elucidate a significant and wholly unknown aspect of scorpion natural history.

Reproduction and Development

Scorpion life-history traits are also reflected in the subtle complexities of their reproductive behaviors. Insemination occurs by indirect transfer of sperm: the male deposits a stalked spermatophore on the substrate and the female subsequently fertilizes herself on it with guidance from the male. These mechanics are simple by arthropod standards, but the courtship "dance" that precedes insemination is an elaborate and evolved process. Mate selection is often the driving force behind increased complexity of courtship behaviors, and this seems to be the case for scorpions. The male risks cannibalization by his larger partner, and the female must make a substantial investment in rearing the few offspring (15–30 for most species) she can produce that year. Under these circumstances, mate choice has increased importance and must be sorted out by sexual selection during courtship.

As described by Benton (chapter 11), scorpion reproductive behavior can be easily studied at different levels. To analyze the proximate (physiological or behavioral) mechanisms of courtship, the investigator can record and quantify several discrete behavioral components (i.e., juddering, pectine sweeps, *promenade aux deux,* headstands, sexual sting, male escape) and begin to determine the chemical and tactile cues regulating timing, expression, and consequence of each. For analysis of the ultimate (evolutionary) forces shaping the behavior, courtship can be studied in a natural ecological setting where, for example, the quality of potential mates or other aspects of courtship are manipulated experimentally. In this way, scorpions can provide a model of reproductive behavior sufficiently complex to be of interest and accessible enough to be analyzed through experimentation.

The time from fertilization to birth can be several months to years (some species store sperm). Most arachnids lay eggs, but scorpion embryos develop internally, attached to the mother's uterine wall where the progress of embryogenesis can be monitored by regular sampling (see Farley, chapter 2). Scorpion embryos make excellent preparations for morphological study by scanning electron microscopy, and Farley has used this technique to describe in detail the morphological transformations of segmental appendages during development. Appendage-related structures, such as the abdominal book lungs and preoral tube, are of particular interest because these organs changed dramatically during the evolution of a terrestrial life history. These evolutionary processes appear to be recapitulated during development and thus may aid in the interpretation of an incomplete fossil record (see Jeram, chapter 16).

The incompletely differentiated young are born live and crawl to the mother's back, where they remain through the first molt. In most species, the second instars disperse to take up independent lives, but in some species, parental care of the brood is extended to subsequent molts. Under conditions of extended contact, offspring and parent begin to forage cooperatively as they defend a common burrow, thus setting the foundation for sociality. Mahsberg (chapter 10) argues that protracted parent–offspring interaction of scorpions in one locality (i.e., brood care in a burrow) is a precedent of full sociality, defined as the permanent cooperation of communicating conspecifics. Thus, behavioral studies of gregarious scorpions may help identify the ecological and physiological processes supporting social behavior.

Physiology and Behavior

Scorpions reproduce and grow slowly to maturity. To reach maturity they must succeed as predators eating a wide variety of prey while avoiding predators and tolerating the physiological and metabolic stresses imposed by their often harsh environments. Evolutionary success under these circumstances implies the existence of finely tuned adaptations, both to maintain physiological homeostasis and to time appropriate behaviors. Thus, for example, a scorpion living in sandy deserts must be adapted to endure extremes of heat and water stresses by day while functioning as a sensitive and nimble predator of flying, running, and burrowing insects by night. Metabolically, the respiratory organs (book lungs) are sufficient to sustain short bursts of activity (i.e., prey capture from ambush) but prohibit sustained pursuit of food as observed in tracheate arachnids (e.g., solpugids) and insects. To their advantage, most scorpions show singularly low rates of metabolism at rest—approximately 25% of the basal metabolic rates of insects of comparable size and the lowest rate among all arthropods except ticks. This unique capacity to conserve metabolic energy also conserves water, an important advantage for desert scorpions. Scorpion cuticle has very low permeability to water because of saturated, long-chain (>30 carbons) hydrocarbon waxes secreted into the epicuticular layer (Hadley, 1994). Respiratory water loss is minimized by the infrequent need to open the spiracles and book lungs for gas exchange at rest. Furthermore, because scorpions are essentially fluid feeders, little water is lost in processing digestive wastes. These metabolic, respiratory, and digestive adaptations combine to make desert scorpions among the most water-efficient animals on the planet and certainly favor their success in hot, dry environments.

Scorpions display a relatively limited repertoire of stereotypical behaviors, and their nervous systems are tractable enough to encourage correlative neurobiological investigations. Neuroethological studies usually require a detailed knowledge of nervous system structure, however, and this is poorly known for scorpions beyond histology. As with other arachnids, the sensory and motor functions of prosomal appendages (i.e., chelicerae, pedipalps, legs, eyes, pectines) are controlled by a complex fusion of neuromeres, called the cephalothoracic nerve mass or "brain" of the animal (see Farley, chapter 2). A ventral, ganglionated nerve cord innervates the abdominal and postabdominal organs (i.e., book lungs, postural musculature, poison gland). The nerve cells of these structures and of peripheral sensory organs (i.e., eyes, pectines, setaform mechanoreceptors) are comparable in size and organization to those of insects, and there is no indication that their cellular functions are distinctive.

A special advantage of scorpions as models for neuroethological studies stems from their sensory specializations for nocturnal life. Scorpions have been most successful as night-active predators of insects, and they display a remarkable array of nonvisual sensory adaptations for orientation to prey in the dark (Brownell, chapter 6). It is noteworthy that most of what we know about sensory systems and behaviors of animals is biased by information from the diurnal, usually visual, species that investigators typically study. Scorpions offer a valuable nocturnal alternative for analysis, especially of mechanical senses (touch, substrate vibration, air currents) to which they are exquisitely responsive. This is not to say that scorpions lack photoreception or vision: As Locket shows here (chapter 3), the small lateral eyes (two to five pairs) lack good image-forming lenses, but they rank among the most sensitive photodetectors known (Fleissner and Fleissner, chapters 4 and 5). The single pair of medial eyes have very good lenses with excellent light-gathering ability, raising the possibility that scorpions see low-resolution images under very low (i.e., starlight) illumination.

Another sensory peculiarity of scorpions is their ability to detect chemical signals on the substrate, especially pheromones involved in mating. Gaffin and Brownell (chapter 7) demonstrate that the unique pectinal organs function similarly to the antennae of insects and crustaceans and are among the most elaborate chemosensory organs of any arthropod. The pectines have highly favorable properties for electrophysiological analysis and should give new insights into the mechanisms of chemosensory behavior.

Scorpions, like all animals, live in environments where conditions fluctuate: temperature, light, availability of food, and predator density all vary in more or less predictable patterns to which activity of the animal must be adjusted for optimal functioning. The strongest of these cycles is circadian, and it is imposed on all cells and tissues by an internal clock. As nocturnal predators typically living in burrows by day, scorpions have evolved one of the sharpest and most accurate photoperiods described for any organism. As described by Fleissner and Fleissner in chapters 4 and 5, the sensitivity and precision of the scorpion's internal clock is sufficiently great that the subtler aspects of circadian regulation, such as feedback effects from locomotor activity, were first accurately revealed and quantified in this animal. Moreover, the neuronal circuitry underlying circadian rhythmicity is being described in exceptional detail, down to the level of neurotransmitters mediating regulation. These informative and comprehensive studies have had a strong impact on the field of circadian biology.

Special Properties

Fluorescence

A most remarkable property of scorpions is their capacity to fluoresce brightly when illuminated by ultraviolet light. It is a mark of our generally poor knowledge of these animals that neither the mechanism nor ecological function of this astonishing trait has been resolved. Fluorescence rarely occurs in terrestrial animals and is nowhere more striking than in scorpions. In *Paruroctonus mesaensis,* a sandy desert species, aromatic compounds in the cuticle absorb maximally in the near-UV (400 nm) and fluoresce in a broad band that peaks in the yellow-green (500–550 nm) range. Other species fluoresce from blue-green to yellow in color, indicating some variance in the family of compounds mediating fluorescence. In at least two families of scorpions, the fluorogenic stubstances appear to be β-carboline and other tryptophan family derivatives, possibly cross-linked to epicuticular proteins (Stachel et al., 1999); these are secreted anew during each molt (first instar scorpions do not fluoresce). Only the lenses of the eyes and tips of the sting and appendages do not fluoresce (see fig. 1.1).

Although the functional significance of scorpion fluorescence is unknown, its practical value as a serendipitous tool to observe a nocturnal animal's behavior and natural history is enormous. In field-oriented studies, a single investigator with a portable blacklight can localize several hundred animals in an evening, essentially all local individuals exposed on the surface at that time. The behavior of most species is much less affected by illumination under UV than under broader spectrum lighting, making it possible to record behavior in the natural setting or in the laboratory. Photography or video recordings of fluorescing scorpions give sharp images of high contrast, even to the level of confocal fluorescence microscopy where all structures lined by epicuticle are inherently labeled in fine detail. One unfortunate consequence of cuticular fluorescence is the devastating advantage it gives to indiscreet commercial (or research) collectors. Of particular concern are species living in specialized habitats, like sand dunes, where continuous "strip harvesting" would soon decimate local populations of these slowly reproducing animals.

Cuticular metals

The chitinous exoskeleton and its articulations are a defining characteristic of the arthropods and a major reason for their unparalleled evolutionary success; the exoskeleton is among the first and most thoroughly studied structures of in-

Figure 1.1 Fluorescence of the Mojave Desert sand scorpion, *Paruroctonus mesaensis,* as seen under black light illumination.

sects. Schofield's recent examination of scorpion cuticle, using nuclear microprobes (chapter 9), uncovered an entirely new characteristic of the exoskeleton—namely, the focal distribution of metals that harden cuticular structure. The cutting, piercing, and abrading tips of structures like the chelicera, sting, and tarsi are reinforced by metal atoms integrated into the chitin structure, giving it greatly improved mechanical strength. Furthermore, the concentration of metal is extraordinarily high, approaching 30% by weight, or well above the amount possible if it were complexed in the form of metalloproteins. Some novel form of metal deposition may be operating in scorpions, and the resolution of this issue could reveal new biochemical or developmental pathways for metabolism and processing of metals in animal tissues.

Radioresistance

Another unusual attribute of the scorpions is their resistance to ionizing radiation. Scorpions were among the few organisms to survive near ground zero during early nuclear testing in the Saharan and North American deserts. The mechanisms of such radioresistance have not been resolved. Goyffon and Roman discuss this phenomenon (chapter 17) and speculate on possible sources of protection and damage repair for irradiated genes, cells, and tissues.

Venoms

The success of scorpions as predators must be attributable in part to their use of venoms. Scorpion venoms are complex mixtures of highly specific toxins, some to immobilize large prey (usually other arthropods) and some to deter predators (commonly vertebrates). Venom toxins are small, rugged proteins that paralyze through their interactions with ion channels in the excitable membranes of nerve and muscle cells. For those using scorpions in research, it is fortunate that only minor components of the venom affect vertebrate (mammalian) nerve excitability. Nevertheless, a few species in the family Buthidae produce potentially lethal venom for humans and pose a particularly serious medical problem where they live among human dwellings or refuse (e.g., *Centruroides* in Mexico, *Androctonus* in North Africa, *Tityus* in South America). While accurate statistics are not available, several hundred human deaths probably occur yearly as a direct result of untreated scorpion stings or indirectly from anaphylaxis induced by antidotes. Unfortunately, these few species have generated an undeserved fear of innocuous species, which far outnumber the seriously poisonous varieties (<2% of total).

A happier application for venoms has been as tools for biochemical isolation and study of ion channels from mammalian nerve and muscle tissues. As discussed by Loret and Hammock (chapter 8), insect-specific toxins from scorpion venoms are being applied to the control of agricultural pests. Genes from scorpions for insect-specific neurotoxins have now been engineered for expression in insect-specific viruses, giving highly selective control against the pest and protection from undesirable side effects on humans, vertebrates, or other nontarget species. Because so few scorpion venoms have been analyzed, basic and applied research should continue to thrive as interest in mechanisms of toxin action, toxin evolution, and use in taxonomy continues to grow.

Scorpions as laboratory animals

Of all the arachnids, scorpions are likely the most easily kept. They are conveniently captured in numbers from dense and resilient populations, and they are largely free of diseases and parasites. Individuals typically survive for years in simple containers and tolerate wide regimes of photoperiod, temperature, and humidity. They require only infrequent feedings of mealworms, crickets, or waxworm larvae, and desert species can live on dry sand without supplementary water beyond that present in food. While these conditions are adequate to hold these animals for several years, subtler aspects of their behavior probably require optimized conditions. Most species are large and conveniently manipulated by hand. Nondangerous species are subdued by grasping and immobilizing the sting segment between the thumb and forefinger, allowing pedipalps and legs to swing free. For toxic species the terminal millimeter of the sting can be clipped to render even these animals harmless. Scorpion pedipalps are generally not strong enough to pinch painfully. One occasional problem is the acquisition of allergies through inhalation of aerosols produced by many scorpions when they are handled frequently. Thus, in almost all practical matters, scorpions make excellent and safe subjects for study, both in the field and laboratory.

LITERATURE ON SCORPIONS

This volume continues the compilation of information about scorpion biology began in *The Biology of Scorpions* (Polis, 1990). For earlier treatises, see Werner

(1934), Millot & Vachon (1949), and Vachon (1952). With the addition of this publication, most of the world's scientific, nontaxonomic literature on scorpions is referenced. Fet et al. (2000) have prepared the complete taxonomic catalog of the world's scorpion fauna; this work should be consulted for correct names, taxonomic references, and geographic ranges of particular species. *Biology of Scorpions* was a broad distillation of factual information about scorpions as organisms—their anatomy, systematics, evolution, and ecology—and it contained summary information about life history, neurobiology, environmental physiology, and toxin biochemistry. The reader is referred to this volume for background information not covered in detail here.

The current volume revisits and updates much of the basic information contained in *Biology of Scorpions.* Moreover, it significantly expands the coverage on reproduction and development (chapter 2), behavior (chapters 5, 10, and 11), sensory neurobiology (chapters 3, 4, 6, and 7), and biogeography (chapters 14 and 15). In each of these areas, scorpions excel as experimental models for basic research, but they also have specialized properties worthy of investigation, which are treated in chapters 8, 9, 17, and 18. Also, fundamental aspects of scorpion ecology and natural history (chapters 12 and 13) and paleobiology (chapter 16) are updated here, with emphasis on how scorpion studies contribute uniquely to broader questions in biology.

Finally, it should be noted that some basic information about the organ systems, anatomy, and physiology of scorpions is not found in either this volume or other literature. Fundamental aspects of circulation, osmoregulation, respiration, metabolism, immunology, and endocrinology are known only through one or a few publications. Biochemistry (other than of venoms), genetics, and cell biology of the group are almost entirely without study in the past half century. It is a challenge to future investigators that these voids should be filled, especially for subject areas where scorpions are likely to yield unique insights by comparison.

References

Fet, V., Sissom, W.D., Lowe, G. and Braunwalder, M.E. (2000). *Catalog of the Scorpions of the World (1758–1998).* New York Entomological Society, New York.

Hadley, N.F. (1994) *Water Relations of Terrestrial Arthropods.* Academic Press, New York.

Millot, J. and Vachon, M. (1949) Ordre des scorpions. *In* Grasse, P. P. (ed.), *Traite de zoologie,* vol. 6, pp. 386–436. Masson, Paris.

Polis, G.A. (ed.) (1990) *Biology of Scorpions.* Stanford University Press, Palo Alto, CA.

Stachel, S.J., Stockwell, S.A., and Van Vranken, D.L. (1999). The fluorescence of scorpions and cataractogeneis. *Chem. and Biol.* 6(8): 531–539.

Vachon, M. (1952) Etudes sur les scorpions. *Publications de l'Institut Pasteur d'Algerie,* Algiers, Algeria.

Werner, F. (1934) Scorpiones, Pedipalpi. *In* Bronn, H.G. (ed.), *Klassen und Ordnungen des Tierreichs,* Band 5, Abteilung 4, Buch 8. Akademische Verlags Gesellschaft, Leipzig, Germany.

2

Structure, Reproduction, and Development

ROGER FARLEY

Scorpions are widely distributed, with representatives on all the continents except Antarctica (Pocock, 1894; Kraepelin, 1905; Lamoral, 1980; Sissom, 1990). They also have an ancient fossil record, which allows some analysis of their evolution and distribution in relation to plate tectonics and other aspects of geology (Sissom, 1990; Jeram, this volume). We have no scorpion fossils older than the Silurian, so information is lacking about the earlier transition from more primitive chelicerates in the Cambrian period (Briggs and Collins, 1988). The Silurian species were similar to modern forms (figs. 2.1, 2.2) except that the former were aquatic with gills and the latter are all terrestrial with booklungs (Kjellesvig-Waering, 1986).

In the absence of fossils, some aspects of scorpion evolution may be inferred from examination of embryos and comparison with structures in related taxa. In this chapter, scorpion reproduction and development are examined in relation to paleontological information. Extant scorpions are viviparous and, unlike many other arthropods, they do not have a separate larval stage with extensive transformation during metamorphosis to the adult (Hjelle, 1990; Polis and Sissom, 1990; Farley, 1999a). The embryos may thereby demonstrate some changes that occurred in the evolution of adult structures of present terrestrial forms. Gould (1977) and Willmer (1990) discuss uncertainties and the caution that must be applied to infer recapitulation from embryonic structures.

The fossil record shows that scorpions migrated from water to land during the Devonian and Carboniferous periods (Selden and Jeram, 1989; Jeram, 1990b, this volume); suggestions of this transition can be seen in the development of embryos. Also, embryos of different taxa have a striking diversity in the timing and sequence of organogenesis, and the maternal reproductive tubules have a variety of specializations for nourishment of the embryos (Farley, 1996, 1998, 1999a). In spite of this developmental diversity, scorpions are much alike at birth.

There has clearly been strong natural selection for efficient maintenance of the embryos inside the mother, although evolution in scorpions is otherwise conservative compared to insects and spiders. We have not yet begun to examine the conditions that favored one mode of embryogenesis over another, but these studies show that scorpions can serve as an excellent model for the influence of selective pressure with viviparity.

To provide perspective for readers unfamiliar with scorpions, this chapter begins with some information about their evolution and phylogeny and an overview of adult features, the end point of the developmental process. For ad-

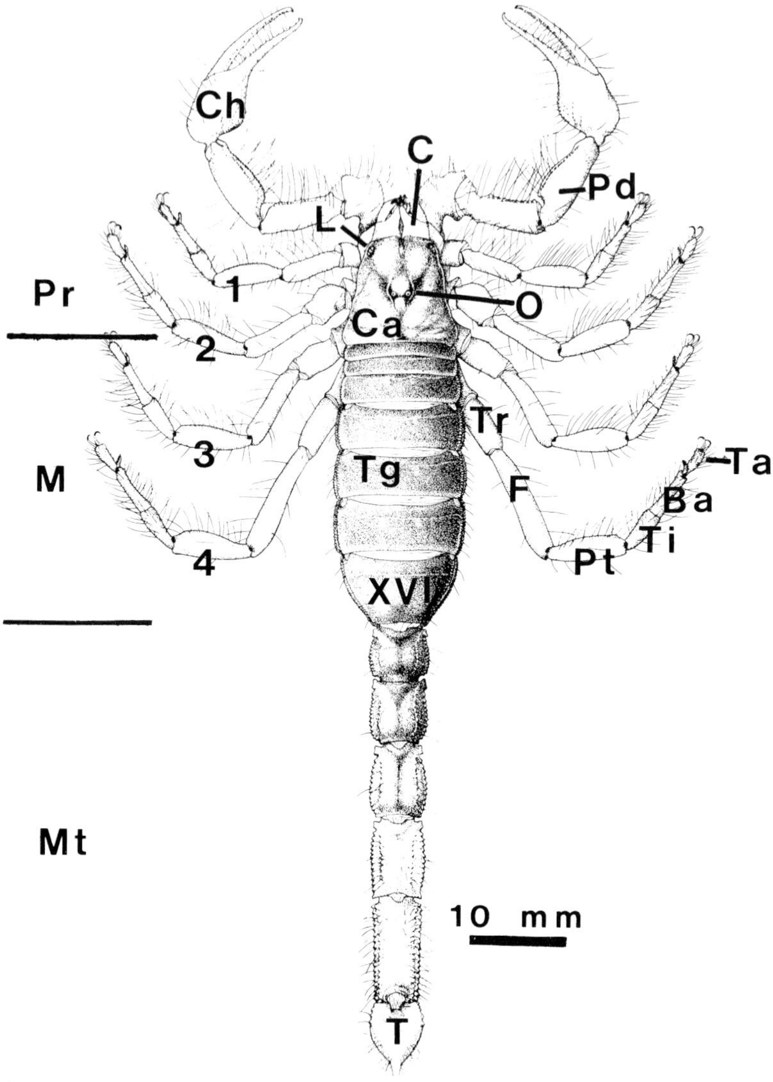

A

Figure 2.1 External features of modern scorpions. Dorsal (A) and ventral (B) views of *Hadrurus arizonensis* (Iuridae). XVI, body segment; B, basal piece; Ba, basitarsus; C, chelicera; Ca, carapace; Ch, chela (pincers); Co, coxa; F, femur; G, genital operculum; L, lateral eyes; M, mesosoma; Mt, metasoma; O, optic tubercle with median eyes; P, pecten;

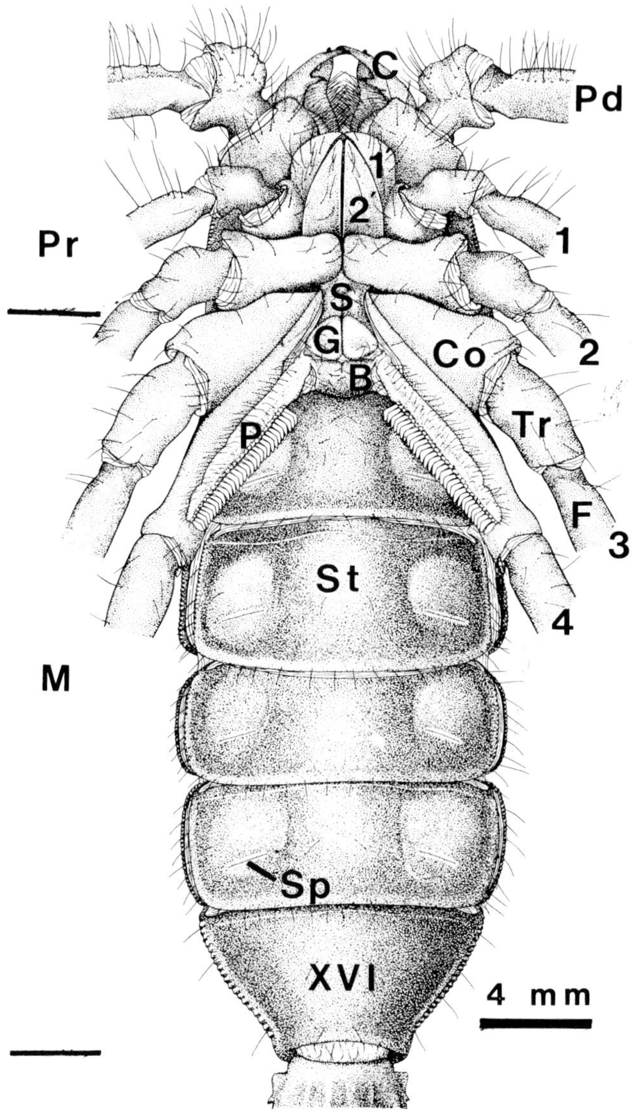

B

Pr, prosoma; Pd, pedipalp; Pt, patella; S, sternum; Sp, spiracle; St, sternite; T, telson with sting; Ta, tarsus; Tg, tergite; Ti, tibia, Tr, trochanter; 1′, 2′, coxapophyses (endites) of first and second walking legs; 1–4, walking legs. From Keegan (1980) with permission (labels added).

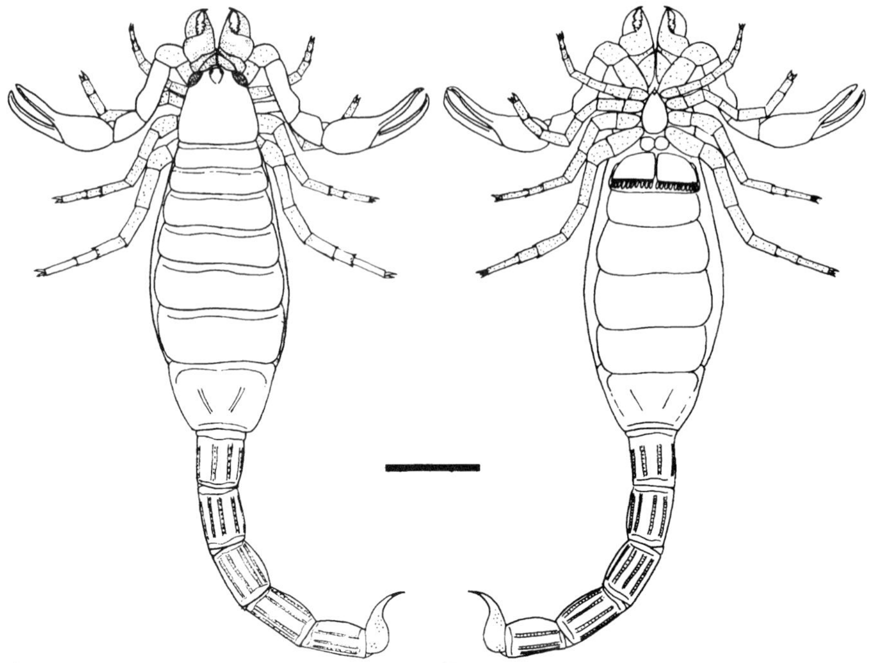

Figure 2.2 External features of Upper Silurian scorpion. Dorsal (A) and ventral (B,C) views of *Proscorpius osborni* (Whitfield), reconstructed from fossils. Large lateral compound eyes and small median eyes are forward on carapace. Prominent chelicerae consist of four segments, the bulbous telson ends in a sting, the pectines are broad and flaplike with teeth, and the ventral mesosoma has five abdominal plates (the first partially covered by

ditional information see Hadley (1990), Hjelle (1990), Sissom (1990), Polis and Sissom (1990), Root (1990), and Farley (1999a).

EVOLUTION AND PHYLOGENY

The arthropods include the major groups (possibly separate phyla) Trilobita (extinct), Crustacea, Uniramia (i.e., insects, myriapods), and Chelicerata (e.g., spiders, scorpions). Sensory antennae and chewing mouthparts (mandibles) are major features of the first three groups (Willmer, 1990), whereas the chelicerates have only one pair of distinctive appendages, the chelicerae, dorsal to the mouth (figs. 2.1–2.3). Included among the chelicerates are the Merostomata (*Limulus, Tachypleus*), Eurypterida (extinct), Pycnogonida or Pantopoda (sea spiders), and Arachnida (Sissom, 1990). Arachnids have five pairs of appendages in addition to the chelicerae. According to Ruppert and Barnes (1994), the class includes 16 extant orders: Uropygi, Amblypygi, Palpigradi, Ricinulei, Pseudoscorpiones, Solifugae, Opiliones, Araneae (spiders), Scorpiones, and seven orders of the subclass Acari (mites and ticks). These taxonomic distinctions are

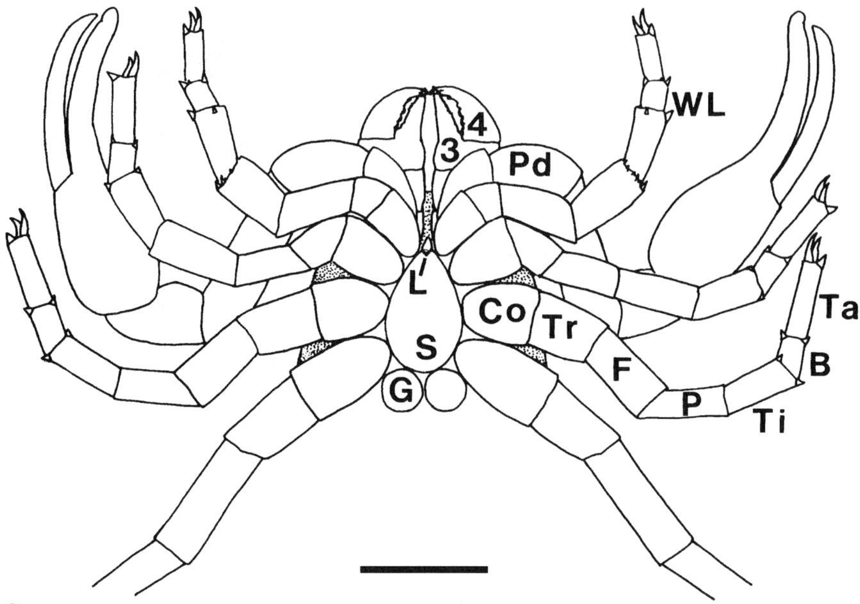

C

pectines). Walking legs are short and tubular with two claws at tip. Scale, 3 mm. (C) No preoral tube is evident in ventral prosoma of this presumed aquatic scorpion. First and second walking legs (WL) have no coxapophyses (compare with figs. 2.1B, 2.3B, 2.17). 3, 4, distal segments of chelicerae; B, basitarsus; Co, coxa; F, femur; G, genital operculum; L, labium; P, patella; Pd, pedipalp; S, sternum; Ta, tarsus; Ti, tibia; Tr, trochanter. Scale, 2 mm. From Kjellesvig-Wearing (1986) with permission (labels added).

useful for extant species, but modifications may be necessary to accommodate fossil Palaeozoic forms (Platnick, 1986; Briggs and Collins, 1988; Selden, 1988).

There was a massive proliferation of arthropods and other groups during the Cambrian period some 500 MYA, an event some authors have labeled as the Cambrian explosion (Valentine et al., 1991). Among the myriad arthropod forms was *Sanctacaris uncata,* which Briggs and Collins (1988) consider an early chelicerate or protochelicerate (fig. 2.4). Chelicerae are lacking in this species, but *S. uncata* had other important features of the class including at least six pairs of appendages on the head. The fossils show 11 segments and a broad telson, which was probably used for stability during swimming. Each body segment has a pair of biramous appendages, a primitive condition thought to have preceded the uniramous condition of recent chelicerates.

Although there are no chelicerae in *Sanctacaris* (Briggs and Collins, 1988), their first pair of head appendages may be cheliceral precursors (fig. 2.4b). In scorpion embryos, the chelicerae initially resemble the other limb-buds (see figs. 2.14, 2.15A, 2.16A), but later move dorsally and away from the leg and pedipalp primordia to a position above the mouth and rostrum where the chelicerae develop chela (see figs. 2.17, 2.18). Fossils of *Sanctacaris* show that it has other

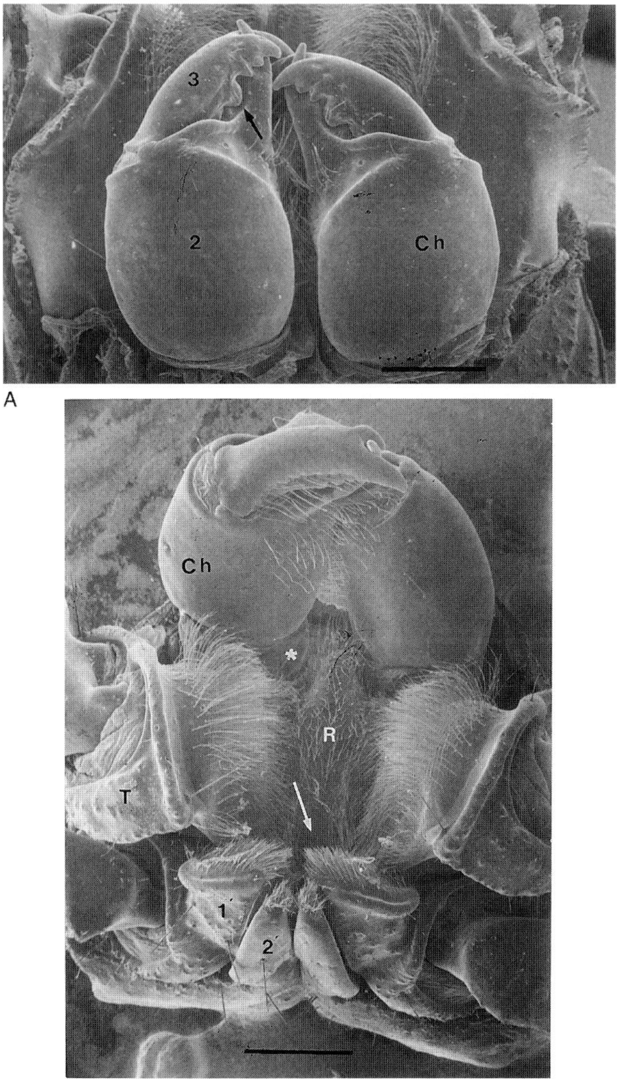

Figure 2.3 Chelicerae consist of three segments and form the roof of preoral tube. Shown is *Hadrurus arizonensis.* (A) Fingers of the chelicerae (Ch) have sensory setae and filtration hairs and teeth (arrow) on opposed surfaces. Divided tip of movable finger (segment 3) allows closure with stationary finger (tip of segment 2). Basal (coxal) segment not shown. (B) Preoral tube formed by bases of anterior appendages. Chelicerae elevated to show basal segment (*) and rostrum (R). Arrow indicates mouth. Filtration hairs and sensory setae line inner surfaces of chamber. T, trochanter of pedipalp; 1′, 2′, coxapophyses (endites) of first and second walking legs.

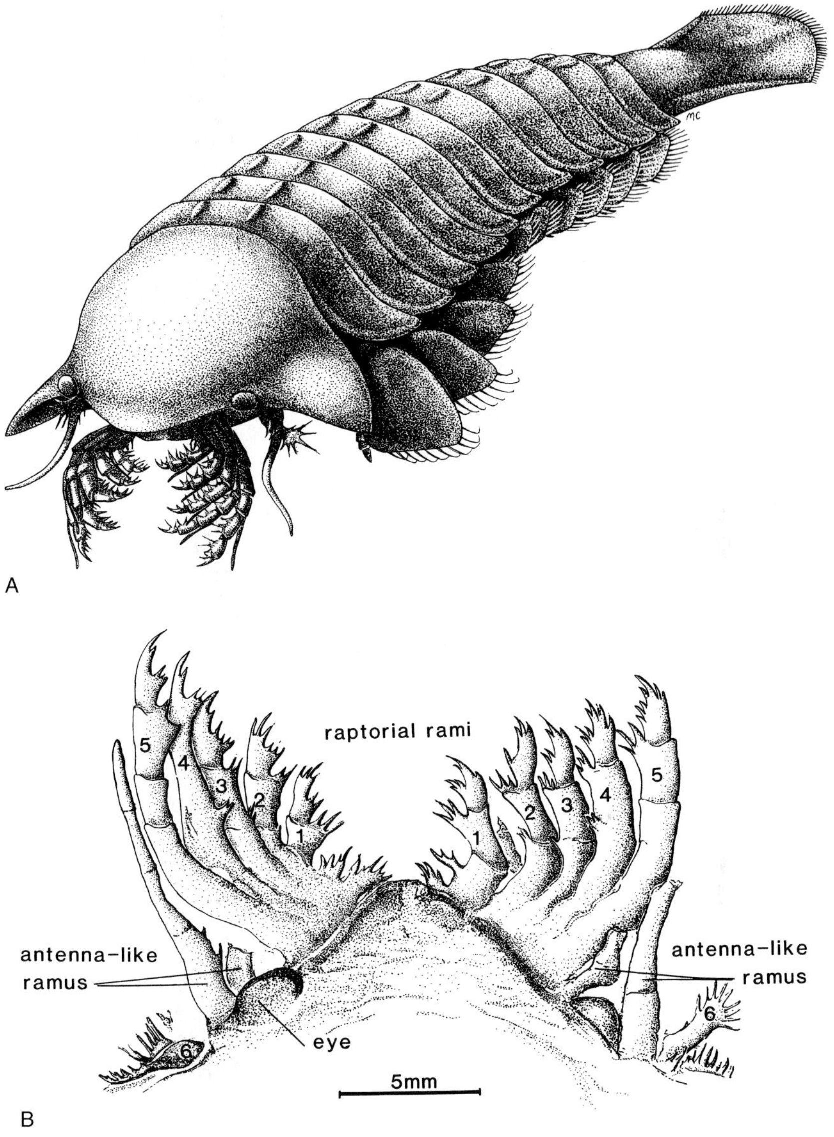

Figure 2.4 Diagrams of *Sanctacaris uncata,* reconstructed from fossils (Middle Cambrian). (A) Chelicerae are not evident, but appendages, tagmata, and other features indicate chelicerate affinity. (B) At least six pairs of biramous appendages (1–6) project beyond head shield. From Briggs and Collins (1988) with permission.

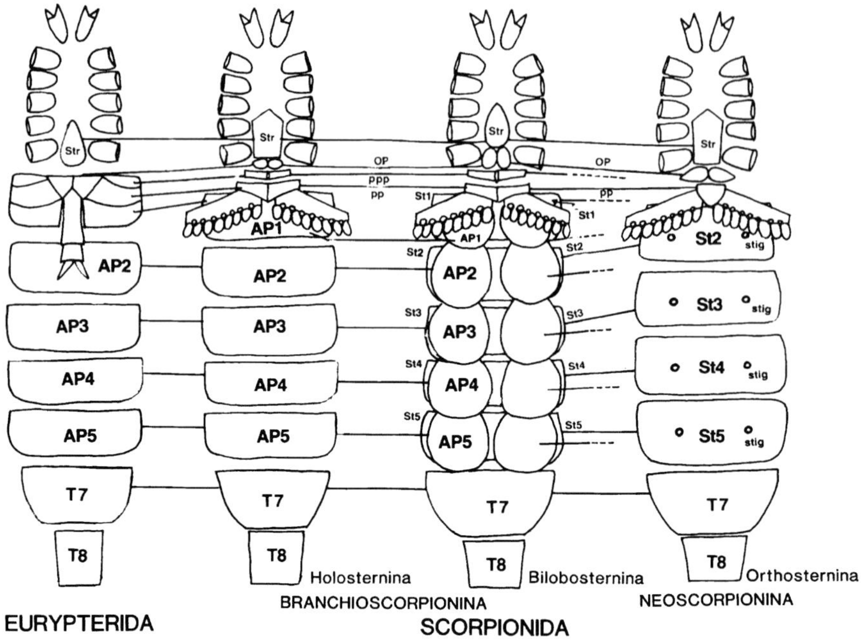

Figure 2.5 Comparison of ventral abdominal plates (AP) in eurypterids and scorpions. The latter have gills (suborder Branchioscorpionina) or book lungs (Jeram, 1990a, this volume) above plates. Some fossil scorpions have reduced abdominal plates (infraorder Bilobosternina) and sternites (St) evident above plates. From studies with scorpion embryos, the Holosternina rather than the Bilobosternina (as proposed by Kjellesvig-Waering, 1986) are most likely the forerunners of Recent scorpions, which have book lungs (suborder Neoscorpionina) and sternites ventrally (Orthosternina). OP, genital operculum; PP, pectinal plate; PPP, pre-pectinal plate; stig, stigmata (spiracle); Str, sternum; T, tergite. From Kjellesvig-Waering (1986) with permission.

features in common with eurypterids, suggesting the former could have been a progenitor for the latter (Briggs and Collins, 1988).

Eurypterids are the oldest chelicerates in which fossils clearly show that chelicerae were present (Plotnick, 1983; Rolfe, 1988; Tollerton, 1989). The phylogenetic relationship of scorpions and eurypterids is uncertain, but they share many features (fig. 2.5) and apparently coexisted in the Paleozoic seas (Kjellesvig-Waering, 1986; Selden, 1988). The eurypterids were very successful during the Ordovician and Silurian (Briggs et al., 1988). Fossils of *Buffalopterus* are 2 m in length, making this the largest known arthropod (Selden, 1984, 1985, 1988). Some species probably spent time on land (Selden, 1984, 1985; Briggs et al., 1988; Shear and Kukalová-Peck, 1990), but the eurypterids did not become fully terrestrial and disappeared by the end of the Permian.

It is unclear why some chelicerates survived among the many arthropods that appeared during the Cambrian explosion, but the adaptability and effectiveness of the chelicerae were probably factors. In extant scorpions, the cheliceral musculature (Vyas, 1974a) and opposing claws (fig. 2.3) enable them to grasp, puncture, and tear prey tissue so it can be further macerated by other mouthparts and solubilized with regurgitated fluid (Farley, 1999a). Surely their transition onto land was key to their survival because, aside from the acarina and pycnogonids, all arachnids are terrestrial inhabitants. Transition onto land appears to have required relatively minor transformations in structure of locomotory appendages and respiratory and sensory organs (Jeram, this volume). The variations in emergent forms of terrestrial chelicerates, as evident in the diverse body plan and size of animals making up the different orders, is reminiscent of the variation seen in insects. The terrestrial radiation suggests that the evolutionary success of the group owes much to the fact that they are among the smaller and more mobile of the arthropod taxa.

Chelicerates certainly reached a large size: some extinct aquatic scorpions, such as *Prearcturus* and *Gigantoscorpio,* were 70–100 cm in length (Størmer, 1963; Rolfe and Beckett, 1984; Briggs, 1985; Kjellesvig-Waering, 1986; Selden, 1988). The large size of these ancient arthropods was probably a response to predation (Vermeij, 1987; Shear and Kukalová-Peck, 1990; Shear, 1991). All the aquatic scorpions became extinct in the Mesozoic, but later than the eurypterids (Briggs et al., 1988). Although aquatic eurypterids and scorpions were abundant and sometimes massive, they were probably vulnerable during molts (Briggs, 1985). Also, fish and other arthropods may have attacked their eggs and juveniles. All surviving scorpions are terrestrial, with the eggs and embryos maintained inside the mother (Francke, 1982) and initial larval stages on her back (Mahsberg, this volume).

Among the largest of present scorpions is the ischnurid *Hadogenes troglodytes* (21 cm, Newlands, 1972) and the scorpionid, *Pandinus imperator,* which is 18–20 cm in length (Vachon, 1953), with some gravid females reaching 65 g in mass. Extant scorpions have a nocturnal, cryptic lifestyle and use their venomous sting for defense and predation on insects, centipedes, spiders, immature scorpions, and other small arthropods (Polis, 1980, this volume). With terrestrialization in the Carboniferous, the retreat to smaller size and cryptic mode probably occurred in response to predation by vertebrates (Jeram, 1990b).

From the fossils, it appears that ancient scorpions were more diverse and successful than present forms. Kjellesvig-Waering (1986) concluded that fossils represent 56 families; Sissom (1990) estimated that there were 11 families in the middle Silurian to lower Devonian and 30 families in the Carboniferous. Today there are 1246 described species and 156 genera belonging to 16 families (Sissom, 1990; Stockwell, 1989, 1992; Fet et al., 1999; Gantenbein et al., 1999). As the oldest fossil scorpions are from the Silurian, more than 400 MYA (Selden and Jeram, 1989), this arthropod lineage had plenty of time to speciate and penetrate into ecological niches. For some reason, they have done much less of this than spiders and insects, which have, respectively, more than 32,000 and 750,000 extant described species (Parker, 1982; Zeh et al., 1989; Ruppert and Barnes, 1994).

Previous authors used adult characteristics to account for differences in arachnid diversity, but Zeh et al.(1989) concluded that the first stages of the life cycle can account for both stasis (e.g., scorpions) and diversity (e.g., spiders). The early development of viviparity in scorpions and some other arachnids may have limited their speciation in comparison with arthropods with resistant or protected eggs. Protected eggs may tolerate more environmental diversity than females sustaining a relatively small number of embryos in utero.

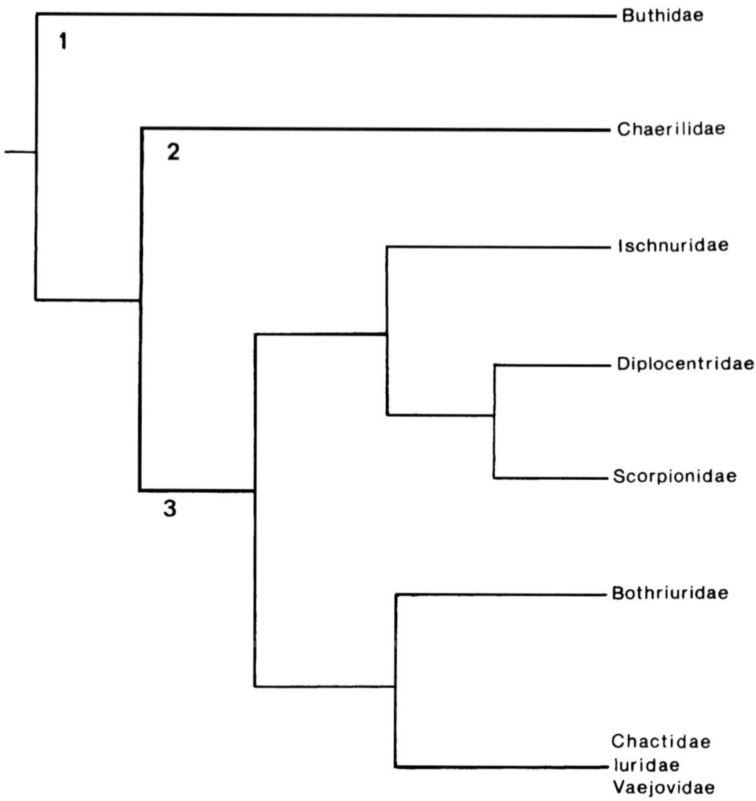

Figure 2.6 Evolutionary branches hypothesized for Recent scorpions: 1, buthid; 2, chaerilid; 3, diplocentroid. This cladogram proposed by Lamoral (1980) and modified by Sissom (1990) is tentative since it shows only 9 of the 16 families, and the interconnecting lines are sure to change with further results. The grouping of the families is based on substantial differences and is not in conflict with recent revisions of systematics base on DNA analysis (Fet et al., 2000). Used with permission.

Extant scorpions are considered to be monophyletic (Sissom, 1990). Their main adaptations enabling survival on land (e.g., book lungs, preoral tube, indirect sperm transfer, viviparity) occur in all present families. Lamoral (1980) proposed a cladogram based on the notion that there were three major branches in the evolution of present families. This division heavily relied on trichobothrial patterns on the pedipalps. Sissom (1990) modified this to include two later identified families, the Ischnuridae and Iuridae (fig. 2.6). This cladogram is helpful in depicting the major groups of scorpion families, but the connecting lines are sure to change with additional information. Buthidae is thought to be the most primitive family, a hypothesis supported by the ontogeny described below. A second evolutionary branch consists of the family Chaerilidae. The third (diplocentroid) clade, which included 7 families on the cladogram of Sissom (1990), currently includes as many as 11 recognized families: Bothriuridae, Chactidae, Euscorpiidae, Diplocentridae, Heteroscorpionidae, Ischnuridae, Iuridae, Scorpionidae, Scorpiopsidae, Superstitionidae, and Vaejovidae (Stockwell, 1992; Lourenço, 1996; Gantenbein et al., 1999). Additional information from morphological, developmental, and genetic studies is needed to clarify relationships among these families (Stockwell, 1989; Fet et al., 2000; Gantenbein et al., 1999).

Many scorpion features were conserved throughout their evolution. Exceptions include some finer morphology (e. g., finger dentition in pedipalps and chela, trichobothrial pattern), variations in venom toxicity (Simard and Watt, 1990; Loret and Hammock, this volume), modifications for the water-to-land transition (Jeram, 1990a, this volume), and maternal and developmental adaptations for nourishment of embryos inside the mother (Anderson, 1973; Polis and Sissom, 1990; Farley, 1996, 1998, 1999a). As described below, maintenance of the embryos is strikingly different in the buthid and diplocentroid branches, and, as yet, we have no information about development in the Chaerilidae (Polis and Sissom, 1990).

For more information about chelicerate evolution and phylogeny, the reader is referred to the following reviews: Petrunkevitch (1955), Firstman (1973), Kraus (1976), Van Der Hammen (1977, 1979, 1982, 1985a,b, 1986a,b), Boudreaux (1979), Weygoldt and Paulus (1979a,b), Shultz (1989), Stockwell (1989; 1992), Sissom (1990), Turbeville et al. (1991), Selden (1993), Jeram (1994a,b), Ruppert and Barnes (1994), Foelix (1996), Punzo (1998), Fet et al. (1999), and Shear (1999).

STRUCTURAL FEATURES

The scorpion body is divided into a cephalothorax or prosoma and the abdomen or opisthosoma (fig. 2.1). The abdomen is divided into an anterior mesosoma and a posterior metasoma or cauda. Internally, the diaphragm, a vertical wall of muscle and connective tissue, separates the prosoma and opisthosoma. The function of this partition is unclear, but it may relate to the different functions and blood pressures of the two body regions (Stewart and Martin, 1974; Foelix, 1996). The prosoma contains muscles associated with locomotion, while the mesosoma has reproductive structures, book lungs, heart, and a large digestive gland (Hjelle, 1990; Farley, 1999a).

Except for some blind troglobiont species (Polis, 1990; Sissom and Cokendolpher, 1998), scorpions have a pair of median eyes mounted on a prominent eye node (ocular tubercle) near the center of the carapace (fig. 2.1A). Two to five lateral eyes are located at the antero-lateral margins of the carapace. There are substantial differences between Recent and Paleozoic scorpions in the shape of the carapace and the position and complexity of the eyes (figs. 2.1, 2.2; Kjellesvig-Waering, 1986; Selden and Jeram, 1989; Locket, this volume). The Silurian scorpion of fig. 2.2 had compound lateral eyes, and the carapace of some fossils has a shovel-like protrusion that was probably helpful for digging in mud.

Hjelle (1990) concluded that scorpions have 21 body segments; some earlier authors reported other numbers because segments are lost or merge during development (McClendon, 1904; Millot and Vachon, 1949; Abd-el-Wahab, 1952; Snodgrass, 1952; Raw, 1957; Keegan, 1980). There is uncertainty about the fate of segments I and II, but the adult prosoma consists of seven segments (III-IX) that can be identified externally. The chelicerae (Fig. 2.3) of segment III have sensory setae (Farley, 1999a), and each ends in a claw (chela) with movable and stationary fingers (Vyas, 1971, 1974a).

The large and sturdy pedipalps (fig. 2.1) of segment IV have claws (chela) that are used to seize and hold prey (Dubale and Vyas, 1968; Vyas, 1971) and that males use to grasp the female chela during courtship (Alexander, 1956, 1957, 1959; Rosin and Shulov, 1963; Polis and Farley, 1979a; Peretti, 1991; Farley, 1999a; Benton, this volume). Sensory receptors along the length of the pedipalps enable them to be used like antennae as the animal moves forward with the pedipalps extended (Shulov and Levy, 1978; Farley, 1999a). The trichobothria are long setae that respond to airborne vibrations and occur only on the pedipalps (Farley, 1999a). They were not present in aquatic scorpions and are apparently a terrestrial adaptation (Selden and Jeram, 1989).

Four pairs of walking legs (fig. 2.1) are associated with segments V–VIII (Vyas, 1971; Couzijn, 1976; Hjelle, 1990). The sclerotized segments of each pedipalp and leg are connected by flexible, nonsclerotized cuticle. Along the length of the legs there are bristles, sensory setae, and spurs; tarsal claws are located at the tip. Modern scorpions (all terrestrial) have stronger legs and different sensory organs than aquatic ancestors (Kjellesvig-Waering, 1986; Selden and Jeram, 1989; Jeram, this volume).

The fused coxae of the third and fourth walking legs form a contiguous plate (fig. 2.1B); internally and dorsal to the plate is the coxal excretory organ (Farley, 1999a). At the ventral midline, the coxal plates are separated by the sternum (figs. 2.1B, 2.2, 2.5), a structure important for distinguishing ancient and recent scorpion families (Petrunkevitch, 1952; Keegan, 1980; Kjellesvig-Waering, 1986; Hjelle, 1990; Sissom, 1990).

Petrunkevitch (1952) noted that the fossils of Silurian scorpions have the original first abdominal segment (IX), but this is not evident in fossils from the Devonian and subsequent periods or in extant adults. Segment IX appears with a limb-bud in scorpion embryos but later disappears (Abd-el-Wahab, 1952; Hjelle, 1990), so the adult mesosoma consists of segments X–XVI.

The mesosoma is the widest part of the scorpion body. Each segment has a dorsal sclerotized tergite and a ventral sclerotized sternite (fig. 2.1). A nonsclerotized intersegmental or arthrodial membrane connects each tergite and sternite to the preceding and following one. At the lateral margin, the tergites and sternites are joined by nonsclerotized pleural membrane (Tandan et al., 1977). The nonsclerotized regions are flexible, allowing movement between sclerites and some swelling of the mesosoma as a result of pregnancy and ingestion of food.

Body segment X has the genital operculum and genital aperture or gonopore (figs. 2.1B, 2.2, 2.5). The genital operculum is a flaplike covering of the aperture, the opening of the reproductive system (Hjelle, 1990). Anterior attachment of the operculum enables the posterior margin to be moved downward for expulsion (male) or attachment (female) of the spermatophore. In the males of some species, two genital papillae extend posteriorly from either side of the gonopore and may have a role in spermatophore deposition (Polis and Farley, 1979b).

The pectinal plate or basal piece and the pectines are associated with segment XI (figs. 2.1B, 2.5; Hjelle, 1990; Farley, 1999a). The pectinal plate constitutes the sternite of that segment, while the pectines are complex sensory structures. Muscles are associated with the pectines at their base (Lankester et al., 1885) and with each tooth of the comb, so the pectines can be retracted dorsally or rotated forward and lowered to form-fitting contact with the substrate. A prepectinal plate (fig. 2.5) is present in some fossil and recent scorpions (Kellesvig-Waering, 1986).

The book lungs with their spiracles are located ventrally in mesosomal segments XII–XV (fig. 2.1B). The spiracles can be closed and thereby reduce water loss (Paul et al., 1987; Fincke and Paul, 1989; Paul and Fincke, 1989; Farley, 1990a,b). Air passes through the spiracles to an atrium and then into the narrow spaces of the book-lung lamellae. Blood is present on the opposite, internal side of the thin lamellar membrane (Vyas, 1974b; Farley, 1990a,b).

The scorpion circulatory system is fairly complex, with hemocyanin for oxygen transport and a network of arteries that carry blood to the tissues (Farley, 1990a,b). Blood collects in veins and sinuses in the ventral mesosoma and flows through the book-lung lamellae and then dorsally to the pericardial sinus and tubular heart. Contraction of the heart forces blood through aortas at the anterior (prosoma) and posterior (metasoma) ends of the heart. The blood flows through small arteries within the organs before collecting again in ventral veins and sinuses (Kästner, 1940a,b; Millot and Vachon, 1949; Farley, 1984, 1985, 1987, 1999a; Farley and Chan, 1985).

The mesosoma of aquatic scorpions had five ventral plates which presumably had gills above them (fig. 2.5; Kjellesvig-Waering, 1986), but extant scorpions have only four pairs of book lungs (fig. 2.1B). Also, there were book lungs above five ventral plates of a Lower Carboniferous terrestrial scorpion (Selden and Jeram, 1989; Jeram, 1990a, this volume). Kjellesvig-Waering (1986) speculated that the anteriormost respiratory segment was lost in the evolution of extant terrestrial species.

The last mesosomal segment (XVI) tapers sharply to join with the first metasomal segment (fig. 2.1). The entire circumference of the five metasomal segments (XVII–XXI) is sclerotized, and their diameter is much smaller than the mesosomal segments. The anus is located at the ventro-posterior end of segment XXI, and distal to this is the telson or sting, which is not regarded as a true segment (Hjelle, 1990).

The cylindrical metasomal segments form a slender, sturdy and flexible shaft that enables the sting (aculeus) to penetrate hard tissues. During develpment, the tip to the sting is hardened by metals that are deposited there by some novel mechanism (Schofield, this volume). The bulbous base of the telson contains the paired venom glands. Contraction of the compressional muscle of each gland forces the venom through two ducts that open near the tip of the aculeus (Farley, 1999a). Loret and Hammock discuss the properties and functions of venom proteins in chapter 8.

SPECIALIZATION FOR FEEDING: PRE-ORAL TUBE

The mouthparts of arachnids vary in form and function but usually enclose a cavity or filter for ingestion of fluid from externally digested prey (Petrunkevitch, 1955). A tubelike chamber in front of the mouth is a characteristic feature of terrestrial scorpions. The tube is formed by the basal segments of the anterior appendages (fig. 2.3B) and enables the scorpion to penetrate the integument of captured prey, macerate and mix prey tissue with regurgitated digestive fluids, and then ingest the partially digested mixture (Srivastava, 1955;

Snodgrass, 1952; Størmer, 1970; Hjelle, 1990; Farley, 1999a). The roof of the tube is formed by the bases of the chelicerae. The inner sides of the pedipalpal coxae and trochanters provide the sides of the tube. The floor is the coxapophyses (maxillary lobes, endites, mesal lobes) that extend forward from the coxal segments of the first and second walking legs. The mouth is evident as a small opening beneath the labrum (rostrum), a laterally compressed structure which projects forward from between the bases of the pedipalps.

The preoral tube was absent in pre-Carboniferous scorpions presumed to be aquatic (figs. 2.2B,C), and its development during the Carboniferous period was probably associated with the transition from aquatic to terrestrial existence (Petrunkevitch, 1952; Størmer, 1970; Rolfe and Beckett, 1984; Kjellesvig-Waering, 1986; Selden and Jeram, 1989; Jeram, 1990b, this volume). Extant scorpions macerate and mix their food with digestive juices before ingestion (Farley, 1999a), and the preoral tube provides a passageway through which food can pass with little spillage or loss from evaporation. In aquatic animals, such external digestion would result in diffusion of digestive juices into surrounding water unless a seal was formed between the captor and prey. Aquatic scorpions probably crushed and ground their food with gnathobasic coxae, much as occurs in xiphosurans (horseshoe crabs) and other arthropods that feed on the bottom (Rolfe and Beckett, 1984). The enlarging coxapophyses of the first and second walking legs of scorpion embryos (fig. 2.17) may reflect this earlier function.

SPECIALIZATION FOR REPRODUCTION: SPERMATOPHORE

The external features of males and females are much alike, yet these animals somehow recognize each other in the dark and suppress their predatory and cannibalistic tendencies long enough to mate. It is likely that pheromones are involved in this process (Polis and Sissom, 1990; Gaffin and Brownell, this volume) along with recognition of substrate vibrations (Brownell, 1977, 1984; Brownell and Farley, 1979a,b,c; Polis and Farley, 1979a) and tactile cues. Scorpions have an elaborate courtship ritual that serves to temporarily reduce their aggressive and defensive behaviors and triggers the release of a spermatophore from the male at a time when the female is present (Francke, 1979; Polis and Farley, 1979a; Benton, this volume).

Benton (this volume) has reviewed details of courtship and mating behaviors; additional information is provided by Polis and Sissom (1990). Initial interactions include juddering (male), mating attacks (female), and clubbing with the tail (either sex). In the later stages, the male uses his pedipalps to grasp those of the female, then pulls and pushes her until a stable object or substrate is found to deposit the spermatophore. This *promenade à deux* lasts from minutes to days, and may include cheliceral massage and sexual sting (male). When a suitable substrate is located, the spermatophore is extruded from the male's genital opening (fig. 2.1B). The spermatophore attaches to the substrate in a nearly vertical position. He then pulls her over it so the tip of the spermatophore contacts her genital pore. She positions and lowers her body on the spermatophore, and in so doing, causes it to open so the sperm are expelled into her reproductive tract (Francke, 1979; Benton, this volume).

The scorpion spermatophore is a cuticular structure formed as mirror-image hemispermatophores inside the paraxial organs (fig. 2.7) of the male reproduc-

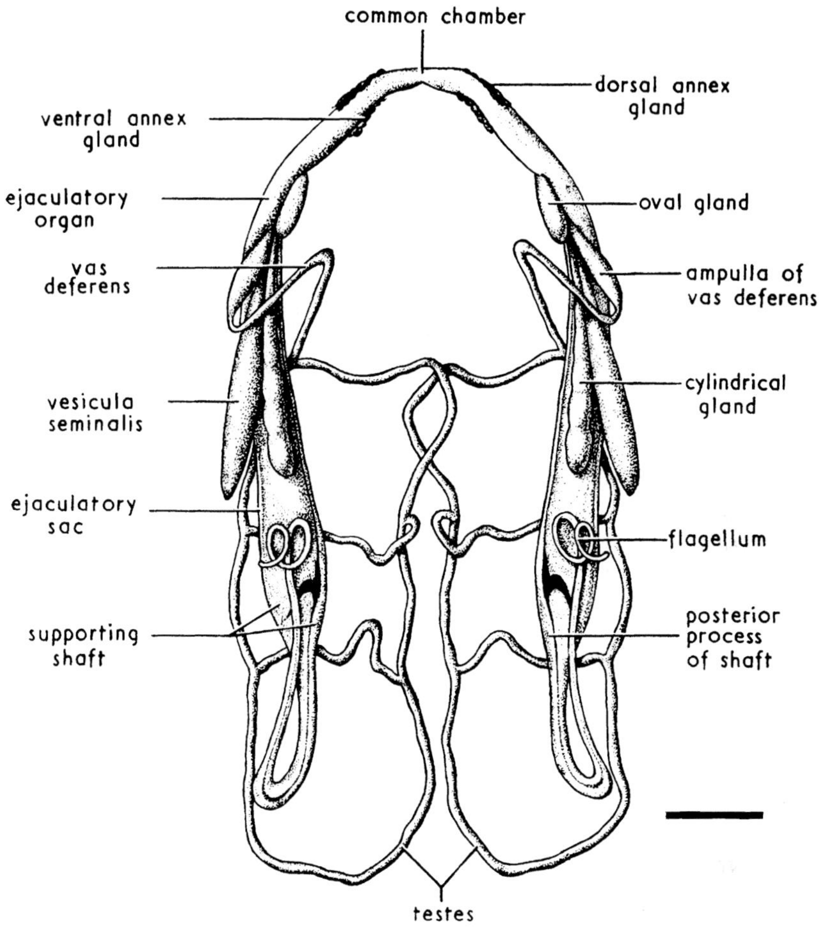

Figure 2.7 Internal reproductive structures in male scorpions. Shown is *Leiurus quinquestriatus* (Buthidae). Sperm produced in testes is stored (vesicula seminalis) and released with each cuticular hemispermatophore. Hemispermatophores mislabeled as supporting shafts by early workers, are formed inside the right and left paraxial organs (ejaculatory organ, sac). The wall of the latter consists of an epithelium, basement membrane, connective tissue, and muscle cells. Contraction of the muscle cells probably ejects the hemispermatophores during mating. Glands (annex, oval, cylindrical) may provide fluid for the sperm and adhesive for fusion of the hemispermatophores at release. Scale, 3 mm. From Abd-el-Wahab (1957) with permission.

tive system (Angermann, 1957; Francke, 1979; Hjelle, 1990; Sissom, 1990). In both males and females, the reproductive structures are in the mesosoma, where the gametes and embryos obtain nutrients from the anterior intestine and massive digestive gland.

The testes in the male (fig. 2.7) and the ovariuterus (fig. 2.8) in the female consist of a network of longitudinal and transverse tubules (Hjelle, 1990). In addition, this system in the male has paired glands, the seminal vesicles where

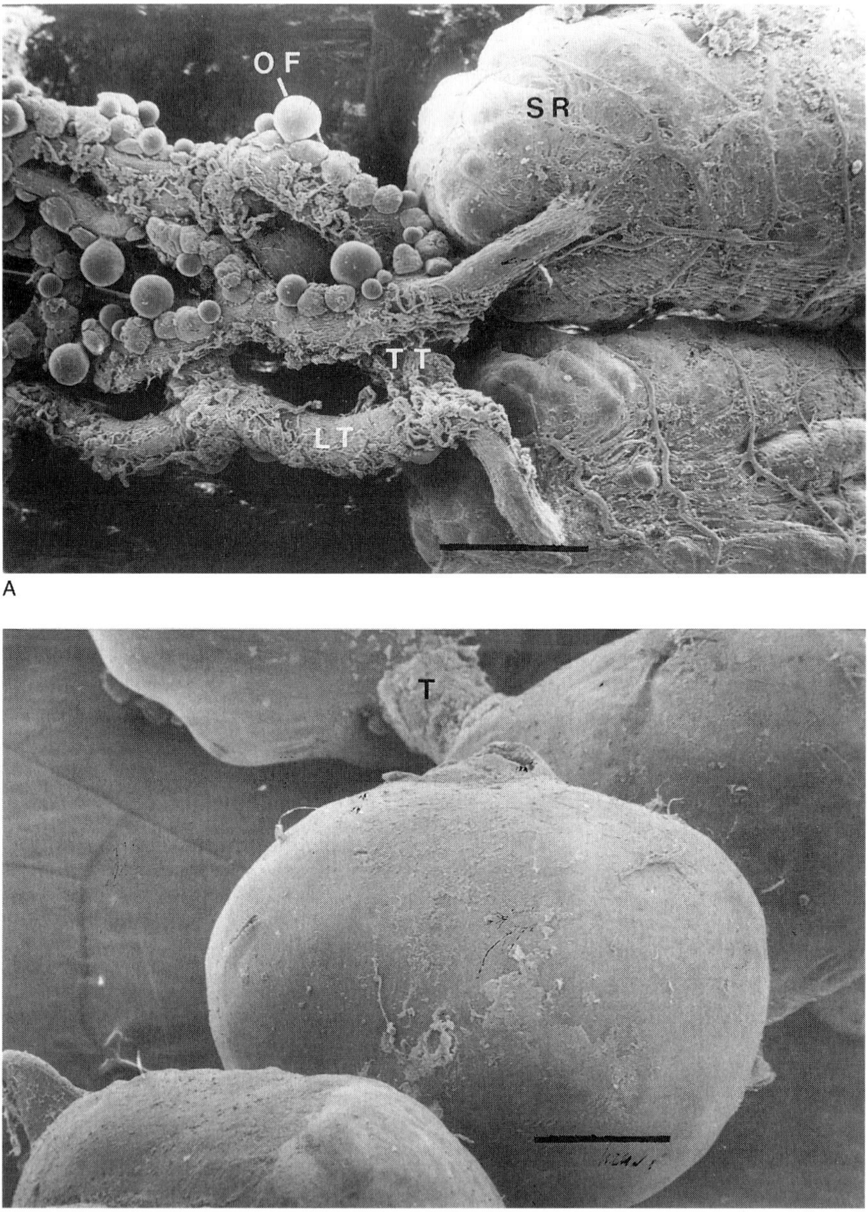

Figure 2.8 Reproductive system of female scorpion. Shown is *Hadrurus arizonensis.* (A) Each ovarian follicle (OF) contains an oocyte. After the ovum is fertilized, the enlarging embryo bulges into the tubule lumen and remains there for further development. LT, TT, longitudinal and transverse ovarian tubules; SR, seminal receptacles. (B) Embryos develop within swollen regions of ovarian tubules (T). With this simple form of apoikogenic development, embryos (see fig. 2.18) apparently absorb maternal nutrients passed through tubule wall. Scale, 1 mm for A and B.

sperm is stored, and the large paraxial organs where the hemispermatophores are synthesized and stored until mating. During extrusion from the genital aperture, the two hemisphermatophores fuse together with sperm enclosed.

The spermatophore consists of a pedicel (base), trunk, capsule with sperm duct, and either a flagellum or bladelike lamina. The Buthidae have a spermatophore with a flagellum at the distal end. The other families lack this and have a spermatophore with a lamina at the distal end (lamelliform). The flagellum and lamina have an important role in spreading the female genital operculum (fig. 2.1B) so the sperm is inserted into the gonopore as the spermatophore is bent (Francke, 1979; Benton, this volume). There is evidence of sperm storage within the female reproductive organs (Kovoor et al., 1987).

Depending upon species, the gestation period ranges from several months to more than a year (Polis and Sissom, 1990). The embryos develop inside the maternal ovariuterus, receive nourishment from her tissues, and are released live from her gonopore (Francke, 1982). Litter size varies from 1 to 105 (Polis and Sissom, 1990). The newborn ascend onto the mother's back and continue development until they disperse after the first molt and take up an independent existence (Mahsberg, this volume). Larval stages are defined by four to six more molts before reaching sexual maturity. Longevity typically ranges from 2 to 8 years but is known to be considerably longer (up to 25 years) in some species (Polis and Sissom, 1990; Mahsberg, this volume).

Animals use a variety of mechanisms to transfer viable sperm. Information about spermatophores in relation to other types of sperm transfer and impregnation in arthropods is provided by Alexander and Ewer (1957), Abalos and Baez (1963), Alexander (1964), Weygoldt (1969, 1975, 1990), Savory (1971, 1977), Schaller (1971, 1978), Cloudsley-Thompson (1976, 1988), Little (1983) and Proctor (1998). In water, the spermatophore prevents sperm loss by dilution; in terrestrial environments the spermatophore reduces loss from desiccation (Schaller, 1978). The reliability of transfer and the chances that a given spermatophore sperm packet will reach a female oviduct depends on communication, substrate, behavior and other circumstances associated with transfer.

Scorpions have indirect sperm transfer because the spermatophore is placed on the substrate before it contacts the female. This occurs in some other Recent orders of Arachnida—the Uropygi, Schizomida, Amblypygi, Acari, and Pseudoscorpiones—and may be the primitive condition from which the transfer methods of other arachnids were derived (Weygoldt, 1975, 1990). Courtship does not occur in all animals with indirect sperm transfer. Weygoldt (1969) described three types of behavior involved with this transfer mode: (1) release of spermatophore without the female present (females may detect it through pheromone markers), (2) release with the female present but with little or no interaction, and (3) as in scorpions, release after mating behavior in which the male assists the female to take up the spermatophore. The latter appears to be relatively efficient and probably contributes to the survivability of scorpions in environments with low density of potential conspecific mates. Before each spermatophore is released, success of impregnation is greatly facilitated by the courtship ritual. All three of the above transfer modes are exhibited among pseudoscorpion species, suggesting stages in the evolution of spermatophore transfer (Ruppert and Barnes, 1994).

Although scorpions differ in the details of the courtship, the *promenade à deux* is present in all species observed (Polis and Sissom, 1990). This behavior and the mechanism of indirect sperm transfer are characteristic of the order and might also have been present in pre-Carboniferous aquatic ancestors. Fossils of these scorpions (fig. 2.2) do not have specialized appendages, copulatory organs, or other structures for direct sperm transfer to the female (Kjellesvig-Waering, 1986). The flaplike structure of the pectines in some fossil scorpions (fig. 2.2B) suggests use as stabilizers during swimming (Størmer, 1963), but the proximity of the pectines with the genitalia raises the possibility of earlier reproductive functions. Genital appendages were present in eurypterids (Tollerton, 1989; Braddy and Dunlop, 1997), and they are used for brood pouches in some extant arthropods (Weygoldt, 1969). Kjellesvig-Waering (1986) speculated that the pectines might initially have been used to scoop a depression for egg deposition and external fertilization.

GAMETOGENESIS

Spermiogenesis

Parthenogenesis was reported in some scorpion species (Makioka and Koike, 1985; Polis and Sissom, 1990; Makioka, 1992a, 1993; Lourenço and Cuellar, 1994, 1995; Farley, 1999a), but most reproduction is sexual. Some external sexual dimorphism is common in scorpions. Males and females differ most prominently in reproductive behavior (Polis and Sissom, 1990; Benton, this volume; Gaffin, this volume) and in the structure and function of the tubular network attached to the genital aperture. The genetic basis for these sexual differences is unknown, as are the developmental mechanisms that bring them about.

The reproductive tubules have been a site of striking evolutionary change, especially among females, as described below. The research in males shows there are taxonomic differences in the number of chromosomes, differentiation of sperm cells, and the ultrastructure of the mature sperm (Alberti, 1983). It is likely that these differences can be helpful in identifying evolutionary relationships, but so far only a few species have been examined. The studies on scorpion spermatogenesis should be extended and compared with results from other invertebrates where more data are available (Baccetti, 1970, 1978, 1987; Franzén, 1970; Baccetti and Afzelius, 1976; Jamieson, 1987).

Descriptions of the number and distribution of chromosomes during meiosis in male scorpions were provided by numerous authors including Sokolow (1913), Sato (1936), Tuzet (1938), De Tolédo Piza (1941, 1943a,b, 1944, 1956), Brieger and Graner (1943), Rhoades and Kerr (1949), Grénin (1957, 1961), Bedi (1962), Sharma et al. (1962), Venkatanarasimhaiah and Rajasekarasetty (1964), Yoshida and Toshioka (1964), Koch (1977), Newlands and Martindale (1980), and Shanahan and Hayman (1990).

A diploid number of six was reported for the buthid *Tityus bahiensis* (Brieger and Graner, 1943; De Toledo Piza, 1943a,b, 1944). Buthids examined by Shanahan and Hayman (1990), and by Goyffon and Ramon (this volume) have diploid numbers of 12–16, while some Scorpionidae had 30–175 (Grénin, 1957). The evolutionary significance of this striking pattern is unclear. The variation may be due to chromosome fusion and fragmentation (De Toledo Piza, 1943a; Srivastava and Uma, 1961) and interspecific mating. Buthids appear to have fewer chromosomes and are considered the most primitive among Recent families (Kraepelin, 1905; Lamoral, 1980; Sissom, 1990). Because only a few chromosomes contain the genetic information necessary for development and adulthood in some species, multiple copies and a substantial amount of unused genetic material may be present in other scorpions with larger chromosome numbers.

Complex chromosomal figures, possibly from reciprocal translocations, were observed at first metaphase in some scorpion species (De Toledo Piza, 1941, 1956; Sharma et al., 1959; Srivastava and Uma, 1961). The possibility of chromosomal fusion and fragmentation is supported by observations of variable chromosome numbers in animals of the same species but from different localities (De Toledo Piza, 1943a; Srivastava and Uma, 1961). These within-species differences might also result from interspecies mating. The males of the scorpionid *Heterometrus* (formerly *Palamnaeus*) *longimanus* were reported to be heterogametic, with X and Y chromosomes (Srivastava and Uma, 1961). Meiosis was found to be achiasmatic in several species (Srivastava and Uma, 1961; Shanahan and Hayman, 1990).

Within the testes, some cells of the germinal epithelium differentiate into large, spherical spermatogonia (Sokolow, 1913; Abd-el Wahab, 1957; Jespersen and Hartwick, 1973). These occur individually or in groups in the testicular inner layer. The spermatogonia differentiate into primary spermatocytes. Large numbers of these can be seen within testicular compartments or spermatocysts as they undergo the first meiotic division to produce secondary spermatocytes. Division of the latter results in spermatids, which gradually undergo transformation to spermatozoa.

Spermatid transformation is a remarkable process in which much of the cytoplasm is expelled, and the remaining components are rearranged (Jespersen and Hartwick, 1973; Farley, 1999a). The result is a highly specialized and motile sperm cell with the minimum components necessary for passage to the ovum and fertilization. During transformation, the spherical spermatid nucleus becomes slender and elongate with very dense chromatin. Microtubules appear not to be involved in the movement that occurs during nuclear shaping in the iurid *Hadrurus arizonensis* and the vaejovid *Vaejovis spinigerus* (Phillips, 1974).

Spermatid transformation in scorpions occurs without supportive or Sertoli-like cells. Within the lumen of the testis, bundles of mature spermatozoa appear to be held together by an electron-dense ground substance (Alberti, 1983). Sperm length was reported to be 275 μm in *Hadrurus* (Jespersen and Hartwick, 1973), 170 μm in *Centruroides* (Riess et al., 1977), and 90–210 μm in *Tityus* (Da Cruz Landim and Ferreira, 1973).

There are species differences in the fine structure of scorpion spermatozoa, but a general pattern is consistent (Alberti, 1983). Each spermatozoon has a head, middle piece, and end piece. The head consists of the acrosomal complex and nucleus. The middle piece has two centrioles arranged linearly just posterior to the nucleus and at the anterior end of the flagellum. The flagellum contains the axoneme, a regular arrangement of microtubules. The anterior two-thirds of the flagellum is surrounded by a cytoplasmic collar with mitochondria and smooth endoplasmic reticulum. The flagellum extends posteriorly beyond the end of the cytoplasmic collar, so that the end piece consists of free flagellum.

Differences within and between species were reported in the flagellar axoneme of scorpion spermatozoa. Nine pairs of peripheral microtubules is characteristic, but taxonomic differences are reported for the axoneme center, which may have one or two microtubules or an electron dense rod (André, 1963; Robison, 1970; Hood et al., 1972; Da Cruz Landim and Ferreira, 1973; Jespersen and Hartwick, 1973; Riess et al., 1977; Esponda and Taffarel, 1980; Alberti, 1983). The significance of these differences is unclear. Motility was observed in spermatozoa with a central rod in the axoneme (Robison, 1970; Hood et al., 1972; Riess et al., 1977).

Wilson (1916a,b, 1931) described in scorpions two different mechanisms by which mitochondria in the primary spermatocytes are distributed to the spermatids. In the buthid *Centruroides exilicauda* (formerly *C. sculpturatus*), the mitochondria fuse into a ring. During the first meiotic division, the mitochondrial ring is pulled apart so that a half-ring moves toward each pole. Each half-ring straightens into a rod, which then divides so that each secondary spermatocyte receives two mitochondrial rods. Each rod divides again during the next meiotic division, so that the spermatid receives two rods. Each of these mitochondrial rods is one-eighth of the original mitochondrial ring.

During transformation of the spermatid of *Centruroides,* the two mitochondrial rods join to form a ring (*nebenkern*) just posterior to the nucleus. The two mitochondria then elongate posteriorly within the cytoplasmic collar that surrounds the flagellum in the middle piece of the spermatozoa.

In the ischnurid *Opisthacanthus elatus,* the mitochondria do not form a ring in the primary spermatocyte (Wilson, 1916a,b, 1931). The separate mitochondria are randomly distributed during the cell divisions so that the spermatids receive approximately an equal number. These mitochondria also come together in a ringlike pattern just posterior to the nucleus. Each mitochondrion then elongates posteriorly within the cytoplasmic collar.

These two divergent mechanisms of mitochondrial distribution may have evolutionary and phylogenetic significance, as the first was observed only in buthids and the second is present in other families. Mitochondrial fusion with ring formation in the primary spermatocyte was reported for the buthids *Centruroides* (Wilson, 1916a,b, 1931, Wilson and Pollister, 1937) and *Buthus* (Tuzet, 1938). Random distribution of separate mitochondria from the primary spermatocyte was observed in the ischnurid *Opisthacanthus,* the iurid *Hadrurus,* the vaejovid *Vaejovis,* the chactid *Euscorpius,* and the scorpionid *Heterometrus* (Wilson, 1916a,b, 1931; Bhattacharya and Gatenby, 1924; Nath, 1924; Gatenby and Bhattacharya, 1925; Bedi, 1962).

Electron microscopy supports Wilson's (1916a,b, 1931) observations that mitochondrial distribution is distinctive during spermatogenesis in the Buthidae. There are two elongate mitochondria in the middle piece of buthid spermatozoa (Da Cruz Landim and Ferreira, 1972, 1973; Hood et al., 1972; Alberti, 1983). Four to nine mitochondria are present in the middle piece of species of Vejovidae and Chactidae (André, 1959, 1963; Hood et al., 1972; Jespersen and Hartwick, 1973).

Oogenesis

Within the mesosoma, the network of ovarian or ovariuterine tubules on each side leads to an ipsilateral enlarged oviduct and seminal receptacle (fig. 2.8A). These two ducts connect with a single enlarged genital chamber which has an opening (gonopore) covered externally by a flaplike genital operculum (fig.

2.1B). The ovariuterus is so named because the inner wall has a germinal epithelium that gives rise to the oocytes (Laurie, 1891; Awati and Tembe, 1956; Mathew, 1962; Farley, 1996, 1998), and these tubules are also the site of fertilization and embryonic development.

In *Euscorpius italicus,* the inner epithelium of the ovariuterus consists of columnar cells with oval nuclei and clear, faintly granular protoplasm (Laurie, 1890). At intervals along the length of the tubules, one cell (oocyte) begins to increase in size. At first, no sign of its presence is evident on the outside of the tube, but as the oocyte increases in size, it pushes outward, surrounded by trophic follicle cells (fig. 2.8A). The enlarging oocyte remains as a protuberance of the ovariuterine tubule, but the inner surface of the oocyte is open to the lumen of the ovariuterus. Presumably, sperm in the ovariuterine tubules can reach the egg while it is in the protuberance.

This initial passage of the egg out of the ovariuterine tubules occurs in all scorpions. It is probably important for the movement of sperm throughout the length of the tubular network. In some families (Buthidae, Bothriuridae, Chactidae, Iuridae, Vaejovidae; fig. 2.6), the fertilized ovum or morula passes back into the lumen of the ovariuterus or the embryo gradually enlarges and bulges into the lumen where development continues (Laurie, 1890; Abd-el-Wahab, 1953; Awati and Tembe, 1956; Francke, 1982; Polis and Sissom, 1990; Farley, 1996, 1998, 1999a; Warburg and Rosenberg, 1996). Apparently, cells in the ovariuterine wall

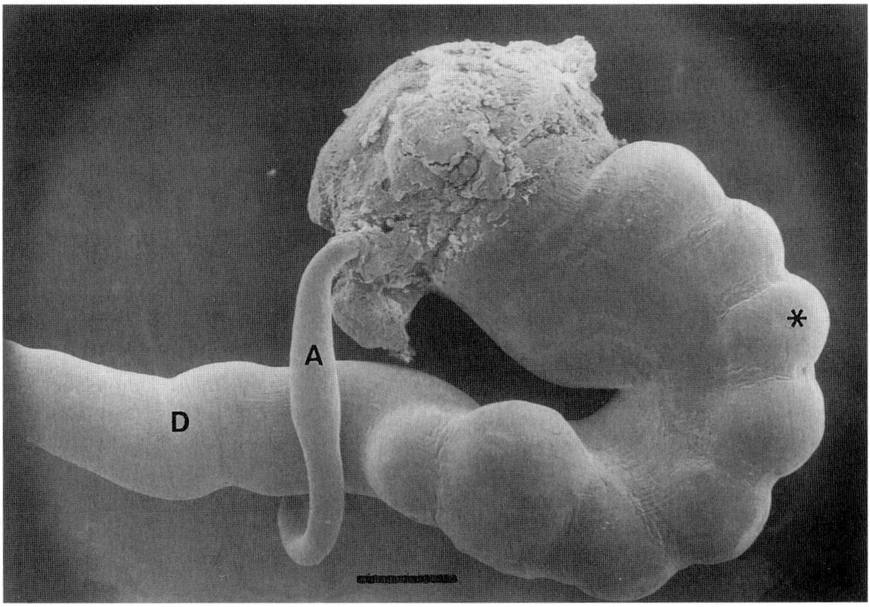

Figure 2.9 Specialized diverticulae of ovarian tubules. Katoikogenic embryogenesis in *Pandinus imperator* (Scorpionidae). Embryo head is oriented toward distal end of diverticulum (D) where appendix (A) apparently passes maternal nutrients to enteron. Diverticulum is swollen (*) by embryo's dorsal processes (eight pairs) which may aid gas exchange. Scale, 1 mm.

transfer maternal nutrients into the lumen, where they are absorbed through the integument of the embryo (Toolson, 1985). Laurie (1896b) used the word, "apoikogenic" ("away from home") to identify these scorpions because the embryo does not develop in an individual diverticulum. In fig. 2.8B, the ovariuterus of the iurid *H. arizonensis* is swollen with embryos.

In the Diplocentridae, Ischnuridae, and Scorpionidae (fig. 2.6), each fertilized ovum remains in a protuberance or diverticulum of the ovariuterine tubule. As the embryo increases in size, the diverticulum lengthens, and, distally, it develops a specialized structure (appendix) that transfers nutrients from the maternal hemolymph to the mouth of the developing embryo (fig. 2.9). This type of development was termed "katoikogenic" ("at home") by Laurie (1896b).

EMBRYOGENESIS

Although other features are associated with each of the two general types of scorpion embryogenesis, Laurie's (1896b) classification is based on the place where the embryo developed—i.e., within the lumen of the ovariuterine tubule (apoikogenic) or in a diverticulum of the tubule (katoikogenic). These terms will be used here, but alternatives may be necessary eventually to accommodate developmental diversity in species not yet examined.

As described below, apoikogenic development is considered primitive because it is less complex and closer to the pattern of embryogenesis thought to occur in ancestral aquatic chelicerates (Anderson, 1973; Yoshikura, 1975). Katoikogenic scorpions have a developmental pattern with many of the features of other primitive chelicerates, but these scorpions have specialized ovariuterine structures (diverticulum, appendix). In addition, the anterior appendages, mouth, and digestive tract differentiate early, enabling the embryos to ingest maternal nutrients.

Aquatic ancestral chelicerates probably released their eggs into seawater, where they needed protective membranes and large amounts of yolk for the developing embryo (Anderson, 1973; Yoshikura, 1975). Such membranes (amnion, serosa) surround the embryos of extant apoikogenic scorpions. Membrane retention and the development of the embryos in the ovariuterine tubules with little modification of the latter is relatively close to the ovoviparity that probably occurred in ancestral aquatic forms.

As pointed out by Laurie (1896a,b), apoikogenic scorpions may have eggs with a large, moderate, or small amount of yolk. The apoikogenic embryos eventually become much larger than the fertilized egg (Francke, 1982), even though the egg is initially swollen with yolk. It is likely, therefore, that the developing embryo absorbs nutrients from the mother, reminiscent of nutrient absorption from the surrounding seawater or brood pouch in ovoviparous aquatic ancestors.

Absorption of nutrients by apoikogenic embryos is probably accomplished by the surrounding membranes and embryonic integument because the mouth and digestive tract develop late (Anderson, 1973; Farley, 1998, 1999a). Isolated apoikogenic embryos of the buthid *C. exilicauda,* with their outer membranes intact, accumulate labeled leucine from the surrounding medium and incorporate the leucine into protein (Toolson, 1985). Presumably, this also occurs inside the mother, and all present scorpions are viviparous regardless of the amount of yolk initially present (Francke, 1982).

In katoikogenic scorpions, the oocyte is relatively small in size and has little yolk. The fertilized egg begins development at a location near the apex of the ovariuterine diverticulum (fig. 2.9). Embryogenesis continues with the embryonic head oriented toward the appendix at the distal end of the diverticulum. The appendix has large, specialized maternal cells that apparently absorb nutrients from the hemolymph and transport them to channels in the central core of the appendix. The nutrients are conducted to the mouth of the embryo (Laurie, 1891, 1896a,b; Pflugfelder, 1930; Mathew, 1948, 1956, 1962; Subburam and Gopalakrishna Reddy, 1989). The early differentiation of the mouth, chelicerae, and anterior digestive tract (Mathew, 1956; Anderson, 1973; Yoshikura, 1975) is an embryonic adaptation that occurs in correlation with the development of specialized maternal structures for feeding the embryo.

Apoikogenic Embryos

In the species examined so far, embryogenesis occurs either within the lumen of the ovariuterine tubules (apoikogenic) or in a diverticulum (katoikogenic). Initial studies (Laurie, 1891, 1896a,b) and reviews (Anderson, 1973; Yoshikura, 1975; Polis and Sissom, 1990) suggested that each of these forms of development is consistent among scorpion taxa. We now know there is much diversity, as described in this and the following sections. Additional variations will probably be discovered as more species are examined; for example, nothing is known about development in the chaerilid evolutionary branch (fig. 2.6; Polis and Sissom, 1990). Also, no experiments demonstrate the transfer of maternal nutrients to apoikogenic or katoikogenic embryos, so function is inferred from morphological studies.

Embryogenesis in Paruroctonus mesaensis

The vaejovid scorpions *Paruroctonus mesaensis* are abundant in deserts of the southwestern United States and are commonly used for experimentation. The embryos are frequently used as examples in this and an earlier review (Farley, 1999a). Reproduction is seasonally synchronized (Polis and Farley, 1979b); gravid females are easily collected and maintained in the laboratory, so embryogenesis can be followed from mating (early summer) to birth 10–14 months later (late summer of following year). Development is apoikogenic, but maternal ovariuterine structures are more complex than those described in other apoikogenic species.

The reproductive system of female *P. mesaensis* has a network of tubules (fig. 2.10) with an attached cylindrical layer of cells. The tubules (ovariuterus) have a lumen for passage of sperm and embryos as in other species (Hjelle, 1990). The cylindrical layer, attached and parallel to the ovariuterine tubules, is continuous with the distal, nutritive end of a broad chamber (uterus) where the embryo develops (Farley, 1996, 1998, 1999a). In transverse section, the trophic layer has no lumen but consists of loosely packed cells. It may serve as a conduit for nutrients similar to the elongate appendix in katoikogenic scorpions (see figs. 2.9, 2.12; Mathew, 1956). Cells in the trophic layer encircle the ovariuterine tubules, possibly inducing a sustentacular state in the latter when embryos are present (Farley, 1998).

The inner lining of the uterus in *Paruroctonus mesaensis* has an epithelium

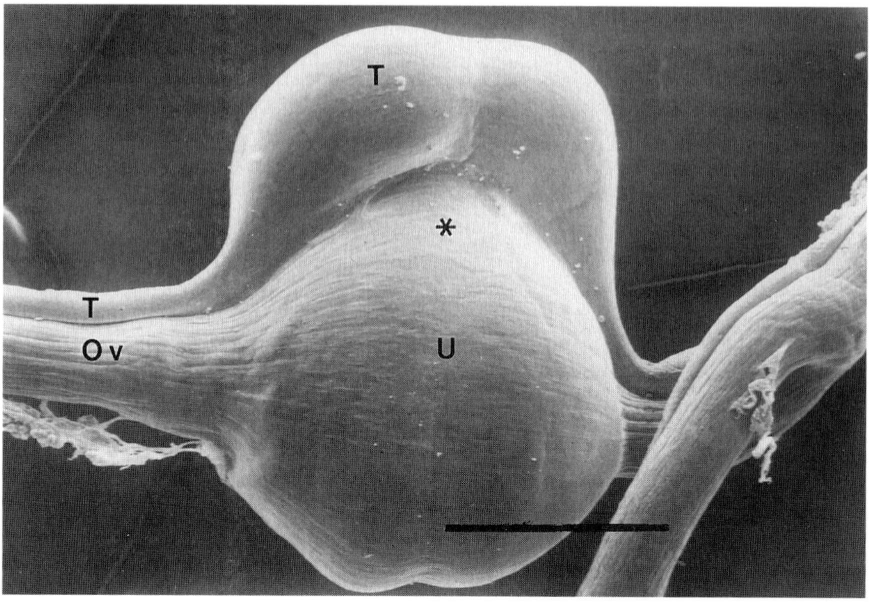

Figure 2.10 Complex apoikogenic development in *Paruroctonus mesaensis* (Vaejovidae). Embryos develop within a specialized region (U, uterus) of ovarian tubules (Ov). Embryo and a presumed trophic cell mass (*) are enclosed within serosa inside uterus (see fig. 2.11). The embryo apparently absorbs nutrients after it reaches the uterus from a trophic network (T) attached to the uterus and ovarian tubules. Scale, 1 mm.

that probably transfers nutrients to the lumen, where they can be absorbed by the egg or embryo. In addition, a bun-shaped mass of cells develops at the distal end of the uterus (figs. 2.10, 2.11; Farley, 1996, 1998, 1999a). The cell mass is continuous with the trophic network and is probably nutritive for the embryo. Only the thin amnion membrane initially surrounds the embryo, but eventually it also becomes enclosed by a thicker serosa. In an early stage of development, the serosa extends upward from the embryo and surrounds the bun-shaped mass along with the embryo (fig. 2.11A). Thus, the cell mass (a follicular placenta; Farley, 1996) is in a position to transfer to the embryo maternal nutrients absorbed in the trophic network (Farley, 1998, 1999a).

The developing embryo is pressed against the trophic cell mass, but there are no intervening tubules or blood vessels. This was evident when the serosa was ruptured with forceps and the embryo was pulled away from the nutritive cell mass (fig. 2.11B). Apparently, embryos adsorb nutrients through the integument, as occurs in the apoikogenic buthid *C. exilicauda* (Toolson, 1985). As the embryo enlarges in the uterus, it compresses the follicular placenta and extends along the length of the ovariuterine tubule well beyond the confines of the initial uterus. That specialized chamber and its placenta are probably important in the early stages of embryogenesis but are no longer needed after development of the circulatory system in the embryo (Farley, 1996, 1998, 1999a).

The ovariuterine tubules, developing embryos, and specialized feeding structures are intermingled among the lobes and branches of the massive digestive

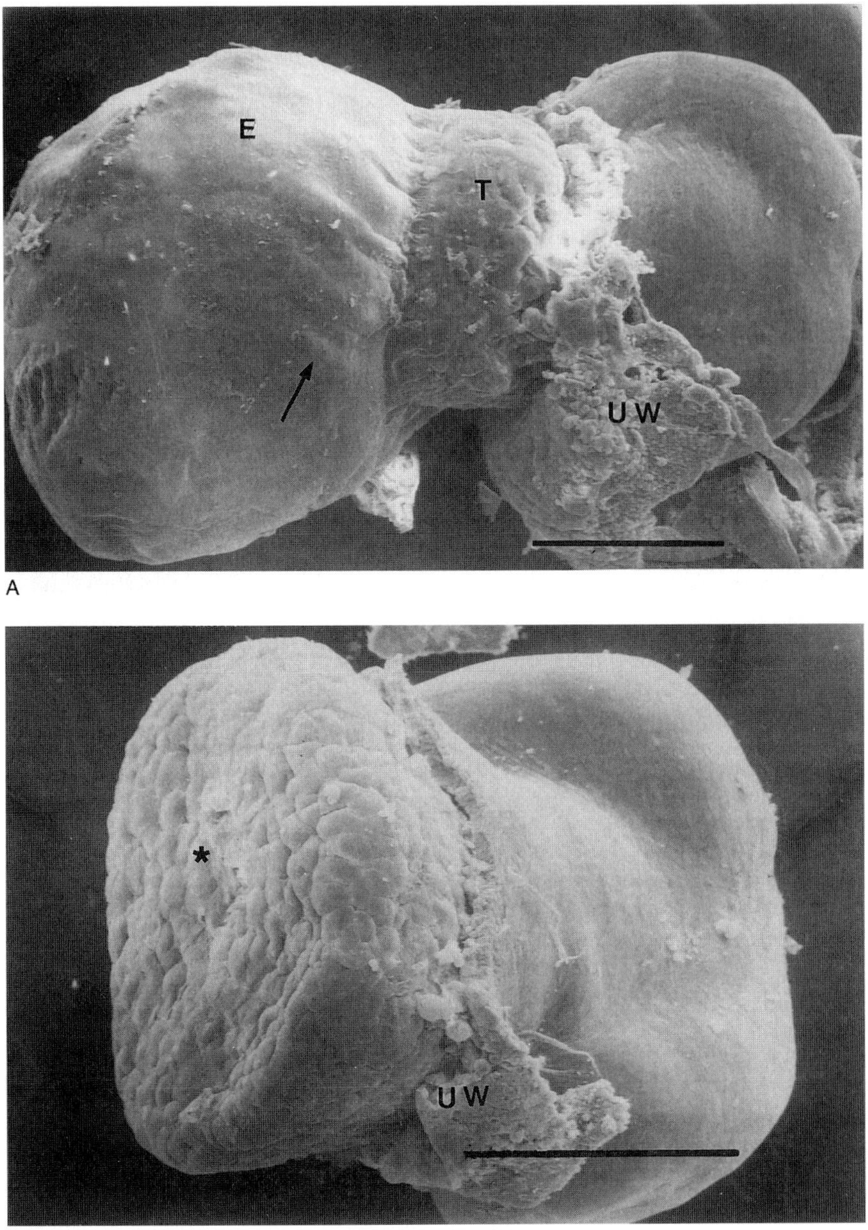

A

B

Figure 2.11 Embryo and trophic cell mass inside uterus of *Paruroctonus mesaensis.* (A) Uterine wall (UW) retracted to show embryo (E) and trophic cell mass (T) enclosed within serosal membrane. Appendages (arrow) of embryo ventrum can be seen through membrane. Embryo left side abuts against cell mass. Scale, 1 mm. (B) A mass of presumed trophic cells (*) is evident after removal of serosal membrane and embryo. Embryo abuts these cells without interconnecting vasculature. Scale, 0.5 mm.

gland in the mesosoma. This occurs in all scorpions and may facilitate nourishment of the embryos. In a preliminary study with a few specimens, Mathew (1948) reported that the digestive gland of gravid females of *Heterometrus* (formerly *Palamnaeus*) *scaber* decreased from 15% of body weight at the beginning of embryogenesis to 7% at the end.

The origin of the trophic layer in female *P. mesaensis* was described by Farley (1996, 1998). The trophic cells are apparently derived from those initially close to the enlarging oocyte. In katoikogenic scorpions, follicle cells are thought to be the precursor of the nutritive cells of the appendix (Mathew, 1956). Rather than transfer maternal nutrients to the egg for storage as yolk, follicle cells and their descendants may pass nutrients directly to the developing embryo and thereby eliminate the metabolic processes associated with yolk.

Embryogenesis in Lychas tricarinatus

Lychas tricarinatus is apoikogenic like other buthids, but it develops an unusual embryonic feeding structure (Mathew, 1960) and thus provides another example of diversity among apoikogenic families. In *Lychas,* the dorsal surface of the embryo develops a stalk with an enlarged globular body at its top. This globular body is thought to be derived from trophic follicle cells. The stalk and globular body protrude into the lumen of an evagination of the ovaruterine wall. At this site, the latter wall is thickened by additional cells in the inner epithelium adjacent to the protrusion from the embryo.

The embryonic globular body and stalk have a central lumen that is continuous with the lumen of the embryonic digestive tract (Mathew, 1960). The morphology suggests that nutrients are secreted by specialized cells in the ovariuterine epithelium, and then the nutrients pass into the embryonic globular body and stalk and from there pass directly into the embryonic digestive tract. In contrast to the specialized feeding mechanisms of katoikogenic species (see next section), the dorsal structure in *Lychas* develops much later, after the embryo is completely segmented. This delay is correlated with late development of the mouth and digestive tract, as occurs in the embryos of other apoikogenic scorpions.

Katoikogenic Embryos

Appendix and embryonic structures

At the apex of the diverticulae, there are species differences in the structures and mechanisms that presumably transfer nutrients to the mouths of embryos with katoikogenic development. The small size and complexity of these apical structures has caused difficulty in interpretation, especially because observations were done mainly with the light microscope. Also, the embryo and apical diverticulum change during gestation, so a sequential series is needed.

In the scorpionids *Scorpio, Heterometrus, and Opistophthalmus* and the Ischnurid *Chiromachus* (formerly *Ischnurus),* the chelicerae enclose a specialized mass of secretory cells (axial cord, teat) that leads directly to the mouth (Laurie, 1891; Vachon, 1950; Mathew, 1948, 1956, 1960). The chelicerae differentiate early, and the distal end has cuticular plates that may facilitate nutrient transfer. In some species (e.g., the scorpionid *Urodacus novaehollandiae),* the

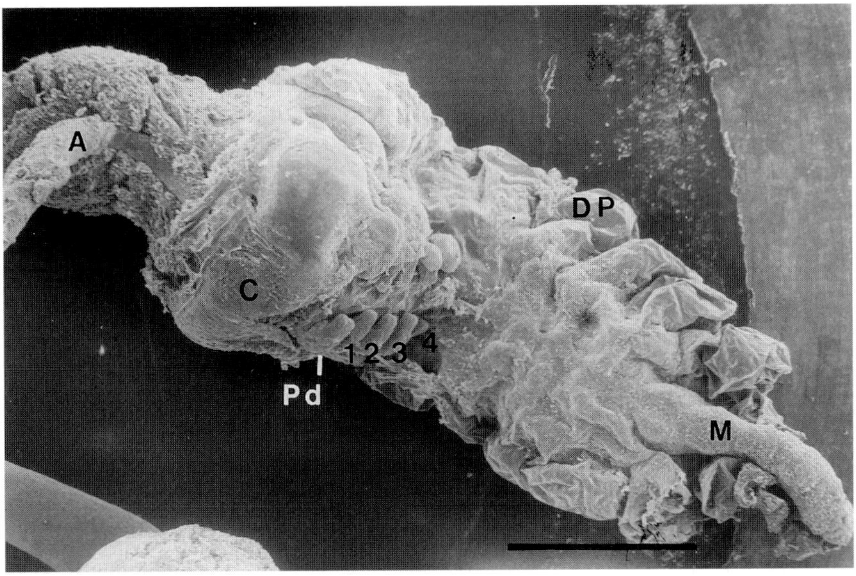

Figure 2.12 Early embryo of *Pandinus imperator,* ventral view, katoikogenic development. Appendix (A) apparently delivers maternal nutrients directly to enteron. Thin-walled dorsal processes (DP) may aid gas exchange. Pedipalp (Pd) and walking legs are evident on one side. No metasomal (M) ventral flexure occurs as in apoikogenic embryos (see figs. 2.15, 2.16A, 2.17B, 2.19). C, cerebral/cheliceral lobe. Scale, 1 mm.

central cord of the appendix terminates close to the distal end of the chelicerae, rather than between the chelicerae (Laurie, 1896b). The two chelicerae are modified to form a groove through which nutrients may pass to the mouth. In embryos of *H. scaber,* Mathew (1948) observed regular pharyngeal movements, which he interpreted as sucking nutrients from the teat into the mouth.

In the ischnurid *Liocheles* (formerly *Hormurus) australasiae,* Laurie (1896b) thought that the embryonic pedipalps, rather than the chelicerae, hold the feeding cord. This was considered a misinterpretation by Pflugfelder (1930), who concluded that the chelicerae enclose the cord. Early stages of development in this species are described by Makioka (1992b).

In the scorpionid *Urodacus abruptus,* a bun-shaped mass of cells forms the appendix at the apex of the diverticulum (Mathew, 1968). Modified chelicerae grasp a buttonlike cone at the base of the appendix and provide a groove for passage of nutrients into the embryo's mouth. Other katoikogenic scorpions have a long and slender appendix at the tip of the diverticulum. Fig. 2.12 shows an early embryo and appendix of *Pandinus imperator.* In this species, the proximal end of the appendix connects early to the anterior end of the embryo. The digestive tract forms just after gastrulation as in other katoikogenic scorpions, and the presumed feeding tube leads directly into the primordial enteron well before the appearance of limb-buds.

Embryogenesis in Opistophthalmus

The genus *Opistophthalmus* is thought to be katoikogenic like other scorpionids (Polis and Sissom, 1990), but Laurie (1896c) described species with unusual mechanisms for embryonic nourishment. The embryos of *Opisthophthalmus capensis* and a related species develop with the head oriented toward the distal end of the diverticulum, but maternal nutrients are apparently absorbed by two outgrowths at the anterior end of the embryo, rather than being ingested. One tubelike structure extends like a probocis from a region just anterior to the mouth. Another protrusion arises from the dorsal carapace, between or behind the median eyes. These outgrowths are filled with connective tissue and blood vessels and are not associated with the mouth, digestive tract, or pumping pharynx.

The chelicerae of these embryos are substantially modified, with distal plates and enlarged joints (Laurie, 1896c). In addition, paired bladderlike structures (dorsal processes) extend from the dorsal surface of both mesosomal and metasomal segments.

Dorsal processes

Although the above developmental diversity apparently evolved for nutrient transfer, the dorsal processes are a striking example of embryonic modification that may facilitate gas exchange. The dorsal processes are thin-walled, bladderlike structures that form on the dorsal opisthosoma as the embryo grows. They were described in Diplocentridae and Scorpionidae (Mathew, 1956; Rosin and Shulov, 1963), but enlarged mesosomal tergites (dorsal ridges, see fig. 2.18) may have a similar function in other families (Farley, 1999a). The dorsal processes and ridges disappear before birth or shortly thereafter. In embryos of *Pandinus imperator,* the cuticle of the dorsal processes is very thin (Farley, 1999a) and frequently collapses during preparation for microscopy (fig. 2.12). The enlarging processes stretch and thin the overlying diverticulum wall (fig. 2.9), probably facilitating gas exchange between the maternal fluids and embryo.

Mathew (1956) suggested that, as the embryo increases in size, and before the embryonic circulatory system develops, there is an unfavorable surface-to-volume ratio so that special respiratory structures are needed. When the circulatory system becomes functional, many of the blood sinuses are sufficiently close to the integument (Kästner, 1940a,b; Millot and Vachon, 1949; Farley, 1990a), so dorsal processes may be no longer necessary.

CLEAVAGE AND EARLY EMBRYONIC DEVELOPMENT

The pattern and sequence of development in scorpions is correlated with the maternal structures and mechanisms (apoikogenic, katoikogenic) for maintenance of the embryos. These differences in cleavage and the earliest stages of embryogenesis raise the possibility that the developmental divergence occurred very early, possibly before terrestrialization in the Devonian and Carboniferous periods. A more likely possibility is that katoikogenic development, including the early embryonic modifications described below, is a secondary mode that originated after viviparity and as a result of the diverse environmental conditions confronted upon terrestrialization.

Cleavage

In scorpion eggs with apoikogenic development, the type of cleavage (total or polar discoidal) is correlated with the amount of yolk. Subsequent stages appear to be similar throughout this group (Anderson, 1973); the embryo develops from a polar germinal or primordial disc, regardless of how much yolk was present in the egg. Katoikogenic scorpions produce small eggs with little or no yolk, so cleavage is total, and subsequent development involves the entire embryonic mass rather than a polar germinal region.

Polar discoidal cleavage occurs in the yolky eggs of apoikogenic scorpions such as *Euscorpius carpathicus, Leiurus quinquestriatus,* and *Euscorpius italicus* (Laurie, 1890; Brauer, 1894, 1895; Abd-el Wahab, 1954). These eggs are about 400 μm in diameter, and no cleavage furrow passes into or through the yolky mass of the egg. Rather, cleavage is confined to a yolk-free polar cap where the initial zygote nucleus is located. Anderson (1973) suggested that the polar discoidal cleavage of yolky eggs is a primitive condition and has been lost in apoikogenic and katoikogenic scorpions that have eggs with little or no yolk. He also noted that polar discoidal cleavage in scorpions differs from the polarized total cleavage in other arachnids and xiphosurans.

Total cleavage occurs initially in the smaller eggs of apoikogenic scorpions (e.g., *P. mesaensis*). Soon, however, the cell divisions occur mainly in a localized region of the egg, and a germinal disc is formed (fig. 2.13), as in the yolky eggs of *Euscorpius* and *Leiurus.* Subsequent development of all these embryos occurs at the germinal disc.

Total cleavage of the small (about 400 μm) eggs in katoikogenic scorpions was described by Pflugfelder (1930) for *Liocheles australasiae* and by Mathew (1956)

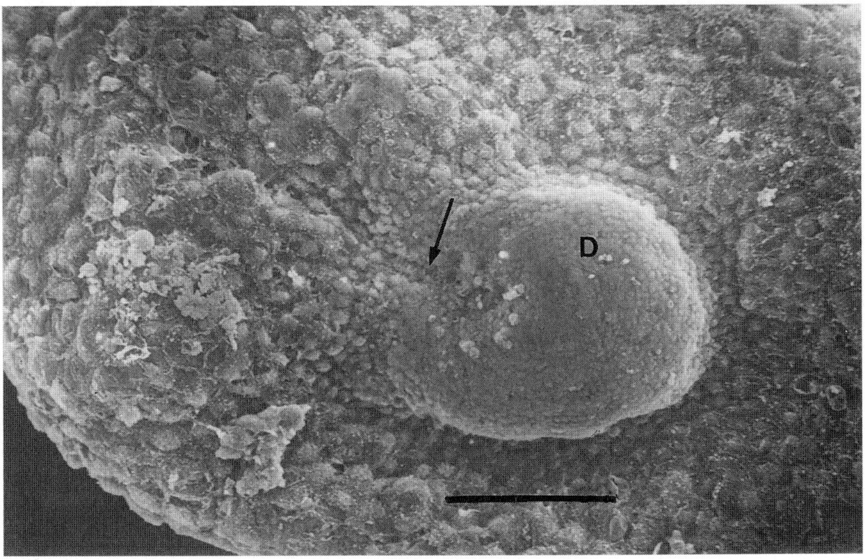

Figure 2.13 Germinal disc of *P. mesaensis.* Apoikogenic development, ventral view. The disc (D) gives rise to the embryo body. Indentation (arrow) may be site of inward cell migration during gastrulation. Scale, 0.1 mm.

for *Heterometrus scaber.* During cleavage in the latter, the follicle cells withdraw slightly and leave a space except at the distal end of the diverticulum. These cells then completely surround the developing embryo, forming an embryonic capsule. The follicle cells at the distal end of the diverticulum proliferate and form a plug, which apparently transfers maternal nutrients to the embryo (Mathew, 1956).

Blastula

In apoikogenic scorpions with yolky eggs (e.g., *Leiurus* and the chactid *Euscorpius),* a discoblastula is formed in which most of the embryo is yolk, but one pole of the egg has a germinal disc or embryonic primordium which consists of a dense mass of cells (Laurie, 1890; Brauer, 1894, 1895; Abd-el Wahab, 1952; Anderson, 1973; Yoshikura, 1975; Farley, 1998). This cell mass gives rise to the main body of the embryo as well as the amnion and serosa (Anderson, 1973). In its initial stages, the germinal disc consists of a single layer of cells. Eventually, some of the cells (yolk cells, vitellophages) push inward into the yolk where they probably assist in the dissolution of yolk platelets (Laurie, 1890; Yoshikura, 1975). There is no indication that these cells give rise to any of the tissues of the embryo (Farley, 1998).

In those apoikogenic scorpions with smaller eggs, a hollow blastula (coelobastula) is formed. The eggs of *Paruroctonus mesaensis* and some other apoikogenic species have apparently reduced their yolk with the development of an efficient mechanism for embryonic feeding. Subsequent embryogenesis is typical of that in apoikogenic embryos. A blastodisc is formed, with a germinal center (fig. 2.13) that gives rise to the main body of the embryo (Farley, 1998, 1999a).

After total cleavage, a hollow blastula develops from the small egg of the katoikogenic scorpion *Heterometrus scaber* (Mathew, 1956). Subsequent differentiation involves the entire embryonic mass, rather than a localized germinal center.

Gastrulation

There is uncertainty about how gastrulation occurs in scorpions with the resulting formation of endoderm and mesoderm. Early workers observed in histological sections a primitive groove or keel in the germinal disc of apoikogenic embryos derived from yolky eggs with polar blastodiscs. This raised the possibility that there might be an inward migration of cells in the groove, somewhat like the primitive streak in vertebrates (Laurie, 1890; Brauer, 1895; Abd-el Wahab, 1954). The resulting gastrula in these animals consists of a yolk-filled region dorsally and a small, three-layered germinal disc ventrally. In the germinal disc of *P. mesaensis* a small depression is commonly seen (fig. 2.13). This may be a site of inward cell migration during gastrulation (Farley, 1998), but cell tracer studies are needed since gastralation can occur by other means in invertebrates (Wilmer, 1990).

In the ischnurid, *Liocheles,* gastrulation was thought to occur by delamination from the ectodermal layer (Pflugfelder, 1930). Mathew (1956) observed in *Heterometrus scaber* a slight depression in the blastula germ disc, where there is an ingrowth of a double row of cells into the blastocoelic cavity. This unipolar ingression of cells soon fills the cavity of the blastula, and mesodermal and endodermal layers are differentiated. Gastrulation in these katoikogenic embryos involves the entire embryonic mass (stereogastrula), because little or no yolk is present.

Embryonic Membranes

The formation of the amnion and serosa probably differs among scorpion species, as suggested by Abd-el Wahab (1954). In histological sections of the apoikogenic embryos of the buthid *Leiurus quinquestriatus,* he concluded that the two membranes arise as an amnio-serosal fold in the ectoderm at the anterior margin of the germinal disc. This fold gradually spreads and encloses the entire germinal disc as well as the yolk-filled dorsal part of the embryo. The embryo continues development within these membranes until after birth, when they are removed.

In katoikogenic embryos, the embryonic membranes may be vestigial and transitory, if they develop at all. A thin, transparent amnion was dissected from the early limb-bud embryo of *P. imperator* in fig. 2.12. Mathew (1956) reported that persisting polar bodies from the egg of *H. scaber* give rise to a trophamnion which temporarily surrounds the embryo. The function of the amnion and serosa is unknown, but their retention in apoikogenic embryos and loss or reduction in katoikogenic embryos suggest the membranes transport nutrients in the former but are supplanted by the appendix in the latter.

Early Stomodeal Invagination in Katoikogenic Embryos

In *Heterometrus scaber,* an invagination for the mouth begins immediately after gastrulation and the formation of the solid stereogastrula cell mass (Mathew, 1956). There is a transitory return of blastocoel space, as the stomodeal invagination continues to lengthen from the anterior end of the embryo. The mouth and digestive tract develop at a much later stage in apoikogenic embryos (see figs. 2.16A, 2.17; Farley, 1999a).

The anterior end of katoikogenic embryos is always oriented toward the distal end of the diverticulum where the appendix forms (fig. 2.9). This rotation of the embryonic axis perpendicular to the ovariuterine tubule is another important difference from apoikogenic embryos. Apoikogenic embryos are oriented parallel to the tubule, and they develop either within it (e.g., *Lychas tricarinatus, Hadrurus arizonensis;* fig. 2.8B) or in a specialized region of it (e.g., *Paruroctonus mesaensis;* figs. 2.10, 2.11). As a result of orientation during gestation, katoikogenic embryos are born tail-first, whereas apoikogenic embryos are born either head- or tail-first (Francke, 1982).

In *Heterometrus* (Mathew, 1956), the stomodeal invagination continues to elongate, extending about two-thirds the length of the embryo when it is about 160 μm long and 100 μm wide. At this stage, the embryo is enclosed within a trophamnion (possibly derived from polar cells) and an embryonic capsule of appendix cells derived from the follicle cells of the egg (Mathew, 1956). The trophamnion and embryonic capsule disappear, and another embryonic membrane develops. This appears to be made up of trophic cells from the appendix. The morphology suggests that nutritive material is secreted from the appendix into the space surrounding the embryo, and then the material flows into the stomodeal invagination, even though the digestive tract (mesenteron) is still only a baglike structure with no hindgut or anal opening. Oral feeding may thus begin very early in katoikogenic embryos, before any other organs are differentiated.

Although the digestive tract and digestive glands continue to differentiate in embryos of *Heterometrus,* the proctodeum does not appear until the embryo is 12–14 mm long, is segmented, and has appendages (Mathew, 1956). The proctodeum begins as an ectodermal invagination from the ventral side of the last metasomal segment. It extends upward and joins with the distal end of the mesenteron.

Segmentation, Embryonic Flexure, and Dorsal Closure

In apoikogenic scorpions, the embryonic primordium of the blastodisc becomes oval shaped and eventually segments into three parts (Laurie, 1890; Brauer, 1894, 1895; Abd-el Wahab, 1952): the primordial cephalic lobe, a cheliceral segment (body segment III), and a growth zone. As the germinal disc continues to elongate, the growth zone proliferates succeeding segments in an anterior-to-posterior direction, starting with the segments of the pedipalps and walking legs (body somites IV–VIII; Farley, 1998, 1999a). This stage is shown in fig. 2.14 in embryos of *P. mesaensis.*

In his review of chelicerate embryology, Anderson (1973) used the xiphosuran *Tachypleus tridentatus* (aquatic, oviparous) as a primitive example that is probably close to the ancestral chelicerate condition. The pattern of segment formation described above for scorpions is similar to that of *Tachypleus* and, as Anderson suggested, this pattern was retained in scorpions in spite of ovariuterine and embryonic changes for viviparity. Xiphosurans and scorpions are unique among chelicerates in having growth-zone proliferation of both prosomal and opisthosomal (mesosoma and metasoma) segments. In other chelicerates, the prosoma and mesosoma emerge directly from the blastoderm, and the growth zone gives rise only to the metasomal segments.

As in scorpions with yolky eggs and apoikogenic embryos, the xiphosuran blastula develops a germinal band (primordium of the embryo) at one pole (Anderson, 1973). A gastral groove develops early in the longitudinal midline of the germ band, where the groove eventually gives rise to the stomodeum and other parts of the digestive tract. A gastral groove is thought not to occur in scorpion embryos; the digestive tract in apoikogenic embryos develops late (Laurie, 1890; Brauer, 1895) by proliferation from the growth zone and invagination of ectoderm for the stomodeum and proctodeum.

Subsequent development in apoikogenic embryos is similar to that in *Tachypleus* and primitive spiders (Anderson, 1973; Yoshikura, 1975). The embryonic primordium is somewhat U-shaped (dorsal flexure) and consists of a band of tissue ventral to the yolk mass (fig. 2.15A). Yolk is retained dorsally, especially at the anterior end. Prosomal segments with limb-buds differentiate first, followed by mesosomal segments which also have limb-buds. The embryo develops a ventral flexure as the metasoma elongates (figs. 2.15, 2.16A, 2.17B, 2.18, 2.19). This type of flexure is considered more primitive than dorsal flexure (Laurie, 1890; Brauer, 1894, 1895; Abd-el Wahab, 1952; Yoshikura, 1975).

In embryos of *P. mesaensis,* the newly formed metasomal segments are about the same width as the mesosomal segments (fig. 2.15). The embryo thus has a stage like *Sanctacaris uncata* (fig. 2.4A) where the abdomen (opisthosoma) narrows posteriorly but is a single unit (Briggs and Collins, 1988) without separation into pre- and postabdomen as occurs in later scorpion embryos (figs. 2.16A, 2.17B, 2.18, 2.19) and adult scorpions and eurypterids (Kjellesvig-Waering,

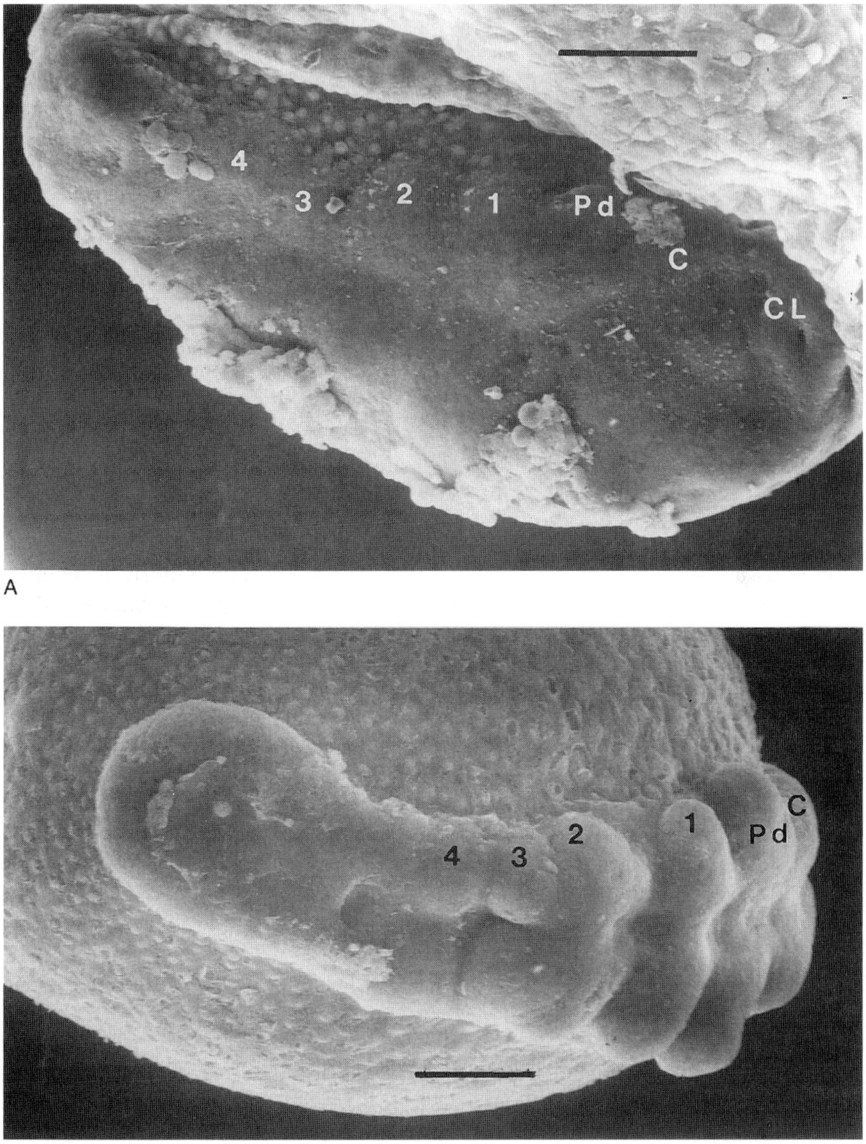

A

B

Figure 2.14 Early limb buds in embryos of *P. mesaensis.* Appendages become identifiable as segments form in an anterior-to-posterior direction (ventral views). (A) Prosomal limb buds are barely discernible. (B) Later stage. Limb buds of anterior legs are more prominent than later emerging posterior ones. C, chelicera; CL, cerebral lobe; Pd, pedipalp; 1–4, walking legs. Scale, 0.1 mm for both A and B.

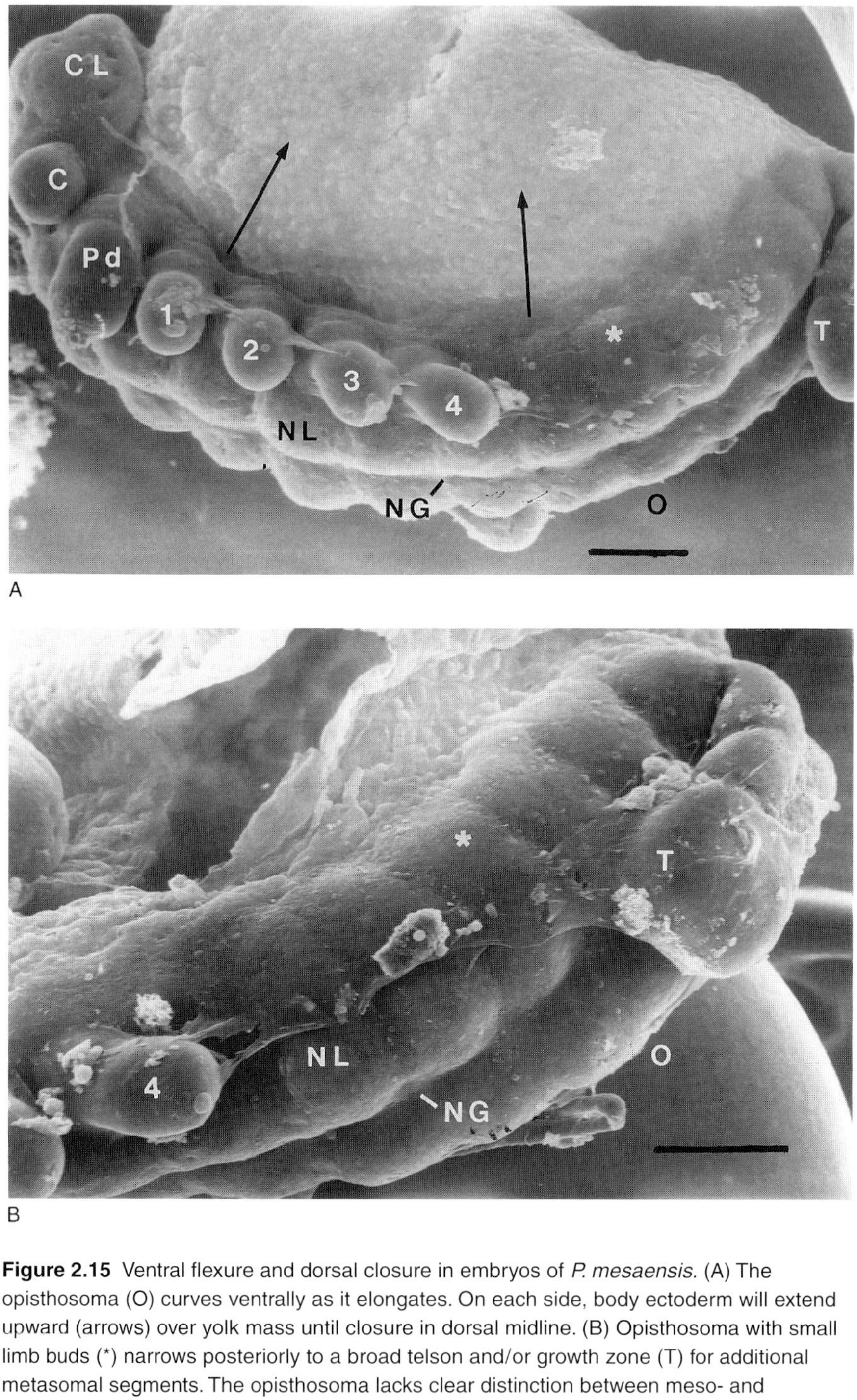

Figure 2.15 Ventral flexure and dorsal closure in embryos of *P. mesaensis.* (A) The opisthosoma (O) curves ventrally as it elongates. On each side, body ectoderm will extend upward (arrows) over yolk mass until closure in dorsal midline. (B) Opisthosoma with small limb buds (*) narrows posteriorly to a broad telson and/or growth zone (T) for additional metasomal segments. The opisthosoma lacks clear distinction between meso- and metasoma. C, chelicera; CL, cerebral lobe; NG, neural groove; NL, neural lobe; Pd, pedipalp; 1–4, walking legs. Scale, 0.1 mm for both A and B.

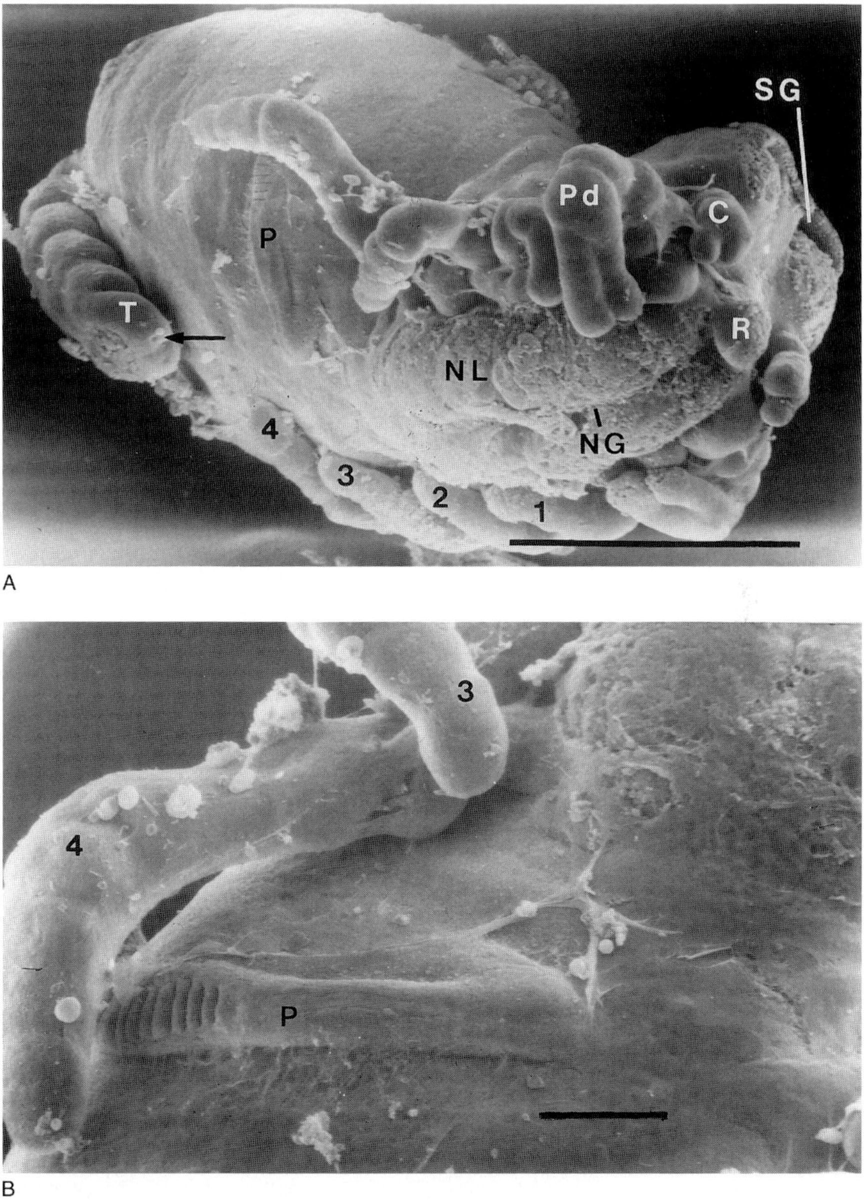

Figure 2.16 Embryos of *P. mesaensis* with features reminiscent of ancestral aquatic existence. (A) Anteroventral view. Broad telson (T) with two transitory, distal indentations (arrow), short and tubular walking legs (1–4) and paddlelike pectines (P). Chelicerae (C), now lateral to rostrum (R), will move dorsally and form roof of preoral tube (figs. 2.3B, 2.17). Walking legs shift medially (fig. 2.17B) with differentiation of nerve cord and reduction of neural lobes (NL). NG, neural grove; Pd, pedipalp; SG, semilunar groove (invagination for medial eyes and protocerebrum) in cerebral lobe. Scale, 1 mm. (B) Ventral view. Distal end of pecten (P) at this stage has transverse striations and is broader than proximal stalk. Pecten later becomes ribbonlike with striations along most of its length (figs. 2.17B, 2.19). 3,4, walking legs. Scale, 0.1 mm.

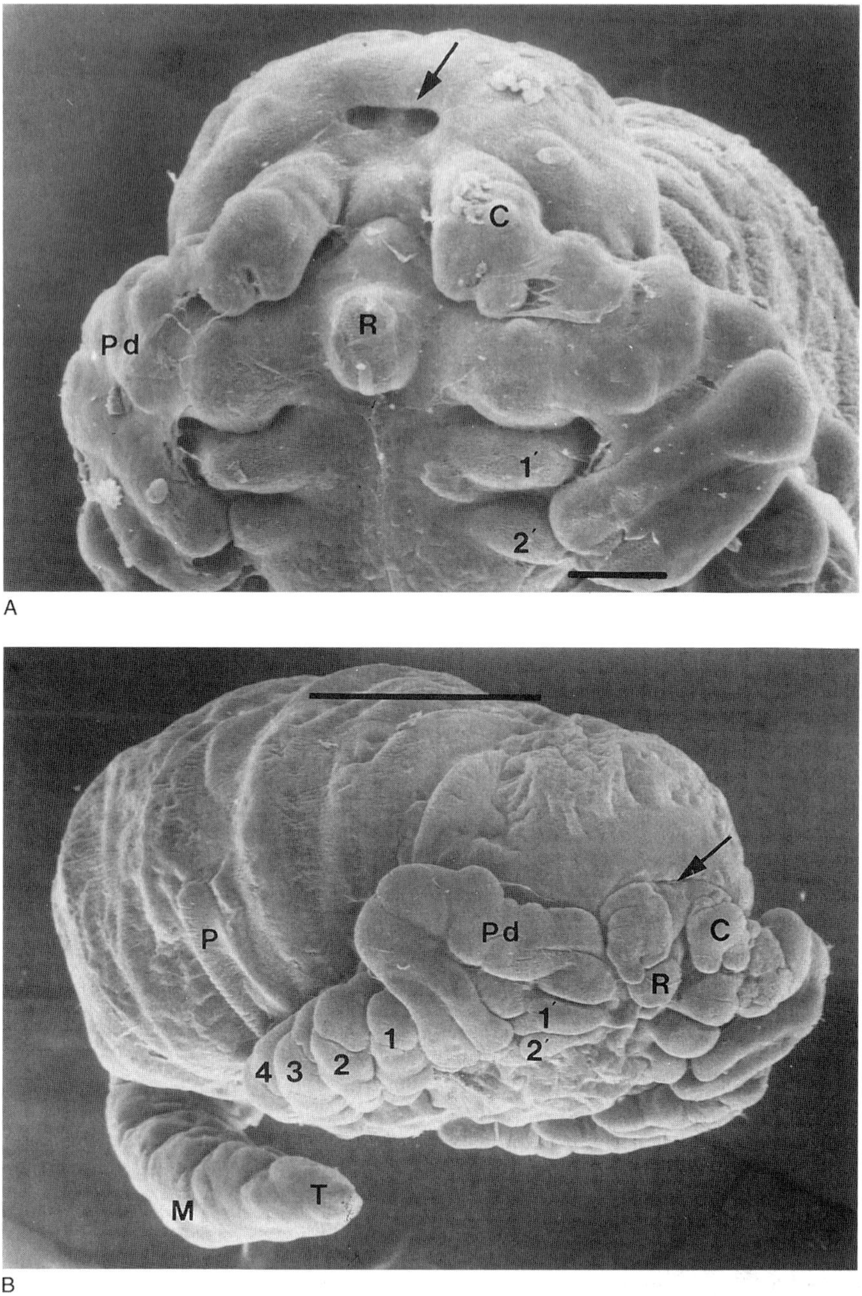

Figure 2.17 Formation of preoral tube in embryos of *P. mesaensis*. Primordial preoral tube appears as chelicerae (C) move dorsally above rostrum (R) and coxapophyseal lobes (1′, 2′) enlarge at bases of first and second walking legs. Coxae and trochanters of pedipalps (Pd) form sides of preoral tube. Anterior views. (A) Arrow indicates openings for invaginations for medial eyes and protocerebrum. Scale, 0.1 mm. (B) Later stage. Arrow indicates closed invagination site. Ventrally flexed metasoma (M) is narrowed and distinct from mesosoma. Telson (T) tapers distally. Tubular legs protrude ventrally beneath body; they will later extend laterally. Pectines (P) are ribbonlike with striations along their length. Abdominal plates not yet present in mesosoma. 1–4, walking legs. Scale, 1 mm.

1986; Tollerton, 1989). The secondary morphological demarcation of mesosoma and metasoma in scorpions reflects emergence of functional specialization (i.e., reproduction, digestion, circulation, gas exchange in the mesosoma and venom delivery in the metasoma) (Farley, 1998, 1999a).

As each body segment demarcates ventrally, the primordial embryonic ectoderm begins to extend laterally and dorsally, overlying the extraembryonic ectoderm that encloses the yolk (Anderson, 1973; Yoshikura, 1975). Eventually, the embryonic ectoderm from each side meets in the dorsal midline of the segment. This stage is termed dorsal closure, and its anterior-to-posterior progression allows early differentiation of the heart and circulatory system for distribution of oxygen and nutrients throughout the embryo (Anderson, 1973; Farley, 1998). In figure 2.15A, the primordial embryonic ectoderm is just starting to extend dorsally in the embryo of *P. mesaensis.* In fig. 2.16A, dorsal closure is completed for the full length of the embryo.

In some primitive spiders (Liphistiomorphae) with a broad and lobular metasoma, the yolk is translocated postero-ventrally as the anterior segments achieve dorsal closure (Anderson, 1973; Yoshikura, 1975; Foelix, 1996). In association with yolk translocation, the embryonic germ band undergoes inversion, (i.e., assumes an inverted U shape). Dorsal closure eventually occurs around the yolk

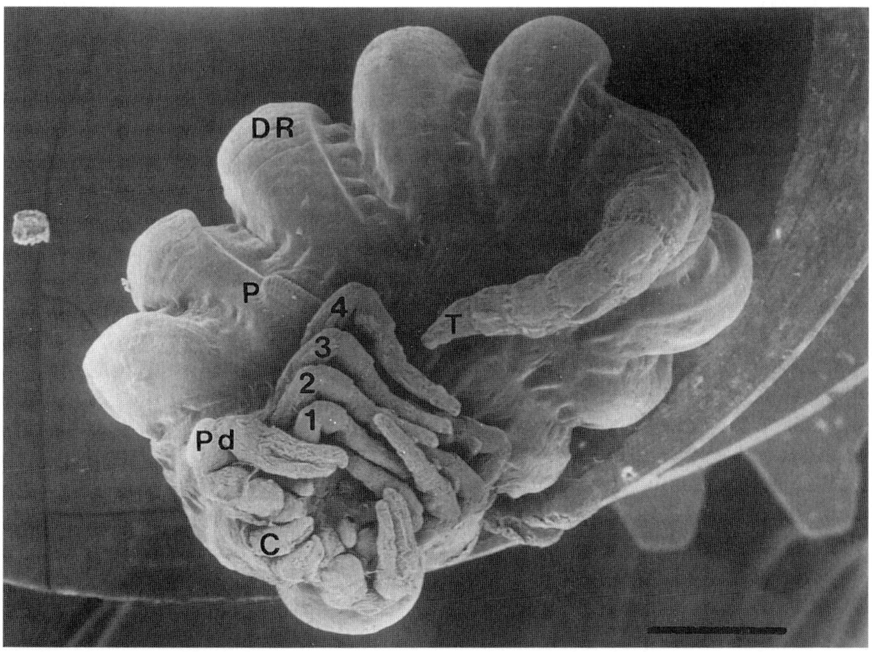

Figure 2.18 Embryo leg structure reminiscent of ancestral aquatic existence. Short, tubular, digitigrade legs protrude ventromedially in early embryo of *Hadrurus arizonensis.* Legs later extend laterally, become flattened (femur, patella) and plantigrade—changes that may be terrestrial adaptations. Prominent dorsal ridges (DR) may aid gas exchange during development. Telson (T) is tapered; abdominal plates are not yet present in mesosoma. C, chelicerae; P, pectine; Pd, pedipalp; 1–4, walking legs. Scale, 1 mm.

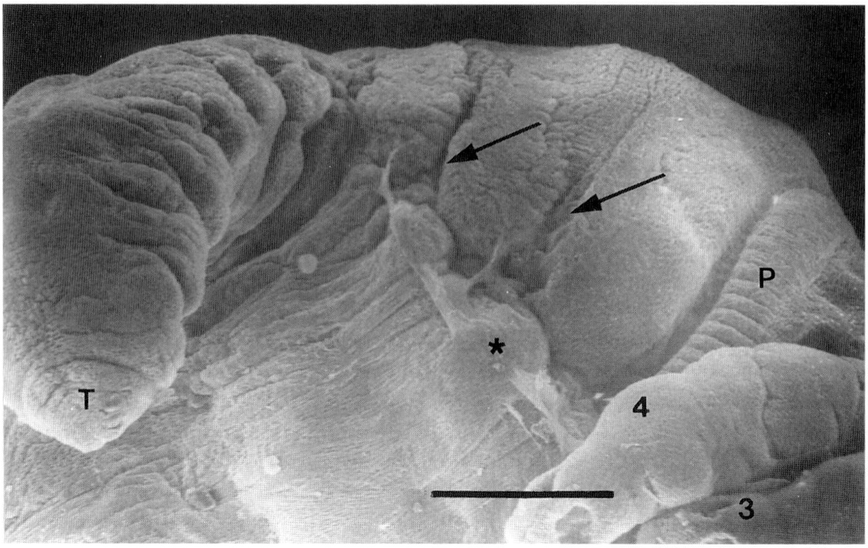

Figure 2.19 Ventral mesosoma of embryo of *P. mesaensis*. Ventral view. The lateral edge of the ventral mesosoma has transverse invaginations (arrows) that delineate the segments, but these invaginations have not yet reached the mid-region (*) where segmental boundaries are still unclear. Demarcation of segments in the ventral mesosoma is followed by formation of segmental structures: abdominal plates (fig. 2.20A), sternites (fig. 2.20B), spiracles and book lungs. More information about these in later stages is provided by Farley (1999a,b). The longitudinal invagination just lateral to the asterisk (*) may be associated with the dorso-ventral muscle band attached at this site in each segment. Pectines (P), ribbonlike with transverse striations along their length, are starting to separate from the body wall at the posterior edge of the pectine. The telson (T) is tapered but does not yet have the specialized features of an aculeus. 3, 4, walking legs. Scale, 0.1 mm.

in the metasomal segments. Yolk translocation and inversion do not occur in scorpions, but the elongating metasoma curves beneath the body in apoikogenic embryos (figs. 2.15, 2.16A, 2.17B, 2.18, 2.19) and in the early stages of some katoikogenic embryos.

Ventral flexure of the metasoma occurs in embryos of xiphosurans and primitive arachnids (Anderson, 1973). Apoikogenic scorpions retain this primitive condition throughout embryogenesis, even though the tail must be curved dorsally for use after birth (Farley, 1998). Adult scorpions hold their metasoma straight out or curved dorsally but never bend it ventrally as in apoikogenic embryos. The structure and interconnection of metasomal segments differ substantially between embryos and adults. The embryo metasoma consists of a series of cylinders of decreasing diameter (figs. 2.17B, 2.18, 2.19), while the adult segments make contact at cuticular pivot points that allow dorsal but not ventral flexure (Bowerman, 1972a,b). In katoikogenic scorpions, the embryo is confined within a tubular diverticulum, so the elongating metasoma is held straight posteriorly or assumes a slight dorsal curvature (figs. 2.9, 2.12; Mathew, 1956).

ORGANOGENESIS

Chelicerae, Pedipalps, and Legs

Van Der Hammen (1985a,b) suggested that the legs of chelicerates are derived from the lobopodia of primitive species, and the three segments of the chelicerae of extant adults may indicate the ancestral condition. In the primitive Cambrian chelicerate *Sanctacarus* (fig. 2.4A), the trunk appendages are lobopodia, while the head appendages are slender and elongate and consist of multiple segments (fig. 2.4A,B; Briggs and Collins, 1988).

Except for variations in claw size and shape, the pedipalps of fossil and present scorpions are much alike (Kjellesvig-Waering, 1986). Some Silurian species, however, had chelicerae with four segments (fig. 2.2C) whereas extant species have only three (fig. 2.3). Some of the former chelicerae were longer and could probably bend sufficiently to pass food directly to the mouth. It is likely this was advantageous for aquatic bottom-feeders, so the four-segment condition may have been a derivation of the basic three segments proposed by Van Der Hammen (1985a,b, 1986a,b). The shorter, three-segment chelicerae of present forms are used to grasp and macerate food for external digestion (Farley, 1999a). Mathew (1956) reported only three segments in the embryos and adults of *H. scaber,* and only three segments occur in embryonic chelicerae in *P. mesaensis.*

Vachon (1945) reported that the pedipalps and walking legs of chelicerate embryos first divide into four segments, and this may be a primitive condition from which the seven-segment legs of present forms developed. The four embryonic segments were designated as protocoxa or prototrochanter, protofemur, prototibia, and prototarsus. A four- or five-segment condition is evident in the embryos of figs. 2.16 and 2.17 (Farley, 1998).

There is uncertainty about which features are aquatic adaptations in the legs of scorpion fossils, so it may be difficult to discern transitory and recapitulated aquatic structures in the developing legs of scorpion embryos. Although eurypterids had some legs adapted for swimming, it appears that aquatic scorpions used their legs to walk, climb, and burrow in the substrate as in present terrestrial forms (Kjellesvig-Waering, 1986; Selden and Jeram, 1989; Jeram, 1990b, this volume). Some Paleozoic scorpions had eurypteroid legs with a digitigrade tarsal region (fig 2.2) that terminated in a single spine, in contrast to the plantigrade tarsi (fig. 2.1A) with claws or spurs in present species (Kjellesvig-Waering, 1986). Størmer (1963, 1970, 1976) suggested that the former condition is advantageous for movement in water, whereas the latter indicates terrestrialization. Others have concluded that the variety of intermediates among fossils makes tarsal form an uncertain indicator of habitat (Kjellesvig-Waering, 1986; Selden and Jeram, 1989; Jeram, this volume).

The legs of terrestrial species must be stronger to sustain the animal's weight without the buoyancy of water, and the eurypteroid legs of Paleozoic scorpions were short and tubular compared with extant scorpions with long legs and flattened femur and patella (Kjellesvig-Waering, 1986). Also, the legs of aquatic scorpions protruded directly from beneath the body, supporting the weight above (fig. 2.2). Without the buoyancy of water, terrestrial scorpions developed a hanging stance (fig. 2.1) with the legs protruding laterally from the body (Van Der Hammen, 1985a,b). In *P. mesaensis,* the legs are initially widespread and separated by the medial neural lobes and groove (figs. 2.15, 2.16; Farley, 1998). The latter disappear with differentiation of the nerve cord, and then the legs protrude more directly beneath the embryo (fig. 2.17). Lateral leg projection occurs at a later stage. Digitigrade tarsi and tubular legs directly beneath the body are evident in the early embryo of *H. arizonensis* in fig. 2.18.

In embryos of *P. mesaensis,* the segments (V–VIII) and limb-buds for the four walking legs appear abruptly (fig. 2.14) and well before the first indications of pectines in the third mesosomal segment (XI; Farley, 1998). In the apoikogenic embryos of *Leiurus quinquestriatus* (Abd-el Wahab, 1952), a limb-bud appears briefly and then disappears from the first mesosomal segment (IX). A transitory ridge (segment unknown) was observed just anterior to the primordial pectine (segment XI) in embryos of *P. mesaensis* (Farley, 1998, 1999a). In some micrographs, the transitory structure appeared to merge with the anterior edge of the pectine.

In the second mesosomal segment (X), the gonopore and the paired covering lobes (genital opercula) develop late in embryogenesis (Abd-el Wahab, 1952). These external reproductive structures have not yet appeared in advanced embryos of figs. 2.17 and 2.19. Pectines emerge at an earlier stage from the third mesosomal segment (XI), and in *P. mesaensis,* they undergo transformations in structure that suggest ancestral functions different from that in extant species (see below).

Cephalic Lobes in Apoikogenic Embryos

The anterior end of apoikogenic embryos enlarges to form the right and left cephalic lobes (Anderson, 1973; Farley, 1998). The earliest stages of the cephalic lobes in embryos of *P. mesaensis* can be seen in figs. 2.14A and 2.15A. With further expansion, each lobe forms a transverse indentation that separates the anterior precheliceral lobe from a posterior cheliceral segment (Anderson, 1973). In some species, a curved groove (semilunar) develops in each precheliceral lobe (fig. 2.16A). These invaginations give rise to the median eye and protocerebral hemispheres (brain). Postero-ventrally, the semilunar grooves merge at the enlarging labral protuberance (rostrum), just anterior to the stomodeal invagination.

The cephalic lobes develop precociously in apoikogenic embryos (Laurie, 1890; Brauer, 1894, 1895; Abd-el Wahab, 1952; Anderson, 1973), and they are prominent in spider embryos. In the katoikogenic embryos of *Heterometrus scaber,* the mouth and anterior digestive tract develop very early; large and distinct cephalic lobes do not appear (Mathew, 1956).

Preoral Tube

At an early stage in the scorpion embryo, the area in front of the mouth is bent strongly upward. The chelicerae migrate to a position above the mouth, and the pedipalps and mouth move to a frontal position (Abd-el Wahab, 1952; Størmer, 1970; Farley, 1998, 1999a), forming the preoral chamber or tube (fig. 2.3B), which extends forward and enables scorpions to digest food partially before ingestion.

The preoral tube is formed by the basal segments of the anterior appendages (Srivastava, 1955; Snodgrass, 1952; Venkateswararao, 1967; Størmer, 1970; Farley, 1999a), and in apoikogenic embryos, these anterior structures differentiate late. In *P. mesaensis,* they begin to take a tubelike position at about the same time as the differentiation of some other structures that may be associated with terrestrial existence. Figure 2.16A shows the cheliceral lobes lateral to the area of presumptive stomodeum. At a later stage, the chelicerae are dorsal to the rostrum and mouth (fig. 2.17). The pedipalps have moved to either side of the mouth, where their coxae and trochanters eventually form the sides of the preoral tube. Coxapophyses (endites) appear initially as small lobes at the base of the first and second walking legs (fig. 2.17A). The coxapophyses increase in size (fig. 2.17B) and extend forward to become the floor of the preoral tube (fig. 2.3B).

Pectines, Early Stages

Ancient scorpions are similar in structure to the extinct eurypterids except for the development in the former of pectines, chelae on the pedipalps, and a telson with a sharp aculeus to deliver venom (fig. 2.5; Kjellesvig-Waering, 1986). The pectines appear fully developed in early Silurian scorpions (fig. 2.2B; Kjellesvig-Waering, 1986). Because these scorpions were probably aquatic, it is possible that the pectines functioned differently from in present terrestrial species. In fossils of aquatic scorpions, there is considerable variation in the size and shape of the pectines, but the basic comblike structure is similar to the pectines in surviving species.

There has been much speculation about the function of the pectines in ancient and present scorpions (Cloudsley-Thompson, 1955). Proposals included equilibrium, audition, fanning the book lungs, auxiliary respiratory organ, detection of ground vibrations, grooming, and a clasping organ during mating. We now know they are elaborate mechano- and chemoreceptor organs with a massive and topographically-organized input to the central nervous system (Brownell, 1991, this volume; Gaffin et al., 1991; Gaffin and Brownell, 1990, 1997a,b, this volume).

The pectines have complex muscles at their base, allowing substantial flexibility of movement (Lankester et al., 1885). This and the flaplike structure of pectines in some species (e.g., *Gigantoscorpio willsi*) suggest use as fins to provide stability during swimming (Wills, 1960; Størmer, 1963). A similar function is attributed to some specialized appendages of eurypterids. The pectines are currently sensory, and they may also have relayed sensory information in the aquatic environment. As pointed out above, the proximity of the pectines and genital opercula raises the possibility that the pectines once had a reproductive function.

In the apoikogenic embryos of *P. mesaensis,* the pectines undergo successive structural transformations, suggesting that they had different functions in succeeding ancestral generations (Farley, 1998, 1999a). Unlike anterior appendages, pectines do not appear with a limb-bud structure that protrudes from the body. Rather, the entire length of the pectinal limb-buds remains attached to the external surface of the mesosoma. At a later stage, transverse grooves or slits appear at the ends of the pectines (fig. 2.16). This results in a paddlelike structure, narrow proximally and broad distally, suggesting an ancestral function to maintain balance in water and/or to circulate water over eggs. These paddlelike pectines continue to remain attached to the body surface as more transverse slits develop along their length and the pectines assume a long, ribbonlike structure (figs. 2.17B, 2.19). This, along with a regular pattern of transverse grooves, suggests a respiratory and/or sensory (rheotactic, olfactory) function for pectines of this form.

Pectines and Structures for Terrestrialization

In *P. mesaensis,* several structures presumed to be terrestrial adaptations appear at about the same time in the embryo (Farley, 1998, 1999a). The structures of the preoral tube become prominent and move into a position in front of the mouth (fig. 2.17). The telson changes from a broad and bilobed structure (figs. 2.15, 2.16A) to one that is tapered distally (figs. 2.17B, 2.19). Eventually, the pectines start to separate from the external mesosomal body surface, and the transverse grooves in the pectines become spaces between teeth that are attached anteriorly but free posteriorly.

The early embryonic stages of pectines in *P. mesaensis* do not resemble any of the pectines in fossil aquatic or terrestrial scorpions (fig. 2.2B; Kjellesvig-Waering, 1986). Fossil pectines vary substantially in shape and number of teeth, but all have a comblike structure with teeth attached anteriorly and free posteriorly. This suggests that these early embryonic stages, if they indeed recapitulate ancestral adult forms, represent pectines from aquatic ancestors that preceded the Silurian species for which we have fossils.

In embryos of the iruid *Hadrurus arizonensis,* the pectines have a ribbon-like stage (fig. 2.18) much like that in *P. mesaensis* (figs. 2.16A,B, 2.17B, 2.19; Farley, 1999a), where the pectine has transverse striations and remains attached to the body wall. In the former species, the pectine separates and forms a comb-like structure (fig. 2.20A,B) at an earlier stage of development than in *P. mesaensis.* In both species, the initial transverse striations disappear and teeth are separated at the posterior edge of the pectine.

In embryos of the scorpionid *Pandinus imperator,* the pectine does not undergo a prolonged ribbon-like stage, but rapidly forms an appendage with posterior teeth. In all the species in which pectine development was studied with the SEM, the external covering of the teeth is thin and smooth in comparison with the surface layer on other parts of the embryo. There is initially no external indication of receptors on the teeth. While abdominal plates are still present in the embryo and before sternites and book lungs are formed, the first few receptors appear on the ventral surface of each tooth. The receptors are initially at various stages of development, suggesting a sequence from a small opening in the tooth surface to a cuticular nodule, cylinder, or tapered sensillum with an opening in the center. The openings are unprotected and 200–400 nm in diameter (Farley, unpublished observations).

These developmental studies suggest that the thin-walled pectinal teeth initially had a respiratory function, and then developed chemoreceptive function while the animals were still in an aquatic environment. Large numbers of peg sensilla with an elongate slit at the tip (Gaffin and Brownell, this volume) develop later, apparently as a terrestrial adaptation. Much variation in the length, shape and density of pectinal sensilla was observed in an electron microscopic study of adult scorpions from all major groups and representative environments (Fet and Brownell, in preparation).

Respiratory Structures

The development of book lungs enabling colonization of land was considered by Kjellesvig-Waering (1986) to be the most important evolutionary change in

Figure 2.20 Early formation of abdominal plates and sternites. Embryos of *Hadrurus arizonensis.* Ventral view. (A) Arrow indicates the outer margin of flap-like abdominal plate. Bilateral depressions (*) in each segment are the attachment site of dorso-ventral muscles. This region may also provide epidermal primordium for the book lungs which form just lateral to the depression in segments XII–XV. Pectines (P) have teeth; distal ends are separated from body wall. (B) Later, the mesosoma becomes broad and flat. The overlapping ventral sclerites are broad and more like sternites rather than flaps. The texture of the epidermis is modified at bilateral depression sites (*), suggesting some invagination of tissue. At the latero-posterior edges of these epidermal scars, oval-shaped spiracles gradually appear in later stages. Scales, 0.5 mm.

scorpions. He proposed that this should be the basis for division of the order into two suborders: Branchioscorpionina (extinct aquatic species with gills) and Neoscorpionina (present and former species with book lungs). He provided some evidence that Paleozoic scorpions had gills located in the space above cuticular plates that were attached at the anterior margin of the segment, but unattached posteriorly (fig. 2.5). This apparently allowed water to pass above each plate, where presumably some type of gill was located. A fossil Carboniferous scorpion *Paraisobuthus duobicarinatus* showed evidence of spongy gill tracts above abdominal plates (Kjellesvig-Waering, 1986). These gills resemble those described in eurypterids. The Devonian scorpion *Waeringoscorpio* had filamentous gills that extended laterally from the ventral mesosoma (Størmer, 1970; Brauckmann, 1987; Selden and Jeram, 1989). External gills have not been seen in scorpion embryos.

In some scorpion fossils, the abdominal plates are smaller with ventral body wall above them (fig. 2.5). Kjellesvig-Waering (1986) hypothesized that there was gradual plate reduction in ancient scorpions, leaving sternites above. Fossil scorpions with small, bilobed plates (infraorder Bilobosternina) were proposed as the precursors of present terrestrial scorpions with sternites and book lungs (suborder Neoscorpionina). This proposal was not supported in recent studies with scorpion embryos (Farley, 1999b). They have transitory holostern abdominal plates (fig. 2.20A) with no evidence of plate reduction.

Abdominal plates designated as protolobosternous were found to cover book lungs in a new scorpion genus from the Lower Carboniferous in Scotland (Selden and Jeram, 1989; Jeram, 1990a, this volume). Thus, plates on the ventral surface of the mesosoma do not necessarily indicate gills and aquatic existence as proposed by Kjellesvig-Waering (1986). These scorpions had five pairs of book lungs and no spiracles (Jeram, 1990a, this volume). Air could apparently enter the book lungs along the posterior margin of each ventral plate. The earliest fossil with attached spiracles (four pairs) and presumably book lungs is *Palaeopisthacanthus* from the Upper Carboniferous (Vogel and Durden, 1966). Most recently, Shear et al. (1996) discovered well-preserved impression fossils of book lungs in cuticle fragments of Lower Devonian mesoscorpions, strong evidence that the group was part of the terrestrial macroarthropod community at this much earlier time.

Lankester (1885) proposed that book lungs of scorpions were derived from book gills like those in *Limulus.* This was supported by Little (1983), who concluded that book lungs are relatively inefficient organs for gas exchange and are unlikely to have originated except as modification of preexisting book gills in aquatic animals. At an early stage of development, the book lung invagination of scorpion embryos has three or four folds reminiscent of book gills (Laurie, 1890; Abd-el Wahab, 1952; Mathew, 1956; Anderson, 1973). Book lungs of the fossil scorpion described by Jeram (1990a) and Shear et al. (1996) also have components similar to the book gills in *Limulus.* Kjellesvig-Waering (1986) doubted that xiphosuran book gills are homologous with book lungs because the former are appendages outside the body wall, whereas the latter are invaginations of the body wall.

Scorpion embryos were examined with the possibility they can add to our understanding of the mesosomal changes that occurred with ancestral terrestrialization (Farley, 1999b). Transitory abdominal plates and spiracles are formed in the embryos. These results and observations from scorpion fossils provide evidence that terrestrialization was a complex process with a diversity of respiratory structures and eventual persistence of the adaptations of modern species (Rolfe, 1985; Kjellesvig-Waering, 1986; Farley, 1999b; Jeram, this volume).

In some embryos of *C. exilicauda*, the abdominal plates have ruffles at the posterior margin, suggesting excurrent channels for water (Farley, unpublished observations). These reflected areas of the plate allowed examination of the tissues in the chamber above the plates. Delicate branching filaments were seen, and in a later stage, the surface layer had vertical striations in the anterior wall of the chamber above the plates. Striations like these would be expected at an early phase in the formation of book lung lamellae. The striations were followed by transverse folds or lamellae that extend into the plate chamber from the body wall above.

In embryos of *C. exilicauda,* and *H. arizonensis,* bilateral spiracles are formed at the latero-posterior margin of the abdominal plates (Farley, unpublished observations). Sections are needed to determine if these spiracles lead to the chamber above the plate or to a respiratory structure within the flap of the plate. Structures that might be considered respiratory were not seen when the space just inside the spiracles were examined with the SEM at high magnification.

In later stages, the posterior spiracles at the plate margin disappear as larger, oval-shaped spiracles form at the more anterior position seen in sternites of adult scorpions (Farley, 1990a,b). Marginal spiracles or gill slits are described in the abdominal plates of some fossil scorpions (Wills, 1947; Kjellesvig-Waering, 1986). Those in the Devonian assemblage from Gilboa, New York are thought to be stem group members of the clade leading to modern scorpions (Jeram, this volume).

Rather than reduction and loss of abdominal plates (Kjellesvig-Waering, 1986), some authors proposed that the plates fuse with the body wall to become sternites (Wills, 1947; Størmer, 1963, 1976; Selden and Jeram, 1989; Jeram, this volume). This notion is supported in studies with scorpion embryos. When abdominal plates first appear, the mesosoma is rounded in transverse section, and the plates occupy the ventral mid-region of each segment (fig. 2.20A). The abdominal plates may grow wider and/or cuticle is added bilaterally at the plate margins to produce broad, flat sternites (fig. 2.20B).

With the transition from abdominal plates to sternites, the faint outline of oval spiracles forms in the epidermis at the adult position. There appears to be little tissue invagination at this site in embryos of *C. exilicauda* (Buthidae), *P. mesaensis* (Vaejovidae) and *H. arizonensis* (Iuridae), suggesting that part of the epidermal primordia for the book lungs comes from some other site in these animals. The bilateral depressions just medial to the spiracles are a possibility (figs. 2.19, 2.20A,B; Farley, 1999b), but tissue sections are needed for clarification. At least part of the depression and associated epidermal scar is the attachment site of the dorso-ventral muscle band (Kästner, 1940a,b; Millot and Vachon, 1949; Farley, 1990b).

There may be differences among scorpion families in the mode of formation of book lungs and associated spiracles; previous workers noted that adult spiracles and book lungs differ in structure among scorpion families (Laurie, 1896a,b; Berteaux, 1889; Pavlovsky, 1926). These morphological distinctions were proposed as a basis for taxonomy, but other characters gained acceptance (Sisson, 1990; Stockwell, 1989, 1992; Jeram, 1994a,b; Fet et al., 1999; Gatenbein et al. 1999).

Telson and Venom Gland

As pointed out earlier, the scorpion body consists of 21 segments (Hjelle, 1990), with the telson not considered a true segment. Sections through the conical telson of 9-mm embryos of *Heterometrus scaber* (Mathew, 1956) show that glandular vesicles and the aculeus (sting) have begun to differentiate. In the broad basal part of the telson, mesodermal cells surround two sacs. These are the precursors for the glandular epithelium and compressional muscles. Two ectodermal invaginations near the tip of the telson eventually join with the glandular vesicles to form the ducts.

The vesicles, ducts, and compressional muscles of the telson are clearly differentiated by the 12-mm stage (Mathew, 1956). The embryos are about 19 mm at birth, and the sting is still conical rather than sharp. When disturbed, the larvae make jabbing movements with their metasoma even though the sting is too blunt to be functional. After the first molt when larvae separate from the mother, the aculeus is pointed and probably functional.

In early limb-bud embryos of *P. mesaensis,* the ventral surface of the telson shows two transitory invaginations that may eventually form the glandular and/or muscular part of the sting. Later, the telson appears as a broad structure, demarcated from the metasomal segments (figs. 2.15, 2.16A). At this stage, the telson appears suitable for swimming and, as the tip has two shallow invaginations, suitable for secreting substances into the environment. These two widespread invaginations in the tip of the lobular telson eventually disappear and do not seem to be precursors of the paired openings that eventually develop for venom at the tip of the aculeus (figs. 2.17B, 2.19).

Nervous System

Neural groove and neural lobes

As reviewed by Anderson (1973), the development of the nervous system follows a consistent pattern among the various orders of Chelicerata. Ventrally in each segment, the embryonic ectoderm proliferates and thickens on either side of a midline neural groove (Laurie, 1890) so that paired ganglionic rudiments are formed. These rudiments are located just medial to the limb-buds in those segments that proliferate appendages (figs. 2.15, 2.16A). Eventually, transverse and longitudinal strands of cells connect the rudiments so that a nerve band with paired ganglionic and interganglionic components extends along the length of the embryo. This eventually becomes the ventral nerve cord, consisting of a composite ganglionic mass and seven successive posterior ganglia. With differentiation of the nerve cord, the neural groove and lobes disappear (fig. 2.17B).

Anterior neuromeres

Authors have disagreed about the segmentation and differentiation of neuromeres in the anterior end of the scorpion embryo (Root, 1990). The confusion has occurred especially with regard to the number of precheliceral segments. From one to three were reported, with succeeding neuromeres and segments numbered accordingly.

As reviewed by Weygoldt (1985), the primitive arthropod head is thought to have a presegmental region (acron) with a ganglion (archicerebrum), but no appendages. This is followed by three segments: prosocephalon, deutocephalon, and tritocephalon. In chelicerates, the cheliceral and pedipalpal neuromeres represent, respectively, segments III (tritocerebrum) and IV. The brain consists of a tritocerebrum and a protocerebrum, the deutocerebrum (antennal lobe of mandibulate arthropods) having been lost or fused with the protocerebrum.

The segmental origin of the scorpion protocerebrum is uncertain. In the 5-mm embryo of *H. scaber,* Mathew (1956) observed a pair of coelomic sacs of a precheliceral segment. He thought these mesodermal cells might give rise to stomodeal muscles. Presumably the neuromeres of this and possibly another precheliceral segment contribute to the scorpion brain. Precheliceral coelomic sacs were also observed in the embryos of other arachnids (Weygoldt, 1985).

Stomatogastric nervous system

The structure and development of the stomatogastric nervous system in spiders was described by Legendre (1953, 1954a,b, 1956, 1958a,b, 1959, 1971, 1985) and has some features similar to scorpions. In *H. scaber* (Mathew, 1956), the labrum first appears when the embryo is about 3 mm long. From strands of mesoderm just above the cheliceral segment, cells pass into the labrum (rostrum) and give rise to the labral muscles. These muscles and those of the pharynx are innervated by the labral (pharyngeal) nerve from the frontal ganglion (McClendon, 1904; Henry, 1949; Farley, 1990c, 1999a). Presumably, this nerve also has afferent fibers from sensory receptors in the labrum and pharynx.

Brain

In scorpions, the paired rudiments of the cerebral hemispheres (brain, supraesophageal ganglion) are derived from invaginated dorsal ectoderm which is completely separate from the ganglionic rudiments of the ventral nerve cord. The brain forms from bilateral invaginations in the precheliceral lobe (figs. 2.16A, 2.17), whereas the ventral nerve cord develops by invagination and proliferation from the neural groove and neural lobes (figs. 2.15, 2.16A) in the ventral midline of the animal (Brauer, 1895; Laurie, 1890; McClendon, 1904; Henry, 1949; Abd-el Wahab, 1952; Mathew, 1956; Anderson, 1973). The cerebral invaginations differ among species and between apoikogenic and katoikogenic embryos.

Large cerebral lobes form early in the embryos of apoikogenic embryos such as *Euscorpius carpathicus* (Brauer, 1895), *E. italicus* (Laurie, 1890), *Centruroides vitiatus* (McClendon, 1904); *Leiurus quinquestriatus* (Abd-el Wahab, 1952) and *P. mesaensis* (figs. 2.14A, 2.15A; Farley, 1998, 1999a). On each side, a shallow indentation separates the cheliceral lobe (posterior) from the precheliceral lobe (anterior). In *E. carpathicus* (Brauer, 1895) and *P. mesaensis,* a curved groove (semilunar) forms in each cerebral lobe (fig. 2.16A). The medial portion of the two curved ectodermal invaginations gives rise to the medial eyes; the cerebral hemispheres of the brain are derived from other parts of the groove (fig. 2.17). The semilunar grooves extend postero-ventrally to the enlarging labrum (rostrum).

In *Centruroides* (McClendon, 1904) and *Leiurus* (Abd-el Wahab, 1952), no semilunar grooves form, but there is an oval-shaped cerebral invagination in each precheliceral lobe. This invagination gives rise to the median eyes and cerebral hemispheres.

In the 4-mm embryo of *Heterometrus scaber* (Mathew, 1956), large cerebral lobes do not appear. Dorsally, a mass of cerebral neuroblasts form in the right and left precheliceral lobes, just anterior to the stomodeum. Each primordial cerebral mass consists of three distinct groups of cells, but the author may have incorrectly interpreted this as the neuromere of a single precheliceral segment. Posterior to the stomodeum is the anterior end of the band of paired neuromeres, which become the ventral nerve cord.

Later stages

At the 4-mm stage in *Heterometrus,* the anterior end of the ventral band of neuroblasts has begun to extend antero-dorsally on either side of the stomodeum (Mathew, 1956). Some of these cells eventually join with the cerebral lobes, while others contribute to the innervation of the chelicerae. This pattern is typical of chelicerates and other scorpions (Anderson, 1973). Embryonic tissue from a postoral cheliceral segment moves around the mouth to give rise to preoral chelicerae. The paired neuromeres of this segment give rise to the cheliceral nerves and become the anterior, cheliceral part of the brain.

The major components of the nervous system are evident in the 6-mm embryo of *H. scaber* (Mathew, 1956). Each of the paired cerebral lobes has three smaller lobes, and these gradually increase in size. The posterior lobes on each side join just above the stomodeum and become most of the brain. The medial lobes form dorsal hollow vesicles which become associated with the medial eyes. The anterior lobes fuse with the first neuromere from the ventral chain, and some of these cells give rise to cheliceral nerves.

Behind the cheliceral segment (III) in the 6-mm embryo (Mathew, 1956), the succeeding pairs of neuromeres of the ventral chain enlarge and are connected by transverse commissures. Eventually, these fuse into the composite subesophageal ganglionic mass. Later, the next succeeding neuromere also joins with the subesophageal mass. Six neuromeres remain as separate abdominal ganglia (three mesosomal, three metasomal). The fourth metosomal ganglion is larger than the preceding abdominal ones and is formed by fusion of the last two neuromeres into a single terminal ganglion.

No trace of a pair of pregenital neuromeres (IX) was observed in the embryos of *H. scaber* (Mathew, 1956). This differs from some other species in which the pregenital neuromeres become part of the adult subesophageal ganglionic mass, even though the segment disappears (McClendon, 1904; Buxton, 1913, 1917).

During development in the mesosoma, neuromeres apparently move forward from their respective segments. As a result, the segmental nerves from the subesophageal and three free ganglia extend backward at an angle to innervate the corresponding segment (Farley, 1990b). Electrophysiological studies have not been done to determine which segmental tissues are controlled by each ganglion, but there are probably species differences because variations are reported in the position of ganglia with respect to mesosomal segments (Laurie, 1896a; McClendon, 1904; Buxton, 1913, 1917; Mathew, 1956; Babu, 1965; Farley, 1985, 1990b).

In some scorpions, workers have reported that the neuromeres in each segment typically have two pairs of segmental nerves. These were called by various names: hemal and neural, external and internal, anterior and posterior (McClendon, 1904; Hanstron, 1923). Usually the anterior pair, if present, is smallest, while the posterior nerves are large and appear to provide most of the segment's innervation. In other species, only one pair of segmental nerves was described from each neuromere (Buxton, 1913, 1917; Henry, 1949; Babu, 1965; Farley, 1985, 1990b). The anterior and posterior pairs may have merged, or one of them may have disappeared during development.

Median eyes

Arthropod eyes can be classified on the basis of the number of cellular layers (Lankester and Bourne, 1883; Mark, 1887). This is an important distinction, not only because it is a factor in the function of the eye, but also because it reflects the pattern of development. The median eyes in the scorpion (fig. 2.1A) have a layer of cells (vitreous body) in front of the retina, but the vitreous is lacking in the lateral eyes (Lankester and Bourne, 1883; Locket, this volume). The vitreous layer is continuous with the hypodermis of the cuticle that surrounds the eye. Lankester and Bourne did not observe the thin, cellular layer (postretinal membrane) proximal to the retina of both the median and lateral eyes. They proposed that the median eyes should be called diplostichous (two-layered) and the lateral eyes monostichous.

In embryological studies of scorpions (Graber, 1879; Patten, 1886; Mark, 1887), the number of cellular layers of the eyes was related to the pattern of development. The postretinal cellular layer proximal to the retina was observed, and it was concluded that the median eyes with vitreous body and retina are actually three-layered or triplostichous. The lateral eyes without a vitreous body are diplostichous. Parker (1886), Laurie (1890), Police (1907), and Scheuring (1913) supported these observations in subsequent studies in scorpion embryos and adults.

In *Centruroides* (Parker, 1886) and *Euscorpius italicus* (Laurie, 1890), each median eye develops from a saclike invagination of ectoderm (semilunar groove, cerebral invagination) at the anterior end of the embryo. Each sac has a lateral component that gives rise to part of the brain and a median component which becomes part of the ipsilateral median eye (figs. 2.16A, 2.17; Farley, 1998). The cerebral sacs from each side join medially to connect the two sides of the brain.

As a result of the invagination of ectoderm, three primordial layers contribute to the median eye (Mark, 1887; Parker, 1886; Laurie, 1890). The outer covering of the embryo eventually secretes the cuticle and becomes the hypodermis of the integument. In the region of the eye, cells in this layer (lentigen; Grenacher, 1880) presumably secrete the specialized lens cuticle and then remain in place as the vitreous body in front of the retina (Lankester and Bourne, 1883; Police, 1907; Bedini, 1967). Presumably, these cells also secrete a new vitreous body just before ecdysis.

The two cellular layers (retina and postretina) beneath the vitreous body are derived from the saclike invagination of ectoderm (Parker, 1886; Laurie, 1890). The wall of the invaginated sac of ectoderm consists of an upper layer that gives rise to the retina and a lower layer that becomes the postretinal membrane. The cavity of the sac gradually becomes thinner and then disappears as the upper retinal layer and lower postretinal layer merge.

Mark (1887) pointed out that the retina of the median eyes is formed from cells that are inverted during ectoderm invagination. In the retinas of adult scorpions, the rhabdomeres of retinular cells are directed toward the light (everted), whereas inverted retinular cells are present in the retinas of some spider ocelli (Foelix, 1996). This raised speculation about a possible change in polarity of scorpion retinular cells at some point during development.

Parker (1886) reported that early in the embryo, the optic nerve fibers appear to originate from the outer ends of these cells. Scheuring (1913) thought the fibers came from the middle of these cells. Spherical, brilliantly refringent bodies (phaospheres) were observed in the retinal layer in association with the receptor cells (Graber, 1879; Mark, 1887; Lankester and Bourne, 1883; Parker, 1886; Locket, this volume). These structures were thought to represent retinular cell nuclei (Patten, 1886; Scheuring, 1913) or rhabdomeres degenerating as a result of the change from embryonic to adult polarity. In electron micrographs, the phaosphere appeared as a dense, granular organelle, and neither of the earlier hypotheses was considered applicable (Bedini, 1967).

Lateral eyes

The lateral eyes (fig. 2.1A) lack a vitreous layer and so consist only of a retina and postretinal membrane (diplostichous). In the embryo, the lateral eyes form from a thickening of the hypodermis rather than by invagination as in the median eyes (Lankester and Bourne, 1883; Parker, 1886; Laurie, 1890). On the external surface of the embryo, ocular areas first appear as pigmented tracts on either side of the cephalic lobe. Sections through this region show a thick cellular layer, apparently resulting from proliferation of the hypodermis (Parker, 1886).

Although adult *Centruroides* have only four lateral eyes, five eye-cup depressions appear initially in the outer surface of the thickened hypodermis. Cells in this layer differentiate into a retina continuous with the surrounding hypodermis (Parker, 1886). Because there is no saclike invagination of ectoderm, the receptor cells of the retina are not inverted. Axons originate from the receptor cell inner surface in the embryo and remain in that position in the adult.

There is uncertainty about the cells (lentigen) that secrete the lens in the embryo and during the molt. Parker (1886) did not take sections at an appropriate stage to observe formation of the lens. He proposed that the lens is secreted by a continuous layer of hypodermal cells which are later retracted to the periphery of the lateral eye. At the outer edges of the adult retina, he identified perineural cells, which may be derived from a lentigen layer that covered the retina during secretion of lens cuticle. These cells should be examined during the molt to see if they migrate in front of the retina, secrete the new lens, and then retreat to the periphery after ecdysis.

DISCUSSION

Modern scorpions look much like Silurian species except for modifications for terrestrialization (Kjellesvig-Waering, 1986). The sixteen surviving families (Fet et al., 1999) share many important features. These include segmentation and appendages, four pairs of book lungs, preoral tube, pectines, venomous sting, filiform-flagellate spermatozoa, networks of ovariuterine tubules and testes, viviparity, spermatophore, and indirect sperm transfer. In addition, they share behaviors that appear to unite the order: the *promenade à deux* during courtship, the upward climbing of newborn larvae onto the dorsal surface of the mother, where they remain until after the first molt (Stahnke, 1966; Polis and Sissom, 1990), and the maternal tolerance and protection of the young during this initial period (Polis and Lourenço, 1986; Mahsberg, 1990).

The common characters of extant scorpions raise many questions about the origin of these features in scorpion history. What were the transitional steps and conditions that led to the present characters, which of these characters began among aquatic ancestors, and which developed with terrestrialization? Any attempt to answer these questions is necessarily speculative because the fossil record is scanty in spite of unusual preservation (Jeram, this volume). The present review shows that examination of scorpion embryos can be helpful, but observations must be interpreted with caution because development among extant families was modified by natural selection.

Bipolar Pattern of Development

The bipolar diversity in embryogenesis and maternal nourishment of embryos is a striking exception to the otherwise conservative pattern of scorpion evolution. Although it appears the families can be designated as apoikogenic or katoikogenic (Polis and Sissom, 1990), results in the preceding sections show a range of diversity within each of the two broad patterns of development.

The dichotomy in scorpion development is evident in the very first stages, in the amount of yolk included in the egg and in the pattern of cleavage and embryogenesis (Farley, 1998, 1999a). This suggests the divergence occurred very early in scorpion history, or that the katoikogenic pattern developed secondarily as a result of natural selection to bypass yolk and provide more direct stomodeal nourishment of the embryo. An argument for the latter notion is that the main adaptations for terrestrialization (e.g., book lungs, preoral tube) are present in extant scorpions whether apoikogenic (e.g., Buthidae, Iuridae, Vaejovidae) or katoikogenic (e.g., Diplocentridae, Scorpionidae). Also, the diverse apoikogenic and katoikogenic modifications described above are adaptations for survival in utero and must therefore have evolved after viviparity.

Viviparity has arisen at least 40 times in fish, and embryos and gravid females show a wide range of specializations for this mode of reproduction (Wourms, 1981; Wourms et al., 1988; Grove and Wourms, 1991; Lombardi, 1993). Some comparison with scorpions is helpful because ancestral scorpions were aquatic, and viviparous fish include such a large number of species (420 chondrichthyans, one actinistian, 510 teleosts) that some evolutionary and environmental factors can be discerned.

Oviparity is considered the most primitive condition of fish (Wourms et al., 1988), but some species also display embryoparity, where the egg is retained internally and some embryogenesis occurs before release. This intermediate condition between oviparity and viviparity may have occurred in scorpions; some postpartum differentiation occurs in the first instar (Polis and Sissom, 1990).

Some viviparous fish exhibit lecithotrophy, where embryos derive nutrition only from yolk (Wourms, 1981; Wourms et al., 1988). This condition is considered more primitive in comparison with matrotrophy, where maternal nutrients are supplied to the embryo throughout gestation. Francke (1982) concluded that matrotrophy occurs in scorpions because the newborn are much larger than fertilized eggs. The early blastula in *P. mesaensis* is about a half millimeter in diameter, whereas the prosoma/mesosoma of the immature embryo of fig. 2.17B is about three times as long.

A dry weight comparison of scorpion eggs and newborn is needed (with adjustment for energy loss during gestation; Wourms et al., 1988) to verify that scorpion viviparity is, indeed, matrotrophic. This mode is advantageous because development is not constrained by a limited store of yolk, and young can be released at an advanced stage better able to cope with the environment. Scorpion first instars are mobile but not predatory, relying on stored nutrients transferred to the embryos during gestation (Polis and Sissom, 1990).

Matrotrophous fish have a remarkable range of adaptations for embryo nourishment (Wourms et al., 1988). These include integumentary absorption by the embryo, oophagy and adelphophagy (ingestion of sibling yolk or embryos), yolk-sac placenta, and placental analogs. In the placental analogs, trophic vasculature is elaborated at a diversity of locations within the embryo and mother. Trophic blood vessels are not reported for scorpion embryos, even though there is striking modification of maternal reproductive tubules (figs. 2.8–2.11), embryo structures (figs. 2.12, 2.14–2.20), and sequence of differentiation (Farley, 1996, 1998, 1999a). Circulatory specializations remain a possibility in tissues and species not yet examined, as a complex microvasculature is present in scorpion muscles, brain and ganglia (Lane, 1981; Lane et al., 1981; Farley, 1990a,b). The numerous placental structures in viviparous fish suggest a greater adaptive capability for their circulatory system as compared with scorpions.

Development and Scorpion Distribution

The ancient origin of scorpions requires consideration of their evolution and distribution in relation to geologic history. A major feature of the Paleozoic era (240–570 MYA) was the merging of the continents into a supercontinent, Pangaea (Bambach et al., 1980; Boucot and Gray, 1983; Scotese et al., 1985; Erwin, 1993). Large regions of Pangaea were tropical, and animals could have dispersed before the supercontinent fragmented and new landmasses separated in the early Mesozoic.

The continental blocks that merged to form Pangaea were not the same as the continents that separated and exist today (Bambach et al., 1980). The Paleozoic continents were Gondwana, Laurentia, Baltica, Siberia, Kazakhstania, and China. Europe, North America, and much of Asia made up the single mass of Laurentia. Gondwana consisted mainly of the present southern continents. A major feature for the 240 million years of the Mesozoic and Cenozoic was the breakup of Pangaea and the dispersal of the new continents along with the animals and plants residing on them.

Aquatic and amphibious scorpions were present in Silurian and Devonian periods (Rolfe, 1980), and there is fossil evidence of terrestrial scorpions in the Carboniferous (Vogel and Durden, 1966; Jeram, this volume). Thus, scorpions were present on the Pangaean supercontinent and were probably able to disperse widely so that they came to occupy most of the continents that separated from Pangaea in the Mesozoic era. Nearly all the known Paleozoic and Carboniferous fossil scorpions are from North America and Europe, regions that were in Laurasia in the northern part of Pangaea (Kjellesvig-Waering, 1986; Sissom, 1990). Rolfe (1982) described a similar situation for some other arachnids and suggested that pre-Pangaean sea barriers may account for the distribution.

The Buthidae have 71 genera and 525 species and thus constitute the largest and most widespread of the present families (Fet et al., 1999). Lamoral (1980) and Sissom (1990) proposed that the buthids are the most primitive and ancient family (fig. 2.6), having originated in Laurasia and then spread to Gondwana. As described in previous sections, the buthids have a distinctive pattern of mitochondrial distribution in spermiogenesis, large yolky eggs, meroblastic cleavage, a discoblastula, and the relatively primitive apoikogenic pattern of embryonic development. All the species of significant toxicity to humans are within the Buthidae (Keegan, 1980; Habermehl, 1981; Simard and Watt, 1990; Loret and Hammock, this volume).

In the cladogram (fig. 2.6) proposed by Lamoral (1980) and Sissom (1990), the chaerilids and diplocentroids represent two additional evolutionary branches. The family Chaerilidae (1 genus and about 21 species; Fet et al., 1999) may have originated in Laurasia but eventually became isolated in the South Asia. The pattern of embryonic development is not currently known in the chaerilids (Polis and Sissom, 1990), so they should be examined. The hypothesized isolation of this group may have resulted in a distinctive form of katoikogenic or apoikogenic development (e.g., the complex apoikogenic development of *Paruroctonus mesaensis* described above), or possibly some modification of the bipolar designation will be necessary.

The Ischnuridae, Diplocentridae, and Scorpionidae all have katoikogenic development (Polis and Sissom, 1990); Lamoral (1980) proposed that they are monophyletic with a common ancestor. The Ischnuridae have a Gondwanaland distribution with representatives in Australasia, Africa, South America, and India (Lourenço, 1989; Stockwell, 1989; Sissom, 1990). The Diplocentridae and Scorpionidae may have had a common ancestor (Lamoral, 1980). The former are distributed in the Middle East (Arabia, Egypt, Israel) and North and South America. The Scorpionidae are thought to have originated in eastern Gondwanaland and are now in Africa, Australia, India, and Southeast Asia (Sissom, 1990).

The Bothriuridae, Chactidae, Iuridae, and Vaejovidae may represent retention of apoikogenic development in the diplocentroid branch (Sissom, 1990; Polis and Sissom, 1990). The iurid *Hadrurus arizonensis* has development like buthids (figs. 2.8, 2.18, 2.20), whereas the vaejovid *Paruroctonus mesaensis* has a form of apoikogenic development (figs. 2.10, 2.11, 2.13–2.17, 2.19) that suggests elaboration and modification of the basic pattern (Farley, 1996, 1998, 1999a). Integumentary absorption of maternal nutrients is likely by embryos of both species in contrast to oral feeding in embryos of more derived katoikogenic scorpions (figs. 2.9, 2.12). Francke and Soleglad (1981) and Sissom (1990) provide information about the distribution of the four families in this group.

In his extensive studies on scorpions in South America, Lourenço (1992, this volume), suggested that reproductive strategy is a factor for distribution in addition to historic geologic and climatic conditions. In the Antilles Islands, the location of *k*-selected or equilibrium species (MacArthur and Wilson, 1967; Pianka, 1970; Polis and Farley, 1979b, 1980; Polis, 1990) is related to the earliest colonizers of the islands and was affected by Tertiary geography and Pleistocene climatic changes. This group of scorpions includes some katoikogenic diplocentrids and ischnurids and some apoikogenic buthids. The distribution of more opportunistic, *r*-selected buthids (*Centruroides, Isometrus*) is based on relatively recent ecological factors.

K- and *r*-selected reproductive strategies relate to the partitioning of energy into somatic and reproductive tissues and activities (Pianka, 1970). Lourenço's (1992) observation in the Antilles raises the possibility that katoikogenic (and complex apoikogenic development such as that in *P. mesaensis;* Polis and Farley 1980) are favored by environmental conditions that lead to *k*-selected features, while simple apoikogenic development is retained in conditions requiring a more opportunistic strategy.

These studies indicate that reproductive and developmental modes can be advantageously studied in relation to long- and short-term environmental and geologic conditions. We have only begun comparative studies on scorpion embryogenesis, but it is clear that a diversity of adaptations for viviparity has emerged within the families. How the environment modifies viviparity through natural selection is an important question for which scorpions are unusually suited for further investigation.

Acknowledgments I thank Morice A. Izmane for assistance in collecting animals, Dr. W. David Sissom for species identifications, and the editors, reviewers, and Dr. Sissom for helpful suggestions for improving the manuscript. This research was supported by intramural grants from the University of California.

References

Abalos, J.W. and Baez, E.C. (1963) On spermatic transmission in spiders. *Psyche* 70: 197–207.

Abd-el-Wahab, A. (1952) Some notes on the segmentation of the scorpion, *Buthus quinquestriatus* (H. & E.) *Proc. Egypt. Acad. Sci.* 7: 75–91.

Abd-el-Wahab, A. (1953) The female genital system of the scorpion, *Buthus quinquestriatus* (H. & E.). *Bull. Zool. Soc. Egypt* 11: 1–8.

Abd-el-Wahab, A. (1954) The formation of the germ layers and their embryonic coverings in the scorpion *Buthus quinquestriatus* (H. & E.) *Proc. Egypt. Acad. Sci.* 10: 110–118.

Abd-el-Wahab, A. (1957) The male genital system of the scorpion, *Buthus quinquestriatus. Q.J. Microsc. Sci.* 98: 111–122.

Alberti, G. (1983) Fine structure of scorpion spermatozoa (*Buthus occitanus* (Amoreux), Buthidae, Scorpiones) *J. Morphol.* 177: 205–212.

Alexander, A.J. (1956) Mating in scorpions. *Nature* 178: 867–868.

Alexander, A.J. (1957) The courtship and mating of *Opisthophthalmus latimanus. Proc. Zool. Soc.* (London) 128: 529–544.

Alexander, A.J. (1959) Courtship and mating in the buthid scorpions. *Proc. Zool. Soc.* (London) 133: 145–169.

Alexander, A.J. (1964) The evolution of mating behavior in arthropods. *Symp.R. Entomol. Soc. Lond.* 12: 78–94.

Alexander, A.J. and Ewer, D.W. (1957) On the origin of mating behavior in spiders. *Am. Nat.* 91: 311–317.

Anderson, D.T. (1973) *Embryology and Phylogeny in Annelids and Arthropods.* International Series of Monographs in Pure and Applied Biology, Zoology Division, vol. 50. Pergamon, New York.

André, J. (1959) Etude au microscope électronique de l'évolution du chondriome pendant la spermatogénése du scorpion *Euscorpius flavicaudis. J. Ultrastruct. Res.* 2: 288–308.

André, J. (1963) Some aspects of specialization in sperm. *In* Mazia, D. and Tyler, A. (eds.) *General Physiology of Cell Specialization,* pp. 91–115. McGraw-Hill, New York.

Angermann, H. (1957) Über Verhalten Spermatophorenbildung und Sinnesphysiologie von *Euscorpius italicus* Habst. und Verwandten Arten. (Scorpiones, Chactidae). *Z. Tierpsychol.* 14: 276–302.

Awati, P.R. and Tembe, V.B. (1956) *Buthus tamulus* (Fabr.) *The Indian scorpion: morphology, anatomy and bionomics.* Zoological Monographs, no. 2. University of Bombay, Bombay.

Babu, K.S. (1965) Anatomy of the central nervous system of arachnids. *Zool. Jahrbücher Abt. Anat. Ontog.* 82: 1–154.

Baccetti, B. (1970) *Comparative Spermatology.* Academic Press, New York.

Baccetti, B. (1978) Ultrastructure of sperm and its bearing on arthropod phylogeny. *In* Gupta, A.P. (ed.) *Arthropod Phylogeny,* pp. 609–644. Van Nostrand Reinhold, New York.

Baccetti, B. (1987) News on phylogenetical and taxonomical spermatology. *In* Mohri, H. (ed.) *New Horizons in Sperm Cell Research.* pp. 333–348. Japan Scientific Societies Press/Gordon and Breach Science Publishers, Tokyo/ New York.

Baccetti, B. and Afzelius.B.A. (1976) The biology of the sperm cell. *Monogr. Dev. Biol.* 10: 1–254.

Bambach, R.K., Scotese, C.R. and Ziegler, A.M. (1980) Before Pangaea: the geographies of the Paleozoic world. *Am. Sci.* 68: 26–38.

Bedi, U. (1962) Studies on the male germ cells of scorpions *Palamnaeus bengalensis* and *Palamnaeus fulvipes,* with particular reference to the morphology and cytochemistry of the cytoplasmic inclusions. *Res. Bull. Panjab Univ. Sci.* 13: 213–225.

Bedini, C. (1967) The fine structure of the eyes of *Euscorpius carpathicus* L. (Arachnida, Scorpiones). *Arch. Ital. Biol.* 105: 361–378.

Berteaux, L. (1889) Le poumon des Arachnides. *La Cellule* 5: 255–316.

Bhattacharya, D.R. and Gatenby, J.B. (1924) Spermatogenesis of an Indian scorpion. *Nature* 113: 858.

Boucot, A.J. and Gray, J. (1983) A Paleozoic Pangaea. *Science* 222: 571–581.

Boudreaux, H.G. (1979) *Arthropod Phylogeny with Special Reference to Insects.* John Wiley & Sons, New York.

Bowerman, R.F. (1972a) A muscle receptor organ in the scorpion postabdomen. I, The sensory system. *J. Comp. Physiol.* 81A: 133–146.

Bowerman, R.F. (1972b) A muscle receptor organ in the scorpion postabdomen. II, Reflexes evoked by MRO stretch and release. *J. Comp. Physiol.* 81A: 147–157.

Braddy, S.J. and Dunlop, J.A. (1997) The functional morphology of mating in the Silurian eurypterid, *Baltoeurypterus tetragonophthalmus* (Fisher, 1839). *Zool.J. Linn. Soc.* 121: 435–461.

Brauckmann, C. (1987) Neue Arachniden-Funde (Scorpionida, Trigonotarbida) aus dem westdeutschen Unter-Devon. *Geol. Palaeontol.* 21: 73–85.

Brauer, A. (1894) Beiträge zur Kenntnis der Entwicklungsgeschichte des Skorpions, I. *Z. wiss. Zool.* 57: 402–432.

Brauer, A. (1895) Beiträge zur Kenntnis der Entwicklungsgeschichte des Skorpions, II. *Z. wiss. Zool.* 59: 351–433.

Brieger, F.G. and Graner, E.A. (1943) On the cytology of *Tityus bahiensis* with reference to meiotic prophase. *Genetics* 28: 269–274.
Briggs, D.E.G. (1985) Gigantism in Palaeozoic arthropods. *Spec. Pap. Palaeontol.* 33: 157.
Briggs, D.E.G. and Collins, D. (1988) A middle Cambrian Chelicerate from Mount Stephen, British Columbia. *Palaeontology* 31: 779–798.
Briggs, D.E.G., Fortey, R.A. and Clarkson, E.N. K. (1988) Extinction and the fossil record of arthropods. *In* Larwood, G.P. (ed.) *Extinction and Survival in the Fossil Record,* pp. 171–209. Systematics Association Special vol. 34. Carendon Press, Oxford.
Brownell, P.H. (1977) Compressional and surface waves in sand used by desert scorpions to locate prey. *Science* 197: 479–482.
Brownell, P.H. (1984) Prey detection by the sand scorpion. *Sci. Am.* 251: 86–97.
Brownell, P.H. (1991) Organized central mapping of chemosensory systems in two arachnid orders (Scorpionida, Solpugida). *Am. Zool.* 31: 84A.
Brownell, P.H. and Farley, R.D. (1979a) Detection of vibrations in sand by tarsal sense organs of the nocturnal scorpion, *Paruroctonus mesaensis. J. Comp. Physiol.* (A) 131: 23–30.
Brownell, P.H. and Farley, R.D. (1979b) Prey localizing behaviour of the nocturnal desert scorpion, *Paruroctonus mesaensis:* orientation to substrate vibrations. *Anim. Behav.* 27: 185–193.
Brownell, P.H. and Farley, R.D. (1979c) Orientation to vibrations in sand by the nocturnal scorpion *Paruroctonus mesaensis:* mechanism of target localization. *J. Comp. Physiol.* (A) 131: 31–38.
Buxton, B.H. (1913) Coxal glands of the arachnids. *Zool. Jahrbücher Suppl.* 14: 231–282.
Buxton, B.H. (1917) Notes on the anatomy of arachnids. *Zool.J. Morphol.* 29: 1–31.
Cloudsley-Thompson, J.L. (1955) On the function of the pectines of scorpions. *Ann. Mag. Nat. Hist.,* ser. 12, 8: 556–560.
Cloudsley-Thompson, J.L. (1976) *Evolutionary Trends in the Mating of Arthropoda.* Meadowfield, Shildon, Durham, UK.
Cloudsley-Thompson, J.L. (1988) *Evolution and Adaptation of Terrestrial Arthropods.* Springer-Verlag, Berlin.
Couzijn, H.W. C. (1976) Functional anatomy of the walking-legs of Scorpionida with remarks on terminology and homologization of leg segments. *Netherlands J. Zool.* 26: 453–501.
Da Cruz Landim, C. and Ferreira, A. (1972) Sperm differentiation in the scorpion *Tityus bahiensis* (Perty). *Caryologia* 25: 125–135.
Da Cruz Landim, C. and Ferreira, A. (1973) The spermatozoa of *Tityus bahiensis* (Perty) Scorpion-Buthidae. *Cytologia* 38: 187–194.
De Tolédo Piza, Jr., S. (1941) Chromosomes with two spindle attachments in the Brazilian scorpion (*Tityus bahiensis* Perty). *J. Hered.* 32: 423–426.
De Tolédo Piza, Jr., S. (1943a) Meiosis in the male of the Brazilian scorpion, *Tityus bahiensis. Revista Agric., São Paulo* 18: 249–276.
De Tolédo Piza, Jr., S. (1943b) The usefulness of the spindle fibre for moving the chromosome. *Am. Nat.* 74: 433–462.
De Tolédo Piza, Jr., S. (1944) A case of spontaneous end-to-end permanent union of two non-homologous chromosomes in the Brazilian scorpion *Tityus bahiensis,* accompanied by irregularities in pairing. *Revista Agric., São Paulo* 19: 133–147.
De Tolédo Piza, Jr., S. (1956) The chromosomes of *Rhopalurus* (Scorpiones-Buthidae). *Can. Entomol.* 89: 565–568.
Dubale, M.S. and Vyas, A.B. (1968) The structure of the chela of *Heterometrus* sp. and its mode of operation. *Bull.S. Calif. Acad. Sci.* 67: 240–244.

Erwin, D.H. (1993) *The Great Paleozoic Crisis: Life and Death in the Permian.* Columbia University Press, New York.

Esponda, P. and Taffarel, M. (1980) Ultrastructure of the flagellum in sperms of the scorpion *Tityus bahiensis. Cell Biol. Int. Rep.* 4: 549–554.

Farley, R.D. (1984) The ultrastructure of hemocytopoietic organs in the desert scorpion, *Paruroctonus. Tissue Cell* 16: 577–588.

Farley, R.D. (1985) Cardioregulation in the desert scorpion, *Paruroctonus mesaensis. Comp. Biochem. Physiol.* (A) 82: 377–387.

Farley, R.D. (1987) Postsynaptic potentials and contraction pattern in the heart of the desert scorpion, *Paruroctonus mesaensis. Comp. Biochem. Physiol.* (A) 86: 121–131.

Farley, R.D. (1990a) Regulation of air and blood flow through the booklungs of the desert scorpion, *Paruroctonus mesaensis. Tissue Cell* 22: 547–569.

Farley, R.D. (1990b) Functional organization of the respiratory and circulatory systems in the desert scorpion, *Paruroctonus mesaensis. Acta Zool. Fenn.* 190: 139–145.

Farley, R.D. (1990c) The retrocerebral complex and the innervation of the digestive tract in the desert scorpion, *Paruroctonus mesaensis. Am. Zool.* 90: 119A.

Farley, R.D. (1996) Formation of maternal trophic structure for embryos of *Paruroctonus mesaensis.* (Scorpionidae: Vaejovidae). *Revue Suisse de Zoologie* Vol hors série 1: 189–202.

Farley, R.D. (1998) Matrotrophic adaptations and early stages of embryogenesis in the desert scorpion *Paruroctonue mesaensis* (Vaejovidae). *J. Morph.* 237: 1–25.

Farley, R.D. (1999a) Scorpiones. *In* Harrison, F.W. and Foelix, R.F. (eds.) *Microscopic Anatomy of Invertebrates,* vol. 8A. Chelicerate Arthropoda, pp. 117–222. Wiley-Liss, New York.

Farley, R.D. (1999b) Ventral mesosomal changes in embryos from three scorpion families: Iuridae, Buthidae, and Vaejovidae. *J. Arachnol.* 27: 123–128.

Farley, R.D. and Chan, D.J. (1985) The ultrastructure of the cardiac ganglion of the desert scorpion, *Paruroctonus mesaensis* (Scorpionida: Vaejovidae). *J. Morphol.* 184: 231–252.

Fet, V. and Brownell, P.H. (1998) Morphological variation in the pectoral sensory organ of scorpions. In *Abstracts,* XIV International Congress of Arachnology (M. Van der Merve, N.I. Platnick, and P. Sierwald (eds.) p. 16.

Fet, V., Sissom, W.D., Lowe, G., and Braunwalder, M.E. (2000) *Catalog of the Scorpions of the World (1758–1998).* New York Entomological Society, N.Y.

Fincke, T. and Paul, R. (1989) Book lung function in arachnids III. The function and control of the spiracles. *J. Comp. Physiol.* (B) 159: 433–441.

Firstman, B. (1973) The relationship of the chelicerate arterial system to the evolution of the endosternite. *J. Arachnol.* 1: 1–54.

Foelix, R.F. (1996) *Biology of Spiders.* Oxford University Press, New York.

Francke, O.F. (1979) Spermatophores of some North American scorpions (Arachnida, Scorpiones). *J. Arachnol.* 7: 19–32.

Francke, O.F. (1982) Parturition in scorpions (Arachnida, Scorpiones): a review of ideas. *Rev. Arachnolo.* 4: 27–37.

Francke, O.F. and Soleglad, M.E. (1981) The family Iuridae Thorell. *J. Arachnol.* 9: 233–258.

Franzén, A. (1970) Phylogenetic aspects of the morphology of spermatozoa and spermiogenesis. *In* Baccetti, B. (ed.) *Comparative Spermatology,* pp. 29–46. Academic Press, New York.

Gaffin, D.D. and Brownell, P.H. (1990) Electrophysiological studies of the pectinal chemosensory system of the scorpion. *Chem. Senses* 15: 579.

Gaffin, D.D., Wennstrom, K.L. and Brownell, P.H. (1991) Water detection in the desert sand scorpion *Paruroctonus mesaensis. Am. Zool.* 31: 85A.

Gaffin, D.D. and Brownell, P.H. (1997a) Response properties of chemosensory peg sensilla on the pectines of scorpions. *J. Comp. Physiol.* A 181: 291–300.
Gaffin, D.D. and Brownell, P.H. (1997b) Electrophysiological evidence of synaptic interactions within chemosensory sensilla of scorpion pectines. *J. Comp. Physiol.* A 181: 301–307.
Gantenbein, B., Fet, V., Largiader, C.R., and Scholl, A. (1999) First DNA phylogeny of *Euscorpius* Thorell, 1876 (Scorpiones, Euscorpiidae) and its bearing on taxonomy and biogeography of this genus. *Biogeographica* 75: 49–65.
Gatenby, J.B. and Bhattacharya, D.A. (1925) Note on the cytoplasmic inclusion in the spermatogensis of the Indian scorpion *Palamnaeus bengalensis* C. Koch. *La Cellule* 35: 251.
Gould, S.J. (1977) *Ontogeny and Phylogeny.* Belknap Press, Cambridge, MA.
Graber, V. (1879) Über das unicorneale Tracheatenauge. *Arch. Mikrosk. Anat.* 17: 58–93.
Grenacher, H. (1880) Ueberdie augen einiger Myriapoden. *Arch. Mikrosk. Anat.* 18: 415–467.
Grénin, H.A. (1957) Contribution a la connaissance cytologique des scorpions: les chromosomes de *Pandinus imperator* Koch. *Rev. Suisse Zool.* 64: 349–353.
Grénin, H.A. (1961) Contributions a la connaissance cytologique des scorpions. Les chromosomes de *Buthus occitanus* Amor. *Vie Milieu* 12: 89–96.
Grove, B.D. and Wourms, J.P. (1991) The follicular placenta of the viviparous fish, *Heterandria formosa.* I. Ultrastructure and development of the embryonic absorptive surface. *J. Morphol.* 209: 265–284.
Habermehl, G. (1981) *Venomous Animals and Their Toxins.* Springer-Verlag, Berlin.
Hadley, N.F. (1990) Environmental physiology. *In* Polis, G.A. (ed.) The *Biology of Scorpions,* pp. 321–340. Stanford University Press, Palo Alto, CA.
Hanstrom, B. (1923) Further notes on the central nervous system of arachnids: scorpions, phalangids and trap-door spiders. *J. Comp. Neurol.* 35: 249–274.
Henry, L. (1949) The nervous system and the segmentation of the head in a scorpion (Arachnida). *Microentomology* 14: 121–126.
Hjelle, J.T. (1990) Anatomy and morphology. *In:* Polis, G.A. (ed.) *The Biology of Scorpions,* pp. 9–63. Stanford University Press, Palo Alto, CA.
Hood, R.D., Watson, O.F., Deason, T.R. and Benton, Jr., C.L.B. (1972) Ultrastructure of scorpion spermatozoa with atypical axonemes. *Cytobios* 5: 167–177.
Jamieson, G.M. (1987) A biological classification of sperm types, with special reference to annelids and molluscs and an example of spermiocladistics. *In* Mohri, H. (ed.) *New Horizons in Sperm Cell Research,* pp. 311–332. Japan Science Society Press, Toyko.
Jeram, A.J. (1990a) Book-lungs in a Lower Carboniferous scorpion. *Nature* 343: 360–361.
Jeram, A.J. (1990b) When scorpions ruled the world. *New Scientist* 126: 52–55.
Jeram, A.J. (1994a) Scorpions from the Viséan of East Kirkton, West Lothian, Scotland, with a revision of the infraorder Mesoscorpionina. *Trans. Roy. Soc. Edinburgh: Earth Sci.* 84: 283–299.
Jeram, A.J. (1994b) Carboniferous orthosterni and their relationship to living scorpions. *Palaeontol.* 37: 513–550.
Jespersen, A. and Hartwick, R. (1973) Fine structure of spermiogenesis in scorpions from the family Vejovidae. *J. Ultrastruct. Res.* 45: 366–383.
Kästner, A. (1940a) Arachnida. *In* Kükenthal, W. and Krumbach, T. (eds.) *Handbuch der Zoologie,* vol. 3, pp. 97–116. de Gruyter, Berlin.

Kästner, A. (1940b) Scorpiones. *In* Kükenthal, W. and Krumbach, T. (eds.) *Handbuch der Zoologie,* vol. 3, pp. 117–220. de Gruyter, Berlin.
Keegan, H.L. (1980) *Scorpions of Medical Importance.* University Press of Mississippi, Jackson.
Kjellesvig-Waering, E.N. (1986) A restudy of the fossil Scorpionida of the world. *Palaeontol. Am.* 55: 1–287.
Koch, L.E. (1977) The taxonomy, geographic distribution and evolutionary radiation of Australo-Papuan scorpions. *Rec.W. Austral. Mus.* 5: 83–367.
Kovoor, J., Lourenço, W.R. and Muoz-Cuevas, A. (1987) Conservation des spermatoides dans les voies génitales des femelles et biologie de la reproduction des scorpions (Chélecérates). *C.R. Acad. Sci.* (Paris) 10: 259–264.
Kraepelin, K. (1905) Die geographische Verbreitung der Skorpione. *Zool. Jahrbücher* 22: 321–364.
Kraus, O. (1976) Zur phylogenetischen Stellung und Evolution der Chelicerata. *Entomol. Germanica* 3: 1–12.
Lamoral, B.H. (1980) A reappraisal of suprageneric classification of Recent scorpions and their zoogeography. *In Proceedings of the Eighth International Congress of Arachnology,* Vienna, pp. 439–444.
Lane, N.J. (1981) Invertebrate neuroglia-junctional structure and development. *J. Exp. Biol.* 95: 7–33.
Lane, N.J., Harrison, J.B. and Bowerman, R.F. (1981) A vertebrate-like blood-brain barrier with intraganglionic blood channels and occluding junctions, in the scorpion. *Tissue Cell* 13: 557–576.
Lankester, E. (1885) A new hypothesis as to the relationship of the lung-book of *Scorpio* to the gill-book of *Limulus. Q.J. Microscop. Sci.* 25: 339–342.
Lankester, E.R., Benham, W.B. S. and Beck, E.J. (1885) On the muscular and endoskeletal systems of *Limulus* and *Scorpio,* with some notes on the anatomy and generic characters of scorpions. *Trans. Zool. Soc. Lond.* Part 10, 11: 311–390.
Lankester, E.R. and Bourne, A.G. (1883) The minute structure of lateral and central eyes of *Scorpio* and of *Limulus. Q.J. Microscop. Sci.* 23: 177–212.
Laurie, M. (1890) The embryology of a scorpion (*Euscorpius italicus*). *Q.J. Microsc. Sci.* 31: 105–141.
Laurie, M. (1891) Some points in the development of *Scorpio fulvipes.Q. J. Microsc. Sci.* 32: 587–597.
Laurie, M. (1896a) Notes on the anatomy of some scorpions and its bearing on the classification of the order. *Ann. Mag. Natl. Hist.* 17: 185–193.
Laurie, M. (1896b) Further notes on the anatomy and development of scorpions and their bearing on the classification of the order. *Ann. Mag. Natl. Hist.* 18: 121–133.
Laurie, M. (1896c) Some newly-hatched specimens and a late embryo of *Opisthophthalmus. Proc. R. Phyisol. Soc. (Edinburgh)* 13: 162–166.
Legendre, R. (1953) Le systéme sympathique stomatogastrique (organe de Schneider) des araigneés du genre *Tegenaria.C. R. Hebdom. Séances Acad. Sci.* (Paris), sér. D, 237: 1283–1285.
Legendre, R. (1954a) Sur la présence de cellules neurosécrétrices dans le systéme nerveux central des Aranéides. *C.R. Acad. Sci.* (Paris) 238: 1267–1268.
Legendre, R. (1954b) Données anatomiques sur le complexe neuroendocrine rétrocérébral des Araéides. *Ann. Sci. Nat. Zool. Biol. Anim.* 16: 419–426.
Legendre, R. (1956) Sur la genése du pont stomodeal chez les araignées. *Bull. Soc. Zool. France* 81: 314–323.
Legendre, R. (1958a) Sur la structure du complexe endocrine rétrocérébral de l'araignée mygalomorphe *Scodra (Stromatopelma) calceata* Fabr. (Theraphosidae). *C.R. Hebdom. Séances Acad. Sci.* (Paris), sér. D, 246: 3671–3674.
Legendre, R. (1958b) Contributions a l'étude du systéme nerveux des aranéides. *Ann. Biol.* 34: 193–223.

Legendre, R. (1959) Contribution a l'étude du systeme nerveux des aranéides. *Ann. Sci. Nat. Zool. Biol. Anim.* 1: 339–473.
Legendre, R. (1971) Etat actuel de nos connaissances sur le développement embryonnaire et la croissance des araignées. *Bull. Soc. Zool. France* 96: 93–114.
Legendre, R. (1985) The stomatogastric nervous system and neurohemal organs of the spider *Argiope aurantia* (Lucas). *J. Morphol.* 141: 77–93.
Little, C. (1983) *The Colonization of Land: Origins and Adaptations of Terrestrial Animals.* Cambridge University Press, Cambridge.
Lombardi, J. (1993) Evolution and diversity of maternal and embryonic adaptations in viviparous fishes. *Am. Zool.* 33: 59A.
Lourenço, W.R. (1989) Retablissement de la famille des Ischnuridae, distincte des Scorpionidae Pocock, 1893, à partir de la sous-famille des Ischnurinae Pocock, 1893 (Arachnida, Scorpiones). *Rev. Arachnol.* 8: 159–177.
Lourenço, W.R. (1992) Les peuplements des scorpions des Antilles: facteurs historiques et écologiques en association avec les stratégies biodémographiques. *Stud. Neotrop. Fauna Environ.* 27: 43–62.
Lourenço, W.R. (1996) Faune de Madagascar. 87. Scorpions (Chelicerata, Scorpiones). Muséum National d'Histoire Naturelle, Paris.
Lourenço, W. R and Cuellar, O. (1994) Notes on the geography of parthogenetic scorpions. *Biogeogrphica* 70(1): 19–23.
Lourenço, W.R. and Cuellar, O. (1995) Scorpions, scorpionism, life history strategies and parthenogenesis. *J. Venon. Anim. Toxins* 1: 52–62.
MacArthur, R.H. and Wilson, E.O. (1967) *The Theory of Island Biogeography.* Princeton University Press, Princeton, NJ.
Mahsberg, D. (1990) Brood care and family cohesion in the tropical scorpion *Pandinus imperator* (Koch) (Scorpiones: Scorpionidae). *Acta Zool. Fenn.* 190: 267–272.
Makioka, T. (1992a) Reproductive biology of the viviparous scorpion, *Liocheles australasiae* (Fabricius) (Arachnida, Scorpiones, Ischnuridae) II. Repeated pregnancies in virgins. *Invert. Reprod. Develop.* 21: 161–166.
Makioka, T. (1992b) Reproductive biology of the viviparous scorpion, *Liocheles australasiae* (Fabricius) (Arachnida, Scorpiones, Ischnuridae) III. Structural types and functional phases of the adult ovary. *Invert. Reprod. Develop.* 21: 207–214.
Makioka, T. (1993) Reproductive biology of the viviparous scorpion, *Liocheles australasiae* (Fabricius) (Arachnida, Scorpiones, Ischnuridae) IV. Pregnancy in females isolated from infancy, with notes on juvenile stage duration. *Invert. Reprod. Develop.* 24: 207–212.
Makioka, T. and Koike, K. (1985) Reproductive biology of the viviparous scorpion, *Liocheles australasiae* (Fabricius) (Arachnida, Scorpiones, Scorpionidae). I. Absence of males in two natural populations. *Invert. Reproduct. Develop.* 8: 317–323.
Mark, E.L. (1887) Simple eyes in arthrodods. *Bull. Mus. Comp. Zool.,* 13: 4–105.
Mathew, A.P. (1948) Nutrition in the advanced embryos of *Palamnaeus scaber. Proc. Natl. Acad. Sci.* (India) B 27: 111–118.
Mathew, A.P. (1956) Embryology of *Heterometrus scaber. Bull. Central Res. Inst.* (University of Travancore) 1: 1–96.
Mathew, A.P. (1960) Embryonic nutrition in *Lychas tricarinatus. J. Zool. Soc. India* 12: 220–228.
Mathew, A.P. (1962) Changes in the structure of the ovarian and diverticular mucosa in *Heterometrus scaber* (Thorell) during a reproductive cycle. *In: Proceedings of the First All-India Congress of Zoology,* 1959, pp. 100–111, pt. 2.
Mathew, A.P. (1968) Embryonic nutrition in *Urodacus abruptus* Pocock. *J. Anim. Morphol. Physiol.* 15: 162–167.

McClendon, J.F. (1904) On the anatomy and embryology of the nervous system of the scorpion. *Biol. Bull.* 8: 38–55.

Millot, J. and Vachon, M. (1949) Ordre des scorpions. *In* Grassé, P.P. (ed.) *Traité de Zoologie,* vol. 6, pp. 386–436. Masson, Paris.

Nath, V. (1924) Cell inclusions in the gametogenesis of scorpion. *Nature* 114: 52.

Newlands, G. (1972) Ecological adaptations of Kruger National Park scorpionids (Arachnida: Scorpiones). *Koedoe* 15: 37–48.

Newlands, G. and Martindale, C. (1980) The buthid scorpion fauna of Zimbabwe-Rhodesia with checklists and keys to the genera amd species, distributions and medical importance (Arachnida: Scorpiones). *Z. Angewandte Zool.* 67: 51–77

Parker, G.H. (1886) The eyes in scorpions. *Bull. Mus. Comp. Zool.* 13: 173–208.

Parker, S.P. (1982) *Synopsis and Classification of Living Organisms,* vol. 2. McGraw-Hill, New York.

Patten, W. (1886) Eyes of mollucs and arthropods. *Mittheilungen Zool. Station Neapel* 6: 542–756.

Paul, R. and Fincke, T. (1989) Book lung function in arachnids II. Carbon dioxide release and its relations to respiratory surface, water loss and heart frequency. *J. Comp. Physiol.* B 159: 419–432.

Paul, R., Fincke, T. and Linzen, B. (1987) Respiration in the tarantula *Eurypelma californicum:* evidence for diffusion lungs. *J. Comp. Physiol.* B 157: 209–217.

Pavlovsky, E.N. (1926) Studies on the organization and development of scorpions. 5. The lungs. *Quart. J. Microsc. Sci.* 70: 135–146.

Peretti, A.V. (1991) Comportamiento de apareamiento de *Zabius fuscus* (Thorell) (Buthidae, Scorpiones). *Bol. Soc. Biol. Concepcion. Chile* 62: 123–146.

Petrunkevitch, A. (1952) Macroevolution and the fossil record of Arachnida. *Am. Sci.* 40: 99–122.

Petrunkevitch, A. (1955) Arachnida. *In* Moore, R.C. (ed.) *Treatise on Invertebrate Paleontology, Part P, Arthropoda 2,* pp. 42–162. University of Kansas Press, Lawrence.

Pflugfelder, O. (1930) Zur Embryologie des Skorpions *Hormurus australasiae* (F.). *Z. wiss. Zool.* 137: 1–29.

Phillips, D.M. (1974) Nuclear shaping in the absence of microtubules in scorpion spermatids. *J. Cell Biol.* 62: 911–917.

Pianka, E.R. (1970) On *r*- and *K*-selection. *Am. Nat.* 104: 592–597.

Platnick, N.I. (1986) Seminar: establishment of phyletic seriations. *Acta X Congr. Intl. Arachnol. Soc. 1986* 2: 55–57.

Plotnick, R.E. (1983) Patterns in the evolution of the eurypterids. (Ph.D. dissertation), University of Chicago.

Pocock, R.I. (1894) Scorpions and their geographical distribution. *Nat. Sci.* 4: 353–364.

Police, G. (1907) Sugli occhi dello scorpione. *Zool. Jahrbücher Abteilung Anat. Ontog.* 25: 1–70.

Polis, G.A. (1980) The effect of cannibalism on the demography and activity of a natural population of desert scorpion. *Behav. Ecol. Sociobiol.* 7: 25–35.

Polis, G.A. (1990) Ecology. *In* Polis, G.A. (ed.) *The Biology of Scorpions,* pp. 247–293. Stanford University Press, Palo Alto, CA.

Polis, G.A. and Farley, R.D. (1979a) Behavior and ecology of mating in the cannibalistic scorpion, *Paruroctonus mesaensis* Stahnke (Scorpionida: Vaejovidae). *J. Arachnol.* 7: 33–46.

Polis, G.A. and Farley, R.D. (1979b) Characteristics and environmental determinants of natality, growth and maturity in a natural population of the desert scorpion, *Paruroctonus mesaensis* (Scorpionida: Vaejovidae). *J. Zool.* 187: 517–542.

Polis, G.A. and Farley, R.D. (1980) Population biology of a desert scorpion: survivorship, microhabitat and the evolution of life history strategy. *Ecology* 61: 620–629.

Polis, G.A. and Lourenço, W.R. (1986) Sociality among scorpions. *Proc. 10th Intl. Congr. Arachnol., Jaca* 1: 111–115.

Polis, G.A. and Sissom, W.D. (1990) Life history. *In* Polis, G.A., (ed.) *The Biology of Scorpions,* pp. 161–223. Stanford University Press, Palo Alto, CA.

Proctor, H.C. (1998) Indirect sperm transfer in Arthropods: Behavioral and evolutionary trends. *Annual. Rev. Entomol.* 43: 153–174.

Punzo, F. (1998) *The Biology of Camel-spiders* (Arachnida, Solifugae). Kluwer Academic, Boston.

Raw, F. (1957) Origin of Chelicerates. *J. Paleontol.* 31: 139–192.

Rhoades, M.M. and Kerr, W.E. (1949) A note on centromere organization. *Proc. Natl. Acad. Sci. USA* 39: 129–132.

Riess, R.W., Barker, K.R. and Biesele, J.J. (1977) The sperm and its formation in the scorpion *Centruroides vittatus. La Cellule* 72: 259–264.

Robison, W.G., Jr. (1970) Unusual arrangements of microtubules in relation to mechanisms of sperm movement. *In* Baccetti, B. (ed.) *Comparative Spermatology,* pp. 311–320. Academic Press, New York.

Rolfe, W.D. I. (1980) Early invertebrate terrestrial faunas. *In* Panchen, A.L. (ed.) *The Terrestrial Environment and the Origin of Land Vertebrates,* pp. 117–157. Academic Press, London.

Rolfe, W.D. I. (1982) Ancient air breathers. *Field Mus. Nat. Hist. Bull.* 53: 12–16.

Rolfe, W.D. I. (1985) Early terrestrial arthropods: a fragmentary record. *Phil. Trans. R. Soc. Long.* B 309: 207–218.

Rolfe, W.D. I. (1988) Early life on land—the East Kirkton discoveries. *Earth Sci. Conserv.* 25: 22–28.

Rolfe, W.D. I. and Beckett, C.M. (1984) Autecology of Silurian Xiphosurida, Scorpionida, Cirripedia and Phyllocarida. *Spec. Pap. Palaeontol.* 32: 27–37.

Root, T.M. (1990) Neurobiology. *In* Polis, G.A. (ed.) *The Biology of Scorpions,* pp. 341–413. Stanford University Press, Palo Alto, CA.

Rosin, R. and Shulov, A. (1963) Studies on the scorpion *Nebo hierichonticus. Proc. Zool. Soc.* (London) 140: 547–575.

Ruppert, E.E. and Barnes, R.D. (1994) *Invertebrate Zoology,* Chapt. 13, Chelicerates, pp. 617–677. Saunders College Publishing, New York.

Sato, I. (1936) On the chromosomes in spermatogenesis of the scorpion *Buthus martensii* Karsch. *Zool. Mag., Tokyo* 48: 956–957.

Savory, T.H. (1971) *Evolution in the Arachnida.* Merrow Publishing, Watford.

Savory, T.H. (1977) *Arachnida.* Chapt. 13, Scorpions, pp. 115–124. Academic Press, London.

Schaller, F. (1971) Indirect sperm transfer by soil arthropods. *Annu. Rev. Entomol.* 16: 407–446.

Schaller, F. (1978) Significance of sperm transfer and formation of spermatophores in arthropod phylogeny. *In* Gupta, A.P. (ed.) *Arthropod Phylogeny,* pp. 587–608. Van Nostrand Reinhold, New York.

Scheuring, L. (1913) Die Augen der Arachniden. *Zool. Jahr. Abt. Anat. Ontog.* 33: 553–636.

Scotese, C.R., Van Der Voo, R. and Barrett, S.F. (1985) Silurian and Devonian base maps. *Phil. Trans. R. Soc., London* B 309: 57–77.

Selden, P.A. (1984) Autecology of Silurian eurypterids. *Autecology of Silurian Organsisms. Spec. Pap. Palaeontol.* 32: 39–54.

Selden, P.A. (1985) Eurypterid respiration. *Phil. Trans. R. Soc. London* B 309: 219–226.

Selden, P.A. (1988) The arachnid fossil record. *Br. J. Entomol. Nat. Hist.* 1: 15–18.

Selden, P.A. (1993) Fossil arachnids—recent advances and future prospects. *Mem. Queensland Mus.* 33: 389–400.

Selden, P.A. and Jeram, A.J. (1989) Palaeophysiology of terrestrialisation in the Chelicerata. *Trans. R. Soc. (Edinburgh) Earth Science* 80: 303–310.

Shanahan, C.M. and Hayman, D.L. (1990) Synaptonemal complex formation in male scorpions exhibiting achiasmate meiosis and structural heterozygosity. *Genome* 33: 914–926.

Sharma, G.P., Parshad, R. and Handa, R. (1962) Meiosis in two species of *Palamnaeus* (Scorpiones—Scorpionidae). *Res. Bull. Punjab Univ.* 13: 85–89.

Sharma, G.P., Prashad, R. and Joneja, M.G. (1959) Chromosome mechanism in the males of three species of scorpions (Scorpiones-Buthidae). *Res. Bull. Punjab Univ.* 10: 197–207.

Shear, W.A. (1991) The early development of terrestrial ecosystems. *Nature* 351: 283–289.

Shear, W.A. (1999) Introduction to Arthropoda and Cheliceriformes. *In* Harrison, F.W. and Foelix, R.F. (eds.) *Microscopic Anatomy of Invertebrates,* vol. 8A Chelicerate arthropoda, pp. 1–19. Wiley-Liss, New York,

Shear, W.A., Gensel, P.G., and Jeram, A.J. (1996) Fossils of large terrestrial arthropods from the Lower Devonian of Canada. *Nature* 384: 555–557.

Shear, W.A. and Kukalová-Peck, J. (1990) The ecology of Paleozoic terrestrial arthropods: the fossil evidence. *Can.J. Zool.* 68: 1807–1834.

Shulov, A. and Levy, G. (1978) Venoms of Buthidae. A, Systematics and Biology of Buthidae. *In* Bettini, S. (ed.) *Arthropod Venoms. Handbook of Experimental Pharmacology,* vol. 48, pp. 309–317. Springer-Verlag, Berlin.

Shultz, J.W. (1989) Morphology of locomotor appendages in Arachnida: evolutionary trends and phylogenetic implications. *Zool.J. Linn. Soc.* 97: 1–56.

Simard, J.M. and Watt, D.D. (1990) Venoms and toxins. *In* Polis, G.A. (ed.) *The Biology of Scorpions,* pp. 414–444. Stanford University Press, Palo Alto, CA.

Sissom, W.D. (1990) Systematics, biogeography and paleontology. *In* Polis, G.A. (ed.), *The Biology of Scorpions,* pp. 64–160. Stanford University Press, Palo Alto, CA.

Sissom, W.D. and Cokendolpher, J.C. (1998) A new troglobitic scorpion of the genus *Typhlochactas* (Superstitionidae) from Veracruz, Mexico. *J. Arachnol.* 26: 285–290.

Snodgrass, R.E. (1952) *A Textbook of Arthropod Anatomy.* Comstock, Ithaca, NY.

Sokolow, I. (1913) Untersuchungen uber die spermatogenese bei den Arachniden I. Ueber die spermatogenese der skorpione. *Arch. Zellforsch.* 9: 399–432.

Srivastava, D.S. (1955) Maxillary processes and mechanism of feeding in scorpions. *Saugar Univ.J.,* pt. 2, 1: 85–91.

Srivastava, M.D.L. and Uma, A. (1961) Absence of chiasmata and formation of a complex chromosomal body in the spermatogenesis of the scorpion *Palamnaeus longimanus. Caryologia* 14: 63–77.

Stahnke, H.L. (1966) Some aspects of scorpion behavior. *Bull.S. Calif. Acad. Sci.* 65: 65–80.

Stewart, D.M. and Martin, A.W. (1974) Blood pressure in the tarantula, *Dugesiella hentzi. J. Comp. Physiol.* 88: 141–172.

Stockwell, S.A. (1989) *Revision of the phylogeny and higher classification of scorpions (Chelicerata)* (Ph.D. dissertation). University of California, Berkeley.

Stockwell, S.A. (1992) Systematic observations on North American Scorpionida with a key and checklist of the families and genera. *J. Med. Entomol.* 29: 407–422.

Størmer, L. (1963) *Gigantoscorpio willsi, a New Scorpion from the Lower Carboniferous of Scotland and its Associated Preying Microorganisms.* Skrifter av det Norske Videnskaps-Akademi, Oslo I, Matematisk-Naturvidenskapelig Klasse, n.s., 8.

Størmer, L. (1970) Arthropods from the Lower Devonian (Lower Emsian) of Alken-an-der-Mosel, Germany. Part I, Arachnida. *Senckenbergiana Lethaea* 51: 335–369.

Størmer, L. (1976) Arthropods from the Lower Devonian (Lower Emsian) of Alken-an-der-Mosel, Germany. Part 5: Myriapods and additional forms, with general remarks on fauna and problems regarding invasion of land by arthropods. *Senckenbergiana Lethaea* 57: 87–183.

Subburam, V. and Gopalakrishna Reddy, T. (1989) Functional morphology of the appendix of the viviparous scorpion, *Heterometrus fulvipes* (Koch) (Arachnida: Scorpionidae). *Intl. J. Insect Morphol. Embryol.* 18: 329–333.

Tandan, B.K., Somadder, K., Kuman, P., Churchman, L.C., Rastogi, V.L. and Gupta, P.D. (1977) The structure of the pleural cuticle of *Heterometrus leiurus* (Arachnida: Scorpionida). *Biol. Mem.* 2: 222–255.

Tollerton, Jr., V.P. (1989) Morphology, taxonomy and classification of the order Eurypterida Burmeister, 1843. *J. Paleontol.* 63: 642–657.

Toolson, E.C. (1985) Uptake of leucine and water by *Centruroides sculpturatus* (Ewing) embryos (Scorpiones, Buthidae). *J. Arachnol.* 13: 303–310.

Turbeville, J.M., Pfeifer, D.M., Field, K.G. and Raff, R.A. (1991) The phylogenetic status of arthropods, as infered from 18S rRNA sequences. *Mol. Biol. Evol.* 8: 669–686.

Tuzet, O. (1938) Sur la spermatogénése de *Buthus occitanus* (Amor.). *Arch. Zool. Exp. Genet.* 80: 335–351.

Vachon, M. (1945) L'appendice arachnidien et son évolution. Note préliminaire. *Bull. Soc. Zool. France* 69: 172–177.

Vachon, M. (1950) Remarques préliminaires sur l'alimentation les organes Chélicériens le biberon et la tétine de l'embryon du scorpion: *Ischnurus ochropus* C.L. Koch (Scorpionidae). *Arch. Zool. Exp. Genet.* (Paris) 86: 137–156.

Vachon, M. (1953) The biology of scorpions. *Endeavor* 12: 80–89.

Valentine, J.W., Awramik, S.M., Signor, P.W. and Sadler, P.M. (1991) The biological explosion at the precambrian-cambrain boundary. *Evol. Biol.* 25: 279–356.

Van Der Hammen, L. (1977) A new classification of Chelicerata. *Zool. Med.* 51: 307–319.

Van Der Hammen, L. (1979) Comparative studies in Chelicerata I. The Cryptognomae (Ricinulei, Architarbi and Anactinotrichida). *Zool. Verh.* 174: 1–62.

Van Der Hammen, L. (1982) Comparative studies in Chelicerata II. Epimerata (Palpigradi and Actinotrichida). *Zool. Verh.* 196: 1–70.

Van Der Hammen, L. (1985a) Comparative studies in Chelicerata III. Opilionida. *Zool. Verh.* 220: 1–60.

Van Der Hammen, L. (1985b) Functional morphology and affinities of extant Chelicerata in evolutionary perspective. *Trans. R. Soc. (Edinburgh): Earth Sci.* 76: 137–146.

Van Der Hammen, L. (1986a) Comparative studies in Chelicerata IV. Apatellata, Arachnida, Scorpionida, Xiphosura. *Zool. Verh.* 226: 1–52.

Van Der Hammen, L. (1986b) On some aspects of parallel evolution in Chelicerata. *Acta Biotheor.* 35: 15–37.

Venkatanarasimhaiah, C.B. and Rajasekarasetty, M.R. (1964) Contribution to the cytology of Indian scorpions. Chromosomal behavior in the male meiosis of *Palamnaeus gravimanus. Caryologia* 17: 195–201.

Venkateswararao, P. (1967) Feeding aparatus in the scorpion *Heterometrus fulvipes. Proc. Indian Acad. Sci.* 66: 1–19.

Vermeij, G. (1987) *Evolution and Escalation.* Princeton University Press, Princeton, NJ.

Vogel, B.R. and Durden, C.J. (1966) The occurrence of stigmata in a Carboniferous scorpion. *J. Palaeontol.* 40: 655–658.

Vyas, A.B. (1971) Adaptive diversification of the prosomatic appendages in the scorpion *Heterometrus fulvipes. Ann. Zool.* 7: 65–80.

Vyas, A.B. (1974a) The cheliceral muscles of the scorpion *Heterometrus fulvipes. Bull. S. Calif. Acad. Sci.* 73: 9–13.

Vyas, A.B. (1974b) Scanning electron microscopy of book-lungs of the scorpion *Heterometrus fulvipes. Entomol. News* 85: 88–91.

Warburg, M.R. and Rosenberg, M. (1996) A different oogenetic pattern in a US apoikogenous scorpion, *Vaejovis spinigerous* (Scorpiones, Vaejovidae), *Tissue and Cell* 28: 751–757.

Weygoldt, P. (1969) *The Biology of Pseudoscorpions.* Harvard University Press, Cambridge, MA.

Weygoldt, P. (1975) Die indirekte Spermatophorenubertragung bei Arachniden. *Verh. Deutsch. Zool. Ges.* 67: 308–313.

Weygoldt, P. (1985) Ontogeny of the arachnid central nervous system. *In* Barth, F.G. (ed.) *Neurobiology of Arachnids,* pp. 20–37. Springer-Verlag, Berlin.

Weygoldt, P. (1990) Arthropoda—Chelicerata: Sperm transfer. *In* Adiyodi, K.G. and Adiyodi, R.G. (eds) *Reproductive Biology of Invertebrates,* vol. IV, part B, pp. 77–119. Wiley, New York.

Weygoldt, P. and Paulus, H.F. (1979a) Untersuchungen zur Morphologie, Taxonomie und Phylogenie der Chelicerata.I. Morphologische Untersuchungen. *Z. Zool. Syst. Evolutionsforsch.* 17: 85–116.

Weygoldt, P. and Paulus, H.F. (1979b) Untersuchungen zur Morphologie, Taxonomie und Phylogenie der Chelicerata. II. Cladogramme und die Entfaltung der Chelicerata. *Z. Zool. Syst. Evolutionsforsch.* 17: 177–200.

Willmer, P. (1990) *Invertebrate Relationships: Patterns in Animal Evolution.* Cambridge University Press, Cambridge.

Wills, L.J. (1947) A monograph of British Triassic scorpions. Part I. *Palaeontogr. Soc. Monogr.,* vol. 100.

Wills, L.J. (1960) The external anatomy of some Carboniferous "Scorpions", Part 2. *Palaeontology* 3: 276–382.

Wilson, E.B. (1916a) The distribution of sperm-forming material in scorpions. *J. Morphol. Physiol.* 52: 429–467.

Wilson, E.B. (1916b) The distribution of the chromosomes to the spermatozoa in scorpions. *Proc. Natl. Acad. Sci. USA* 2: 321–324.

Wilson, E.B. (1931) The distribution of sperm-forming materials in scorpions. *J. Morphol. Physiol.* 52: 429–467.

Wilson, E.B. and Pollister, A.W. (1937) Observations on sperm formation in the Centrurid scorpions with especial reference to the Golgi material. *J. Morphol.* 60: 407–433.

Wourms, J.P. (1981) Viviparity: the maternal-fetal relationship in fishes. *Am. Zool.* 21: 473–515.

Wourms, J.P., Grove, B.D. and Lombardi, J. (1988) The maternal-embryonic relationship in viviparous fishes. *In* Hoar, W.S. and Randall, D.J. (eds) *Fish*

Physiology, vol. XI, *The Physiology of Developing Fish,* part B, *Viviparity and Posthatching Juveniles,* pp. 1–134. Academic Press, San Diego, CA.

Yoshida, Y. and Toshioka, S. (1964) Studies on spermatogenesis in scorpions. *Arachnol. Soc. E. Asia* 19: 1–4.

Yoshikura, M. (1975) Comparative embryology and phylogeny of Arachnida. *Kum.J. Sci.* 12: 71–142.

Zeh, D.W., Zeh, J.A. and Smith, R.L. (1989) Ovipositors, amnions and eggshell architecture in the diversification of terrestrial arthropods. *Q. Rev. Biol.* 64: 147–168.

3

Eyes and Vision

ADAM LOCKET

Scorpions, compared to some other arthropods such as spiders and insects, cannot be regarded as highly visual animals in the sense of using image-forming eyes to their full potential. This is not to say that their eyes are useless: Fleissner and Fleissner (this volume) and their colleagues have shown that scorpion's eyes are extremely sensitive to light, corresponding well to their nocturnal habits. Even where they are abundant, scorpions, with the exception of a few species (e.g., *Pandinus*) are seldom seen by day. At night they usually respond to sudden illumination with inaction, in strong contrast with the abrupt burst of activity evoked by a small air puff, mediated through wind-sensitive hairs, particularly trichobothria (Hoffman, 1967; Linsenmair, 1968; Ignat'yev et al., 1976). Scorpions use the direction of light sources to calibrate their anemomenotactic orientation (Linsenmair, 1968). Dube and Fleissner (G. Fleissner, personal communication) have shown patterns of activity that are entrained or triggered by light. In most cases scorpions are not visual feeders; sit-and-wait at the burrow mouth (Williams, 1966; Shorthouse, 1971; Eastwood, 1978a,b; Polis, 1979; Shachak and Brand, 1983) or locating prey by mechanoreception while roaming (Brownell, 1977, 1984; Brownell and Farley, 1979a,b,c) are more usual modes of foraging. Some cave scorpions lack eyes altogether and manage to find enough food to attain quite large sizes (W. D. Sissom, personal communication).

Though the structure of scorpion eyes was studied in the nineteenth and early twentieth century (e.g., Lankester and Bourne, 1883; Parker, 1891; Police, 1907; Scheuring, 1913), little further work was done on them until the 1960s. Machan (1966, 1967), Bedini (1967), Belmonte and Stensaas (1975), and Fleissner and co-workers (Fleissner and Schliwa, 1977; Schliwa and Fleissner, 1977, 1979, 1980; Fleissner and Siegler, 1978; Fleissner and Fleissner, 1985, 1988) have since published ultrastructural information, much of it correlated with electrophysiology. Root (1990) summarized the previous work in the context of scorpion neurobiology.

The pattern of eye distribution is rather uniform among recent, as opposed to fossil, scorpions. Typically there is a pair of median eyes, located on the dorsal surface of the carapace, and an array of lateral eyes on its anterolateral margin. The number of lateral eyes varies from two to five, depending on species, and they are almost always present as a row, with larger eyes anteriorly. Occasionally the number of lateral eyes varies, even between the two sides in an individual, but there is always a single pair of median eyes if the scorpion has eyes at all (fig. 3.1).

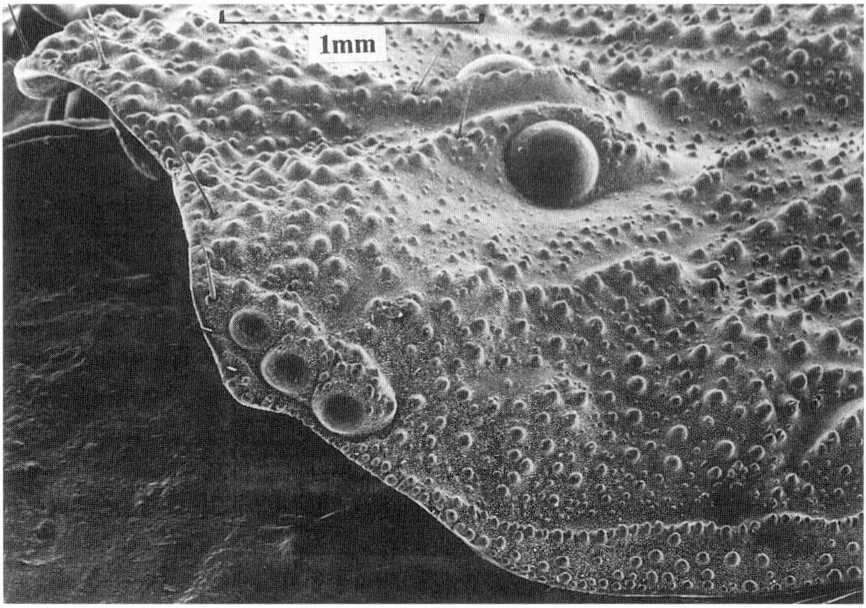

Figure 3.1 Carapace of scorpion showing position of median and lateral eyes.

In some cave-dwelling species the eyes are either reduced or absent altogether. Mitchell (1968, 1972) recorded the eyeless chactid genus *Typhlochactas* from caves in Mexico, and later Mitchell and Peck (1977) recorded another eyeless species of the same genus, *T. sylvestris,* which occurs in forest leaf litter. Francke (1977, 1978) described diplocentrid cave scorpions, of which *Diplocentrus anophthalmus* has no eyes, *D. mitchelli* markedly reduced eyes, and *D. cueva* small but apparently normal eyes. Lourenço and Francke (1985) reviewed cavernicolous scorpions, and Lourenço (1995) raised the number of known taxa to 14.

Vision is usually regarded as the sense mediated by eyes, but light sensitivity is not limited to eyes alone. Zwicky (1968, 1970) showed by behavioral and neurophysiological experiments that the metasoma (tail) of an Australian scorpion, *Urodacus,* is light sensitive, and that the effect was not due to heat or infrared. The action spectrum of sensitivity was from long UV to blue-green, with a maximum at 478 nm and no response to green, yellow, or red. Geethabali and Rao (1973), investigating the metasomal light sense of *Heterometrus* species from India, found a main peak, at 568 nm, and a smaller one at 440 nm. Geethabali (1976) further reported that the metasomal ganglia and nerves are directly light sensitive and that there is considerable crossing of information within the nerve cord. The precise nature of the metasomal photoreceptor remains unclear, but its existence should certainly be considered when dealing with photoreception by the intact scorpion.

Ranging in size from a few millimeters to tens of centimeters, scorpions are a convenient size to keep in numbers and to work on. They are easily sedated by chilling or anesthetized with carbon dioxide and tough enough to survive surgery well. The cuticle is readily cut with scissors, and pieces cut away can be replaced and sealed with wax, or the animal's own hemolymph, making survival experiments possible. The median eyes, the optic nerves and laminae, and the upper parts of the brain are all within reach after a small piece of carapace is excised, and the work of the Fleissners' group shows what can be achieved by recording from and excising parts of the visual system. The lens, though fairly hard, is easily penetrated by an entomological pin which may be sealed in place for electroretinogram (ERG) recordings, or a different electrode may be inserted through the hole. Indifferent electrodes are easily inserted beneath the carapace or tergites.

There is now, largely thanks to the Fleissners' laboratory, a sound body of knowledge on the scorpion visual system, and there is a large literature on related groups, particularly *Limulus* and spiders (see Barlow and Chamberlain, 1980, and Land, 1985, respectively for references). There are still, however, plenty of problems to be solved. The lamina, differently placed in Australian scorpionids than in most other examples, and discussed further below, is an example, and details of the central visual connections are not well known.

Most workers on scorpion vision have done their experiments and fixations by day, understandably since scientists are largely diurnal. Scorpions, with few exceptions, are not; active mainly by night, they spend the daylight hours in retreats or burrows (Cloudsley-Thompson, 1956; Wuttke, 1966). Night, as the Fleissners have shown convincingly (Fleissner, 1972; Fleissner and Fleissner, 1988), is when scorpion eyes are most sensitive and when the shielding pigment in the retinula cells migrates clear of the rhabdoms. Not only is night their active time, but they have well-marked circadian rhythms of sensitivity and pigment migration which persist even in constant darkness. Fixations or experiments on vision designed to reflect the "natural" state, when the scorpion would normally be using its eyes, should therefore be carried out not only in the dark, but also at night.

Many routine experimental methods are suitable for use with scorpions, which are large enough to dissect without undue difficulty, though a dissecting microscope is all but essential. Fixation of eyes and brain for light microscopy may include routine fixatives such as buffered formalin, formal calcium, or Bouin's fluid. Synthetic aged alcoholic Bouin (Gregory, 1980) and Carnoy's fluid are suitable for Bodian silver staining. Golgi preparations made with various mixtures containing glutaraldehyde, formaldehyde, and potassium dichromate, followed by 0.75% silver nitrate work with varying degrees of success. Fixation for electron microscopy and 1-μm plastic sections in routine phosphate-buffered glutaraldehyde/paraformaldehyde mixtures with postosmication work well, and the addition of sucrose to adjust tonicity may be helpful. The hard cuticle and lens, which fail to impregnate well with most embedding media, including the epoxy resins used for electron microscopy, can be a nuisance in histological work, though sometimes the soft tissues can be separated from the cuticle during processing. Other special techniques are discussed in context. Best results usually depend on trial and error, particularly if the species has not been investigated before. An excellent collection of neuroanatomical methods is found in Strausfeld and Miller (1980).

OPTICAL PROPERTIES OF SCORPION EYES

Visual Fields

Land (1985) has shown that the visual fields of those spiders which have a retinal tapetum, notably lycosids, may be measured accurately by observing the

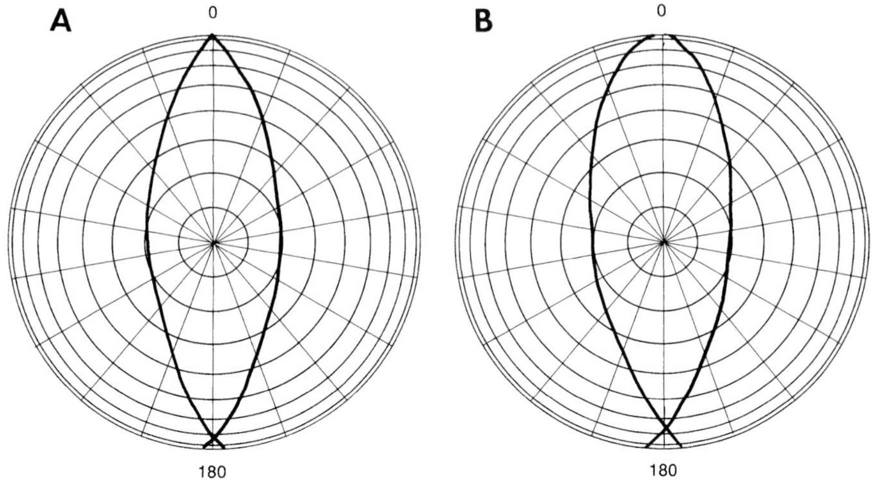

Figure 3.2 Visual fields of the median eyes. Both *Urodacus yaschenkoi* (A) and *Lychas* spp. (B) have all round vision at least down to the horizontal, with a strip of binocular vision above. Graph represents a hemisphere above the scorpion, with the scorpion positioned at its center.

tapetal reflection. A perimeter that moves a light and telescope around the spider enables one to plot the latitudes and longitudes where the reflection disappears.

The visual fields of scorpions are less easily measured optically, as the scorpion has no tapetum. Scheuring (1913) gives estimates derived from shining lights into the eyes and observing eyeshine, and Marschall-Pauly and Fleissner (1988) have recorded visual fields from single cells. Approximate measurements can be made using a simple tilting-table goniometer. A cover slip stuck to the front lens mount of a dissecting microscope at approximately 45° to a light source will reflect light into the eyes, illuminating the retina. The light reflected can be seen, though it is not bright, particularly in the lateral eyes, making it hard to determine an accurate end point. The goniometer table is rotated with the tilt set at a known angle and the extent of the eye shine measured. The results may be plotted onto polar graph paper, showing a representation of a hemisphere in space, seen from the zenith and with meridians marked. The scorpion would be at the center of the hemisphere. Observation of a restrained scorpion shows that the field extends below the horizontal in some regions, but this has not been measured because limbs of the scorpion or the apparatus itself usually get in the way.

The fields of the median eyes of *Urodacus yaschenkoi*, a scorpionid, and of a *Lychas* species (Buthidae), are shown in fig. 3.2. Each eye sees rather more than half a hemisphere, so there is complete coverage round the animal, with a region of binocular vision, as much as 40° wide, in a strip above.

Optics of Lens

The lenses of median and lateral eyes are similar, though the median eyes are always larger. Lenses of an adult *Urodacus yaschenkoi* are about 0.5 mm diameter, somewhat smaller than the 0.8 mm in a large *Lycosa* spider, but are still large for an ocellus type eye in a terrestrial invertebrate. Formed from modified cuticle, the outer surfaces of the lenses are notably smooth, with a strong, regular curvature. Fleissner and Fleissner (personal communication) have observed that the lenses in the buthids *Androctonus australis* and *Buthus occitanus* remain clean even when the rest of the scorpion is dusty. They have found a small brush of setae beneath the femur of the pedipalp which works across the ocular eminence when the scorpion assumes a resting position, brushing dust from the surface of the median eyes. This brush is present in many buthids and some other scorpions.

Removed from the animal or seen in exuviae, the high transparency of the lens is apparent, unlike the cloudy or opaque cuticle with which it is contiguous. The inner surface of the lens is more strongly curved than the outer, thus the whole lens, though thick, is not truly spherical. The inner projecting portion can easily be split away from the more superficial part, along planes of cleavage due to the laminated structure of cuticular derivatives.

The optical properties of the scorpion eye can be investigated using Homann's (1928) hanging drop method (fig. 3.3A), coupled with histological studies. The hanging drop method, developed by Homann for use with spiders, and used by Land (1985), who gives details of the method, provides simple and elegant evidence of the focal length and quality of the image of the eye. Comparable measurements have been made on *Urodacus yaschenkoi.* The image formed in the drop by the lens can be inspected and measured by eyepiece micrometer, as can the diameter of the deep surface of the lens, immersed in the drop (fig. 3.3B).

The aperture of the eye, familiar as the F-number in photography, can be found by dividing the focal length of the lens by its diameter. The F-number for *Urodacus,* about 0.75, shows that though small, the eye is "fast" in photographic terms and well suited to its function in dim light.

The hanging drop method also enables one to assess the image quality, particularly by using a grid of lines as the object, which will reveal any distortions caused by aberrations. Such observations show a surprisingly good and distortion-free image, with little if any divergence or convergence of the lines from center to periphery (fig. 3.3C). Some degree of regular astigmatism may be seen if the lens is not viewed along the optic axis. This astigmatism is hard to avoid because the piece of cuticle bearing the lens is asymmetrical and often hangs askew in the surface of the drop. Carricaburu (1968) found only a small degree of irregular astigmatism, which could be due to damage to the lens surfaces or remnants of tissue. Coupled with the short focal length, the good image quality suggests that the lens is inhomogeneous, as are the lenses of fishes, which also have a short focal length. This suggestion is supported by the observation that the outer surface of the lens is considerably more resistant to mechanical damage than the softer inside portion.

Carricaburu (1968) used a different approach to examine the optics of *Androctonus australis,* measuring the distance between the posterior pole of the lens and an image formed in jelly on a slide, using the calibrated fine adjustment of the microscope, and correcting for refractive index. He measured refractive indices of 1.536 in the anterior part of the lens nucleus and 1.531 in the relatively homogeneous posterior part of the lens. Believing that the retina was in direct contact with the posterior surface of the lens, Carricaburu concluded that the median eye is strongly hypermetropic (far-sighted). But as Fleissner and Fleiss-

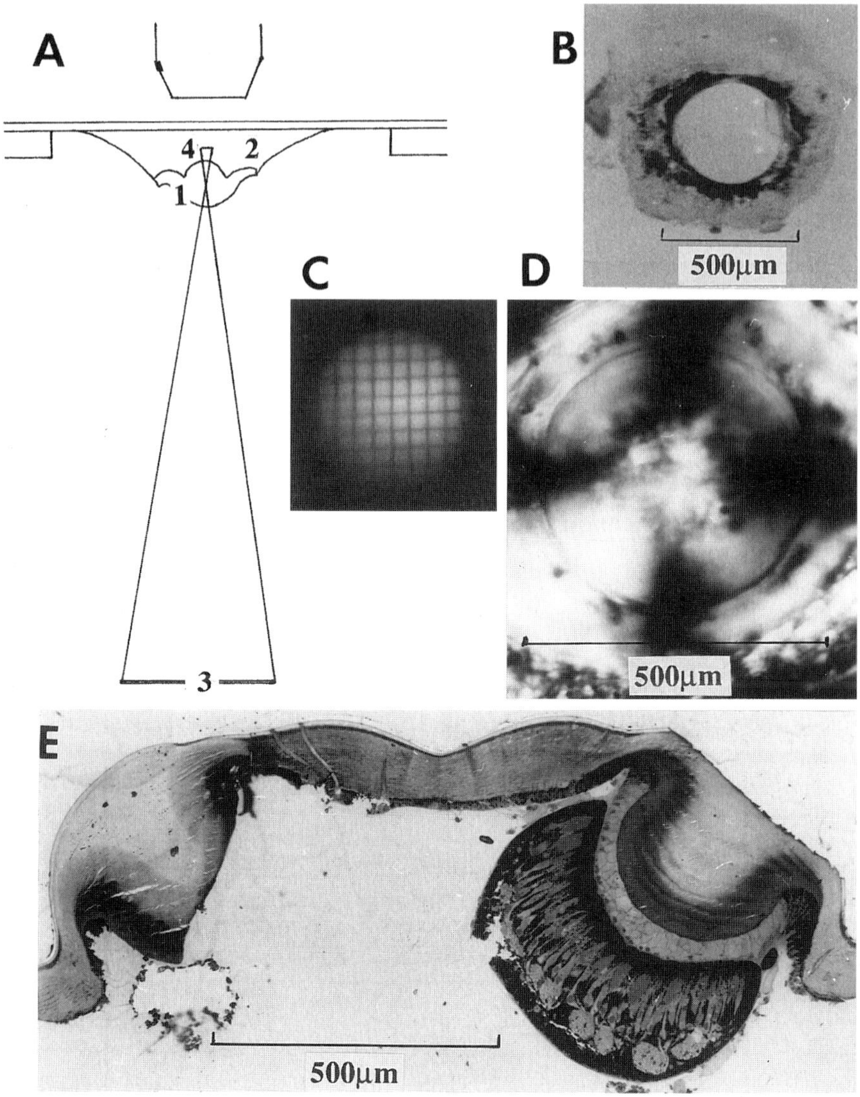

Figure 3.3 Optics of the median eyes of *Urodacus yaschenkoi* as measured by the hanging drop preparation. (A) Arrangement for hanging drop preparation. Lens (1) is suspended in a drop of saline (2) from a cover slip. The lens views an object (3) and forms an image of the object in the drop (4). This can be measured with the microscope and the focal length calculated. (B) Median eye lens and surrounding cuticle photographed in a hanging drop. The diameter is easily measured. (C) Image of a grid pattern formed in a hanging drop by median eye lens. The image is remarkably free from aberrations. (D) Trans-illumination of lens by polarized light shows darkened cross pattern characteristic of birefringent spherical lenses. (E) Histological section through both median eye lenses. Slices were cut from inner (right) and outer (left) surfaces before fixation.

ner (1985) point out, Carricaburu appears to have overlooked the presence of the vitreous, though this is clearly shown in the drawings of early authors (e.g., Lankester and Bourne, 1883; Scheuring, 1913).

The intact lens, or equatorial sections of it, are birefringent, as Carricaburu (1968) reported. The birefringence is well shown in a hanging drop preparation (fig. 3.3D), but the microscope condenser is required, and a straight tube microscope with achromatic objectives are best because prisms in the light path and fluorite components may affect polarized light.

The orientation of the cross pattern visible by polarized light depends on the orientation of the polarizer. Retinal analyzers would convert this to intensity differences over large areas of retina and could well account for the polarization sensitivity of scorpions demonstrated behaviorally by Brownell and Weber (personal communication). Many insects are sensitive to polarized light (see Rossel, 1989, for review), and the analyzer may be the orientated visual pigment molecules in the rhabdomeral microvilli, which are themselves orientated. Certainly the microvilli in scorpion rhabdoms (*vida infra*) are in rather uniform arrays, and in many cases rhabdomeres of the different visual cells have differently orientated microvilli. Such orientation could account for the observed polarization sensitivity, but the question is at present open.

Figures for lens diameter, focal length, and F-number are given in table 3.1. The variation in values for F-number may be due to the difficulty of measuring an image less than 50 μm long accurately with an eyepiece micrometer and a low-power objective, which is needed to provide sufficient working distance. Movements of the lens in the drop also reduce accuracy of measurements.

The resolution of an eye may be limited by the quality of the lens or of the retina. In the hanging drop, one can resolve the image of a scale of millimeters quite plainly, and at the distance used, 120 mm, 1 mm subtends 28′, so the lens resolves at least this figure. This compares well with the figures of 15–55′ of arc obtained by Carricaburu (1968), who measured resolution by observing test patterns at different distances on an optical bench.

The F-number derived from hanging drop experiments allows us to predict where the image will fall in histological sections of the eye. From this ratio and the lens diameter, the focal length can be found, and from this and the dimensions of the receptors, angular resolution may be calculated. In an adult *Urodacus yaschenkoi* the lens is about 375 μm diameter, and the F-number 0.75. Tangential sections of retina show that the vitread ends of the rhabdoms are separated by about 12 μm, and therefore subtend about 2.7°. Clearly, resolution is limited by the retina, not the lens.

In contrast to the median eyes, most accounts of the lateral eyes note that the lens is close against the retina and that there is no vitreous forming a clear zone between the two. Hanging drop preparations of lateral eye lenses are not easy to make, as the eyes are small and there is cuticle of the doublure close against their deep surface. Also, the lens protrudes obliquely from the carapace, so that it will not lie in the surface film of the drop with its optic axis perpendicular to the surface. Given these constraints, an image formed by the lens can be viewed, and it turns out to be located surprisingly close to or at the posterior surface of the lens. Such an image will be projected onto the rhabdoms, despite the absence of a clear zone. Though the lens forms a fair image, the lateral eye rhabdoms are fused to form a network, discussed below, which is probably incapable of resolving the image it receives.

Table 3.1 Optical parameters of the median eyes of *Urodacus yaschenkoi*

Carapace, mm	4.8	4.9	5	6	7	7	7.6	7.7	8.5	10.5	10.5	10.5
Lens diam, μm	242	250	234	265	296	281	304	312	343	374	452	374
Focal length, μm	187	187	182	229	219	244	234	239	280	276	318	280
F-number	0.77	0.75	0.78	0.86	0.74	0.86	0.77	0.76	0.82	0.74	0.70	0.75

STRUCTURE OF THE EYES

All scorpion eyes, median or lateral, are of the ocellus type, with a single lens condensing light into a retina made up of retinula units. Each of these contains a small number of visual cells, bearing rhabdomeres that contribute to the fused rhabdom and contain the visual pigment. Each retinula also contains a single arhabdomeric cell that makes contact with all the visual cells of its retinula, but which is not itself photosensitive. A further contribution is from efferent neurosecretory fibres (ENSFs), which make contact with the rhabdomes. Pigment cells containing shielding pigment surround the retina, except where the optic nerve fibres emerge, and other pigment cells lie between the retinulae. Shielding pigment granules are also present in the retinula cells.

Median Eyes

The median and lateral eyes are of similar but not identical structure; the median eyes are the more prominent and will be described first. The ocular tubercle that bears the median eyes is situated in the midline of the carapace (fig. 3.1), where its position along this line is of taxonomic use in some cases. The lens, continuous with and differentiated from the cuticle, forms a rounded prominence on the surface that grades continuously into its surroundings without a sharp boundary. The interior of the lens is much more sharply distinguished from its surroundings by a groove that separates the prominent inward bulge of the lens from the relatively thin surrounding cuticle. The lens, the only part of the cuticle that is not fluorescent, is further distinguished by being transparent and free of cuticular pores, most noticeable in those species with darkly tanned cuticle, such as *Heterometrus* spp. In sections the continuity of the lens with the adjacent cuticle is clearly seen, and like that cuticle the lens shows regular laminations, oriented with the equatorial plane. There are more laminae in the lens than in the cuticle surrounding it, and it is possible to split an isolated lens along these laminae, when the deep part comes free from the surround. In some histological sections the superficial part of the lens stains red like the cuticle, but with a pale region extending into the substance of the lens from the outer surface; the deep portion of the lens stains blue. If a slice is cut from the inner or outer surface before fixation, the staining pattern is altered (fig. 3.3E).

When the outer face of the lens is removed, the pale region stains uniformly with the rest of the red portion, suggesting that there is a barrier to penetration of fixatives at the outer surface, breached by the slicing. Cutting away the deep surface brings the red-staining core into direct contact with fixative, but the normal pattern of red and blue persist, suggesting that this is due to an inherent difference in staining properties of the red and blue regions. The nature of this difference remains unknown. Dissection shows that the inner parts of the lens are certainly softer than the outer, but there does not appear to be a central projection of harder material. The entire lens is molted with the rest of the cuticle at ecdysis and can be seen clearly in such molts. It would be interesting, though tedious, to follow the formation of the new lens at ecdysis.

The hypodermal cells at the eyes are specialized in two ways. Those directly beneath the lens become transparent, forming a clear zone or vitreous between the lens and the retina, across which the lens focuses light. At the periphery of the lens the hypodermal cells contain abundant black pigment granules, form-

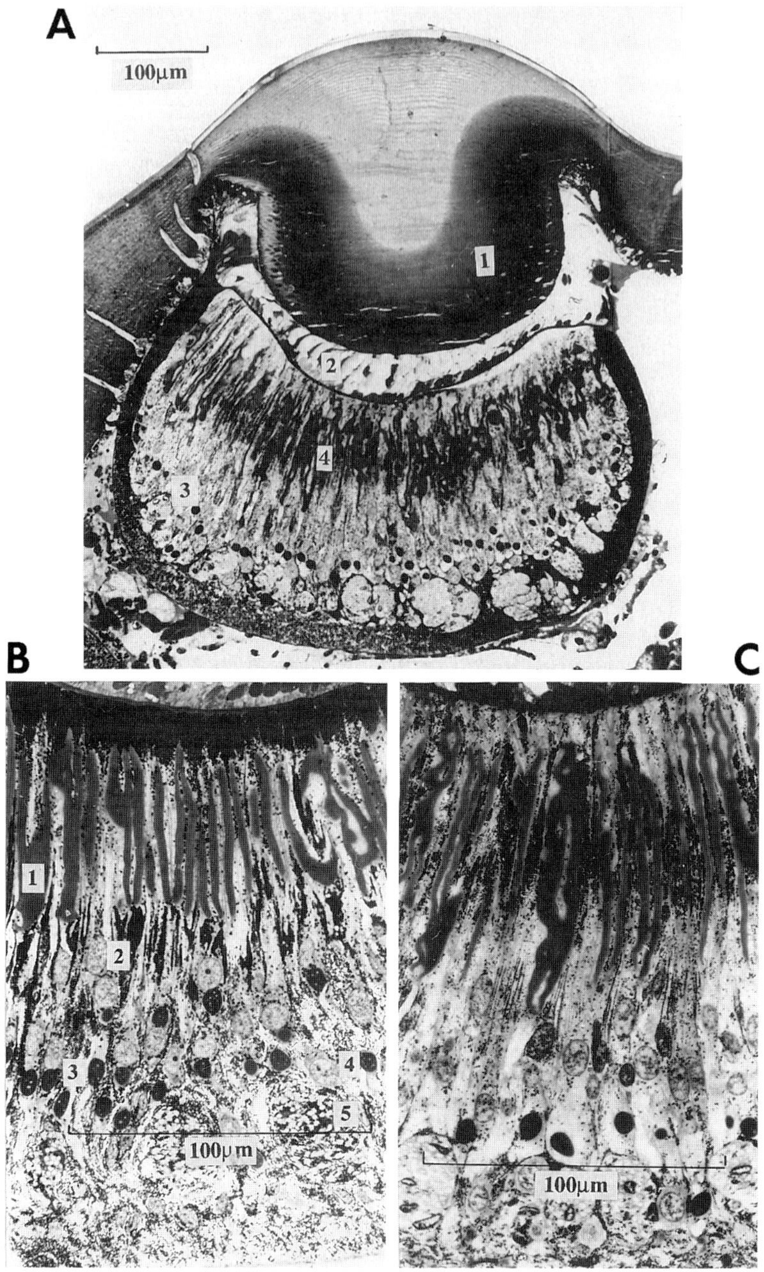

Figure 3.4 Structure of median eyes in night and day states of adaptation. (A) Laminated structure of lens (1) in a night-state eye, with vitreous (2) separating lens from retina (3). Rhabdom layer (4) is exposed to incoming light by pigment moving basally with dark adaptation (1-μm plastic section). (B) Median eye retina fixed in day state. For light-adapted retina, pigment is located between and around rhabdoms, shielding them from light. (C) Median eye retina fixed in night state. Shielding pigment is withdrawn to bases of retinular cells, exposing rhabdoms to incoming light. 1, rhabdoms; 2, retinular cell nuclei; 3, phaospheres; 4, arhabdomeric cell nuclei; 5, nerve fibre bundles.

ing a pigmented limit to the aperture of the lens, analogous to the vertebrate iris. Unlike many vertebrate irises, this aperture does not alter in response to light. Beyond this zone, away from the eyes, the hypodermal cells may or may not be pigmented (fig. 3.4A). The vitreous is separated from the retina by a thin non-cellular preretinal membrane, which extends around the retina to form a postretinal membrane as well. This membrane appears to be complete except where nerve fibres pierce it to form the optic nerve and where they pick up a thin sheath, continuous with the postretinal membrane.

Retina

A "scorpiocentric" view of the scorpion retina would consider two sets of adaptational states: not only dark and light adapted, but night and day states as well (Fleissner and Fleissner, 1988). The normal condition for a scorpion eye is the night state because this is when the animals are active. Illumination is usually low at night, so the night eye is also dark adapted. But the day/night and dark/light cycles are separate, so that the sensitivity—and some structural features—of the eye of a dark-adapted scorpion examined by day will differ from one kept in the dark and examined by night. As Fleissner and Fleissner (1985) point out, most published descriptions (Lankester and Bourne, 1883; Scheuring, 1913; Bedini, 1967; Belmonte and Stensaas, 1975; Locket, 1986) have been on eyes in the day state, though Machan (1968) mentioned the possibility of a diurnal change in sensitivity.

Some features of the retina can be seen in simple squashes of fresh material, perhaps removed from the lens during a hanging drop preparation. Such squashes show the refractile rhabdoms in partial isolation when the abundant pigment that surrounds them is displaced. The rhabdoms do not show birefringence between crossed polaroids. Smaller round, smooth, colorless, refractile (but not birefringent) bodies are also seen. These bodies, not easily disrupted by squashing, are the phaospheres discussed below.

In scorpions of the genus *Urodacus,* the lamina is adjacent to the eye and may be included in the eye preparation, where it is clearly visible as a swelling with a translucent middle (neuropil by conventional microscopy). The fibres leaving the eye and entering the lamina are easily visible, but their cells of origin are obscured by the shielding pigment.

The retina contains about 500–600 functional units, called retinulae (fig. 3.4B,C), in *Androctonus australis* (Fleissner, 1985). Each retinula contains a closed rhabdom, formed by the rhabdomeres of (usually) five visual cells, and separated optically from neighboring rhabdoms by shielding pigment, located mainly within the visual cells. Pigment cells with cell bodies just beneath the postretinal membrane and processes extending vitreally between the visual cells further insulate the visual cells spatially and optically. The arrangement of visual cells in retinulae, also found in spiders and in the ocellar eyes of insects, is reminiscent of the ommatidia in the compound eyes of insects, some crustaceans, and particularly *Limulus* (Fahrenbach, 1975). There is the major difference that the compound eye has an optical system for each ommatidium, whereas the retinulae of ocellar eyes are all arranged beneath a single lens.

The distal portions of the retinula cells bear the rhabdomeres, V-shaped in transverse section and arranged so that those of five (usually) abut to form a fluted, closed rhabdom, star-shaped on section (fig. 3.5A,B). The rhabdomeres extend to the tips of the retinula cells in the night eye, but in the day eye fall short of the preretinal membrane, and are shielded from incoming light by pigment (see Fleissner and Fleissner, 1985, for references).

The rhabdomeres are formed from regular arrays of microvilli, about 80 nm in diameter by up to 2 μm long, projecting from the surface of the cell, their closed tips abutting those of the opposing rhabdomere with no intervening cytoplasm. The microvilli within a rhabdomere are arranged parallel to each other and perpendicular to the surface of the retinula cell from which they protrude, except where the rhabdomere bends from one arm of the rhabdom to the next. The microvilli of two adjoining rhabdomeres are also parallel to each other, but those of different rhabdomeres or parts of rhabdomeres may be at large angles to each other (fig. 3.5B). In tangential sections the rhabdoms appear as "starfish," spanning up to 50 μm; these may be discrete or their arms may interdigitate, depending on the region of the rhabdom cut (fig. 3.5A).

The cytoplasm of the retinula cells contains numerous pigment granules that optically isolate the rhabdoms from each other. These granules undergo marked migrations with day and night adaptation (Fleissner, 1971, 1974), being aggregated in the vitread part of the cell in the day condition, when they effectively screen the rhabdoms from incoming light. At night the pigment withdraws toward the base of the cells and radially away from the rhabdom, leaving it exposed to light.

Close to the visual cell nuclei, which are located deep to the rhabdomeral region, are one or more bodies that Lankester and Bourne (1883) called phaospheres, presumably because they stained darkly. In fresh preparations the phaospheres are as large as the nucleus, transparent, and refractile; they are not affected by lipid solvents, and their histochemistry suggests a glycogen component. By electron microscopy phaospheres are non–membrane-bound aggregations of about 40-nm dense granules. Some phaospheres contain vacuoles that are refractile, according to Bedini (1967), which contain some similar granules and other smaller ones, often in a paracrystalline array (fig. 3.6A).

Opinions have diverged on the occurrence of phaospheres in different taxa. Lankester and Bourne (1883) stated that there were no phaospheres in the buthid *Androctonus,* though they were abundant in the chactid *Euscorpius.* Schliwa and Fleissner (1980) reported phaospheres in the lateral and median eyes of *Androctonus australis.* They are prominent in both lateral and median eyes of the Australian scorpionid *Urodacus yaschenkoi,* but small and sparse, or absent, in *Lychas marmoreus,* a buthid. As Schliwa and Fleissner (1980) remark, their location in the cell is not constant, but usually close to the nucleus, and they are seldom found in the rhabdomeral part.

Found only, but not invariably, in the visual cells, the nature and function of phaospheres remains unknown. Their variability of occurrence suggests that phaospheres may be transient structures. Rhabdomere turnover occurs in spiders (Blest, 1978, 1985) and other arthropods including *Limulus* (see Chamberlain and Barlow, 1984, for references). In *Limulus* the rapid breakdown and recycling of microvilli is associated with the presence of membrane whorls and multivesicular bodies, quite unlike the phaospheres of scorpions. The size of scorpion rhabdomeres does alter with day and night states (Fleissner, 1974), and membrane turnover may well occur. If so, phaospheres may be concerned with such processes, but evidence is lacking.

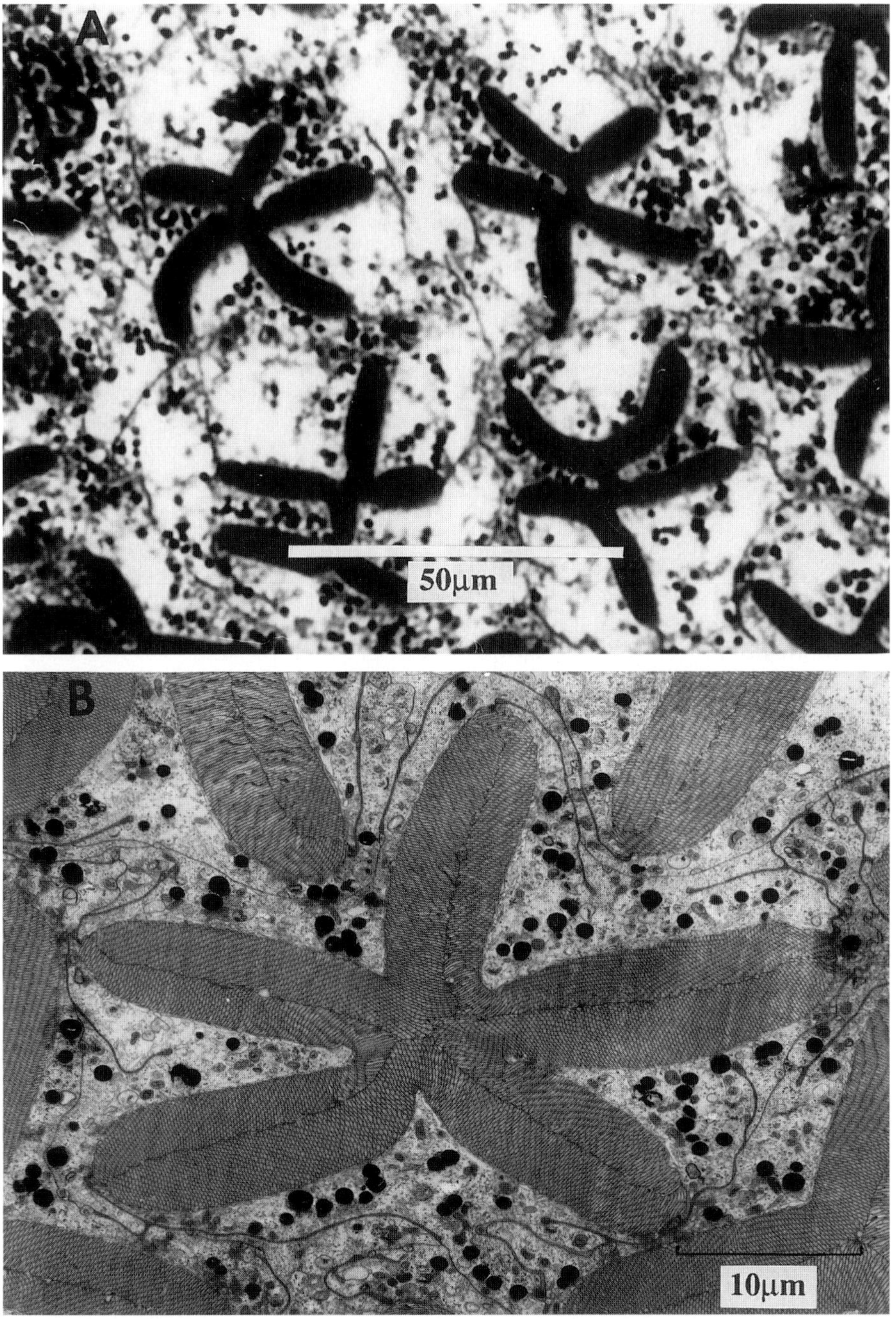

Figure 3.5 Structure of the median eye rhabdoms. (A) Transverse histological section through rhabdoms. Each is formed from rhabdomeres of showing five retinula cells. Size and spacing of rhabdoms limit resolution to about 2.7° of visual angle. (B) Transverse thin section through rhabdomeres shows they are formed by microvilli that abut but do not interdigitate with those of adjacent rhabdomeres. Microvilli of different rhabdomeres make large angles with each other.

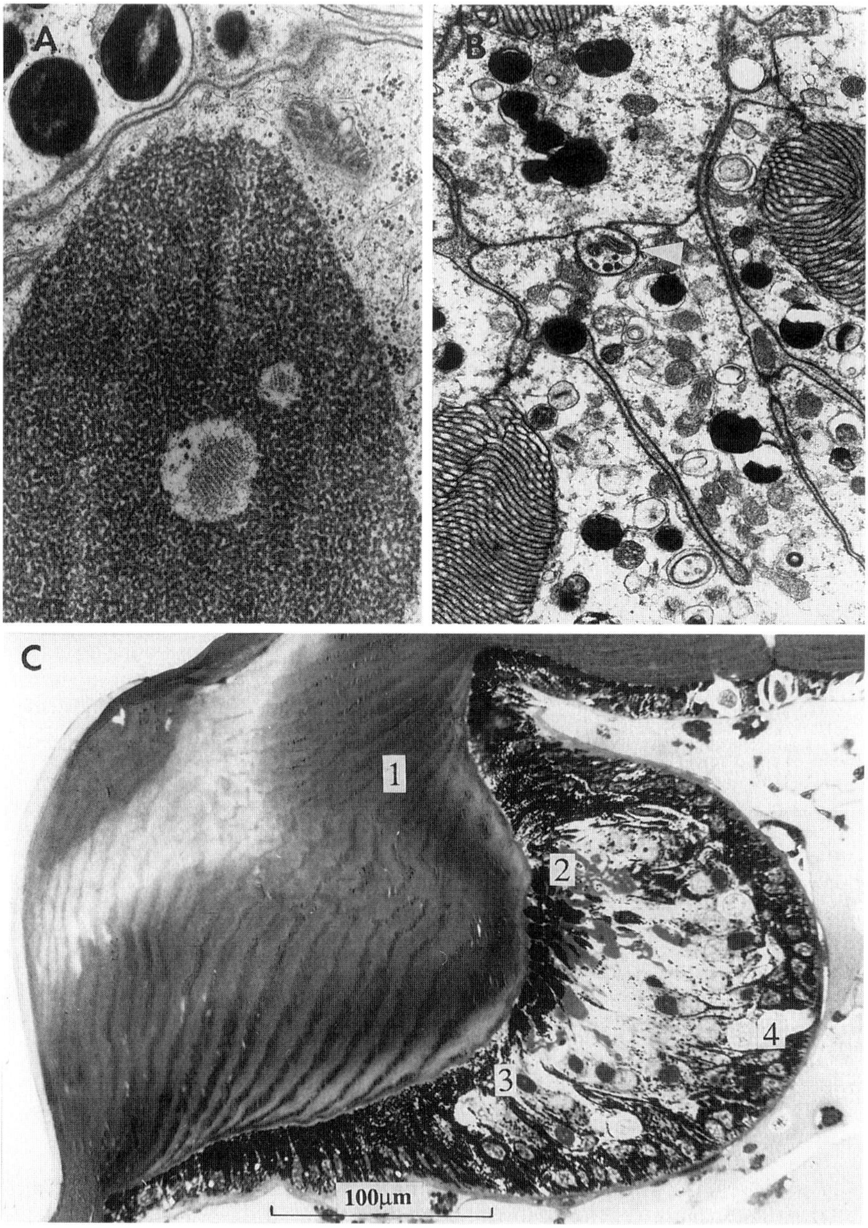

Figure 3.6 Cytological structure of the median and lateral eyes. (A) Phaosphere appears as aggregations of dense granules unbounded by membranes. Vacuoles sometimes contain paracrystalline material. (B) Efferent neurosecretory fibre (arrow) making contact with rhabdomeric cell close to tip of rhabdom. (C) Section through lateral eye showing slightly asymmetrical lens (1) and abutting retina without vitreous. Rhabdomeres (2) are poorly organized. Phaospheres (3) and arhabdomeric cell nuclei (4) are present (1-μm plastic section).

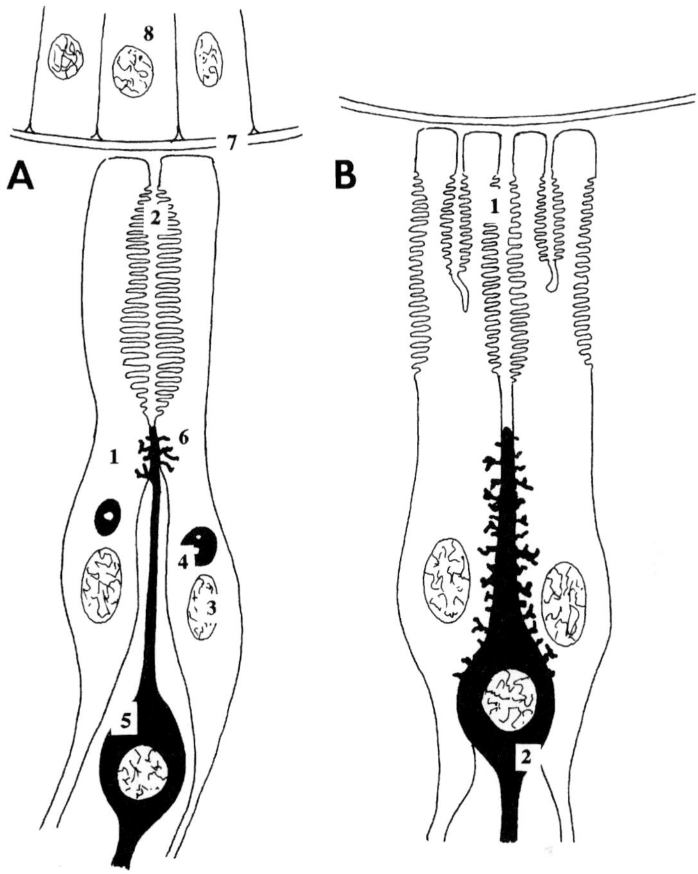

Figure 3.7 Diagram of median and lateral eye retinula cells. (A) Median eye retinula cells (1) form centralized rhabdom (2). Soma of retinular cells contain nuclei (3) and phaospheres (4). Arhabdomeric cell (5) has dendrites (6) contacting retinula cells. 7, preretinal membrane; 8, vitreous cells. (B) Lateral eye retinula cells have dispersed rhabdomeral membrane (1) surrounding distal ends and in clefts. Arhabdomeric cells (2) make contact with few retinula cells, and their dendrites do not extend to the rhabdoms. Adapted from Schliwa and Fleissner (1979).

Associated with each retinula unit is the arhabdomeric cell discovered by Fleissner and Siegler (1978) and described in detail by Schliwa and Fleissner (1979). This cell is devoid of pigment granules and has a round nucleus located deeper than those of the visual cells. The arhabdomeric cell extends a dendrite to the base of the rhabdom and contributes an axon to the optic nerve, but does not form rhabdomeres. The terminal dendrite of these cells was shown by Fleissner and colleagues to end in close association with the base of the closed rhabdom, where projections from the dendrite invaginated the adjacent retinula cell membrane (fig. 3.7A). Arhabdomeric cells, lacking microvilli and not inherently photosensitive, are thus second-order neurons resembling and corresponding functionally (Fleissner, 1985) to the eccentric cells of *Limulus* (Fahrenbach, 1975).

Elements of a third type of neuron are also present in the retina: endings of ENSF (Fleissner and Schliwa, 1977) against visual cells occur, usually near the tips of the rhabdomal ridges. Characterized by their content of dense-cored vesicles, these fibres branch extensively within the retina and may make multiple contacts with visual cells within a retinula or with visual cells of neighboring retinulae. A given visual cell may be multiply innervated by these fibres (fig. 3.6B). The cell bodies of the ENSF lie in the brain (Fleissner and Fleissner, chapter 4, this volume).

Lateral Eyes

Lankester and Bourne (1883) and Scheuring (1913) clearly recognized that, though similar in major respects, the lateral and median eyes also show marked differences. The ultrastructure of the lateral eyes of *Androctonus* was described by Schliwa and Fleissner (1980). Lateral eyes have a cuticular lens, but the quality of this lens is poorer than the lens of the median eye. The retina, smaller relative to the lens than in the median eyes, is in contact with the lens, and there is no vitreous body (fig. 3.6C).

The rhabdomeric cells of the lateral eye retina are distinct from those of the median eye in bearing rhabdomeral material all around their distal portion. This feature produces a more complex rhabdomal network than in the median eyes, where each of the five rhabdomeral cells contacts only two other cells. Schliwa and Fleissner (1980) also showed that the distal faces of the lateral eye rhabdomeral cells are incised by a further rhabdomere-bearing portion, so that the network is more complex distally than basally (fig. 3.7B). They also showed that the microvilli interdigitate in the rhabdoms of lateral eyes, instead of abutting tip to tip as they do in median eyes (fig. 3.5A,B). Such a rhabdomeral network is well suited to enhancing photon catch at the expense of acuity.

In the median eye each retinula of five cells is associated with a single arhabdomeric cell, but the proportion of arhabdomeric to rhabdomeric cells is higher in the lateral eyes. Schliwa and Fleissner (1980) found a retinular/arhabdomeric cell ratio of 2:1 in the larger, anterior eyes and 5:4 in the smaller, posterior eyes. As in the median eyes, the arhabdomeric cell bodies are deeper in the retina than the visual cell nuclei, but the difference in nuclear shape is the same. The arhabdomeric cell dendrites project between the basal parts of the rhabdomeric cells but not as far as the rhabdoms themselves. Their numerous projections penetrate the sheathing pigment cell cytoplasm to make intimate contact with the visual cells.

Pigment and Adaptation

Oriented to receive focused light, the retinal elements in camera-type eyes must be screened from light that does not pass through the lens. In scorpions this is achieved partly by fixed pigment, which does not move in response to light. This pigment is located in pigment cells that line the postretinal membrane and send processes between the visual cells and further into the retina. The pigmented hypodermal cells surrounding the lens and forming a fixed "iris" are described above.

Besides the fixed pigment, the visual cells themselves contain pigment granules that undergo migration, causing marked changes in the screening of the rhabdoms with the day/night and light/dark adaptive states. Scheuring (1913) showed that the pigment granules in the visual cells migrate between the basal and apical regions of the visual cells. In the day state the pigment is between and distal to the rhabdoms, shielding them from incoming light. This pigment causes the reddish reflection from the eyes seen when measuring visual fields (fig. 3.2). At night the pigment migrates toward the base of the visual cells, though in some cases not very far. Within each retinula unit the pigment also migrates centrifugally, widening the aperture through which light can enter the rhabdom. This night state, which is brought about by the circadian clock, is the "natural" state for the scorpion, but before the Fleissners' work this was not recognized, and only the (less sensitive) day state was studied (fig. 3.4B,C).

The withdrawal of pigment at night exposes the tips of the rhabdoms to light through the lens, while, in median eyes, it optically isolates each rhabdom from its neighbors. The radial migration of pigment within the retinula may also contribute to widening the acceptance angle of the single visual cell from about 5° in the day eye to about 16° in the night eye. Change of size of rhabdoms from day to night state is the other parameter controlling the width of acceptance angle (Fleissner and Marschall-Pauly, personal communication). The rhabdoms are optically homogeneous because the dimensions of the microvilli of which they are composed are considerably smaller than the wavelength of light, so light entering the apex of the rhabdom will be conducted throughout its substance, subject to absorption by visual pigment.

Such pigment migrations may affect the resolution of the eye. In tangential sections the starfish-shaped profiles of the rhabdoms interdigitate at certain levels, though there is pigment between them. If such a field receives an image, this is likely to be shared among adjacent rhabdoms in the absence of screening pigment, but if the pigment limits light entry to an array of pupils, each centered over a rhabdom, the latter may be stimulated discretely. The resolution will then be determined by the spacing of these pupils (fig. 3.8).

The pigment migrations within visual cells have been extensively studied by Fleissner (1971, 1972, 1974, 1985) and account for the observed increase in sensitivity of the night-adapted eye. The time course of the effect of these migrations on the ERG due to light adaptation was studied by Machan (1968) and those due to the circadian control by Fleissner (1971, 1974, 1977d). When the optic nerve was cut, the ERG stopped and permanently dropped to the day state, and pigment migration was stopped in the distal day-state position (Fleissner and Fleissner, 1978), indicating that the night position depends on an intact optic nerve.

These circadian adaptational changes, as shown by Fleissner and Schliwa (1977) and Fleissner and Heinrichs (1982), depend on neurosecretory cells in the brain, with fibres that reach the retina via the optic nerves. Cutting the nerve forces the eye to the day state (Fleissner and Fleissner, 1978), but stimulation of the distal end of the cut nerve drives it back to the night state (Fleissner and Fleissner, 1985; Marschall-Pauly and Fleissner, 1988). The neurosecretory fibres, of which about 10 terminal branches reach each retinula unit, make synaptoid contacts with visual cells only. They extend along the cell, and branches may cross to other visual cells in the same or neighboring retinula units. Fleissner and Fleissner (1988) suggest that the transmitter released by these neurosecretory cells is octopamine.

Machan (1968) measured spectral sensitivity by determining the intensity at various wavelengths to elicit a threshold ERG response. She found the spectral

Figure 3.8 Rhabdom pattern in median eye, traced from a micrograph. If pigment between rhabdoms retracts fully, a stimulus (large circle) may excite several rhabdoms, reducing acuity. If isolating pigment only withdraws from rhabdom tips, leaving a "pupil" (small circles) over each rhabdom, acuity will be maximal.

sensitivity maximum of the median eyes to be 493 nm, whereas the lateral eyes showed two maxima, one at 493 nm and another in the UV spectrum. Machan demonstrated the greater sensitivity of the lateral eyes but, though dark adapting the animals, did not state whether the experiments were carried out by night. Fleissner (1968) measured spectral sensitivity by amplitude of the ERG, reaching similar conclusions to Machan: a peak at 490 nm with a lesser peak at 360 nm. Circadian adaptation does not change spectral sensitivity. The shielding pigment acts almost as a neutral density filter (Fleissner, 1971).

Albinism and Eye Structure

Scorpions are not brightly colored but show a wide range of shades from very pale through gray or brown to almost black. Some of the color is due to darkly sclerotized cuticle, but hypodermal pigment also contributes, particularly in the darker examples. Even in very pale species, such as *Leiurus quinquestriatus,* the eye pigment is dark. Many newborn scorpions, including the young of black parents, are colorless except for their eye pigment.

Very pale arthropods, including scorpions, are known from caves, where they may live all their lives in total darkness. Several cave scorpions are eyeless, though others, not totally pale, have reduced eyes (Mitchell, 1968, 1972; Francke, 1977, 1978; Mitchell and Peck, 1977). Albinism, a specific lack of pigment, is known in arthropods, including crabs, insects, and isopods. Albino scorpions were described by Locket (1986) from a normal population of *Urodacus yaschenkoi* from South Australia. These individuals lacked shielding pigment in the eyes as well as elsewhere in the body. Electron microscopy of the eyes showed that the pigment granules were replaced by premelanosomes.

The lens and vitreous were apparently normal, but the retina was not—the rhabdoms were represented by a few irregular microvilli, and mitochondria and ribosomes were much reduced. The phaospheres in the visual cells were disordered, consisting of amorphous dense material, quite unlike the aggregations of regular granules with paracrystalline inclusions in the normal eye. The scorpion studied had been kept alive under laboratory lighting conditions for several weeks before preparation, and it is possible that the degeneration of rhabdoms observed was due to not shielding them from light, rather than representing this condition prior to capture.

PHYSIOLOGY

Physiology of the eyes certainly deals with vision, but this is not the sole function of the photoreceptive system in scorpions. The phenomenon of dark adaptation is well known to zoologists, but Fleissner (1972) established that, in addition to dark/light changes in sensitivity, there are much greater changes associated with night and day, even under constant conditions of illumination. These circadian sensitivity changes have proved a powerful tool for investigating scorpion circadian biology and continue to provide insights into the workings of circadian systems in general (Fleissner and Fleissner, 1988; chapters 4 and 5, this volume).

The circadian changes are not the same for median and lateral eyes. The median eyes increase in sensitivity at night by up to 4 log units (fig. 3.9), but the lateral increase sensitivity by less than 1 (Fleissner, 1975, 1977d). Despite the greater adaptational range of the median eyes, the lateral eyes have the greater absolute sensitivity (Machan, 1966; Fleissner, 1975; see also Fleissner, 1977a,b,c).

The changes in sensitivity are brought about largely by movements of shielding pigment within the visual cells of the retina, exposing the rhabdoms to light through the lens in the night state and shielding them by day. To measure sensitivity, Fleissner (1968) used the amplitude of the ERG. Recording from a platinum electrode embedded in the lens, with an indifferent electrode a few mil-

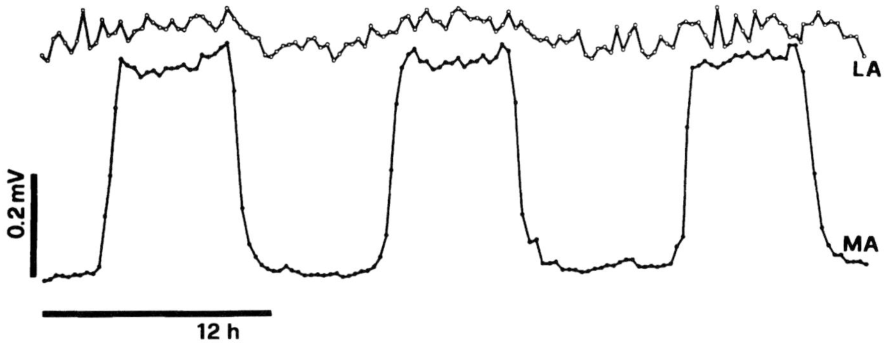

Figure 3.9 ERGs measured by night and by day. Due to circadian rhythm, the amplitude of the median eye ERG (MA) increases by 4 log units at night, and that of lateral eye (LA) by about 1. (G. and G. Fleissner)

limeters away in the prosoma, he found a cornea-negative generator potential, which varied linearly with log of light intensity. Fleissner also determined the spectral sensitivity, which is the same in median and lateral eyes with a peak at about 490 nm. Fleissner (1977c) also measured the absolute threshold of the median and lateral eyes, of which the latter are more sensitive by about 1 log unit when both are in the night state. Maximum sensitivity was attained after 60 days in the dark, when ERGs were elicited from lateral eyes using a source radiance of 1.5×10^{-8} μW $cm^{-2}sr^{-1}$. The threshold of median eyes by night or lateral eyes by day was equal, about 1.5×10^{-7} μW $cm^{-2}sr^{-1}$, but median eyes by day were some 3 log units less sensitive. Fleissner points out that a stone reflecting 0.2 of the light from a clear starry sky would have a radiance of $6 \times 10^{-5}$$\mu$W cm^{-2} sr^{-1}, two log units above median eye threshold, so that a scorpion could orient to stones and their shadows. Linsenmair (1968) demonstrated orientation to a starry sky, and Fleissner's results show that the light from such a sky is well within the physiological range for a scorpion. The fact that full moonlight may be 9 log units above median eye threshold may explain why scorpions are seldom found on the surface under these conditions.

The circadian rhythm persists indefinitely in darkness, but the period length, τ, does not keep perfectly synchronized with external time, a general finding in circadian systems. To attain this synchronization, a zeitgeber stimulus, in the present case a light stimulus, must be fed into the system at the appropriate intervals. Fleissner (1977a,b) showed that light applied to the median or lateral eyes will entrain the rhythm and that the lateral eyes do so at very low intensities of about 2.5×10^{-4} lux.

The circadian signals causing the pigment migration within the visual cells, which is responsible for sensitivity change, was shown (Fleissner and Fleissner, 1977) to be mediated via the optic nerve. Fleissner and Fleissner (1978) showed that cutting the optic nerve between the brain and eye stopped the normal circadian changes of sensitivity and drove the pigment irreversibly to the day state. Efferent neurosecretory fibres had been described in the retina and optic nerve by Schliwa and Fleissner (1977) and Fleissner and Schliwa (1977), and their origin from bilateral groups of 10–20 cells in the protocerebrum was later traced by Heinrichs and Fleissner (1982) and Fleissner and Heinrichs (1982). Fleissner (1983) followed this work by assessing the effects of cutting the pathway at various locations in the brain, confirming the bilateral innervation of both median eyes by the two groups of cells. A full synthesis on the ENSF cells and their connections was given by Heinrichs and Fleissner (1987). Experiments by Bhatti and Fleissner (cited by Fleissner et al., 1987), strongly suggest that octopamine, which mimics the action of the ENSF when applied directly to the retina, is the transmitter released by the ENSF (see also Fleissner and Fleissner, chapter 4, this volume).

The presence of arhabdomeric cells in the retina was reported by Fleissner and Siegler (1978), and their anatomy was fully characterized by Schliwa and Fleissner (1979). Intracellular microelectrode recordings at different depths in the retina (Fleissner, 1985) were correlated with the visual and arhabdomeric cells. Close to the vitreous nonpropagated receptor potentials were obtained quite easily, but deeper penetration, to the depths where the arhabdomeric cells are situated, occasionally gave records of spiking cells. This work refuted the claim of Belmonte and Stensaas (1975), who had recorded from bundles of optic nerve fibres but were unaware of the arhabdomeric cells, that scorpion visual cells produce spikes. Fleissner (1985) also found that visual cell axons end in the lamina, but that the arhabdomeric cell axons go farther, through a chiasma, to end in the medulla.

OPTIC NERVE

The optic nerve is formed by bundles of fibres that pierce the postretinal membrane, joining to form the nerve close to the eye. All the fibres are non-myelinated and are derived from at least three cell types (fig. 3.10A). Most of the fibres arise in the retina, as the axons of visual cells or arhabdomeric cells, but the optic nerve also contains ENSFs that originate in the brain and can be identified in the retina by their content of dense-cored vesicles (Fleissner and Schliwa, 1977). In the nerve only occasional profiles containing such vesicles are seen, and the neurosecretory fibres are not recognizable as a distinct population.

The axons within the nerve are arranged in bundles, separated from each other by septa derived from the optic nerve sheath, which is continuous with the postretinal membrane. Within each bundle the individual fibres are almost completely separated from each other by lamellae derived from glial cells, the nuclei of which are within the bundles.

LAMINA

The connections of the retina to the brain have been studied in the buthid *Androctonus australis* (Fleissner, 1985). In *Androctonus* and many other scorpions, the optic nerve leads, just before it reaches the brain, to a thickening, the first optic neuropil or lamina. Fleissner found, by cobalt and lucifer yellow tracing experiments, that the visual cell axons end by arborizing in the lamina, unlike axons from arhabdomeric cells, which do not end in the lamina but continue to the medulla or second optic neuropil. In the genus *Urodacus,* and in some other scorpions, the lamina is not located at the brain, but is adjacent to the eye.

This difference in position has implications for the fibre makeup of the optic nerve. If the visual cell axons end in the lamina, as they do in *Androctonus,* the optic nerve between lamina and brain should contain only the axons of arhabdomeric and efferent neurosecretory cells and those of any secondary lamina neurons present. The *Urodacus* lamina consists of a mass of neuropil, with a few cells, probably neurons, around the periphery and glial cells, particularly where fibres enter and leave (fig. 3.10C). Fibres can be traced from the retina into the neuropil, but none have been seen to traverse the lamina uninterrupted. The complex neuropil contains synapses with both clear and dense-cored vesicles (fig. 3.10B). Preganglionic fibres make synaptic contacts with small unidentified processes containing hollow vesicles, and processes containing dense cored vesicles make contacts onto the preganglionic fibres. Golgi preparations show arborizations of fibres within the neuropil, with many varicosities along them.

Zanmin Song, working in Locket's laboratory, found that the optic nerve of *Urodacus yaschenkoi* appeared to contain two populations of fibres: one larger

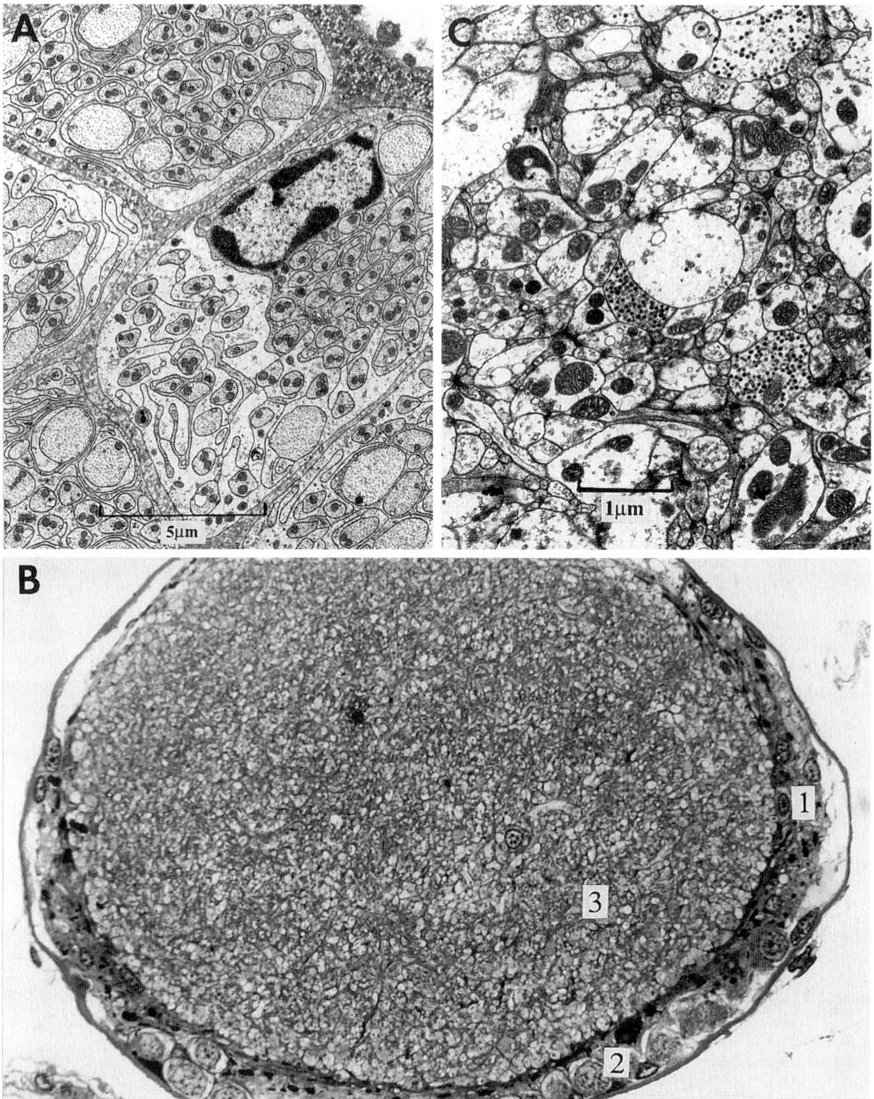

Figure 3.10 Neural and glial processes in optic nerve and lamina of *Urodacus yaschenkoi*. (A) Nerve fibre bundles of optic nerve near retinal origin. Large and small axons are surrounded by glial sheaths. (B) Neuropil within lamina showing many processes containing hollow synaptic vesicles and dense-cored vesicles. (C) Cross-section through lamina, showing glial cells (1) and presumptive neurons (2) surrounding dense neuropil (3).

in diameter (~2.7μm) and one smaller in diameter (~0.7 μm) (fig. 3.10A). Counts of these fibres between eye and lamina gave 4425 prelamina fibres, of which 1200 were small and 3325 large. Postlamina counts gave 1360 fibres: 990 small and 370 large. Some of the fibres are presumably efferent neurosecretory cells from the endings recognized in the retina, though they cannot be recognized as such in the nerve from vesicle content. The reduction in fibre number through the lamina may be explained by visual cell fibres ending there or by convergence within the lamina, but the details of connectivity are at present unresolved (fig. 3.11A, B).

A wider survey of species and further work is required to draw valid conclusions about the detailed connections within the lamina and its implications for the visual system. The possibility that the arhabdomeric cells produce lateral inhibition by connections to visual cell axons within the lamina is an intriguing problem for the future.

Fleissner and Fleissner (1985) describe in *Androctonus* a chiasma, whereby the anterior fibres from the lamina cross to the posterior part of the medulla, and vice versa, between the median eye lamina and medulla and another between the lateral eye lamina and medulla. In *Urodacus* there appears to be such a chiasma between the median eye retina and the lamina beneath it. In the lateral eyes there is no lamina close to the eye.

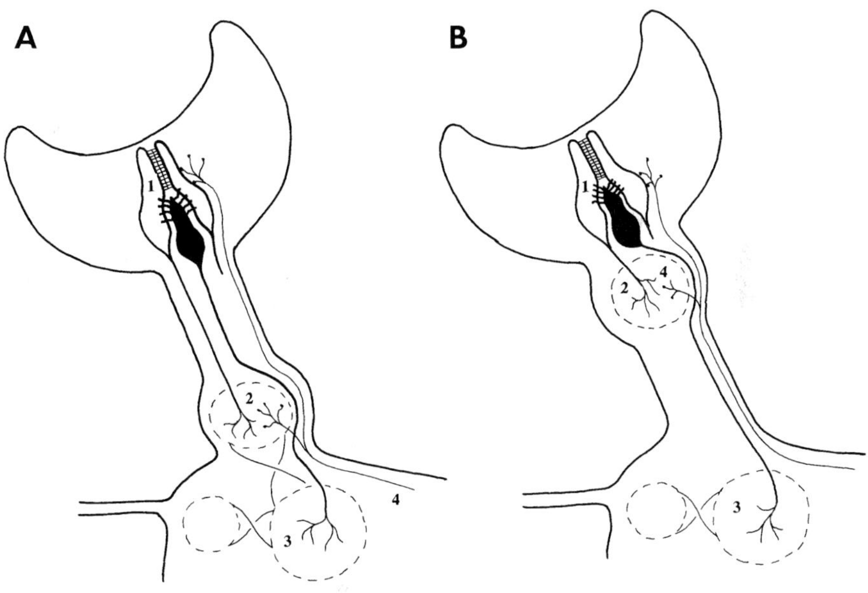

Figure 3.11 Diagram of retinal connections in scorpions. (A) In *Androctonus* (a buthid), axons of retinula cells (1) end in lamina (2) adjacent to supraesophageal ganglion (SEG). Axons of arhabdomeric cells (black) bypass lamina and end in medulla of SEG (3). Neurosecretory fibres (4) pass from SEG to retina, with connections to lamina. (Adapted from Fleissner and Fleissner (1985). (B) In *Urodacus* (a scorpionid) retinula cell axons (1) end in lamina (2) adjacent to the eye. Arhabdomeric cell axons traverse nerve to the medulla (3). Neurosecretory fibres pass to retina, with some (4) possibly connecting to lamina neuropil, as they do in *Androctonus*.

EFFERENT NEUROSECRETORY FIBRE CELLS

The ENSFs demonstrated in the retina by Fleissner and Schliwa (1977) have been shown by tracer experiments to arise in a group of 10–20 cells near each circumoesophageal commissure (Fleissner and Heinrichs, 1982; Heinrichs and Fleissner, 1982, 1987). The fibres bifurcate close to the dorsal surface of the supraoesophageal ganglion, sending a branch forward to the median eye nerve and contributing to optic neuropils within the brain on the way. The branch to the contralateral eye makes contacts with the central body, an integrative centre, as it crosses. These efferent fibres mediate the circadian changes in pigment distribution and consequent sensitivity changes so clearly documented by the Fleissners' group (see Fleissner and Fleissner, this volume).

CONCLUSION

Many biologists will have wished that they could study live examples of extinct forms, and been frustrated that so little, including virtually nothing of soft parts, survives fossilization. In scorpions we have an animal that has come down from the Silurian with apparently little change. The eyes of the early scorpions were different from those now extant—in particular, the lateral eyes of fossil scorpions were sometimes "compound" (Kjellesvig-Waering, 1986). Modern compound eyes are known to be of many types (Nilsson, 1990), and it may be impossible to decide from observation of external features which type is present. The fact that early scorpions had multiple lenses on their lateral eyes does not necessarily mean these eyes were compound in the sense of having one lens for each retinula unit, with the ommatidial inputs integrated either within the eye or by the central nervous system, as in modern compound eyes. The lateral eyes of modern scorpions show a certain amount of plasticity; the number varies between species, and occasionally between individuals of a species. *Androctonus* has five pairs of lateral eyes, *Urodacus* only two. The present-day lateral eyes may be the remains of what were once true compound eyes or, perhaps more likely, reduced groups of separate ocelli. The presence of two median eyes with single lenses goes back as far as the fossil record shows, yet they contain retinulae with closed rhabdoms, similar in many ways to the closed rhabdoms of compound eyes, including those of *Limulus*. Though we cannot study soft parts of the eyes of long-dead scorpions, we can study their modern descendants. In this way, our speculations about their evolution can at least be informed.

References

Barlow, R.B. and Chamberlain, S.C. (1980) Light and a circadian clock modulate structure and function in *Limulus* photoreceptors. *In* Williams, T.P. and Baker, B.N. (eds.) *The Effects of Constant Light on Visual Processes,* pp 247–269. Plenum Press, New York.

Bedini, C. (1967) The fine structure of the eyes of *Euscorpius carpathicus* L. (Arachnida, Scorpiones). *Arch. Ital. Biol.* 105: 361–378.

Belmonte, C. and Stensaas, L.J. (1975) Repetitive spikes in photoreceptor axons of the scorpion eye: invertebrate eye structure and tetrodotoxin. *J. Gen. Physiol.* 66: 649–655.

Blest, A.D. (1978) The rapid synthesis and destruction of photoreceptor membrane by a dinopid spider: a daily cycle. *Proc. R. Soc. Lond.* B 200: 463–483.

Blest, A.D. (1985) The fine structure of spider photoreceptors in relation to function. *In* Barth, F.G. (ed.) *Neurobiology of Arachnids,* pp. 79–102. Springer-Verlag, Berlin.

Brownell, P.H. (1977) Compressional and surface waves in sand used by desert scorpions to locate prey. *Science* 197: 479–482.

Brownell, P.H. (1984) Prey detection by the sand scorpion. *Sci. Am.* 251: 86–97.

Brownell, P.H. and Farley, R.D. (1979a) Detection of vibrations in sand by tarsal sense organs of the nocturnal scorpion, *Paruroctonus mesaensis. J. Comp. Physiol.* A131: 23–30.

Brownell, P.H. and Farley, R.D. (1979b) Prey localizing behaviour of the nocturnal desert scorpion *Paruroctonus mesaensis:* Orientation to substrate vibrations. *Anim. Behav.* 27: 185–193.

Brownell, P.H. and Farley, R.D. (1979c) Orientations to vibrations in sand by the nocturnal scorpion *Paruroctonus mesaensis:* Mechanism of target location. *J. Comp. Physiol.* A131: 31–38.

Carricaburu, P. (1968) Dioptrique oculaire du scorpion *Androctonus australis. Vision Res.* 8: 1967–1072.

Chamberlain, S.C. and Barlow, R.B. (1984) Transient membrane shedding in *Limulus* photoreceptors: control mechanisms under natural lighting. *J. Neurosci.* 4: 2792–2810.

Cloudsley-Thompson, J.L. (1956) Studies in diurnal rhythms. VI. Bioclimatic observations in Tunisia and their significance in relation to the physiology of the fauna especially woodlice, centipedes, scorpions and beetles. *Ann. Mag. Nat. Hist.* (12) 9: 305–329.

Eastwood, E.B. (1978a) Notes on the scorpion fauna of the Cape. III, Some observations on the distribution and biology of scorpions on Table mountain. *Ann. S. Afr. Mus.* 74: 229–248.

Eastwood, E.B. (1978b) Notes on the scorpion fauna of the Cape. IV, the burrowing activities of some scorpionids and buthids (Arachnida, Scorpionida). *Ann. S. Afr. Mus.* 74: 249–245.

Fahrenbach, W.H. (1975) The visual system of the horseshoe crab. *Intl. Rev. Cytol.* 41: 285–349.

Fleissner, G. (1968) Untersuchungen zur Sehphysiologie der Skorpione. *Verh. Deutsch. Zool. Ges.* 61: 375–380.

Fleissner, G. (1971) Über die Sehphysiologie von Skorpionen. Belichtungspotentiale in den Medianaugen von *Androctonus australis* L. und ihre tagesrhythmischen Veränderung(Inaugural-Dissertation). J. W. Goethe-Universität, Frankfurt am Main, Germany.

Fleissner, G. (1972) Circadian sensitivity changes in the median eyes of the North African scorpion *Androctonus australis. In* Wehner, R. (ed.) *Information Processing in the Visual Systems of Arthropod,* pp. 133–139. Springer-Verlag, Berlin.

Fleissner, G. (1974) Circadiane Adaptation und Schirmpigmentverlagerung in den Sehzellen der Medianaugen von *Androctonus australis* L. (Buthidae, Scorpiones). *J. Comp. Physiol.* A91: 399–416.

Fleissner, G. (1975) A new biological function of the scorpion's lateral eyes as receptors of Zeitgeber stimuli. *Proc. VIth Intl. Arachnol. Congr. Amsterdam,* pp. 176–182.

Fleissner, G. (1977a) Entrainment of the scorpion's circadian rhythm via the median eyes. *J. Comp. Physiol.* A118: 93–99.

Fleissner, G. (1977b) Scorpion lateral eyes: extremely sensitive receptors of Zeitgeber stimuli. *J. Comp. Physiol.* A118: 101–108.

Fleissner, G. (1977c) The absolute sensitivity of the median and lateral eyes of the scorpion *Androctonus australis* L. *J. Comp. Physiol.* A118: 109–120.

Fleissner, G. (1977d) Differences in the physiological properties of the median and lateral eyes and their possible meaning for the entrainment of the scorpion's circadian rhythm. *J. Interdiscipl. Cycle Res.* 8: 15–26.

Fleissner, G. (1983) Efferent neurosecretory fibres as pathways for circadian clock signals in the scorpion. *Naturwiss.* 70: 366–367.

Fleissner, G. (1985) Intracellular recordings from spiking and non-spiking cells in the scorpion eye. *Naturwiss.* 72: 46–47.

Fleissner, G. and Fleissner, G. (1977) Steuerung der circadianen Pigmentwanderung in den Medianaugen des Skorpions. *Verh. Deutsch. Zool. Ges.* 1977: 231.

Fleissner, G. and Fleissner, G. (1978) The optic nerve mediates the circadian pigment migration in the median eyes of the scorpion. *Comp. Biochem. Physiol.* A61: 69–71.

Fleissner, G. and Fleissner, G. (1985) Neurobiology of a circadian clock in the visual systems of scorpions. *In* Barth, F.G. (ed.) *Neurobiology of Arachnids,* pp. 351–375. Springer-Verlag, Berlin.

Fleissner, G. and Fleissner, G. (1988) Efferent control of visual sensitivity in arthropod eyes: with emphasis on circadian rhythms. *In* Lindauer, M. (ed.) *Information Processing in Animals,* vol. 5, p 67. Gustav Fischer Verlag, Stuttgart.

Fleissner, G. and Heinrichs, S. (1982) Neurosecretory cells in the circadian clock system of the scorpion *Androctonus australis. Cell Tiss. Res.* 124: 233–238.

Fleissner, G. and Schliwa, M. (1977) Neurosecretory fibres in the median eyes of the scorpion *Androctonus australis. Cell Tiss. Res.* 178: 189–198.

Fleissner, G. and Siegler, W. (1978) Arhabdomeric cells in the retina of the median eyes of the scorpion. *Naturwiss.* 65: 210–211.

Fleissner, G., Stevenson, D. and Fleissner, G. (1987). Circadian systems in invertebrates (including an evolutionary perspective of circadian signal transmission in the visual system). *In* Scharrer, B., Korf, H.-W., and Hartwig, H.-G. (eds.) *Functional Morphology of Neuroendocrine Systems,* pp. 87–104. Springer-Verlag, Berlin.

Francke, O.F. (1977) The genus *Diplocentrus* in the Yucatán peninsula with description of two new troglobites. *Bull. Assoc. Mexican Cave Stud.* 6: 49–61.

Francke, O.F. (1978) New troglobite scorpion of genus *Diplocentrus. Entomol. News* 89: 39–45.

Geethabali (1976) Central course of photic input in the ventral nerve cord of scorpion (*Heterometrus fulvipes*). *Experientia* 32: 345–347.

Geethabali and Rao, K.P. (1973) A metasomatic neural photoreceptor in the scorpion. *J. Exp. Biol.* 58: 189–196.

Gregory, G.E. (1980) The Bodian protargol technique. *In* Strausfeld, N.J. and Miller, T.A. (eds.) *Neuroanatomical Techniques in Insect Nervous System,* chapt. 6, pp. 77–95. Springer-Verlag, New York.

Heinrichs, S. and Fleissner, G. (1982) Anatomie der efferenten neurosekretorischen Zellen im circadianen Oszillatorsystem des Skorpions. *Verh. Deutsch. Zool. Ges.* 1982: 312.

Heinrichs, S. and Fleissner, G. (1987) Neuronal components of the circadian clock in the scorpion *Androctonus australis:* Central origin of the efferent neurosecretory elements projecting to the median eyes. *Cell Tiss. Res.* 250: 277–285.

Hoffmann, C. (1967) Bau und Funktion der Trichobothrien von *Euscorpius carpathicus* L. *Z. Vergl. Physiol.* 54: 290–352.

Homann, H. (1928) Beiträge zur Physiologie der Spinnenaugen. I. Untersuchungsmethoden. II. Das Sehvermögen der Salticiden. *Z. Vergl. Physiol.* 7: 201–269.

Ignat'yev, A.M., Ivanov, V.P. and Balashov, Y.S. (1976) The fine structure and function of the trichobothria in the scorpion *Buthus eupeus*, Scorpiones, Buthidae. *Entomol. Rev.* 55(3): 12–18. [English translation of *Entomologicheskoe obozrenie* 55: 525–532.]

Kjellesvig-Waering, E.N. (1986) A restudy of the fossil Scorpionida of the world. *Palaeontgr. Am.* no. 55.

Land, M.F. (1985) The morphology and optics of spider eyes. *In* Barth, F.G. (ed.) *Neurobiology of Arachnids*, pp. 53–78. Springer-Verlag, Berlin.

Lankester, E.R. and Bourne, A.G. (1883) The minute structure of the lateral and central eyes of *Scorpio* and *Limulus*. *Qt. J. Microsc. Sci.* 23: 177–212.

Linsenmair, K.E. (1968) Anemomenotaktische Orientierung bei Skorpionen (Chelicerata, Scorpiones). *Z. Vergl. Physiol.* 60: 445–449.

Locket, N.A. (1986) Albinism and eye structure in an Australian scorpion, *Urodacus yaschenkoi* (Scorpiones, Scorpionidae). *J. Arachnol.* 14: 101–115.

Lourenço, W.R. (1995) *Chaerilus sabinae*, nouvelle espèce de Scorpion anophthalme des grottes de Matampa en Inde (Scorpiones, Chaerilidae). *Rev. Suisse Zool.* 102: 847–850.

Lourenço, W.R. and O.F. Francke. (1985) Révision des connaissances sur les Scorpions cavernicoles (troglobies) (Arachnida, Scorpiones). *Mém. Biospéleol.* 12: 3–7.

Machan, L. (1966) Studies on structure, electroretinogram and spectral sensitivity of the median and lateral eyes of the scorpion. (Ph.D. dissertation). University of Wisconsin, Madison. [Dissertation Abstracts 27: 1329B].

Machan, L. (1967) Studies on structure and electrophysiology of scorpion eyes. The time course of dark adaptation determined by constant stimulus and constant response methods. *S. Afr. J. Sci.* 63: 512–520.

Machan, L. (1968) Spectral sensitivity of scorpion eyes and the possible role of shielding pigment effect. *J. Exp. Biol.* 49: 95–105.

Marschall-Pauly, A. and Fleissner, G. (1987) Simulation von circadianen Signalen und Einzelzellableitungen in Medianaugen des Skorpions. *Verh. Deutsch. Zool. Ges.* 80: 277–278.

Mitchell, R.W. (1968) *Typhlochactas*, a new genus of eyeless cave scorpion from Mexico (Scorpionida, Chactidae). *Ann. Spéléol.* 23: 753–777.

Mitchell, R.W. (1972) *Typhlochactas elliotti*, a new eyeless cave scorpion from Mexico (Scorpionida, Chactidae). *Ann. Spéléol.* 26: 135–148.

Mitchell, R.W. and Peck, S.B. (1977) *Typhlochactas sylvestris*, a new eyeless scorpion from montane forest litter in Mexico (Scorpionida, Chactidae, Typhlochactinae). *J. Arachnol.* 5(2): 159–168.

Nilsson, D-E. (1990) From cornea to retinal image in invertebrate eyes. *Trends Neurosci.* 13: 55–64.

Parker, G.H. (1891) The eyes in scorpions. *Bull. Mus. Comp. Zool.* (Harvard) 13: 173–208.

Police, G. (1907) Sugli occhi dello scorpione. *Zool. Jahrb. Anat. Abt.* 25: 1–70.

Polis, G.A. (1979) Prey and feeding phenology of the desert sand scorpion *Paruroctonus mesaensis* (Scorpionidae: Vaejovidae). *J. Zool.* 188: 333–346.

Root, T.M. (1990) Neurobiology. *In* Polis, G.A. (ed.) *The Biology of Scorpions*, pp 341–413. Stanford University Press, Palo Alto, CA.

Rossel, S. (1989) Polarization sensitivity in compound eyes. *In* Stavenga, D.G. and R.C. Hardie, (eds.), *Facets of Vision*, pp 298–316. Springer-Verlag, Berlin.

Scheuring, L. (1913) Die Augen der Arachnoiden. *Zool. Jahrb. Anat. Abt.* 33: 553–636.

Schliwa, M. and Fleissner, G. (1977) Efferente neurosekretorische Fasern in den

Medianaugen und im Sehnerv des Skorpions *Androctonus australis. Verh. Deutsch. Zool. Ges.* 70: 230.

Schliwa, M. and Fleissner, G. (1979) Arhabdomeric cells of the median eye retina of scorpions. I. Fine structural analysis. *J. Comp. Physiol.* A130(3): 265–270.

Schliwa, M. and Fleissner, G. (1980) The lateral eyes of the scorpion *Andoctonus australis. Cell Tiss. Res.* 206(1): 95–114.

Shachak, M. and Brand, S. (1983) The relationship between sit-and-wait foraging strategy and dispersal in the desert scorpion *Scorpio maurus palmatus. Oecologia* 60: 371–377.

Shorthouse, D.J. (1971) Studies on the biology and energetics of the scorpion, *Urodacus yaschenkoi* (Birula (1904) (Ph.D. dissertation)). Australian National University, Canberra.

Strausfeld, N.J. and Miller, T.A. 1980. *Neuroanatomical techniques. Insect nervous system.* 496 pp. Springer-Verlag, New York.

Williams, S.C. (1966) Burrowing activities of the scorpion *Anuroctonus phaiodactylus* (Wood) (Scorpionida: Vaejovidae). *Proc. Calif. Acad. Sci.* 34: 419–420.

Wuttke, W. (1966) Untersuchungen zur Aktivitätsperiodik bei *Euscorpius carpathicus. Z. Vergl. Physiol.* 53: 405–448.

Zwicky, K.T. (1968) A light response in the tail of *Urodacus,* a scorpion. *Life Sci.* 7: 257–262.

Zwicky, K.T. (1970) Behavioural aspects of the extraocular light sense of *Urodacus,* a scorpion. *Experientia* 26: 747–748.

4

Neuronal Organization of Circadian Systems

GERTA FLEISSNER
GÜNTHER FLEISSNER

Circadian clock systems are composed of a central pacemaker, or master clock, that generates the subjective time of the organism. In animals this master clock is located in the central nervous system. The endogenous timing signal is conducted via nervous or hormonal pathways to the periphery, where the central reference time is imprinted on different effector systems—for example, leading to circadian modulation of visual sensitivity, locomotor activity, blood pressure, and enzymatic activity. These rhythms may be "phase modulated" in different ways to enable the organism to adjust its sleep and activity to the most appropriate time of day. But the central rhythm generators produce only "circadian" periodicities; that is, the endogenous rhythm regularly deviates from 24 h and may range (i.e., in scorpions) between less than 22 h and more than 27 h. Therefore, all circadian systems can be readjusted to the external light/dark cycle by "Zeitgeber" signals, a feature necessary to guarantee their adaptive value.

In this chapter we describe what is known about the neurobiological basis of the circadian clock system of scorpions. In chapter 3 Locket summarized the optics and histology of the eyes of scorpions and described their circadian changes in sensitivity. This sensitivity rhythm of the median eyes is one of the clearest parameters to represent the "hand" of the central clock. Simultaneous recordings of electroretinogram (ERG) and behavioral parameters (e.g., locomotor activity) offer unique access to the underlying neurobiological mechanisms. Using these rhythms we have been able to analyze the clock system in scorpions down to the cellular level. After considering the different elements of the optic system for Zeitgeber perception, we describe an efferent neurosecretory fiber system that distributes the clock signal throughout the brain, possible candidates for pacemaking neurons, and pathways for feedback signals between the clock-modulated periphery and the pacemaker itself. Feedback is well established in many fields of physiological research, but only lately has it been discovered in chronobiology. The scorpion offers one of the clearest examples for study of feedback mechanisms in a clock system and their functional meaning. This is topic of chapter 5.

For neurobiological studies of the circadian clock system, the scorpion is an ideal experimental animal. According to the time of day, several effector systems may undergo profound structural and physiological changes, anticipating external periodicities. To test whether these rhythms are endogenously controlled by a clock or triggered by external time cues, one must determine whether the rhythms persist under constant conditions of light or darkness, temperature, and humidity. Common laboratory animals such as rats, hamsters, or flies cannot tolerate these conditions nearly as well as scorpions. Many desert scorpions experience constant conditions naturally during the hot season when they remain for weeks at a time in the shelter of their burrows.

There are, of course, areas of chronobiological research where model systems other than scorpions are preferred. These other systems, long and widely used, have dominated some chronobiological studies with an immense body of knowledge and theory to which work on scorpions can hardly contribute. For example, in genetics, studies of numerous mutants in fruit flies, hamsters, rats, and fungi have provided genetic maps of structures and functions in many-fold contexts (for review, see Hall and Rosbash, 1988; Hall, 1990; Green, 1998; Dunlap, 1999). Without neurobiological investigations however, these studies remain incomplete. But neurobiology might be difficult or impossible when the organisms are as small as *Drosophila,* as complex as vertebrates, or have no neurons at all, like *Neurospora.* Likewise, at the subcellular level, studies on isolated pacemaker cells from molluscs such as *Bulla* (e.g., Michel et al., 1992), or single-cell organisms such as *Gonyaulax* (e.g., Roenneberg and Hastings, 1991) have shown biochemical features of pacemaking function (for review, see Block et al., 1993). The small neurons of scorpions and other arthropods do not lend themselves well to this field of research. But within a single cell's response, it is difficult to analyze the different components of a clock system independently. This, however, can be achieved in scorpions.

NEUROBIOLOGICAL RESEARCH ON SCORPIONS

Scorpions have structural features that allow elegant neurobiological investigations that are difficult or impossible in other animals. Scorpions are a convenient size for manipulation and recover from surgery very well. Their brains are accessible from outside without major damage to the animal, and they are large enough for stereotactic microlesion experiments. Nerve back-filling techniques with tracer dyes do not take unduly long, as the pathways are shorter than, for example, in *Limulus.* Histology, including immunohistology, is relatively easy because scorpion tissues can be stained by methods developed for insects and vertebrates without complicated modifications. Scorpion brains have relatively few neurons compared to vertebrates, and fewer neurons are involved in circuits with discrete behavioral functions.

Similarly, the scorpion's circulatory system has important features useful for neurobiological research. The blood–brain barrier (Lane et al., 1981) is tighter against diffusion between body fluids and intracerebral fluids than in other arthropods (e.g., *Limulus;* Harrison and Lane, 1981). This allows the intracerebral ion-milieu to stay constant, even when the extracerebral fluid has become ten times more concentrated after a few days of dessication (Paul and Locket, pers. comm.). Additionally, it enables pharmacological effects of applied drugs or putative neurotransmitters to be distinguished by their actions on, or derivation from, peripheral or central mechanisms. Except for the central nervous system, the scorpion has an open circulatory system like other arachnids (for review, see Foelix, 1992). Arteries and veins are not connected by closed capillary beds; instead, the arteries open within individual organs of the body and blood eventually flows into the hemocoel, where veins return the blood to the dorsal

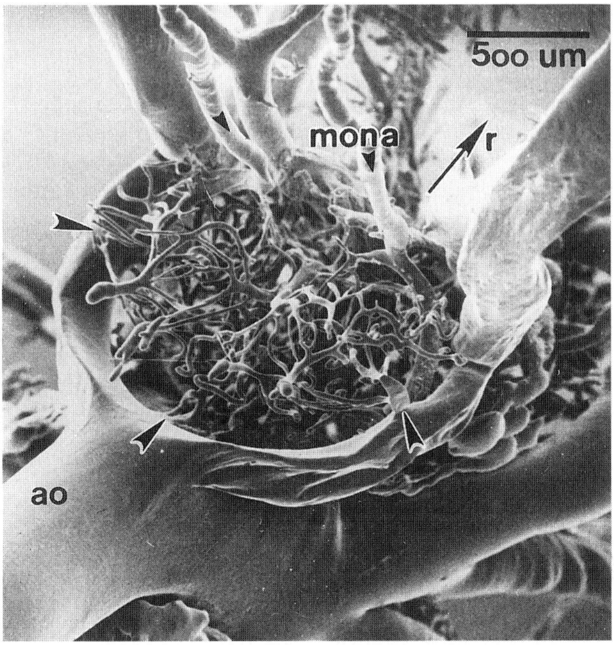

Figure 4.1 Corrosion cast of intra- and extracerebral vascularization as seen by scanning electron microscopy. This dorsal view shows the aorta (ao) branching into smaller arteries which parallel several nerves, including the median optic nerve (mona). Inside the brain a symmetric network of capillaries can be seen. Arrowheads indicate the transition between extra- and intracerebral vessels. This vascular system provides a convenient access for pharmacological manipulation of central neural processing in the scorpion brain; r at arrow points rostrally (after data from Heinrichs and Fleissner, unpublished).

heart through the respiratory book lungs. Within the central nervous system, however, the scorpion has a closed blood system of highly organized capillaries arising from well-defined branches of extracerebral arteries, which penetrate the membrane surrounding the brain (fig. 4.1). These intracerebral capillaries ensure optimal distribution of hemolymph or infused fixatives to complex tracts and neuropils of the brain.

Some scorpions have their eyes in a natural "test-tube" separated from both intra- and extracerebral fluid. In *Androctonus australis,* the primary subject of investigations in our laboratory, the median eye retinae are not directly attached to each other as they are in other nonbuthid scorpions (see chapter 3). Instead, a small transverse tube interconnects the retinae, with a capacity of about 10 nl of hemolymph (fig. 4.2) (Fleissner, 1974). The lumen of this tube has direct access to the retinal cells with no intervening membrane. The eyes are otherwise separated from the body fluid by the blood–brain barrier. Drugs applied to this tube only influence the retinae and do not invade other body tissues or the central nervous system. A clear separation of direct versus indirect effects of applied drugs on retinal functions is thus possible, without the need to injure the

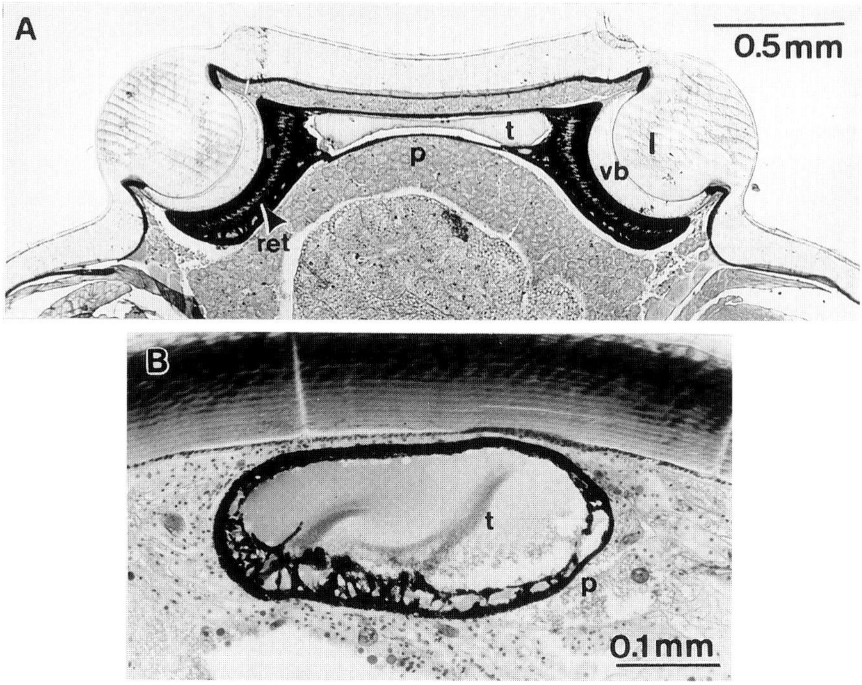

Figure 4.2 The retinas of the median eyes are interconnected by a fluid-filled tube. (A) Transverse and (B) sagittal sections of the dorsal prosoma show the relative position of the median eyes, dorsal cuticle, and eye tube. l, lens; p, postretinal membrane; ret, retina; t, eye tube; vb, vitreous body. Different magnifications indicated by scale bars. Adapted from Fleissner (1974).

retina by injections. This tube also offers an indicator for the correct osmolarity of Ringer solution to use. If *Leiurus*-Ringer of Gilai and Parnas (1970) is injected into the eye tube of *Androctonus,* the tube swells within a few minutes and may even burst, whereas Bowerman's (1976) *Hadrurus*-Ringer is tolerated without distortion.

The optic nerves connecting the eyes to brain are relatively long in adult *Androctonus* (about 1–2 mm) and conveniently accessible for lesion, stimulation, or permanent implantation of electrodes for chronic recordings. At the family level, scorpions differ in the types of visual information conducted to the brain through the optic nerve. In buthids, the nerve contains efferent neurosecretory fibers (ENSFs) and the axons of photoreceptors and arhabdomeric cells (Fleissner, 1985), as it connects the retina to the first optic neuropil (lamina) within the protocerebrum. In some scorpionids the lamina is located outside the brain, and in the Australian scorpionid genus *Urodacus* the lamina is even attached to the retina (see chapter 3). In this arrangement, the optic nerve fibers are axons of arhabdomeric cells, the ENSF axons and the interneurons connecting the first and second optic neuropils.

ZEITGEBER RECEPTORS AND PROCESSING PATHWAYS OF THE TIMING SIGNAL

Circadian clocks are precise but not perfectly accurate: their period lengths deviate from 24 h scale and must be regularly phase-adjusted relative to an external timing reference. In principle, light perceived by any photoreceptor can be used as the timing cue, or Zeitgeber (for review, see Fleissner and Fleissner, 1993). The Zeitgeber receptors and pathways have been studied both to determine their physiological importance to the timing mechanisms and to find the pacemaker, or master oscillator, via these input pathways.

Some features of photoreceptors, which have evolved to improve image-processing ability, detract from time perception. For example, light/dark-adaptation of a photoreceptor optimizes its visual contrast detection by shifting the steepest slope of the intensity–response function according to ambient light. Timing functions of a photoreceptor would be better served by receptors with constant intensity–response relationships that process successive contrast rather than simultaneous contrast of the environment. The latter is an essential feature of photoreceptors that see images. The nonvisual function of "seeing time" is mediated by specialized photoreceptors or by integrative neuronal circuitry. In scorpions both solutions were observed (Fleissner and Fleissner, 1993), suggesting a model of the neuronal mechanism of Zeitgeber perception which can be verified by behavioral and neurophysiological studies (Fleissner and Fleissner, 1998) (see also chapter 5).

Modified Photoreceptors

Scorpion eyes are all composed of similar photoreceptive elements, but the median eyes clearly show more specialization for image formation than the lateral eyes. As summarized in chapter 3, and in fig. 4.3 here, these features include in the median eyes a well-focused dioptric apparatus (fig. 4.3A) and a highly organized retina with receptor cells arranged in retinular units (fig. 4.3B), each with a fused rhabdom and one arhabdomeric cell as first visual interneuron. The lateral eyes are nearly devoid of a dioptric apparatus, as they lack a vitreous body and their corneal lenses are relatively flattened (fig. 4.3C). This varies for different lateral eyes (e.g., lateral eye 3 versus the smaller lateral eye 5; fig. 4.3C) and different scorpion species (e.g., buthids versus scorpionids; see chapter 3). In the lateral eyes there is no optical isolation of retinular units and rhabdoms form a fused network throughout the retina (fig. 4.3D) (Schliwa and Fleissner, 1980). This allows for photon diffusion within the rhabdomeric light guides over the entire retina. Unlike the median eyes, the lateral eyes hardly show circadian adaptation (Fleissner, 1977d). All these features enhance low light perception, making the lateral eyes permanently at least 10 times more sensitive than the median eyes in their most sensitive (night) state (Fleissner, 1977c; cf. fig. 4.12C). The lateral eyes are optically designed like irradiation meters.

Physiological experiments show the lateral eyes are extremely sensitive receptors of Zeitgeber stimuli (less than 10^{-4} cd/m^{-2} sec compared to about 10^{-2} cd/m^{-2} sec necessary for synchronizing via the median eyes; Fleissner, 1977a,b). This means that light intensities close to the absolute threshold of the lateral eye (Fleissner, 1977c) are being perceived and used by the circadian clock.

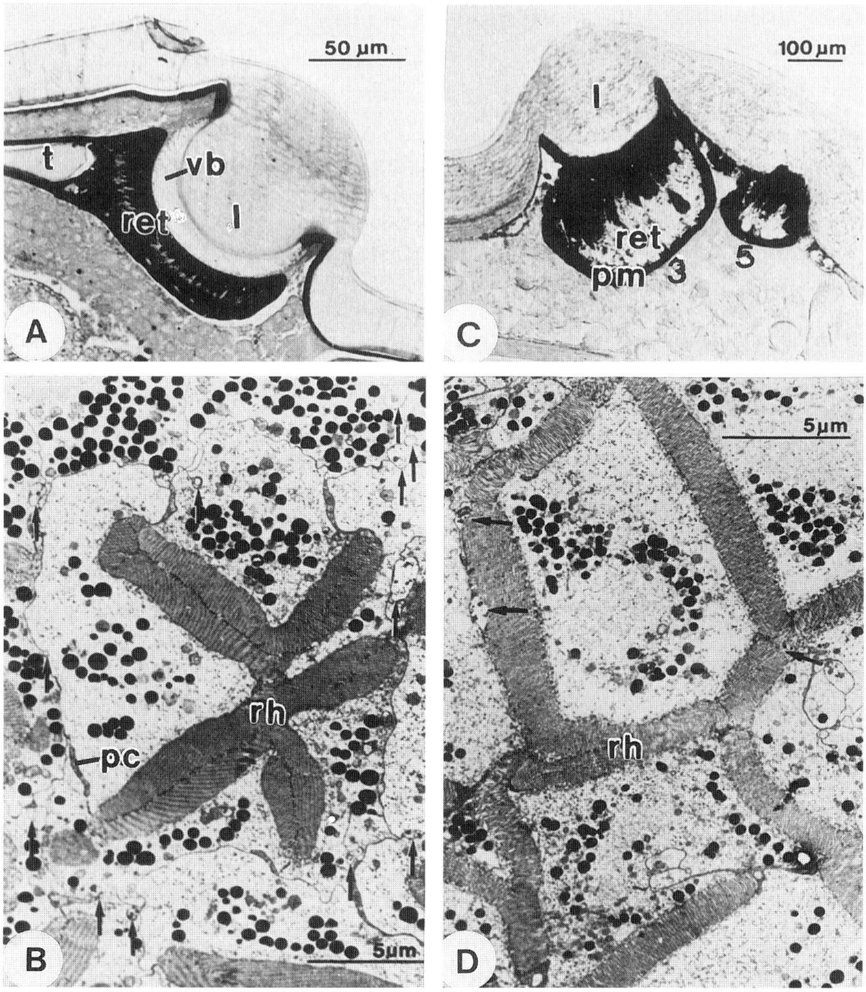

Figure 4.3 Morphological difference between the median and lateral eyes as seen in the light and electron microscope. The median eye (A, B) has features of an image processing (visual) photoreceptor, whereas the lateral eye (C, D) has poor focusing dioptrics and forms a fused network rhabdom. (A and C, light microscopic sections; B and D, electron microscopic sections; l, lens; pc, pigment cell; pm, postretinal membrane; ret, retina; rh, rhabdom; t, eye tube; vb, vitreous body; 3, 5, number of lateral eye. Arrows point to efferent neurosecretory fiber profiles (after Fleissner and Fleissner, 1993).

Neuronal Circuitry

The median eyes of scorpions have a pronounced circadian rhythmicity (compensating for at least 3 log units intensity difference between day and night), and the lateral eyes show little such adaptation. Thus, the scorpion brain is simultaneously informed about the external light conditions from two sources, one unmodified via the lateral eyes and one substantially modified by the clock-controlled "sunglass rhythm" of the median eyes. The light information-processing pathways from median and lateral eyes enter the brain on separate tracts that first converge in the second optic neuropil, or medulla (Fleissner, 1985; Fleissner and Fleissner, 1993). This creates the possibility for neuronal interaction between the two input channels for the purpose of recalculating "photic reality." The interaction of lateral and median eye terminals in the medulla may be similar to that assumed for nonvisual ocelli and image-processing compound eyes in insects: light to the ocelli facilitates perception in the compound eyes (Goodman, 1981).

The neuronal pathways for visual and nonvisual information are also separate within the same eye from the retina onward. The spiking arhabdomeric cells of the median eyes bring visual information directly down to the second optic neuropil, the medulla. The receptor cell axons probably conduct nonvisual information by graded receptor potentials (i.e., successive contrast, rather than simultaneous contrast of environment) to interneurons of the first optic neuropil, the lamina (Fleissner, 1985).

In the lamina, the light input to the system may be modified by the clock because terminals of the efferent fiber system (ENSF, see below) deliver a copy of the circadian signal to the receptor endings and interneurons at that point (fig. 4.4C,G). Ultrastructural investigations show that neurosecretory fibers contact afferent axons before they synapse on their first interneurons (fig. 4.4B; cf. fig. 4.8A). Neurosecretory fibers also synapse on interneurons directly, making an effective channel for circadian modulation of information flow. Terminals of the ENSFs of the lateral eyes are also present in the median eye lamina (Heinrichs, 1985), but there are no endings of the ENSFs in the medulla. The efferent circadian signal (fig. 4.4F), which modulates the sensitivity of the eyes (fig. 4.4A), is thus present where afferent light information via the retinal photoreceptors is processed. Theoretically this information can be used to recalculate the true ambient light conditions (fig. 4.4G), which are masked by circadian adaptation at the retinal photoreceptor level (Fleissner and Fleissner, 1993).

In the scorpion the receptors and neuronal pathways involved in Zeitgeber perception are distinct, though closely associated. This offers the opportunity of analyzing their functional significance one by one and to understand three important features of Zeitgeber perception, found in many organisms, but of which the underlying mechanisms are still unknown. For example, it is commonly observed that combined stimulation of several photoreceptor systems is more effective for circadian timing than stimulation of any one type alone (e.g., sparrow: Zimmerman and Menaker, 1975; horseshoe crab: Horne and Renninger, 1988; scorpion: Luettgen, 1993). It is also commonly observed that light pulses with simulated dawn and dusk periods are better Zeitgeber signals than the abrupt on-and-off pulses typically used in experimentation (Fleissner and Fleissner, 1993; Luettgen, 1993). Perhaps this dawn-and-dusk perception is feasible only by interaction of different photoreceptor systems. Finally, the photic phase-response curve shows a "dead zone" during the subjective day, where light pulses are incapable of shifting the phase of the clock (cf. chapter 5). The dead zone phenomenon appears to be an inherent feature of pacemaker neurons, but peripheral gates may play a reinforcing function, too. Such gating might allow Zeitgeber light pulses to reach the clock only when the proper circadian signal is present in the pathway (Fleissner and Fleissner, 1998).

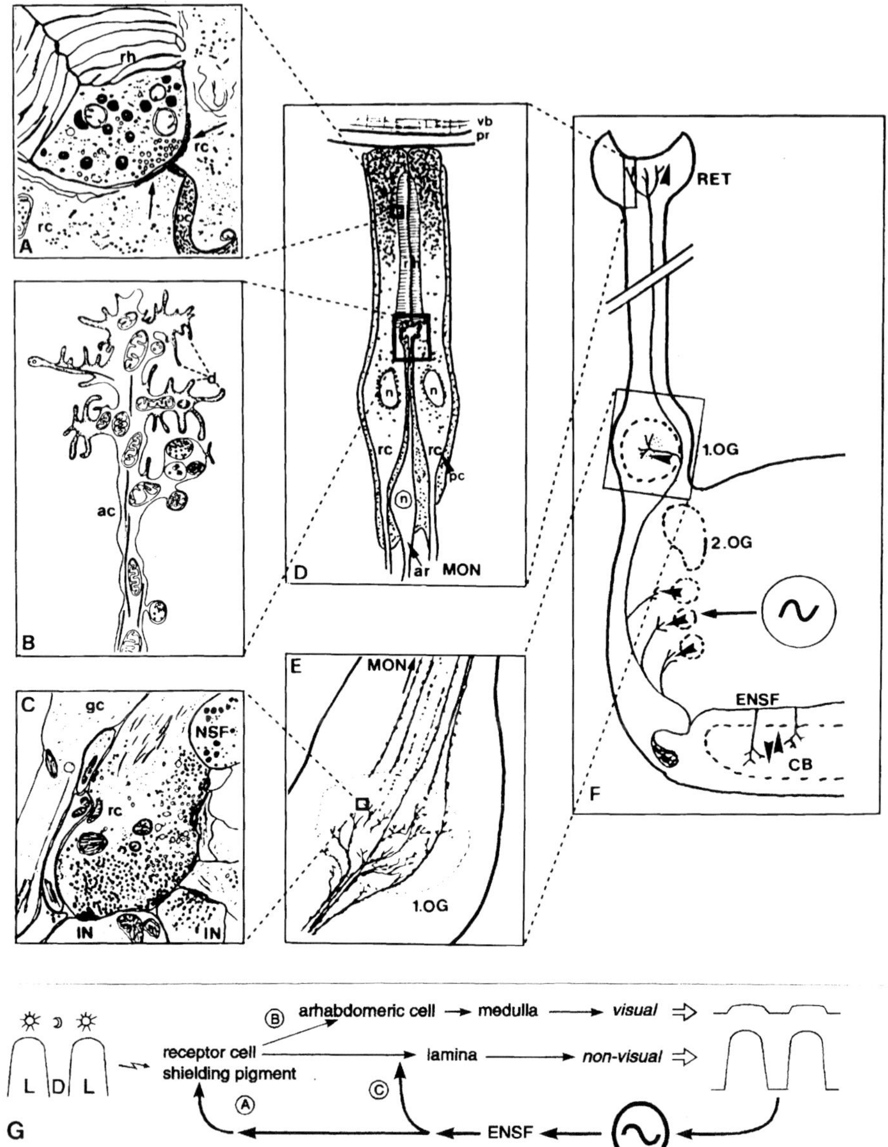

Figure 4.4 Feedback loops in the Zeitgeber signal processing pathways of the visual system of scorpions. (A–C) Drawings of the cytological structures and synaptic contacts between sensory and efferent pathways within the retina (D) and lamina (E). (F) Scheme of neural pathways for central and peripheral input and output (arrows) for efferent neurosecretory fiber (ENSF) processes in protocerebrum. In panel A terminals of the ENSF contact photoreceptor cells (rc) in the retina and control shielding pigment position: rh, rhabdom; arrows indicate postsynaptic specializations. (B) distal dendrite (d) of the arhabdomeric cell (ac) receives input from receptor cells at the base of the rhabdom. (C) visual afferents (rc) form synapses onto interneurons (IN) in the distal part of the lamina. Neurosecretory fibers (NSF) contact these afferents presynaptically (gc, glial cell). (D) Retinular unit from a median eye showing the arrangement of receptor cells, the

EFFERENT NEUROSECRETORY FIBERS AS PATHWAYS OF THE CIRCADIAN SIGNAL

Experiments involving lesioning by electrical stimulation and pharmacological treatments have shown that the circadian adaptation of the eyes is controlled via neuronal signals from the central nervous system (cf. fig. 4.10A; Fleissner, 1974, 1983; Fleissner and Fleissner, 1978, 1988). The clock does not reside in the retina, but the circadian signal is mediated by efferent neurosecretory fibers (ENSF). Analyzing this efferent fiber system and the neurons interacting with it provides at least a partial analysis of the wiring of the circadian clock system.

Morphology of the Efferent Neurosecretory Fibers

We used a variety of techniques to reveal the neuroanatomy of the efferent neurons of origin for the neurosecretory fiber system in *Androctonus.* Electron microscopy of sections through the retina of both median and lateral eyes show many endings of neurosecretory fibers (cf. fig. 4.3B,D; 4.4A), which can be counted and traced throughout the retina (fig. 4.5). Less than 10 ENSF fibers enter the retinular unit of photoreceptors at its proximal end, where the fibers branch repeatedly (fig. 4.5E) to form more than 40 terminals per retinular cross-section. Most of these terminals are located at the basal end of the rhabdom (fig. 4.5D,F), where they make synaptic contacts to photoreceptor cells only (Fleissner and Fleissner, 1988); none extends to the preretinal membrane. In many retinular units one branch of the neurosecretory innervation extends through the center of the fused rhabdom, interposed with the distal dendrite of the arhabdomeric cell. We have looked for the source of these terminals to determine whether they all represent the same type of neurons and which parts of the clock system they connect.

Neuronal pathways from the median and lateral eyes to the brain are revealed by selective back-filling of nerves with neuronal marker dyes (lucifer yellow [LY], propidium iodide [PI], horseradish peroxidase [HRP], cytochrome C [cC], cobalt chloride [$CoCl_2$]) as well as differential and double stainings with different dyes (Heinrichs, 1985). Back-filled structures can then be analyzed at the light microscopic or electron microscopic level (Heinrichs, 1983). The combination of these different methods is necessary, as they all have advantages and disadvantages: Lucifer Yellow is highly sensitive and shows axons down to their terminals within minutes or few hours, but the yellow fluorochrome fades rapidly, and it may pass between cells through gap junctions. As such, its use

---

arhabdomeric cell (ar) and pigment cells (pc). The axons of arhabdomeric cells (ar) and receptor cells (rc) constitute the optic nerve (MON). n, nucleus; vb, vitreous body; pr, preretinal membrane. (E) First lamina neuropil (1.OG) of the median eye and ENSF innervation as marked by contralateral dye application. The ENSF send axonal branches into the lamina. (F) Anatomical model of circadian information flow via the ENSF: the circadian pacemaker (circle) feeds its signal into the ENSF at the level of the higher optic neuropils. The ENSFs distribute circadian information to the effector systems or their control centers. Visual sensitivity is directly modulated via this fiber system. Locomotor behavior may be regulated through ENSF effects on the central body (CB). OG, optic neuropils; RET, retina. (G) Functional model of visual and nonvisual light processing in the median eye system. (After Fleissner and Fleissner (1993)

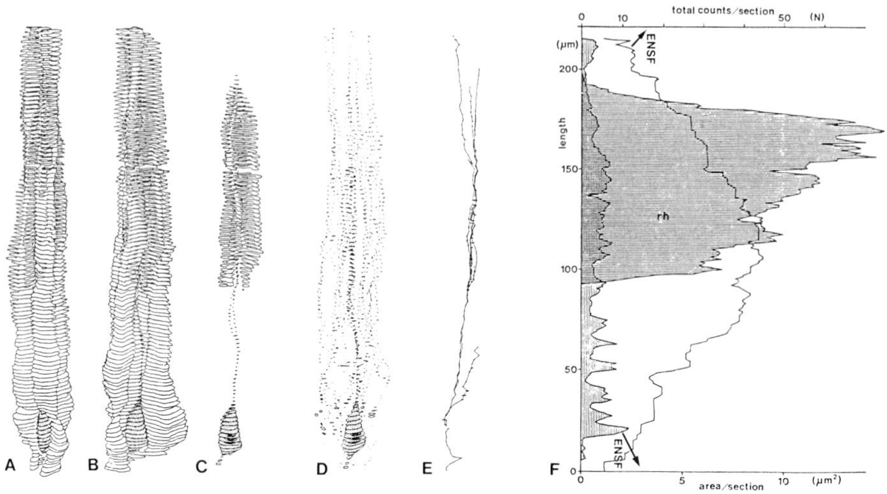

Figure 4.5 Computer-aided reconstruction of a median eye retinula in the day state. (A) The retinular unit is composed of five photoreceptor cells and an arhabdomeric cell (shaded area) centrally located. As the retinula is "dissected" by removing one receptor cell (B) and then all receptor cells (C), the rhabdom and the arhabdomeric cell are revealed. (D) The arhabdomeric cell and the network of all efferent neurosecretory fibers (ENSFs); (E) a single ENSF. Note that the ascending dendrite of the arhabdomeric cell ends where the rhabdom begins, and the ENSFs branch several times along their length. (F) Graphical representation of the number of ENSF profiles (left ordinate), and cross-sectional area of ENSF (right ordinate) and rhabdom (rh) (right ordinate) of this retinular unit, plotted over its longitudinal axis (scale valid for A–E). The number of ENSF cross-sections reaches a maximum at the proximal end of the rhabdom, but the total area remains relatively constant. This is due to the greater number of large varicosities near the posterior retina. After unpublished data from a cooperative project with M. Teckhaus and M. Schliwa.

for camera lucida drawings is limited. Propidium iodide, a red fluorochrome, is used in double and differential stainings, when a second marker is needed in combination with, for example, LY. Horseradish peroxidase is much slower than LY but nearly as sensitive; the granular staining product does not fade and gives a Golgi-like appearance of the neurons. Cytochrome c is used for retrograde labeling similar to HRP, it does not stain neurons trans-synaptically but is slowly transported along the axon (here up to 3 days) and is detectable in the soma region only. Cobalt chloride also gives the stained neurons a Golgi-like appearance; it has been used for whole mounts to evaluate the results in reconstructions from serial sections with different markers, but it may provide unreliable results in fine dendrites.

Back-fills from the median eye retina or cut optic nerve show 10–20 neuronal somata of about 15 μm diameter on either side of the protocerebral ganglion (fig. 4.6) (Fleissner and Heinrichs, 1982). These cell bodies (fig. 4.6D) are located near the ventrolateral border of the supraesophageal ganglion (brain), at the level where it surrounds the esophagus (fig. 4.7D). No other central neurons are filled via the cut optic nerve and all of the marked neurons are of the same type with the same central course and terminal branching pattern (Heinrichs, 1985; Heinrichs and Fleissner, 1987; for review, see Fleissner et al., 1987). Each of these somata sends an ascending axon toward the dorsal rim of the brain (fig. 4.6B,C), where it bifurcates into two branches, both of which run horizontally forward along the lateral aspect of the brain (figs. 4.7D, 4.6A). One branch decussates to the contralateral side, making en route contacts within the central body and the contralateral optic neuropils; the other axon branch passes ipsilaterally along the optic neuropils. Both branches enter the optic nerves and end in the median eye retinae. These cells, their axons, and particularly their terminals are characterized by their content of large, electron-dense vesicles of about 80–100 nm diameter, indicative of neurosecretory function (Fleissner and Schliwa, 1977). Therefore, we refer to this fiber system as ENSF.

Terminal branches arising from these axons can be seen in the central body of the brain, in the optic neuropils 1 (lamina), 3, 4, and 5, and in the retina, but not in the second optic neuropil (medulla) (fig. 4.7). The terminals make chemical synapses with other neurons, as indicated by membrane-associated clear vesicles at contact zones and en-passant synapses along the axons (fig. 4.8) (Heinrichs, 1985; for review, see Fleissner and Fleissner, 1988). Postsynaptic differentiation is rarely seen, but ultrastructural studies of back-filled axons of ENSF neurons demonstrate the direction of information flow.

In the retina (fig. 4.4A) and the lamina (figs. 4.4C, 4.8A) all marked neurosecretory axons are presynaptic terminals, as shown by the presence of clear vesicles within the marked profile (fig. 4.8A). In the lamina, both the photoreceptor terminals and interneurons receive input from the ENSFs. In optic neuropils 3 to 5, all back-filled neurosecretory terminals appear to be postsynaptic (fig. 4.8B), as unstained terminals form synapses with clear vesicles and characteristic membrane thickenings onto the marked profiles. In the central body, both input (fig. 4.8D,E) and output (fig. 4.8C) chemical synapses are associated with marked neurosecretory profiles.

Efferent terminals are present in both lateral and median eye retinae, raising the possibility of two functionally independent subsets of ENSF neurons serving the different eyes. Dyes of different microscopical appearance (e.g., LY and PI) applied to the same type of eye on the two sides (right and left) always reach the same somata and reveal crossover of right and left side ENSF in the central body (fig. 4.9A) (Heinrichs, 1985). Dyes applied to the nerves of lateral and median eyes on the same side stain different somata in the protocerebral population of ENSF cells (fig. 4.9D), showing that there are different ENSF systems for lateral and median eyes (L-ENSF, M-ENSF). Both systems, which probably consist of one type of neuron only, show the same central course parallel to each other (fig. 4.9B), as described for the M-ENSF above. Their somata are located in the same general area, with the L-ENSF somata placed somewhat more caudally. The terminal branches of both systems are closely interlaced in all neuropils to which they contribute (fig. 4.9C), with one important difference: the M-ENSF only contact the median eye lamina (fig. 4.7D), while the L-ENSF innervate both the lateral eye laminae and, by a fine branch, the median eye laminae as well (fig. 4.7E). These L-ENSF branches to the median eye lamina, however, do not extend as far as the median eye retina.

The likely pathways of information processing within the ENSF system are summarized in fig. 4.4F. Both retinae contain terminals of the ENSF neurons of

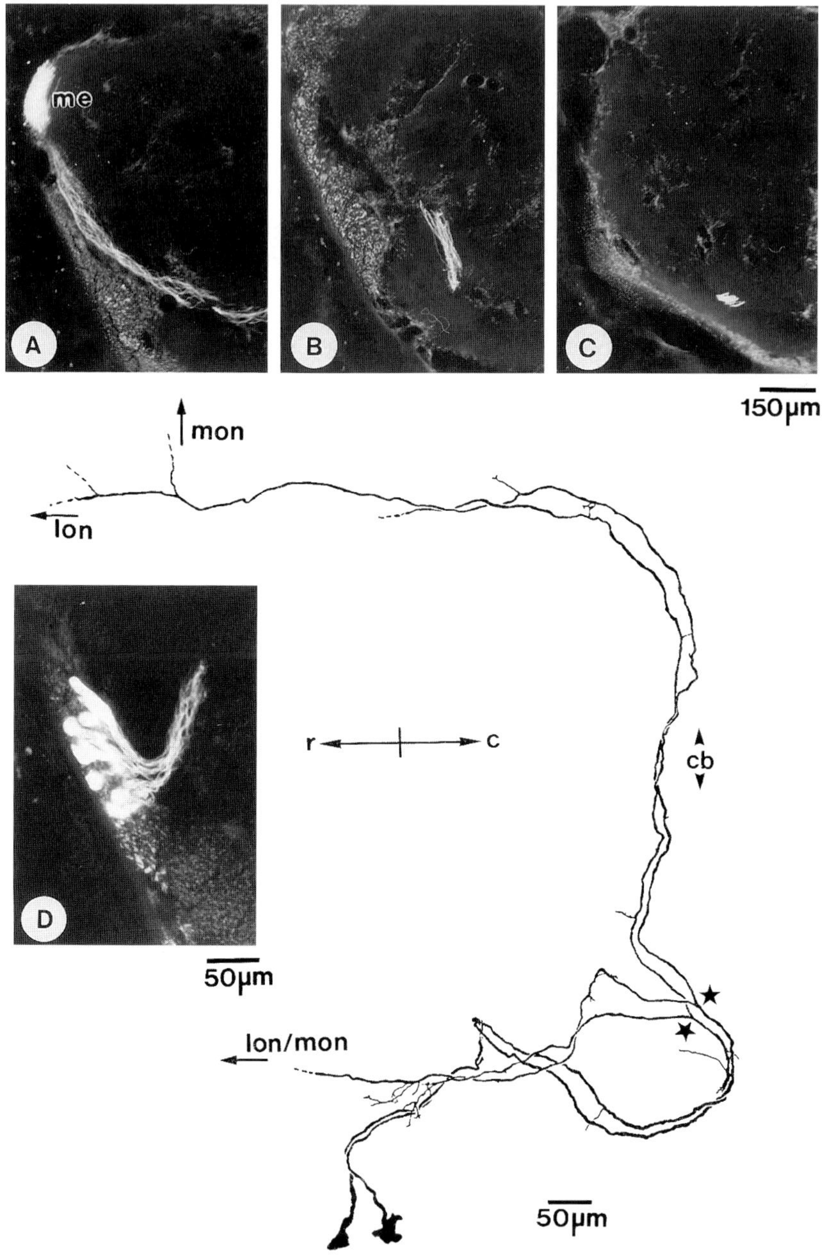

Figure 4.6 Horizontal sections through the scorpion's left protocerebrum showing efferent neurosecretory fiber (ENSF) axonal pathways and location of their somata. The fluorescent dye Lucifer Yellow applied to the cut median eye nerve back-fills (A) a bright spot, the medulla, with terminals of the arhabdomeric cells and a few fluorescing fibers (ENSF) connecting both sides; me, medulla. (B) Near the caudal rim of the brain a fiber bundle descends, concentrates, and (C) turns laterally toward the lateral rim of the brain. (D) A

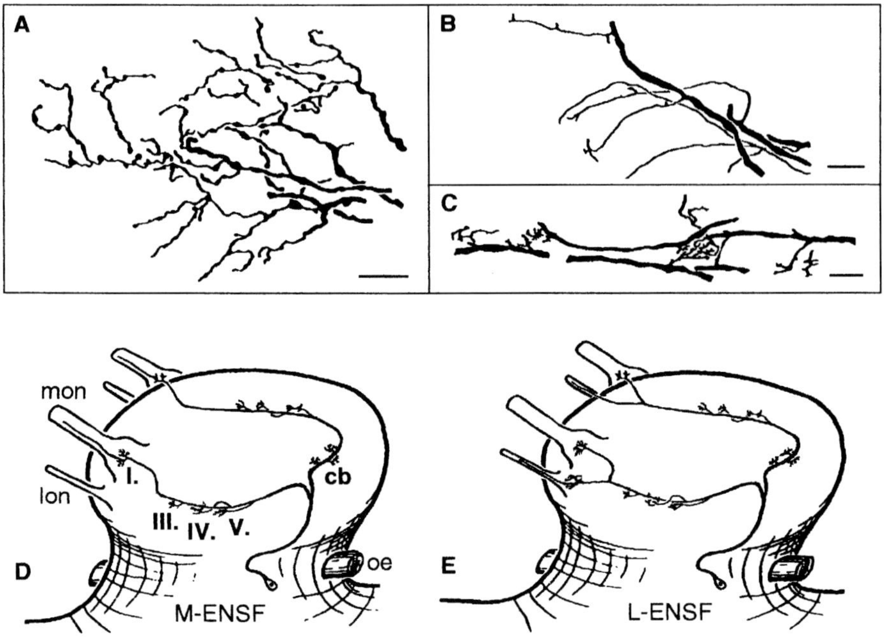

Figure 4.7 Terminal branching and pathway of the efferent neurosecretory fibers (ENSFs). (A–C) Camera lucida drawings of cobalt chloride-labeled terminal branches of the median eye ENSF system: (A) in the lamina, (B) in the higher optic neuropils, and (C) in the central body; supraesophageal ganglion (brain) with half-schematic reconstruction of the ENSF systems of the median (D) and lateral eyes (E). (A–C) all scale bars are 10 μm; cb, central body; lon, optic nerve of the lateral eye; mon, optic nerve of the median eye; oe, esophagus; I, III, IV, V, number of optic neuropils. Data from Heinrichs (1985).

both sides of the brain, but the lateral and median eye system is separate. Information entering the ENSF systems at the level of optic neuropils 3–5 and from the central body is released onto the retinae, the laminae, and the central body.

Functional Significance of the ENSF

The ENSF fiber system and the necessity of efferent innervation of the retina for its circadian regulation of sensitivity were recognized at much the same time

---

higher magnification shows the fibers originate from about 20 neuronal somata. Only these somata are filled by dye application to the proximal stump of the cut optic nerve. Below, a camera lucida drawing of two ENSF axons labeled by horseradish peroxidase application to one lateral eye nerve (projected into horizontal plane). The curvature and bifurcation (stars) of the ascending fibers can be seen. The two branches innervate the lateral eye (lon) and median eye (mon) on both sides and send terminals into the central body (cb). c, caudal; r, rostral. Data from Heinrichs (1985).

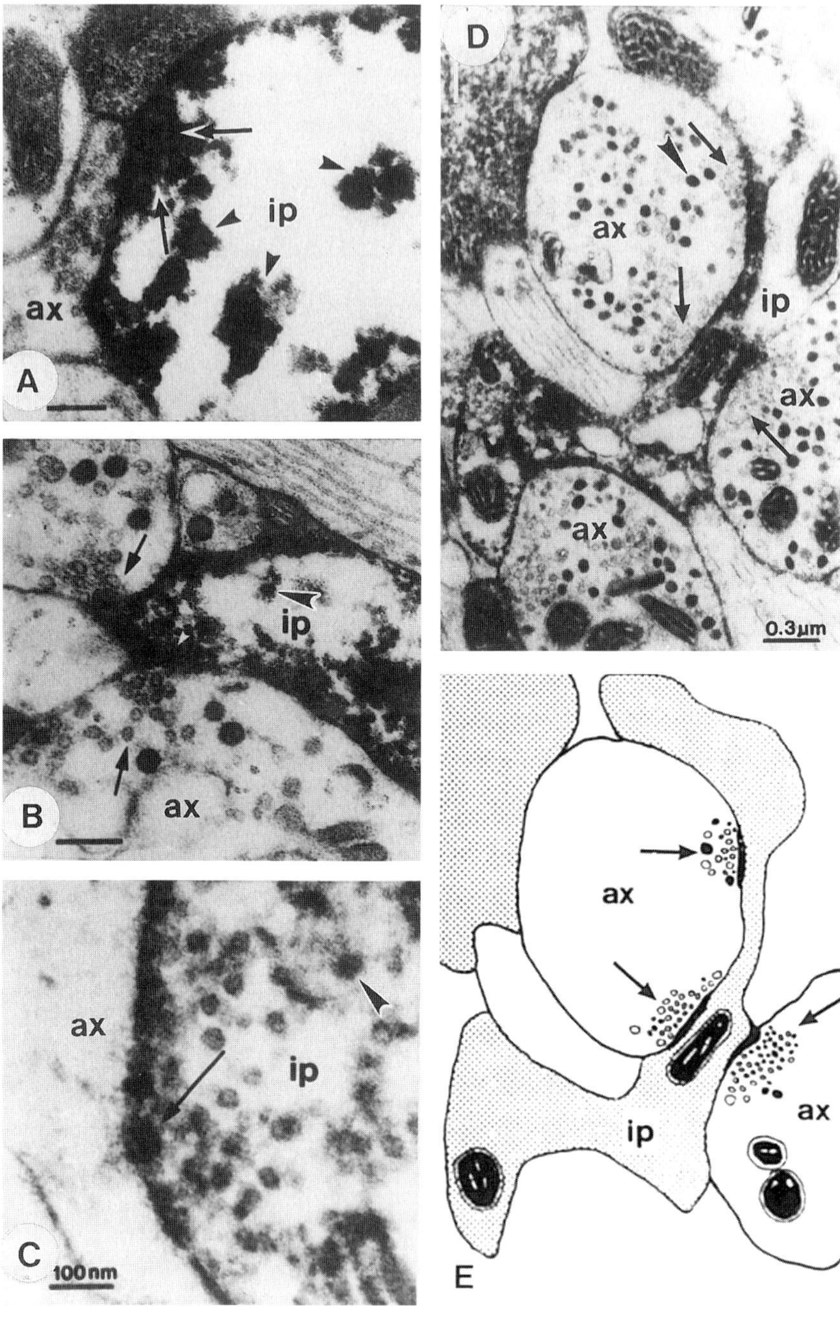

Figure 4.8 Synaptic profiles between efferent neurosecretory fiber (ENSF) terminals and axonal processes within optic and cerebral neuropils. ENSF are labeled by horseradish peroxidase infusion. They contain neurosecretory granules (80–100 nm) (arrowheads) and clear synaptic vesicles (20–40 nm) (arrows) near synaptic densities. The marked fibers form (A) output synapses in the lamina (cf. fig. 4.5A), and (B) input synapses in the higher optic neuropils (3–5) (cf. fig. 4.5B). In the central body (cf. fig. 4.5C) there are both output terminals (C) and terminals with multiple input (D, E). (E) A scheme of the section in panel D. ip, identified profile of the ENSF system; ax, terminal of an axon contacting the ENSF; (A–C) scale bars = 100 nm; D, E) same magnification. Data from Heinrichs (1985).

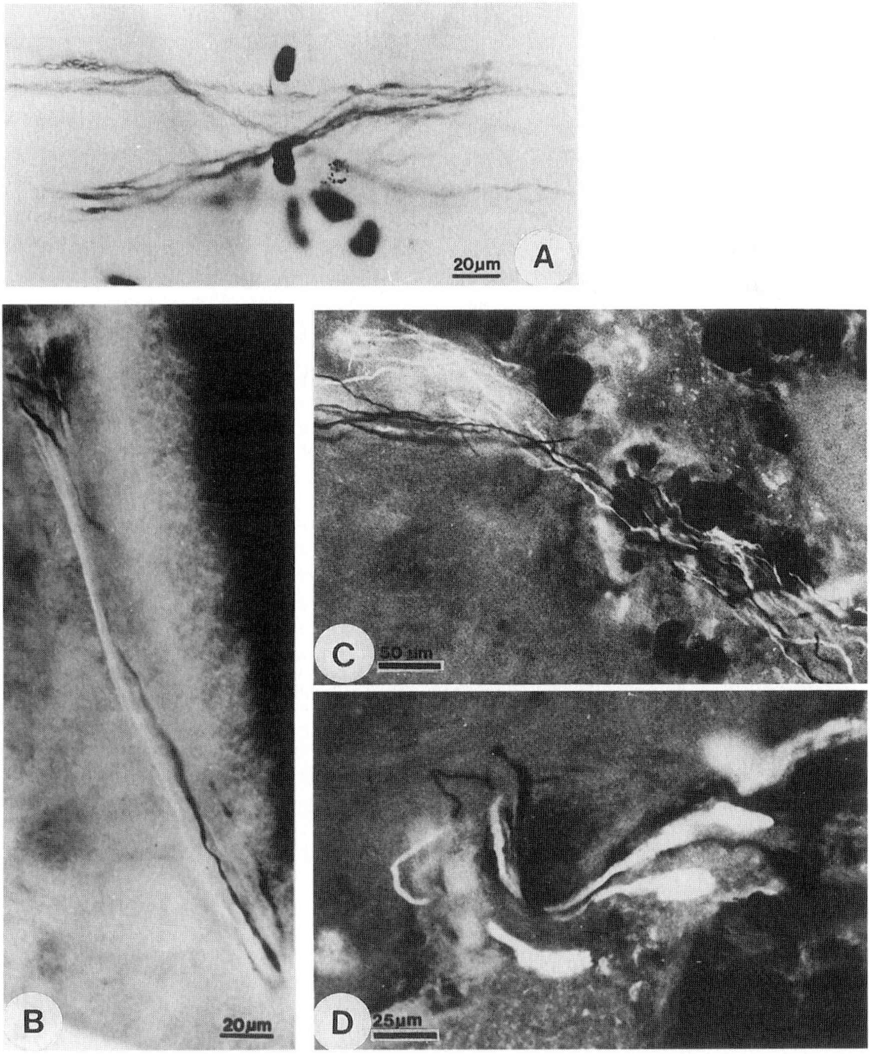

Figure 4.9 Central projections of median and lateral efferent neurosecretory fiber (ENSF) systems. (A) Application of dyes to the right and left median eye nerves shows labeled fibers in the same tracts. In this section the chiasma of right and left median ENSF in front of the central body is shown. (B–D) The median ENSF and lateral ENSF are stained differentially by filling the left lateral eye nerve with horseradish peroxidase (dark reaction product) and the right median eye with Lucifer Yellow. Lateral and median ENSF are separate systems but use the same central pathways (B, ascending branches) and have their terminals (C) in the distal part of higher optic neuropils), and somata (D) are located in the same sites. Data from Heinrichs (1985).

(Fleissner and Fleissner, 1978; Fleissner and Schliwa, 1977). The central course and synaptology of these fibers was elucidated subsequently (Fleissner and Heinrichs, 1982; Heinrichs, 1985; Heinrichs and Fleissner, 1987). Currently, functions of the ENSF are being investigated by disrupting their axonal connections at different sites along the ENSF pathway through the brain. We also seek to find the exact location of the circadian oscillator and its relation to the ENSF, whether it is one clock or a multioscillator system, and whether there are different pacemakers for different effector systems (e.g., locomotion and visual sensitivity).

To investigate the role of the ENSF for circadian control, a series of microlesions was performed (fig. 4.10A,C,E). Before and after the operation, the rhythmic changes of visual sensitivity in different eyes (fig. 4.10B,D,F) and locomotor activity were recorded simultaneously. Histological controls ascertained the site of the lesion (Fleissner and Fleissner, 1978; Fleissner, 1983). In intact animals the circadian changes in sensitivity of the eyes on both sides are precisely synchronized and persist throughout life (Fleissner, 1971, 1972). The rhythm of the animal's locomotor activity follows at identical period length, but may be phase locked at different angles to the eye rhythm (Michel, 1989). This behavior is unlike comparable rhythms in some other arthropods (e.g., beetles), where spontaneous arhythmicity and internal desynchronization between left and right eyes is frequently observed (Koehler and Fleissner, 1978; Fleissner and Fleissner, 1986).

All lesions that interrupt the course of the ENSF system (Fleissner and Fleissner, 1978; Fleissner, 1983) distal to the lamina immediately terminate the rhythm in ipsilateral eyes without affecting the rhythm in contralateral eyes. The photosensitivity of the de-efferented retina decreases to the lower, or "day state" of photosensitivity and remains there indefinitely. This is true even with partial lesion of an optic nerve which induces two patterns of sensitivity in a single retina. Only retinal tissue that retains intact efferent connection to the ipsilateral optic neuropils 3–5 continues to express circadian rhythmicity, while the de-efferented portion shows the chronic "day" configuration of photosensitivity. As lesions are made more proximally along the visual pathway, the disturbances to rhythmicity become more complex and evident in the contralateral eyes (fig. 4.10C,D). Sensitivity rhythms for both eyes may become less pronounced due to an increase of their day-state sensitivity (fig. 4.10D,F). Arhythmicity, manifested as permanent night-state (fig. 4.10F) or day-state (fig. 4.10B) may occur, and mutual inhibitory interactions between left and right side can become evident (fig. 4.10D). Lesions of the median eye hemolymph-vessels have no effect on rhythmicity.

The connection of ENSF axons to their somata is not crucial for the persistence of the circadian rhythmicity (fig. 4.10D). Thus, ENSF axons probably receive the circadian signal from presynaptic elements and mediate it to the effector systems. About 2 weeks after unilateral axotomy, the rhythm in the ipsilateral eye fades as the rhythm in the contralateral eye returns to normal (Fleissner, 1983). This result is also observed when the central course of the ENSF is interrupted proximal to the higher optic neuropils, where the ENSF ascend from their cell bodies. The distal parts of the ENSF are then disconnected from both their somata and their contralateral branches.

It is generally accepted in chronobiological research that circadian pacemaking function is bound or even generated by nuclear factors (Takahashi, 1992).

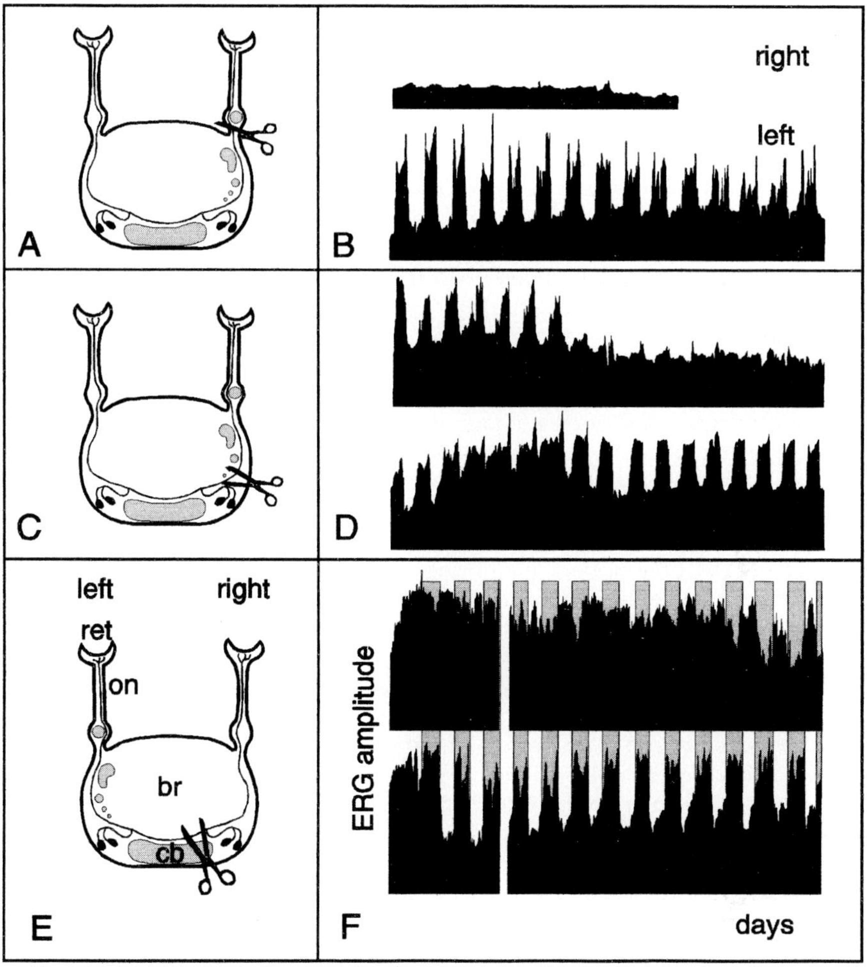

Figure 4.10 Lesion experiments illustrating circadian regulatory function of the efferent neurosecretory fiber (ENSF) system for median eyes. Simultaneous ERG recordings (B, D, F) of right (upper traces) and left median eyes (lower traces) after lesioning (scissors in A, C, E). (A, B) Cutting the visual pathway between optic neuropil 1 and 2 (lamina and medulla) induces permanent day state to the lesioned ipsilateral eye, while the rhythm in the contralateral eye persists. (C, D) Lesions close to or within optic neuropils 3–5 have dramatic and complex effects on ipsilateral and contralateral eye rhythmicity. E,F, lesion in the central body partly obscures rhythmicity and mainly desynchronizes the left and right-side ERG rhythms. The night states of the left eye (shaded area) are drawn in the recording of right median eye to illustrate the desynchronizing effect. br, brain; cb, central body; on, optic nerve; ret, retina. Panel D adapted from Fleissner (1983).

Persistence of rhythmicity in scorpions with the ENSF disconnected from their somata provides evidence that the ENSF cannot be pacemaker neurons. These set of data also show that there must be additional contralateral connections serving mutual inhibitory interaction between left- and right-side effector systems.

By cutting the supraesophageal brain along its midline, the decussating axons of the ENSF are severed in the region of the central body such that the ENSF only innervate their ipsilateral targets. In these "split-brain" animals, both eyes retain rhythmicity but may develop different period lengths (fig. 4.10F). Locomotor rhythmicity is coupled to the rhythm of the side retaining ENSF connection of the central body to the neuropil 3–5 region. Bilateral disconnection between the central body and neuropils 3–5 destroys the rhythmicity of locomotion (Michel, 1989).

From these morphological and behavioral studies of the circadian signal pathway, we postulate that (1) the circadian pacemaker is a bilaterally symmetrical system, (2) the circadian signal is transmitted by an efferent fiber system in the optical pathway, being fed into this fiber system near the higher optic neuropils, (3) the clock system appears to be the same for all eyes on both sides and for locomotor control centers, indicating that the "eye-clock" is a master clock for locomotory rhythms as well, and (4) the circadian signal is available in the lamina, where photoreceptor cells terminate, and can be used for correction of visual input signals when these are modified by the clock-controlled sensitivity rhythm.

The circadian clock system in scorpions, though a multioscillator system, functions as a single clock. This is important for burrowing scorpions when they lose access to external synchronizing stimuli during long periods of inhospitable weather (e.g., during desert summer). The ENSF can best be interpreted as a hard-wired "bus"-system distributing the circadian signal among several users, and are not themselves the circadian pacemakers (Fleissner et al., 1987). It is not yet known whether the ENSF also serve as a coupling pathway for synchronization of the left and right pacemakers or whether a separate oscillator coupling pathway different from the ENSF mediates this function.

Circadian Signals

The clock-controlled efferent "night" signal drives the shielding pigment inside the retinal photoreceptor cells from its distal day-state position proximally into their somata. This exposes the rhabdoms to incident light and increases visual sensitivity by several log units (for review, see chapter 3). Thus, the day-state results from absence of a night signal. The circadian state of retinular cells can be determined either by histology or sensitivity records. Using such information we can study which type of signal is carried along the ENSF and how it is transferred to the effectors.

The circadian signal in the ENSF has both neurochemical and electrical components, which interact to produce their effects. A major advantage of using *Androctonus* for investigating pharmacology of the circadian signal is the possibility of drug application via the eye tube (cf. fig. 4.2), the lumen of which is in direct contact with retinular cells. Such exclusively retinal application could otherwise be achieved only in isolated eyes, which then would lack efferent innervation. Electrical stimulation can be performed with permanently implanted suction-electrodes on the optic nerve, using simultaneous long-term recordings of left- and right-side ERG to monitor stimulation effects and the circadian state of sensitivity.

Neurochemical component

Among the various neurotransmitters we have tested, octopamine is the most specific for conversion of a day-state retina into the night-state (Fleissner et al., 1987). Both the histological and electrophysiological effects of transition from day to night state are evident—namely, withdrawal of shielding pigment (fig. 4.11C) and change in visual sensitivity as measured by ERG amplitude (fig. 4.11A). The higher the concentration of the transmitter, the more proximally the pigment granules migrate, sometimes even down into the optic nerve (fig. 4.11B). Octopamine application also moves the rhabdoms of the photoreceptor cells closer to the preretinal membrane and may even destroy the rhabdoms if the concentration is too high. Octopamine also changes the shape of the ERG and its intensity–response-function to night state (fig. 4.11C). Such an octopamine-induced period of higher ERG amplitude has no phase-shifting effect (fig. 4.11A; see also chapter 5).

In other experiments the effects of octopamine, dopamine, and serotonin (5HT) were compared after drug application to the eye tube or via the hemolymph. These studies showed that octopamine produced the retinal night state in the lowest concentration of the three, with dopamine and 5HT requiring at least 10–100 times higher concentrations to produce similar effects. This can best be interpreted as agonistic but nonspecific cross-reactions with a special set of octopamine receptors (Orchard and Lange, 1987) in the photoreceptor membrane. Antagonists to these transmitters can block both circadian and simulated night states: reserpine, which empties all aminergic vesicles, and phentolamine and clozapin, both specific blockers of octopamine membrane receptor binding sites, all inhibit the increase in photosensitivity. The effects are dose dependent and may last several hours or even days, but the underlying circadian rhythms meanwhile persist undisturbed, only its expression is temporarily blocked (Michel, 1989).

These experiments lead us to conclude that the circadian night state of the eyes requires the release of octopamine, as has been shown for the lateral eyes of *Limulus* (Kass and Barlow, 1984; for review, see Battelle, 1991). These studies also show that the pacemaker function does not interfere with the blocked night state, even when aminergic systems are pharmacologically inhibited. The results do not suggest that the eye tube (cf. fig. 2) naturally carries the circadian signal via hemolymph factors because in split-brain preparations both median eye rhythms run independently on their own period length (fig. 4.10F), although their extracellular space is interconnected by the intact eye tube (Fleissner, 1983).

Electrical component

If octopamine is acting on the photoreceptors in the mode of a conventional neurotransmitter then its release should be triggered by appropriate stimulation of impulse activity in the visual pathway. We observed that electrical stimulation of the distal stump of a cut optic nerve induces the night state in the ipsilateral

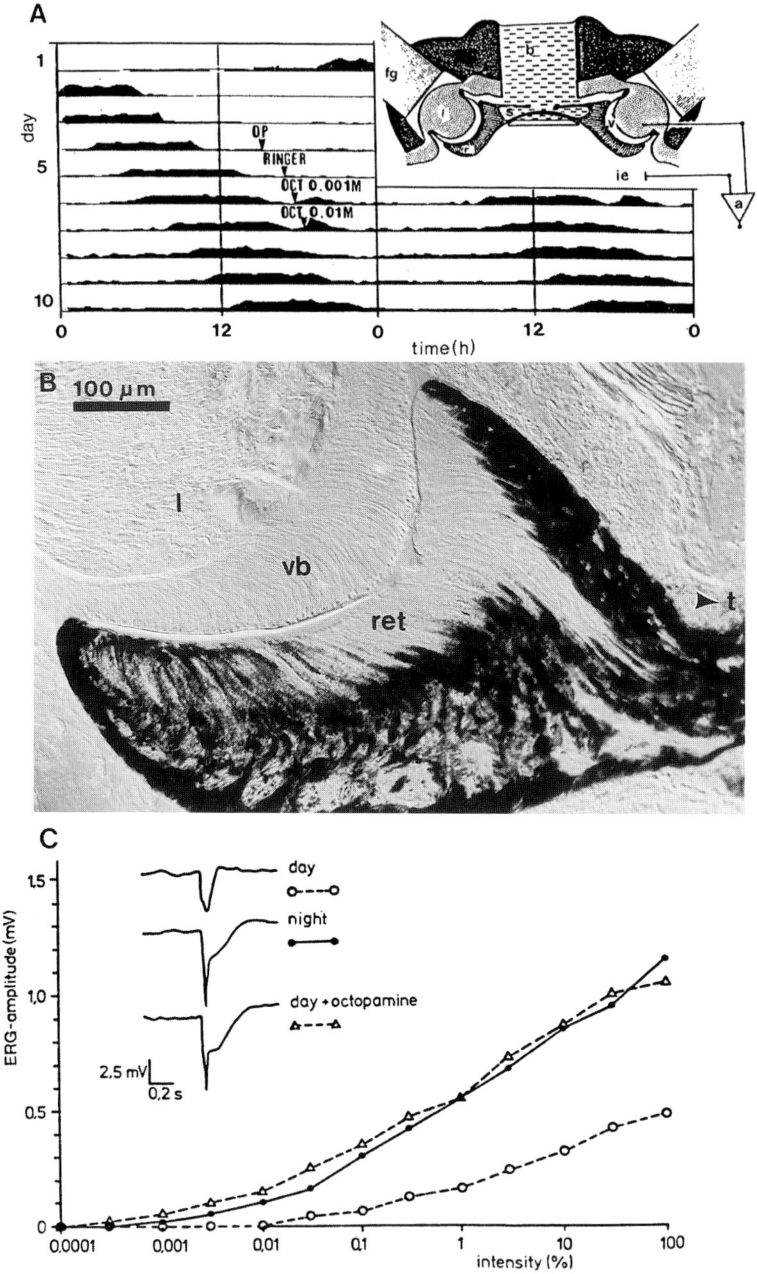

Figure 4.11 Neurochemical simulation of circadian night signal. (A) Continuous recordings of circadian fluctuations in ERG amplitude of a median eye before, during, and after application of physiological saline (Ringer) or saline with octopamine (of different concentrations; OCT 0.001M and OCT 0.01 M; one day of experiment below the other, double plot). Note that octopamine pulses do not phase shift the rhythm. Inset shows

median eye (fig. 4.12A,B; Fleissner et al., 1987), while stimulation of the proximal stump induces this transition in the contralateral eye (Moustafa, 1995). The frequency of stimulation necessary to affect these sensitivity changes is surprisingly low. Maximal sensitivity, sometimes higher than the circadian night state, is induced by pulse frequencies as low as 1 Hz (fig. 4.12C). Both lower and higher stimulation frequencies (Fleissner et al., 1987, 1989) gave lower levels of sensitivity (Fleissner et al., 1987; Marschall-Pauly and Fleissner, 1987). In the lateral eyes the change from day to night state is induced by stimulus frequencies as low as 0.1 Hz; increasing frequency to 1 Hz can superstimulate the lateral eyes to a "200% night state" level of sensitivity (fig. 4.12C). With higher rates of stimulation sensitivity declines again as it does in the median eyes.

The time course of this efferent circadian signal seems to be different in the subjective morning and evening: during subjective dusk the impulse frequency slowly rises to the night level and declines abruptly before dawn (Marschall-Pauly and Fleissner, 1987). This observation is decisive for the analysis of timing by dawn and dusk detection. Based on the histological data we proposed a neuronal model which makes use of the circadian signal as internal copy of the external twilight transitions (see fig. 4.4 and chapter 5, this volume). But as the circadian signal is present only during dusk, not during dawn, synchronization of the rhythm should be possible only with evening twilight. This could be convincingly corroborated in behavioral experiments (Luettgen, 1993; Fleissner & Fleissner, 1998).

Nervous impulses are necessary to evoke the night state in retinal photoreceptor cells, but they are not sufficient: octopamine must be present in the ENSF terminals for this to occur. This can be demonstrated by prolonged electrical stimulation of a cut optic nerve (e.g., for 10 days in the experiment shown in fig. 4.12D). A clearly defined night state begins to break down after about 15 to 20 h of ongoing stimulation. After recovery for several hours without excitation, nerve stimulation again increases photosensitivity but only for about 3 h. With repetition of this intermittent stimulation regime, the artificial night state evoked by electrical stimulation of the optic nerve gradually shortens until the response disappears altogether. This pattern of desensitization during chronic

experimental setup: a small basin (b) was attached above the opened eye tube (s; cf. fig. 4.2), drugs reached the retinas (r) without damage to the photoreceptor layer via the fluid space of the eye tube. a, amplifier for recording ERGs; fg, fiber-optic light guide for testing ERG response to light pulses; l, lens with active ERG electrode inserted into the lens and indifferent electrode (i.e., contacting hemolymph); OP, preparation of basin; v, vitreous body. (B) Cross-section through a day-state median eye after octopamine injection into the eye tube (t at arrowhead; cf. fig. 4.2) to show optomotor effects. The shielding pigment is retracted within the visual cells according to the diffusion gradient of the drug. Drug concentration is highest close to the eye tube; here the pigment retracts into the optic nerve, while the lateral parts of the retina (ret) still show day-state position of shielding pigments; vb, vitreous body. (C) Electrophysiological effects of octopamine application to the median eye. Intensity response functions of day and night states differ by more than 3 log units of sensitivity. The difference in light sensitivity is neutralized when in day state the eye tube is perfused by octopamine. Inset traces show single ERGs recorded for the various treatments (identical light stimulus flashes). Data from Bhatti, unpublished; panels B and C from Fleissner et al. (1987).

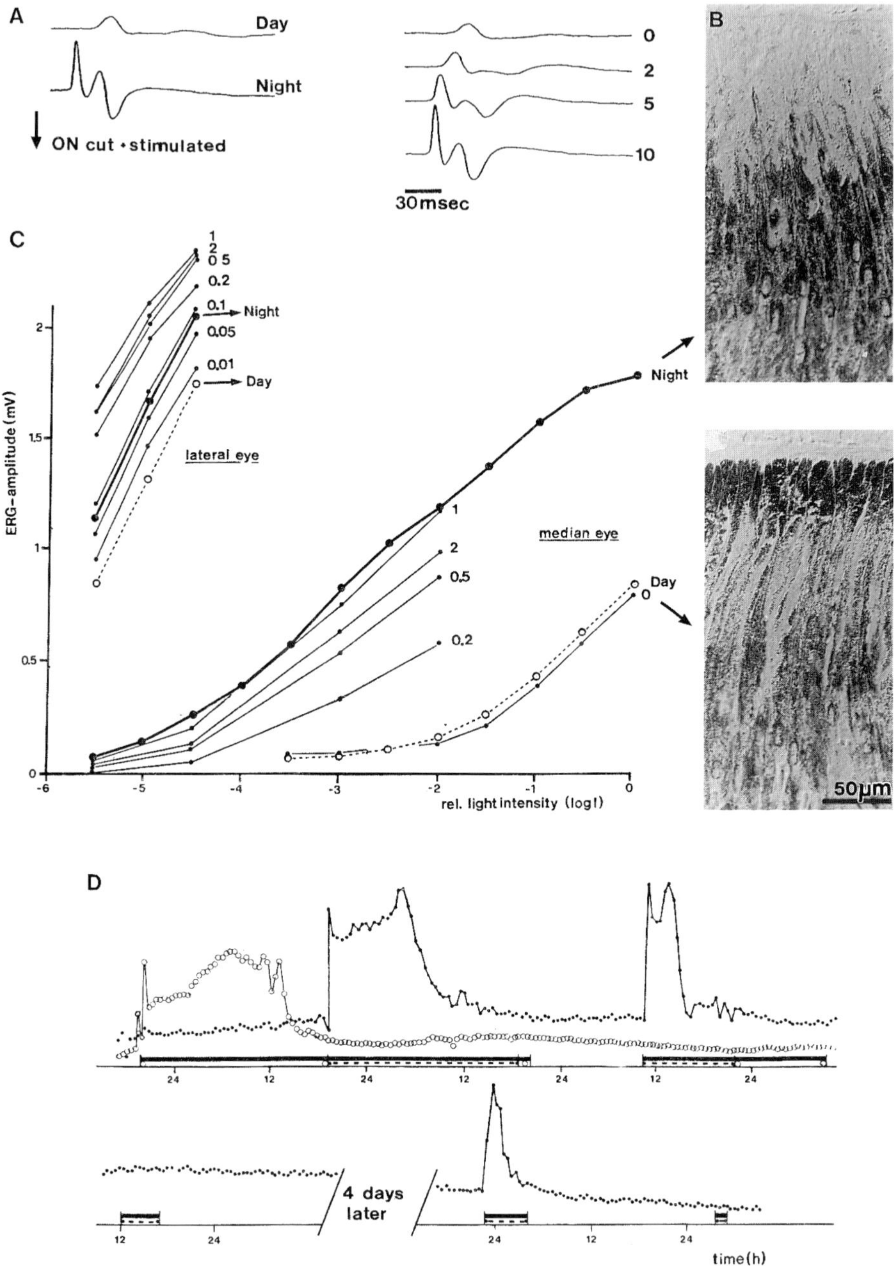

Figure 4.12 Electrophysiological simulation of the circadian night signal. (A) Electrical stimulation (7 Hz, 5 msec, 10 µA) of the de-efferented optic nerve causes transition from day to night patterns of ERG. (Left) ERGs recorded in intact day and night eye; (right) ERG recordings from the same eye after de-efferentiation and subsequent stimulation; numbers indicate minutes after start of stimulation, and bar indicates duration of light flash evoking ERGs. (B) Histological control of electrical stimulation: the retina shows the shielding

stimulation is consistent with depletion of the natural transmitter, probably octopamine, from synaptic terminals of the ENSF. The gradual decline in response is not caused by destruction of terminals or receptor sites within retinal cells (Fleissner et al., 1987).

It is noteworthy that electrical stimulation of a partially severed median eye nerve evokes only partial transition of the retina to the night state, as measured by ERG recording (Marschall-Pauly, personal communication). Histologically, the distribution of night-state and day-state photoreceptor cells in these preparations is not random. There is evidence of a topographical organization, with ENSFs in the lateral part of the optic nerve innervating the ventral half of the retina, and ENSFs within the medial half of the nerve innervating the dorsal retina (Fleissner et al., 1987). Such regional mapping of denervation effects gives further indication that the circadian signal is conducted along discrete axonal fibers and that humoral factors in the brain or hemolymph are not necessary for mediation of clock signals.

In summary, the ENSF discharge impulses at low frequency during the circadian night state, releasing octopamine from their terminals in the retinae. Left- and right-side clocks located in the protocerebrum regulate photosensitivity of both retinae integratively, and dorsal and ventral parts of the retina are controlled by different fibers.

IMMUNOHISTOLOGICAL IDENTIFICATION OF CIRCADIAN PACEMAKER NEURONS AND FEEDBACK PATHWAYS

Having described the basic features of Zeitgeber receptors and the function of the ENSF as a "system bus" for distributing circadian regulation in scorpions, other questions remain. Where exactly are the pacemaker neurons located, and where does Zeitgeber information meet these neurons? Also of interest are the pathways and signals for different steps of feedback in the circadian clock system. Two lines have been followed to investigate this, and both are based on immunocytology.

Our methods for immunohistology of scorpion tissues followed standard protocols developed for different arthropods (Nässel and Elekes, 1985, for anti-5HT

---

pigment is retracted to its proximal night-state position by stimulation (upper); it stays in its distal day state position without efferent signals (lower). (C) Intensity-response functions of lateral (left set of curves) and median eyes (right set of curves) in day (dashed line) and night state (solid line); electrical shocks of different frequencies (numbers indicate stimulus frequency in Hz). The lateral eyes are at least 3 log units more sensitive to light than the median eyes. The circadian variation in sensitivity of the median eyes is greater than in lateral eyes (3–4 log units versus <1 log unit). Most striking is the low frequency of impulses that can shift the day eye to night eye: in median eyes 1 Hz is optimum; in lateral eyes 0.1 Hz stimulation is sufficient to drive day eyes to their night state. (D) Long-term stimulation of left and right median eye nerve forces night state (high amplitude) in de-efferented retinas only for about 15 h, after which the day state (low amplitude) returns. After recovery, electrical stimuli can increase visual sensitivity again, but only for a shorter duration. Panels A, B, and D adapted from Fleissner et al. (1987); panel C from A. Marschall-Pauly, unpublished data.

studies; Siwicki et al., 1988, for anti-PER [period protein] studies), except that longer fixation times (4 h) and higher concentrations (0.1%) of the permeabilizing agent Triton X-100 were used to improve penetration of the sheath surrounding the scorpion cephalothoracic mass. Both cryosections and paraffin sections were used because cryosections gave slightly better marking of the finest terminals and paraffin-block sectioning allows better orientation and reconstruction. To demonstrate immunoreactivity, we used the avidin-biotin methods and Diaminobenzidine (DAB)-staining rather than fluorescent dyes. These modifications are necessary for reliable reconstruction of the marked fiber systems. Using these procedures we could demonstrate neuronal networks as candidates for components of the circadian system and thus lay the groundwork for physiological studies to begin.

Anti-*"per"*-like Immunoreactivity

Mutants of the fruit fly *Drosophila melanogaster* in which the period gene (*per*) is affected exhibit differences in their circadian behavior. The circadian clock of these mutants may run faster or slower than in wild-type flies, or the flies may even show no circadian rhythm, when the *per* gene is not expressed (Konopka and Benzer, 1971). This has been taken to indicate that *per* gene is involved in circadian oscillator mechanisms (for review, see Hall, 1990). The protein PER encoded by the *per* gene has been sequenced, and the full-length molecule or invariant parts have been used to generate anti-PER antibodies (Siwicki et al., 1988).

Such an antibody, developed for fruit flies, selectively labels a discrete set of cells in the scorpion brain, too (Brandes-Frisch et al., 1990), suggesting that the staining is rather specific. But while in flies only nuclei of neurons and glia cells are marked (Siwicki et al., 1988), in scorpions entire neurons stain (fig. 4.13A,B) giving a clear indication of their course through the central nervous system (fig. 4.13D–F) and target areas (fig. 4.13C). These marked fibers seem to be identical to the ENSF or at least contained within the same tract: they can be traced throughout the brain and optic nerves and have varicosities and terminal branches coincident with the placement of back-filled ENSF axons (cf. fig. 4.4, 4.6). Double stainings would confirm the identity of the two systems, but as yet this has not been achieved. Apart from this ENSF-like system, only few neurons medial to the higher optic neuropils show *per*-like reactivity, but their axons and terminal regions have defied tracing, and their putative function remains unknown.

All of our physiological experiments indicate the ENSFs of scorpions are conducting circadian clock signals between pacemaker and effector. They are possibly involved in the signaling between Zeitgeber receptors and the pacemaker as well. However, they are not the pacemaker neurons generating circadian periodicity (see above). The *per*-like fibers in scorpions may also act as signal-conducting pathways between pacemaker and effector and between Zeitgeber input and pacemaker. These suggestions are consistent with the results from comparative studies where *per*-like immunoreactivity was tested: beetles (Brandes-Frisch et al., 1990; Frisch et al., 1996), molluscs (Siwicki et al., 1989), and vertebrates (Garcia-Fernandez et al., 1994). In all these animals *per*-like immunoreactivity demonstrated possible pacemaker sites but pathways of the

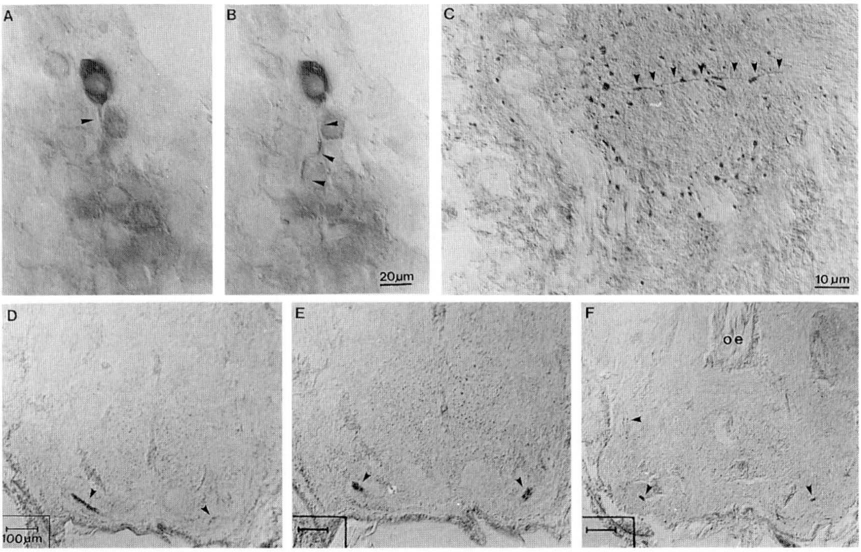

Figure 4.13 *per*-like immunoreactivity in the scorpion's cerebral ganglion. Anti-*per* antibody labels a distinct set of neurons and axonal processes in the scorpion protocerebrum. (A, B) *per*-like reactivity in a soma and its axon (arrows, same section, different focal plane). (C) Varicosities along an axon in the lamina. (D-F) serial sections showing that PER-like fiber system has a central course like that of the efferent neurosecretory fiber system (cf. fig. 4.4).

circadian clock signal as well, especially those between Zeitgeber input and potential circadian oscillator circuitry. This supports the hypotheses that *per* may encode coupling signals rather than, or including, oscillatory functions.

These comparative investigations seem to contradict the results in fruit flies, where *per* was found to interfere with period length and therefore was directly involved in rhythm-generating processes. Based on scorpion studies, we propose a solution: when the findings on this clock-gene and its protein first became known, the circadian clock system was looked upon as a simple system where the Zeitgeber input adjusts the self-sustained pacemaker, which, in turn, controlled the overt rhythm of the effectors. The importance of feedback and its impact on the circadian period length was less clear than it is now. Experiments resulting in changes of period length were then regarded as direct manipulations of the pacemaker itself. But our experiments with scorpions show that manipulation of pathways or coupling signals to and from the clock can also change circadian parameters.

To summarize, *per*-like immunoreactivity in scorpions demonstrates a fiber system, which may well be a signaling pathway of the circadian clock system, but it does not conclusively identify its circadian pacemaker neurons. This is a good example where the findings of molecular biology, dealing only with subcellular compartments, is in need of supplementary studies in model systems where hypotheses can be verified at the organismic level (e.g., by means of neurobiology).

Serotonergic Fiber Systems

As discussed at length in chapter 5, scorpions were one of the first animal model systems to demonstrate at the systemic level feedback connections between several component parts of the circadian clock system (Fleissner and Fleissner, 1992). Most striking was the clear feedback between clock-controlled rhythms in different effector systems and circadian pacemaker function. For example, behavioral hyperactivity of the scorpion changes the interpretation of light stimuli as possible photic Zeitgeber (see chapter 5). To explain this strange phenomenon, we must understand the signal for this "nonphotic photic Zeitgeber" stimulus and determine whether hyperactivity inhibits light-generated signals from reaching the pacemaker. As a first hint toward a possible physiological approach, we found that 5 HT antagonists can block this effect.

There must be a neuroanatomical interface between behavior and the nonvisual information-processing pathway. Studies of other animals (e.g., honeybees: Harris and Woodring, 1992; flies: Fowler et al., 1972; mice: Edgar et al., 1991) suggest that serotonergic neurons are likely to fulfill this function, and such pathways were accordingly sought. Both polyclonal and monoclonal antibodies against 5HT stain fibers, terminal branches, and somata of neurons throughout the brain of scorpions (fig. 4.14A; for review, see Fleissner and Fleissner, 1988). Stained neurons often appear as chains of discrete varicosities instead of continuous axons (fig. 4.14A,B), making it often difficult to follow single-fiber connections. Mainly two sets of 5HT-reactive neurons can be discriminated on each side of the brain.

All optic neuropils uniformly show abundant 5HT immunoreactivity (fig. 4.14A,B). Varicose fibers can be seen along the optic chiasms and side branches of the visual pathway and diverging from here toward a chain of little accessory neuropils parallel to higher optic neuropils (2–5) (fig. 4.14B). This fiber system appears to arise from less than 10 somata bilaterally. No fibers have been seen distal to the lamina. The optic nerves and retinae are free of 5HT immunoreactivity.

In the central body the layers described from conventional staining methods include layers of serotonergic terminals (fig. 4.14D). The caudo-dorsal layers have the densest staining, the rostral layers being more weakly marked. Approximately 20 somata on each side seem to be the neuronal origin of this fiber system. The protocerebral bridge, which forms a contralateral connection between the left and right side, shows horizontal and transverse serotonergic branches to neighboring neuropils (fig. 4.14C). The mushroom bodies and their globuli are nearly free of 5HT reactivity, but a sharp thin rim of marked varicosities surrounds them.

The 5HT-reactive neurons, both in the first optic ganglion and in the central body, show anatomical features expected of feedback pathways influencing circadian behavior. In the lamina, where nonvisual light information is likely processed en route to the pacemaker, 5HT-reactive fibers would have access to ENSF systems and other interneurons signaling circadian information. In the central body there is a convergence of all these pathways with higher centers of

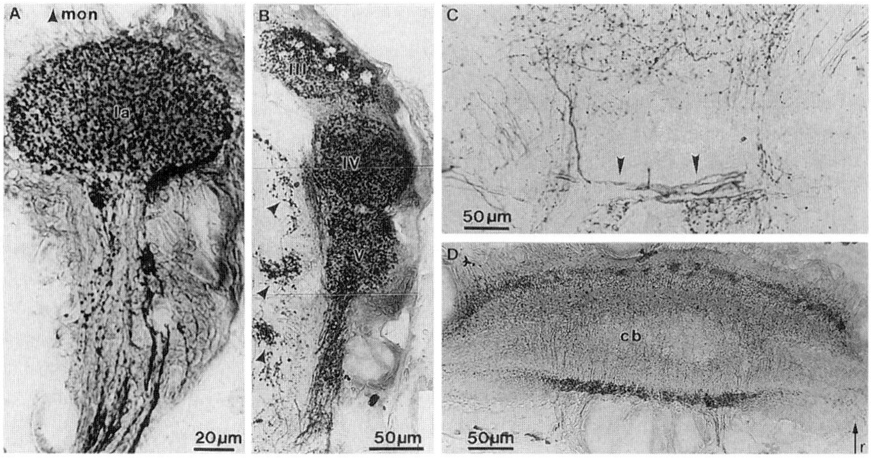

Figure 4.14 Serotonin (5HT) immunoreactivity in the scorpion's visual pathway. (A) Horizontal section through the median eye laminar neuropil showing heavy staining within the neuropil and proximal to it, but not within the optic nerve, (B) the optic neuropils 3–5 (III, IV, V) (right side shown) have a similar intensity of 5HT-staining paralleled by focal staining of small neighboring neuropils (arrowheads). (C) The protocerebral bridge shows serotonergic fibers as contralateral connections (arrowheads). (D) The central body is labeled in distinct layers, with the caudal side more condensed than the rostral part (arrow at r points rostrally for all figures) (from H. Hess and G. Fleissner unpublished data).

locomotory control. Again, terminal branches of the ENSF are well placed to influence information flow through these serotonergic pathways.

CONCLUSIONS

It is apparent that scorpions are indeed good model systems in which to investigate the neurobiology of the circadian clock system. Books and papers dealing with the circadian systems in arthropods include numerous reports on studies of crayfish and insects. Among chelicerates, *Limulus* is better known than the scorpion. Basic knowledge on circadian clocks, theoretical and practical, has been derived from the study of these systems over the years. We now offer good reasons to revise this emphasis: in no other arthropod system have the different components of the circadian clock system been so clearly observed as discrete neuronal entities and manipulated by experimentation as in scorpions. There is information on Zeitgeber receptors and their possible connection to the pacemaker centers controlling circadian rhythms. The efferent pathways conducting the circadian signal are known and the signal itself is described in some detail at different levels within the central nervous system. Good candidates for the site of the circadian pacemaker and feedback pathways are presented here, and a plausible model exists for the assembly and coupling mechanisms of multiple oscillators. This synthesis will continue to improve through concerted application of anatomical, physiological, and behavioral analysis and pharmacology and field work.

SUMMARY

The neuronal elements of the scorpion's circadian clock system are described based on neuroanatomical, neurophysiological, and pharmacological experiments. Zeitgeber perception is mediated by the highly sensitive and non–image-forming lateral eyes and also by the median eyes, which seem to have different pathways for visual information (via the spiking arhabdomeric cells) and for time information (via the nonspiking axons of the photoreceptor cells). Efferent neurosecretory fibers in the visual system with known central course and terminal synapses in optic neuropils and central body conduct the circadian signal from the pacemakers (located near the higher optic neuropils) to the different effector systems. During subjective night, low-frequency electric impulse activity in the ENSFs causes release of the transmitter octopamine from their terminals in the retinae. This stimulates retraction of the shielding pigment in the retinal photoreceptor cells onto the soma and extension of the rhabdoms toward the preretinal membrane, thus increasing photosensitivity by about 3–4 log units. Immunocytology with antibodies against a clock gene product of the fruit fly and against serotonin reveals sets of neurons, which may be in the coupling, and feedback pathways of the circadian clock system.

Acknowledgments We thank Frau Elke Noering for her great efforts to keep the scorpions as happy as possible in our animal house over the years and Karl-Heinz Schmitt who helped everywhere in the lab and in the library. We thank Adam Lockett (Adelaide) for improving our English writing. We are grateful to all our students and co-workers who contributed to the "clock" story for scorpions. We especially acknowledge those students who contributed otherwise unpublished data from their theses to various parts of this work: Susan Bhatti (octopamine pharmacology), Heike Hess (serotonin immunohistology), Stefan Heinrichs (ultrastructure of the ENSF), and Annerose Marschall-Pauly (electrical simulation of the circadian signal). The studies of PER-immunoreactivity were done in collaboration with our late colleague Brigitte Frisch, Brandeis University (supported by grants from the Max-Kade-Foundation to B.F. and the U.S. NIH [GM-33205] to J. C. Hall, Brandeis University). Our research and the many guests working on chronobiology of scorpions in our laboratory were supported by various grants from the Deutsche Forschungsgemeinschaft, the DAAD, and the Humboldt-Foundation. The scorpions used in these studies were collected in Tunesia (with permission of the Ministry of Agriculture, the General Director of Forests), in India (Mani Manuel, Madurai), and in the United States (Deep Canyon Research Station of USC, Riverside). We thank all these institutions for their generous support.

References

Battelle, B.A. (1991) Regulations and retinal functions by octopaminergic efferent neurons in *Limulus*. *Prog. Retinal Res.* 10: 333–355.

Block, G.D., Khalsa, S.B.S., McMahon, D.G., Michel, S. and Geusz, M. (1993) Biological clocks in the retina: cellular mechanisms of biological timekeeping. *Int. Rev. Cytol.* 146: 83–144.

Bowerman, R.F. (1976) Ion concentrations and pH of the hemolymph of the scorpion *Hadrurus arizonensis* and *Paruroctonus mesaensis*. *Comp. Biochem. Physiol.* A54: 331–333.

Brandes-Frisch, B., Fleissner, G., Fleissner, G. and Hall, J.C. (1990) "per"-Reactive neurons in the CNS of scorpion and beetle. *Verh. Deutsch. Zool. Ges.* 83: 630–631.

Dunlap, J.C. (1999) Molecular basis for circadian clocks. *Cell* 96: 271–290.

Edgar, D., Martin, C.E. and Dement, W.C. (1991) Activity feedback to the mammalian circadian pacemaker: influence on observed measures of rhythm period length. *J. Biol. Rhythm* 6: 185–199.

Fleissner, G. (1971) Über die Sehphysiologie von Skorpionen. Belichtungspotentiale in den Medianaugen von *Androctonus australis* L. und ihre tagesrhythmischen Veränderungen (Inaugural dissertation). JWG-University, Frankfurt/Main, Germany.

Fleissner, G. (1972) Circadian sensitivity changes in the median eyes of the North African scorpion *Androctonus australis*. *In* Wehner, R. (ed.) *Information Processing in the Visual System of Arthropods*, pp. 133–139. Springer Verlag, Berlin.

Fleissner, G. (1974) Circadiane Adaptation und Schirmpigmentverlagerung in den Sehzellen der Medianaugen von *Androctonus australis* L. (Buthidae, Scorpiones). *J. Comp. Physiol.* 91: 399–416.

Fleissner, G. (1977a) Entrainment of the scorpion's circadian rhythm via the median eyes. *J. Comp. Physiol.* A118: 93–99.

Fleissner, G. (1977b) Scorpion lateral eyes: extremely sensitive receptors of Zeitgeber stimuli. *J. Comp. Physiol.* A118: 101–108.

Fleissner, G. (1977c) The absolute sensitivity of the median and lateral eyes of the scorpion *Androctonus australis* L. (Buthidae, Scorpiones). *J. Comp. Physiol.* A118: 109–120.

Fleissner, G. (1977d) Differences in the physiological properties of the median and the lateral eyes and their possible meaning for the entrainment. *J. Interdisc. Cycle Res.* 8: 15–26.

Fleissner, G. (1983) Efferent neurosecretory fibers as pathways for the circadian clock signals in the scorpion. *Naturwiss.* 70: 366.

Fleissner, G. (1985) Intracellular recordings of light responses from spiking and nonspiking cells in the median and lateral eyes of the scorpion. *Naturwiss.* 72: 46–48.

Fleissner, G. and Fleissner, G. (1978) The optic nerve mediates the circadian pigment migration in the median eyes of the scorpion. *Comp. Biochem. Physiol.* A61: 69–71.

Fleissner, G. and Fleissner, G. (1985) Neurobiology of a circadian clock in the visual system of scorpions. *In* Barth, F.G. (ed.) *Neurobiology of Arachnids*, pp. 351–375. Springer Verlag, Berlin.

Fleissner, G. and Fleissner, G. (1986) Circadian rhythms in the compound eyes of the carabid beetle *Pachymorpha (Anthia) sexguttata* I. Sensitivity-rhythms and the bilateral circadian oscillator system. *In* den Boer, P.J., et al. (eds.), *Carabid Beetles: Their Adaptations and Dynamics*. pp. 3–17. Gustav Fischer, Stuttgart.

Fleissner, G. and Fleissner, G. (1988) Efferent control of visual sensitivity in arthropod eyes: with emphasis on circadian rhythms. *In* Lindauer, M. (ed.) *Information Processing in Animals*. 5, pp. 1–67. Gustav Fischer Verlag, Stuttgart.

Fleissner, G. and Fleissner, G. (1992) Feedback loops in the circadian system. pp. 79–84. *In* Zatz, M. (ed.) *Discussion in Neuroscience*, FESN Study Group, VIII Circadian Rhythms. Elsevier, Amsterdam.

Fleissner, G. and Fleissner, G. (1993) Seeing time. *In* Wiese, K., Gribakin, F.,

Popov, A.V. and Renninger, G. (eds.) *Sensory Systems of Arthropods,* pp. 288–306. Birkhaeuser, Basel.

Fleissner, G. and Fleissner, G. (1998) Natural photic Zeitgeber signals and underlying neuronal mechanisms in scorpions. *In* Y. Touitou (ed.) *Biological Clocks. Mechanisms and Applications,* pp. 171–180. Elsevier Science, Amsterdam.

Fleissner, G. and Heinrichs, S. (1982) Neurosecretory cells in the circadian clock system of the scorpion *Androctonus australis. Cell Tiss. Res.* 224: 233–238.

Fleissner, G., Heinrichs, S., Marschall-Pauly, A. and Fleissner, G. (1989) Mechanismen und zelluläre Bausteine der circadianen Uhr. *Verh. Deutsch. Zool. Ges.* 83: 630–631.

Fleissner, G. and Schliwa, M. (1977) Neurosecretory fibers in the median eyes of the scorpion *Androctonus australis* L. *Cell Tissue Res.* 178: 189–198.

Fleissner, G., Stevenson, R.D. and Fleissner, G. (1987) Circadian systems in invertebrates (including an evolutionary perspective of circadian signal transmission in the visual system). *In* Scharrer, B., Korf, H.W. and Hartwig, H.G. (eds.) *Functional Morphology of Neuroendocrine Systems. Evolutionary and Environmental Aspects,* pp. 87–104. Springer, Berlin.

Foelix, R.F. (1992) *Biologie der Spinnen.* 2. Aufl. Thieme, Stuttgart-New York.

Fowler, D.J., Goodnight, C.J. and Labrie, M.M. (1972) Circadian rhythms of 5-hydroxytryptamine (serotonin) production in larvae, pupae and adults of *Drosophila melanogaster* (Diptera, Drosophilidae). *Ann. Entomol. Soc.* 6: 138–141.

Frisch, B., Fleissner, G., Fleissner, G., Brandes, C. and Hall, J.C. (1996) Staining in the brain of *Pachymorpha sexguttata* mediated by an antibody against a *Drosophila* clock-gene-product: labeling of cells with possible importance for the beetle's circadian rhythms. *Cell Tissue Res.* 286: 411–429.

Garcia-Fernandez, J.M., Siwicki, K.K. and Foster, R.G. (1994) *per*-like immunoreactive neurons within the brain of *Xenopus* innervate both retina and pineal organ. *In* Moller, M. and Pevet, P. (eds.) *Advances in pineal research.* Vol. 8, pp. 19–28, John Libbey & Co., London.

Gilai, A. and Parnas, J. (1970) Neuromuscular physiology of the closer muscles in the pedipalp of the scorpion *Leiurus quinquestriatus. J. Exp. Biol.* 52: 325–344.

Goodman, L.Y. (1981) Organisation and physiology of the insect dorsal ocellar system. *In* Autrum, H. (ed.) *Comparative Physiology and Evolution of Vision in Invertebrates. C. Invertebrates Visual Centres and Behaviour II,* pp. 201–286. Springer Verlag, Berlin.

Green, C.B. (1998) How cells tell time. *Trends in Cell Biology* 8: 224–230.

Harris, J.W. and Woodring, J. (1992) Effects of stress, age. season and source colony on levels of octopanine, dopamine and serotonin in the honeybee (*Apis mellifera* L.) brain. *J. Insect. Physiol.* 38: 29–35.

Harrison, J.B. and Lane, N.J. (1981) Lack of restriction at the blood-brain interface in *Limulus* despite atypical junctional arrangements. *J. Neurocytol.* 10: 233–250.

Heinrichs, S. (1983) Identification of neurons in the transmission electron microscope after retrograde labelling with the dye Lucifer Yellow. *Mikroskopie* 40: 79–86.

Heinrichs, S. (1985) Neuronale Komponenten der circadianen Uhr des Skorpions. Anatomie und Verschaltung efferenter neurosekretorischer Neurone im Zentralnervensystem. (inaugural dissertation). JWG-University, Frankfurt/Main, Germany.

Heinrichs, S. and Fleissner, G. (1987) Studies on neuronal components of the scorpion's circadian clock: I. Central anatomy of efferent neurosecretory fibers. *Cell Tiss. Res.* 250: 277–285.

Kass, L. and Barlow, R.B. (1984) Efferent neurotransmission of circadian rhythms in *Limulus* lateral eye. I. Octopamine-induced increases in retinal sensitivity. *J. Neurosci.* 4: 908–917.
Koehler, W.K. and Fleissner, G. (1978) Internal desynchronisation of bilaterally organized circadian oscillators in the visual system of insects. *Nature* 274: 708–710.
Konopka, R.J. and Benzer, S. (1971) Clock mutants in *Drosophila melanogaster. Proc. Natl. Acad. Sci. USA* 68: 2112–2116.
Lane, N.J., Harrison, J.B. and Bowerman, R.F. (1981) A vertebrate-like blood-brain barrier, with intraganglionic blood channels and occluding junctions, in the scorpion. *Tissue Cell* 13: 557–576.
Luettgen, M.A. (1993) Entrainment der circadianen Laufrhythmik durch Licht-zeitgeber: Untersuchung biologisch relevanter Lichtparameter am Beispiel der Lokomotionsrhythmik von *Androctonus australis* (Scorpiones, Buthidae) (Inaugural dissertation). JWG-University, Frankfurt/Main, Germany.
Marschall-Pauly, A. and Fleissner, G. (1987) Electrical simulation of a circadian signal in the visual system of the scorpion. *Chronobiol.* 14: 204
Michel, S. (1989) Laufaktivitaet und Augenempfindlichkeit von Skorpionen: Untersuchungen der Kopplung in einem circadianen Multioszillatorsystem. (Inaugural dissertation). JWG-University, Frankfurt/Main, Germany.
Michel, S., Geusz, M.E., Zaritzky, J.J. and Block, G.D. (1993) Circadian rhythm in membrane conductance expressed in neurons. *Science* 259: 239–241.
Moustafa, A. (1995) Das circadiane System des Skorpions: elektrophysiologische Analyse des peripheren und zentralen circadienen Signals in den efferenten neurosekretorischen Fasern des optischen Systems. (Inaugural dissertation). JWG-University, Frankfurt/Main, Germany.
Naessel, D.R. and Elekes, K. (1985) Serotonergic terminals in the neural sheath of the blowfly nervous system: electron microscopical immunocytochemistry and 5, 7-dihydroxytryptamine labelling. *Neuroscience* 15: 293–307.
Orchard, I. and Lange, B. (1987) Octopamine in insects: control of hemolymph lipid and visceral muscle in locusts. *Am. Chem. Soc. Symp.* ser. 356: 136–153.
Roenneberg, T. and Hastings, J.W. (1991) Are the effects of light on the phase and period of the Gonyaulax clock mediated by different pathways? *Photochem. Photobiol.* 53: 525–530.
Schliwa, M. and Fleissner, G. (1980) The lateral eyes of the scorpion *Androctonus australis. Cell Tiss. Res. 206:* 95–114.
Siwicki, K.K., Eastman, C., Petersen, G., Rosbash, M. and Hall, J.C. (1988) Antibodies to the *period* gene product of *Drosophila* reveal diverse tissue distribution and rhythmic changes in the visual system. *Neuron* 1: 141–150.
Siwicki, K.K., Strack, S.T., Rosbash, M., Hall, J.C. and Jacklet, J. (1989) An antibody to the *Drosophila* period protein recognizes ciracdian pacemaker neurons in *Aplysia* and *Bulla. Neuron* 3: 51–58.
Takahashi, J.S. (1992) Circadian clock genes are ticking. *Science* 258: 238–240.
Zimmerman, H.H. and Menaker, M. (1975) Neural connection of sparrow pineal: role in circadian control of activity. *Science* 190: 477–479.

5

The Scorpion's Clock

Feedback Mechanisms in Circadian Systems

GÜNTHER FLEISSNER
GERTA FLEISSNER

An animal's circadian timing system synchronizes its biological functions and behavior with daily fluctuations of important factors in its external environment (reviewed by Aschoff, 1981). As a familiar example, the circadian clock in nocturnal animals, sends them to sleep around dawn and arouses them again around dusk. In principle, a circadian timing system would be composed of four elements as illustrated in fig. 5.1: (1) an oscillator, with a period of about 1 day, (2) a Zeitgeber, or synchronization input for phase adjustment of the oscillator, (3) output machinery for effecting biological functions, and (4) coupling pathways between these components. This simple scheme for organization of circadian systems has, among chronobiologists, become almost dogmatic. It is looked at as a one-way flow of information from the Zeitgeber input (usually photoreceptors) through the oscillator to the clock-driven process (e.g., locomotor activity). The focus of this chapter is to describe the multiple types of feedback flow occurring at the systemic level. Examples of the scorpion's circadian system are used to demonstrate the fundamental significance of these regulatory mechanisms. A schematic overview of these feedback loops is given in fig. 5.13.

As in other systems that oscillate, feedback interactions are often involved as mechanisms for generating the circadian rhythm at a cellular and subcellular level only. Recently this was discovered also at an organismic level where feedback from the effector output (e.g., locomotor activity) to the pacemaker has impact on circadian period length (Hohmann and Fleissner, cited in Mrosovsky et al., 1989; Turek, 1989; Hohmann et al., 1990; Edgar et al., 1991; Fleissner and Fleissner, 1992). Consideration of feedback effects at this level began 15 years earlier when Aschoff et al. (1973) observed that hamsters with access to a running wheel had circadian period lengths different from animals without such access to a wheel and consequently less locomotor activity, but these ideas had been largely forgotten.

The existence of feedback loops at all organizational levels within circadian oscillatory systems is consistent with the ubiquity of servo-control feedback in other regulated processes. While considering locomotor activity as an important determinant of circadian regulation, it is noteworthy that other parameters indicated the existence of feedback loops at the organismic level long ago. Kiesel (1894) first proposed that the photic sensitivity of insect compound eyes is controlled by "endogenous daily rhythms" (i.e., the circadian clock)—a term not then introduced. When, on the other hand, the eyes also function as input for photic Zeitgeber stimuli to the clock, this endogenously controlled adaptation

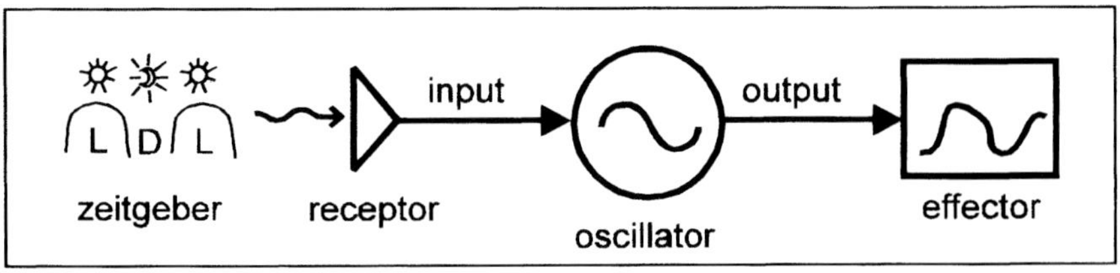

Figure 5.1 Simplified and commonly used flow chart of a circadian clock. Zeitgeber information (L/D/L) originating from periodic change of environmental illumination (L, light; D, dark) at dusk and dawn enters a photoreceptive path as input. Zeitgeber input shifts the phase of the self-sustained oscillator according to the expected phase relation between the circadian rhythm and the Zeitgeber periodicity (see fig. 5.7). The oscillator's output controls effector systems such as locomotor activity or visual sensitivity (see fig. 5.2).

provides a feedback loop through which the clock controls its own input. In the 1970s, the neurobiological basis of this efferent control of visual sensitivity became a subject of intensive research (for arthropods, see Fleissner and Fleissner, 1988), and the scorpion was one of the first animals in which it was shown that circadian rhythms in the eyes feed back on the sensitivity of their Zeitgeber inputs (Fleissner, 1972, 1975, 1977a). Another clear case of an oscillator feeding back to its own Zeitgeber input is found in the behavior of nocturnal animals such as squirrels (DeCoursey, 1986, 1989), rats (Terman et al., 1991), or scorpions (see below): by leaving their burrows at dusk, and reentering them at dawn, these animals expose themselves only to short periods of twilight (see fig. 5.6). Thus, clock-induced locomotor activity produces its own photoperiod, where external day and night are replaced by two "light pulses," each about 1 h long and interrupting continuous darkness at dawn and dusk (skeleton photoperiod).

As this concept of feedback has fundamental value for understanding the organization of circadian systems in general, current research focuses on mechanisms and pathways of locomotory feedback loops. Mammals such as hamsters (Mrosovsky et al., 1992), mice (Edgar et al., 1991), and arthropods, particularly scorpions (Fleissner and Fleissner, 1985, 1992) provide excellent model systems for this field of research. Since the discovery of a "paradoxic phase-response curve " (see below) in our laboratory, the findings from scorpion studies have substantially contributed to knowledge in this field (Mrosovsky et al., 1989). These results have broadened the scope of research on circadian systems (which otherwise remains largely a mammalian enterprise) and stimulated investigations concerning the significance of locomotor activity for the human clock (Schmidt et al., 1990).

A great advantage of scorpions as a model for research on the circadian system is their ability to sustain activity for long periods without food or water (Cloudsley-Thompson, 1956; Hadley, 1990). Chronobiological experiments often continue for several days or weeks, during which time the animal should not be disturbed by feeding. Desert scorpions, such as *Androctonus australis*, are favored experimental animals in our laboratory because they can survive for more than 100 days without food and water (Cloudsley-Thompson, 1956). Due to its residence in deep burrows by day and strictly nocturnal periods of activity, *Androctonus* spends most of its life in darkness; as such, the optimal experimental conditions for circadian research (constant darkness and constant temperature, uninterrupted for weeks by intake of food or water) come close to this burrower's normal living conditions.

Another advantage of the scorpion as a model for circadian experimentation is its sharply defined and obligatory rhythm of photosensitivity in the eyes. This was demonstrated by continuous electrophysiological recordings lasting more than 8 months (Fleissner, 1971; review in Fleissner and Fleissner, 1988). This ERG (electroretinogram) rhythm plays an important role as a model parameter for circadian system analysis because it marks the "central circadian time" of the system (see below). The scorpion is also the first arthropod in which the pathways of the circadian signal controlling visual sensitivity have been described to the level of identifiable neurons (see chapter 4, Fleissner and Schliwa, 1977; Fleissner and Fleissner, 1978; Fleissner and Heinrichs, 1982; Fleissner, 1983). The scorpion is the first creature in which a paradoxic photo–phase-response curve (PRC) has been observed. This observation has finally lead us to the hypothesis of feedback at the level where locomotor activity evidently affects the circadian system in a way not previously understood (Hohmann et al., 1990; Fleissner and Fleissner, 1992).

In this chapter we report five examples of feedback within the circadian system of scorpions. Most of these are concerned with Zeitgeber perception required for correct phase angle between the temporal organization of body functions and the periodicity in its environment.

ERG RHYTHM: THE MARKER FOR SUBJECTIVE TIME

Central Circadian Time

Time of day in our environment as it is caused by the earth's rotation is always unequivocal: at any given moment only one time of day is valid. Because of their complexity as multi-oscillator assemblies (see below), circadian systems are at risk of losing their internal time coherency by internal desynchronization, when Zeitgeber and internal coupling mechanisms are not effective. For example, subjective day and subjective night can then coexist within different parts of the same brain at a given moment (Koehler and Fleissner, 1978). Therefore, they must be phase locked to the 24 h environmental periodicity (day and night) which is the ultimate external reference for a circadian system. It must have its own internal reference timer—the master pacemaker—which finally controls the phase relationship within the entire system (Pittendrigh, 1974; Moore-Ede and Sulzman, 1981). In other words, a "central circadian time" must exist within the organism.

ERG Rhythms

How can one assess this central circadian time from the animal's behavior? Spontaneous locomotor activity of a nocturnal animal is evidence of its "subjective night." However, an animal can be active for many different reasons, and differently motivated activities might then be optimal at different times within the daily cycle. In addition, the response pattern to a given motivation can change during the year, even from nocturnal to diurnal and back. Thus, spon-

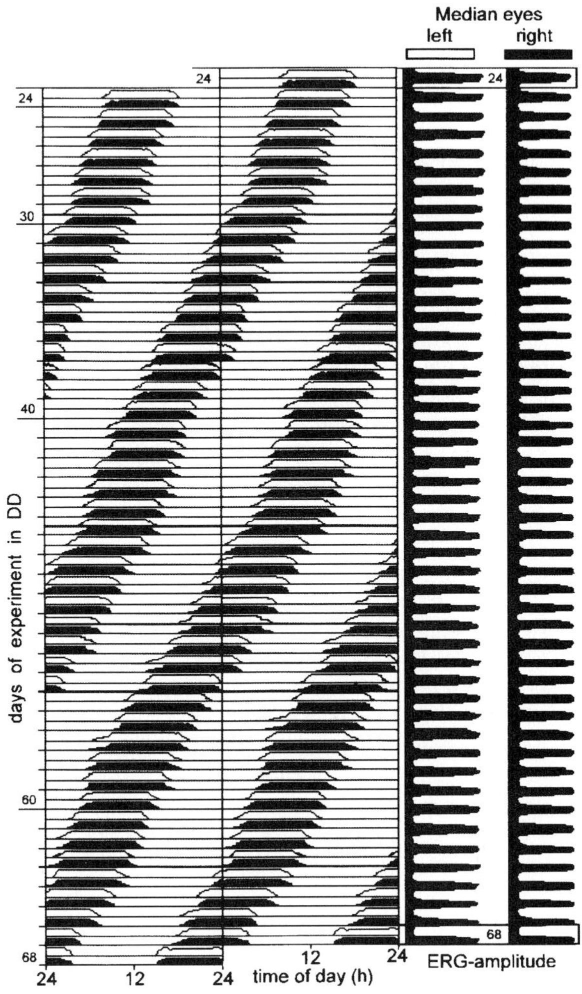

Figure 5.2 Circadian rhythm of ERG in left and right median eyes of the scorpion *Androctonus australis.* Left panel shows phase plot of ERG rhythm for successive days under conditions of continuous darkness and constant temperature (about 28°C) (double plot: each day is plotted twice). The right panel displays a continuous recording of the same ERG responses to a 10-msec light flash of the same intensity applied every 30 min throughout the experimental period. Note that ERG response varies strikingly over 24-h periods: large ERG amplitude indicates subjective night, a period of about 12 h when the eye is approximately 1000 times more sensitive to light than during subjective day. When sensitivity is assayed as ERG response amplitude and plotted on a 48-h scale (left panel), the precise alternation of subjective night and subjective day is easily seen. If consecutive 48-h periods are plotted one below the other, a phase plot is generated from which the variation in period length, τ, and its deviation from 24-h can clearly be recognized and measured. The endogenous rhythm is free running in this example (continuous darkness), and τ is shorter than 24 h. To run exactly at 24 h, τ must be synchronized by Zeitgeber input (see fig. 5.4A). After data from Michel (1989).

taneous activity produces a rather irregular pattern, preventing the experimenter from recognizing the animal's central circadian time precisely and reliably. To analyze circadian organization, it is advantageous to find additional related parameters, such as different types of locomotion, or completely different ones such as body temperature.

Circadian rhythmicity in the eyes, which persists in constant darkness and anticipates the daily changes of light, is among the best indications of central circadian time because visual function is more directly linked to the cyclic appearance of daylight than any other parameter (fig. 5.2; Fleissner, 1972). The clock's effect on pigment cells in the retina shelters the highly sensitive photoreceptors against damaging daylight (Terman et al., 1991), periodically reducing the incident light flux to the receptors by raising thick pigmentary screen, or "internal sunglasses" (review in Fleissner and Fleissner, 1988; see also the scheme in figs. 5.4 and 5.5 and chapters 3 and 4). Exposure to daylight, as it increases and decreases, is the only factor to which the time course of this sunglass-rhythm must accommodate. Consequently, eye rhythms precisely mirror the daily periodicity of the photic environmental fluctuations (fig. 5.3). The amplitude of ERGs is a convenient measure of the eye's state of sensitivity and a preferred indicator of the animal's central circadian time. ERGs are recorded noninvasively through electrodes attached to the cuticular surface of the eye lenses and the body. Scorpions and other arthropods tolerate these miniature wires indefinitely, thus allowing for lifelong and precise quantitative measurements. The scorpion ERG shows an unusually clear and robust circadian rhythmicity, which can be recorded without disturbance for more then a year (Fleissner, 1974; Fleissner and Fleissner, 1985) (fig. 5.2). These attributes made scorpions one of the first experimental models useful for analysis of the neuronal organization of a circadian clock system (see chapter 4; Fleissner, 1971, 1972; Fleissner and Fleissner, 1988). ERG rhythmicity is particularly promising for analyzing the clock because it appears to be more directly connected to the oscillator output than locomotory activities.

FEEDBACK

Feedback from the Oscillator to Its Photic Zeitgeber Input

Circadian rhythms must be synchronized by periodic timing cues, or Zeitgeber stimuli, of which natural photoperiodic fluctuations in the environment are most relevant and easily manipulated (Aschoff, 1981; Pittendrigh, 1981). In scorpions, both median and lateral eyes function as Zeitgeber inputs (fig. 5.4A; Fleissner, 1977a,b). Because the median eyes' sensitivity is modulated by the circadian pacemaker (chapters 3 and 4), a feedback loop is established where the pacemaker controls its own Zeitgeber input (fig. 5.4B; see loop no. 1 in fig. 4.13). The functional significance of this loop for vision is clearly to adjust the dynamic range of photo reception to the actual light conditions. It also offers an important cue for the perception of natural Zeitgeber programs. It may occlude Zeitgeber information during twilight hours due to its sensory compensation of light change during dusk and dawn (sensitivity increases (or decreases) as light decreases (or increases, fig. 5.3B,C). In spite of this obvious disadvantage, this loop may even sharpen the Zeitgeber time course whenever there is a phase mis-

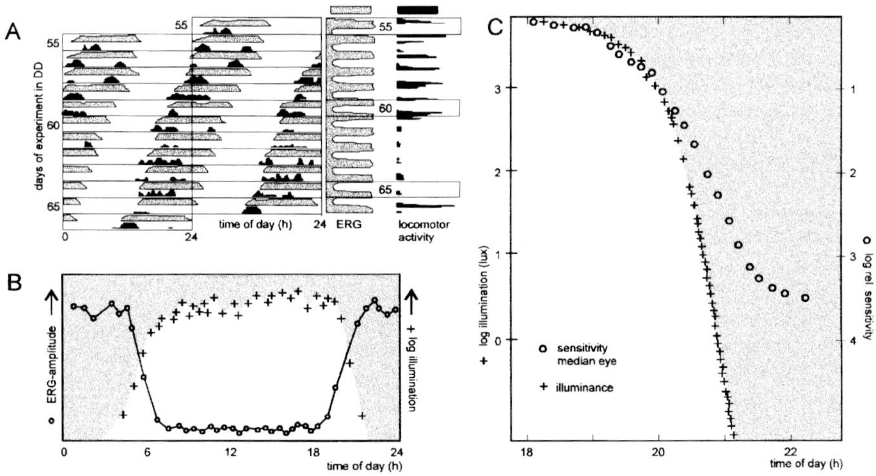

Figure 5.3 The ERG rhythm marks "central circadian time." The ERG rhythm is the best indicator of the scorpion's central circadian time because it mirrors most closely the time course of the change between day and night. (A) Example of simultaneous recording of ERG (stipling) and locomotor activity (black) demonstrates that the ERG amplitude indicates the daily periodicity of illumination more precisely than does the pattern of spontaneous locomotor activity. (Note: right panel of A shows the data over the continuous time scale and at an absolute amplitude scale, whereas the data in the left panel are on a relative scale normalized in each line to the highest value.) Data from Michel (1989). (B) Compensatory changes in ERG amplitude (o) under natural fluctuations of illumination (+) are synchronous with the light regime, even when the daytime light is attenuated to less than 1 mlux: as the light increases at dawn, ERG sensitivity decreases and this process reverses at dusk. Data from Fleissner (unpublished). (C) Inverted plot of median eye sensitivity under continuous darkness (DD) (o) shows how precisely it follows the decrease in illumination (+) at dusk (plotted on the same time and logarithmic intensity scales) (illumination measured in summer at latitude where *Androctonus australis* lives). If direct light adaptation is additionally taken into account, the match would be even more striking. Data from Fleissner (1977c).

match between the circadian rhythm and the Zeitgeber cycle (Fleissner, 1977c). For example, if subjective night starts before dusk, the sensitivity of the median eyes will increase earlier and faster than light intensity decreases. The afferent neural signal to the clock will then report "light is increasing," though it is actually decreasing. In this context note that, while the median eyes are strongly controlled by the circadian clock, sensitivity of the lateral eyes is almost free of circadian modulation. Afferent information from both types of eyes could, if compared at a more central stage of the neural pathway, provide special Zeitgeber information (see chapter 4).

Feedback from the Oscillator to Zeitgeber Processing

The feedback loops (no. 2 in Fig. 5.13) involving the efferent neurosecretory fibers (ENSF, see chapter 4) may supply "reafference" in the afferent pathway

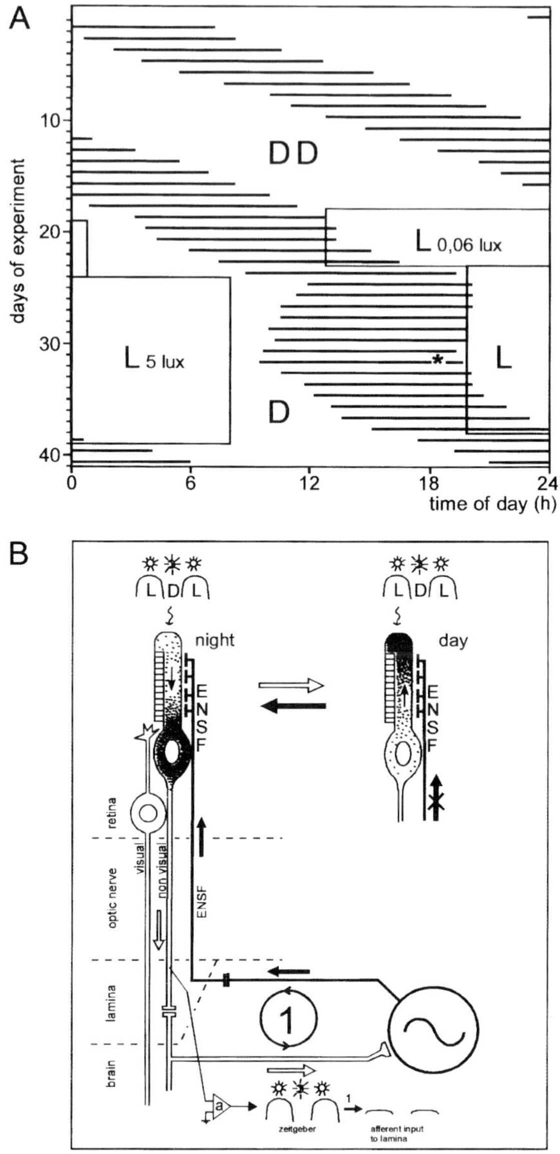

Figure 5.4 Feedback from the circadian oscillator onto its Zeitgeber input via the median eye. (A) The circadian ERG rhythm (black bars represent subjective night) in one median eye, free running in constant darkness (DD) with τ longer than 24 h, is synchronized by light/dark cycle (LD) with L of about 5 lux (days 24–38) via the contralateral median eye; 60 mlux (days 18–23) in this case produced only modest relative coordination. When the contralateral eye was covered by opaque wax (asterisk on day 32), the synchronization was lost. Data from Fleissner (1977a). (B) Schematic representation of a receptor cell in two circadian states (day and night) with its afferent and efferent neuronal connections to the circadian oscillator and the screening pigment in the respective position (proximally

of the median eyes of the scorpion (von Holst and Mittelstaedt, 1950; Fleissner and Fleissner, 1993, 1998). We have seen that the median eyes mediate photic Zeitgeber signals and that the pacemaker changes the photosensitivity of these eyes in a circadian manner. Due to this sensitivity change, the eyes cannot accurately inform the pacemaker about illumination changes throughout the daily cycle. Consequently, adaptation compensates for the slow changes of illumination, which could otherwise give information on the time of day. Direct light/dark adaptation (which is not considered here) may even multiply this compensatory effect. Thus, the visual channel must be more or less "time blind."

Photic information from the environment takes two paths to the brain (Fleissner, 1985). One is via the nonspiking photoreceptors directly into the first optic neuropil, the lamina. The other path enters the nervous system via second order neurons, the arhabdomeric cells, which project to the second optic neuropil or medulla. The ENSF axons, which carry the circadian signal to the eyes, also branch to the lamina, but not into the medulla where visual image processing occurs. Thus, two photic channels may exist in parallel: a visual pathway for image processing and a nonvisual input for Zeitgeber perception.

It is possible to infer a function for feedback loop 2 of Zeitgeber perception: as shown in figure 5.4B, afferent information about the photic environment is masked by ENSF-controlled circadian adaptation of the retina (ERG rhythm) and passes to the lamina. Under these circumstances information about the level of brightness is largely lost. On the other hand, information about the signal driving the circadian adaptation at the retina is also available in the lamina through branches of the ENSF pathway. We hypothesize that appropriate information processing at the level of the lamina may correct the afferent light/dark (LD) signal for the effect of circadian adaptation (fig. 5.5). This feedback loop may thus supply the clock with information about the absolute level of light intensity, information that is principally not available in any visual channel but is needed by the clock to recognize time reference in its photic environment.

Feedback from the Oscillator on the Scorpion's Exposure to Zeitgeber Stimuli

Another example of a feedback loop (no. 3 in Fig. 5.13) becomes evident if we consider the time course of the scorpion's exposure to environmental light throughout the daily cycle. With few exceptions, scorpions are strictly nocturnal animals residing in relatively dark burrows or shelters during daylight hours

during night, distally during day). Feedback loop 1: the oscillator, via the efferent neurosecretory fibers (ENSFs) (black arrows), causes the screening pigment within the receptor cells to migrate proximally, thus allowing the incident light (L/D/L) free access to the rhabdoms (striped part; subjective night). When the efferent clock signal is absent, shielding pigment migrates back to its distal position over the rhabdom and reduces incident photon flux by a factor of 1000 (subjective day). The afferent signal of the external light (white arrow, LD) is thus modulated by the clock. As a result, Zeitgeber information to the pacemaker (cartoon in lower part of the figure) is masked by output from the clock (loop 1). The afferent information then does not fully correspond to the environmental cycle of light and dark.

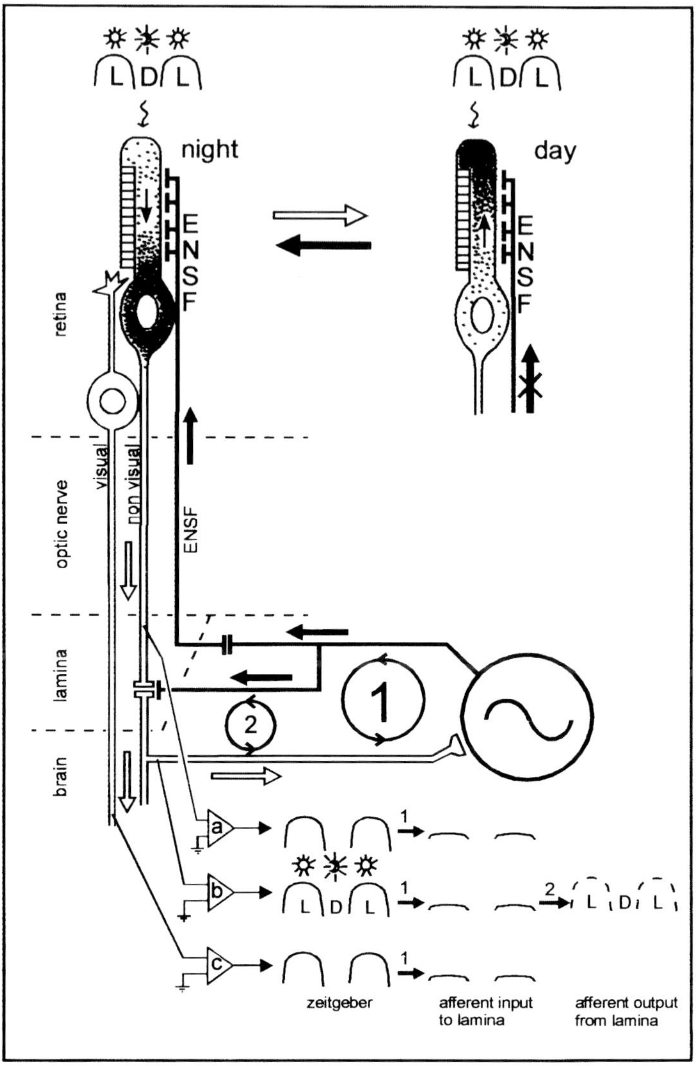

Figure 5.5 Feedback (loop 2) from the central oscillator onto neuronal pathways for Zeitgeber processing. The feedback loop represented in fig. 5.4B is further modified by neural processes (loop 2) from the efferent neurosecretory fibers (ENSFs) carrying a copy of the efferent clock signal to the lamina neuropil (black arrow), putatively a nonvisual photic channel. This second loop could largely nullify the clock-driven Zeitgeber masking effect at the photoreceptor level in the retina. Via this channel a "corrected" LD (light/dark) signal could be available to the central oscillator (dashed line in cartoon b). The visual channel from the arhabdomeric cells in the retina bypass the lamina and hardly reports on the external photic timing signal (cartoon c).

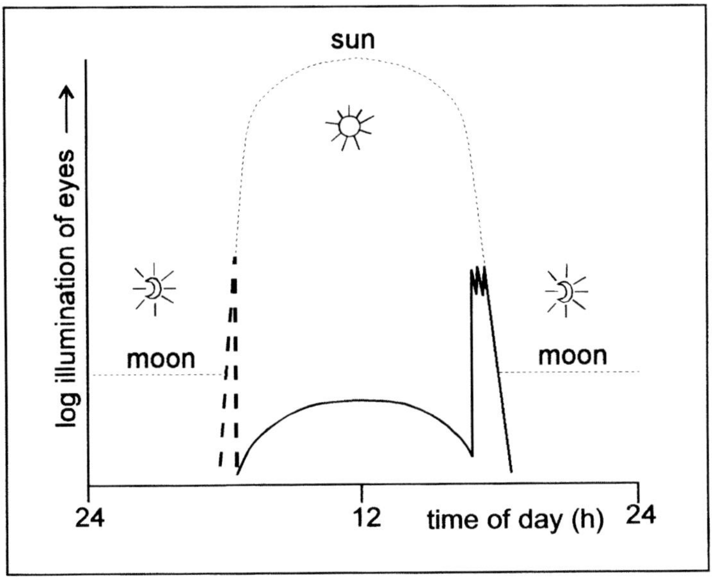

Figure 5.6 Feedback (loop 3) of clock-controlled behavior onto Zeitgeber exposure of the eyes during 24 h. Under simulated natural conditions scorpions experience a different regime of light and dark (solid line) than that observed during day and night outside the burrow (stippled line). During night the scorpion is outside its burrow where it is exposed only to moonlight and starlight (horizontal dotted line). Long before dawn approaches, the scorpion usually retreats into its burrow but may experience some increase in ambient light before reentering the burrow (dashed line). In its shelter it is only exposed to very dim light. At dusk the scorpion wakes up and gradually approaches the exit so that light exposure increases as the scorpion moves and decreases as the night falls. When outside illumination is about 1 mlux, the scorpion leaves its burrow and experiences diminishing illumination until full night. The light exposure produced by this entering and leaving its burrow is the "skeleton photoperiod." After Fleissner and Fleissner (1998).

and only emerging in nearly complete darkness. Infrared time-lapse video recordings (of *Androctonus,* 1993) show that, driven by their circadian clock, the scorpions approach the exit of their dark shelter about 1 h before dusk, exiting later, but not before light outside has decreased to intensities well below 1 lux (ca. 0.04 $\mu W/cm^2$). Thus, the scorpion's eyes are only exposed to light for a shorter time (fig. 5.6). They receive a skeleton photoperiod. The feedback loop is established by clock-controlled behavior, which in turn controls the light exposure of the clock's Zeitgeber input. Though we do not yet know the underlying neuronal wiring, it must be investigated whether the input and output synapses of the ENSFs in the central body (cf. chapter 4) may deliver a reafferent loop to control the synchrony of clock-driven surface activity and perceived light program, similar to loops 1 and 2.

Feedback from Locomotor Activity to the Oscillator

The paradoxic photo–phase-response curve

In keeping with every other circadian clock, the scorpion's circadian system reacts to light/dark stimuli (Zeitgeber) by shifting its phase. Analysis of this effect reveals the well-known photic PRC (fig. 5.7). This PRC has the same shape for

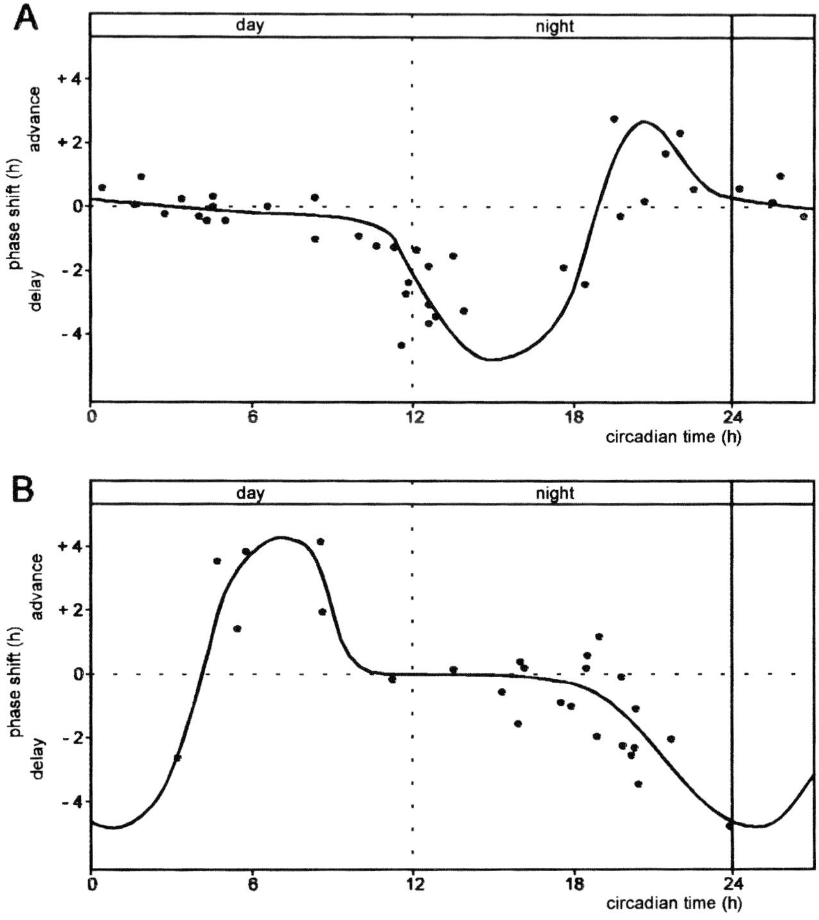

Figure 5.7 Photic phase–response curves (PRC) of the ERG rhythm. Phase shift produced by light pulses breaking continuous darkness (DD) at a certain circadian time (advance and delay) is plotted over circadian time scale (subjective night starts at circadian time 12). (A) Light pulses of about 4 h interrupting continuous darkness produce phase shifts when applied during subjective night: delays in the first half and advances in the second half. These data are from scorpions not responding with locomotor activity to the Zeitgeber light pulse. (B) A "paradoxic" photo-PRC recorded from the same animals under the same experimental conditions and the same light treatments as in panel A above, but the scorpions, aroused by the Zeitgeber light stimuli, showed high locomotor activity during light exposure. After Hohmann and Fleissner, published in Mrosovsky (1989).

all known circadian systems whether of plant or human origin. The scorpion, however, can react in two different ways to a photic Zeitgeber: the usual way is to show phase delay at the beginning, phase advance at the end of subjective night, and no reaction at all to light during subjective day (fig. 5.7A). But this pattern is only seen when the scorpion is resting quietly during light exposure. Sometimes, unpredictably, scorpions begin running under the light treatment, and their response to light then shows a most unusual circadian rhythm: large phase advances when exposed to light during its daytime, no reaction in the first half of the night, and a pronounced phase delay in the second half (fig. 5.7B). This is completely unexpected and a paradoxical way for a circadian oscillator to respond to light; so far it has only been observed in scorpions (Hohmann et al., 1990; Fleissner and Fleissner, 1992). This response, which bears some resemblance to a dark-pulse PRC (Boulos and Rusak, 1982), poses many challenging questions which are under investigation in our laboratory. What lies behind this strange and unique light response? Is it the locomotor activity per se, which only is triggered by light, or is it a mechanism, which impacts on light processing triggered by the locomotor activity?

The hypothesis that light-induced locomotor activity causes the paradoxic PRC is attractive for the following reason: any PRC has a dead zone during which the system does not respond with a phase shift to an adequate stimulus. Logically, synchronization is given as long as the periodically repeated phase-shifting stimulus falls into the dead zone (e.g., light occurs during subjective day, which is the dead zone of the normal photic PRC). Thus, any PRC is homeostatically regulating a constant phase-angle difference between a periodic stimulus and the corresponding rhythm of physiological parameters. If we compare the paradoxic PRC with the normal circadian locomotor pattern of the scorpion (fig. 5.8), we find the activity concentrated within the dead zone of this PRC (fig. 5.9). In other words, the circadian phase of spontaneous locomotor activity and the paradoxic PRC fit well (Fleissner and Fleissner, 1992).

The feedback loop

As a consequence of this fit, the paradoxic photo-PRC may reflect the existence of yet another feedback loop: from the pacemaker's output, shown by the overt rhythm of locomotion, a signal feeds back onto the pacemaker mechanisms, which thereby determines the period length (see no. 4 in Fig. 5.13). Evidence exists of spontaneous and unexplained changes of period length in circadian rhythms (Michel 1989). Knowledge of the paradoxic PRC, however, could provide an explanation for this change of τ, at least in the scorpion. In the example of fig. 5.10, when the scorpion is active in the second half of its subjective night, τ is always longer than without activity in that circadian phase. The paradoxic PRC shows a delay portion during the second half of the night (see fig. 5.7B). This supports the idea that the locomotor activity produces the paradoxic PRC. In functional terms this PRC guarantees that spontaneous locomotor activity in the scorpion occurs at the beginning of subjective night. If this paradoxic PRC reflects the pacemaker's phase response to "misphased" locomotor activity, it should delay the phase when activity occurs in the second half of the night. When this delay is repeated from cycle to cycle, an apparently longer τ will result (fig. 5.10)

If locomotor activity per se causes the phase shift, it should be possible to shift the pacemaker by provoking the scorpion to run at the wrong circadian time us-

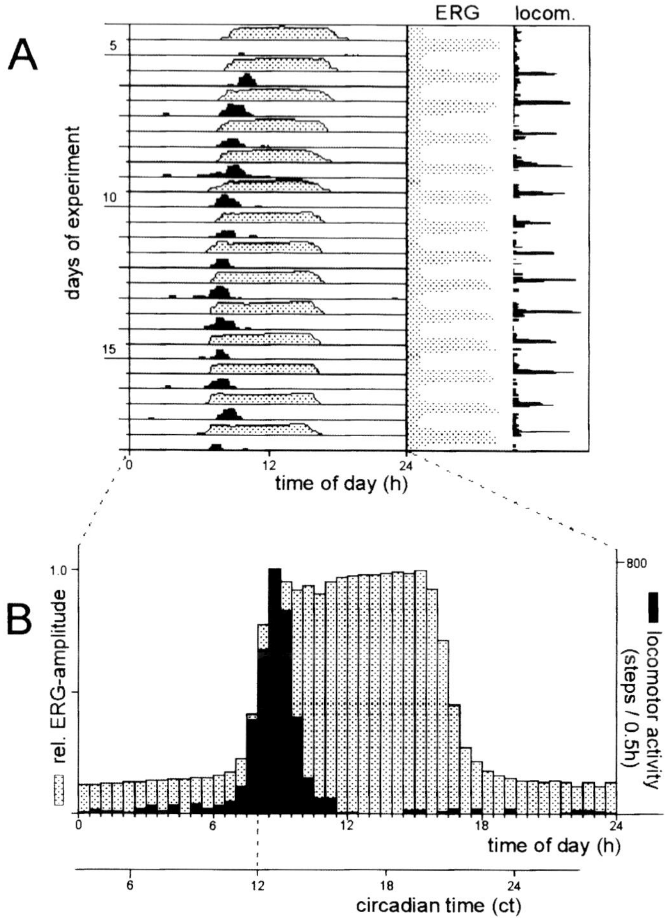

Figure 5.8 Normal distribution of locomotor activity within the circadian cycle. (Upper panel) daily recordings of an undisturbed and unstressed *Androctonus australis* in continuous darkness (DD) show that it is preferentially active only for a few hours at the beginning of each night. (Lower panel) a compilation of these daily activities shows that locomotor rhythm of the animal is synchronous with the beginning of subjective night in its ERG rhythm (an indicator of the animal's central circadian time). After data from Michel (1989).

ing a stimulus other than light. The scorpion can be aroused by blowing puffs of air on it, but this fails to elicit the phase shift. It may be that arousal mechanisms not directly linked to locomotor activity are decisive for the phase reaction of the pacemaker. This view is supported by the observation that scorpions can also show spontaneous activity changes without corresponding τ changes.

The existence of the paradoxic photo PRC and its power to explain spontaneous τ changes is a strong indication for the existence of a feedback loop at this level. Again, the scorpion appears to be a promising experimental preparation for further analysis of this challenging phenomenon, and there is neuro-

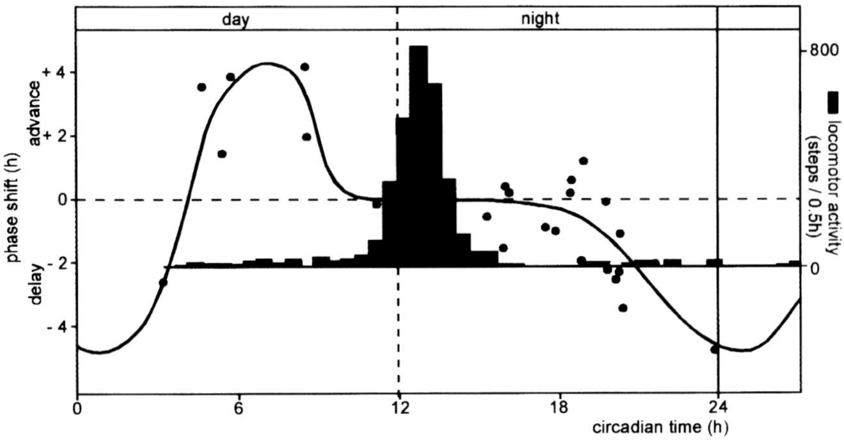

Figure 5.9 The paradox photic phase–response curve (PRC). When the normal pattern of circadian locomotor activity (from fig. 5.8) is superimposed on the phase–response behavior shown in fig. 5.7B, the normal activity pattern fits precisely into the "dead zone" (period of phase shift insensitivity) of this paradoxic PRC. That means that activity occurring outside this circadian time window would, if phase reactive on the basis of this paradoxic PRC, automatically readjust the phase relation between locomotor activity and ERG rhythm (i.e., the central circadian time).

anatomical evidence that serotonin efferents in the photic neuropils may be involved as part of the mechanism (see chapter 4; Fleissner and Fleissner, 1988).

Feedback from Oscillator to Oscillator

Left/right bilateral multioscillator

In addition to their organization as multiple oscillator assemblies with one master and several "slaves" (Pittendrigh, 1974), circadian systems are duplicated bilaterally. The left and right oscillators are equivalent and mutually coupled (Koehler and Fleissner, 1978) so that their mutual entrainment provides a feedback loop between two circadian master oscillators (see no. 5 in fig. 5.13). If coupling is weak between the left and right pacemakers the rhythms separate, running on different τ's and resulting in internal desynchronization. This is seen often in beetles (Koehler and Fleissner,1978; Fleissner and Fleissner, 1986), but scorpions are tightly coupled at this level, always showing internal synchrony between left and right. For example, application of different Zeitgeber stimuli to the two median eyes does not uncouple their oscillations, and ERG rhythms on both sides remain equal (Fleissner, 1974, 1977a). However, when the ENSF are disconnected by surgery in the region of the central body (see chapter 4; Fleissner, 1983; Michel, 1989), the oscillators on both sides of the brain become uncoupled and free-run at different period lengths (fig. 5.11). Changes occur in the circadian pattern as well, as subjective night is extended at both ends at the cost of subjective day.

The difference of the left and right rhythms (comparing the coupled with the

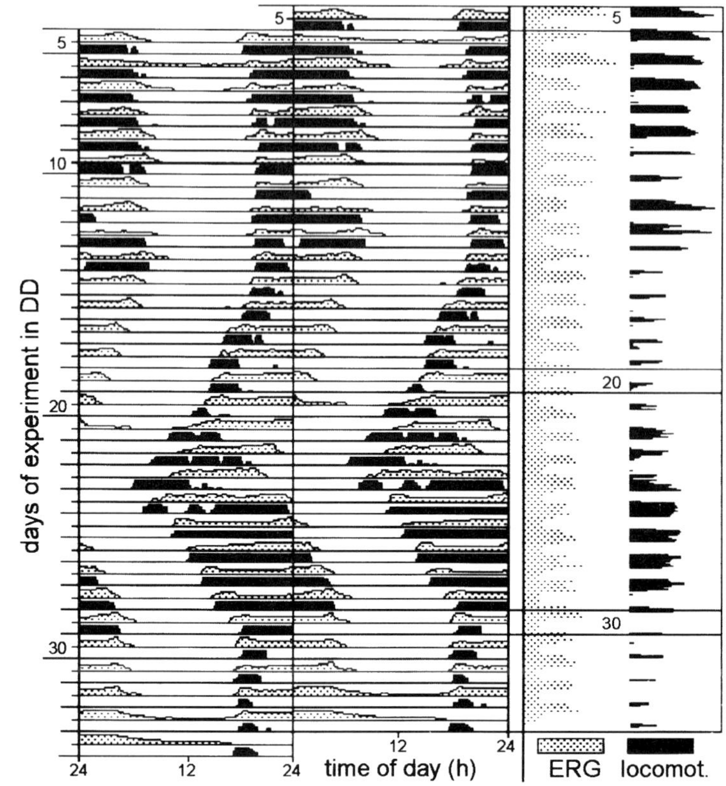

Figure 5.10 Spontaneous changes of τ during constant darkness are correlated with changes in the circadian pattern of locomotor activity (loop 4). Circadian ERG rhythm was used to measure τ simultaneously with locomotor activity over more than 30 days of continuous darkness (DD). The left part of the graph clearly shows that the appearance of high activity in the second half of subjective night (i.e., nearly continuous running during the whole subjective night on days 5–10, 13–14, 24–28) was correlated with longer τ than when the animal rested during late night (days 11, 12, 15–23, 30–34). During the latter, the paradoxic phase–response curve predicts a delay of the period (see figs. 5.7B and 5.9). Data from Michel (1989).

uncoupled stage) shows the effect of feedback due to the mutual interaction between the two pacemakers.

Is morning/evening a matter of left and right?

Under free-running conditions of constant darkness, the relation between length of subjective day and night occasionally varies in a predictable way: the night (normally 9–12 h long) gradually shortens until it reaches a critical point and begins to lengthen again back to its normal duration (fig. 5.12). After 2 weeks this process may be repeated. There are at least two remarkable features of this behavior: first, the changes in the ERG amplitude of both median eyes during shortening and lengthening are identical, even in "noisy" parts of the rhythm.

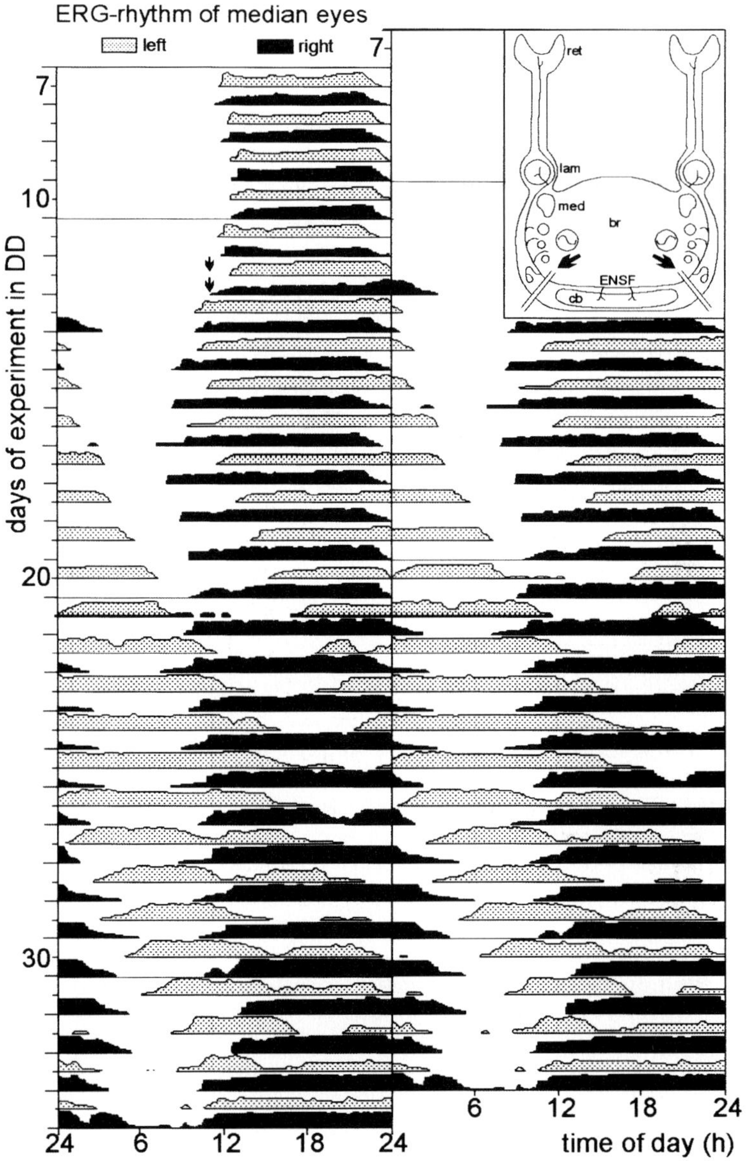

Figure 5.11 Internal desynchronzation between left and right circadian ERG rhythms in constant darkness (DD) after lesioning the eye's neurosecretory fiber (ENSF) system. When the ENSFs are cut (arrows on day 12) bilaterally close to the central body (see inset), the subjective night state increases in length as subjective day decreases. The rhythms of the two eyes uncouple and free run on different τ's. Presumably, mutual feedback (loop 5) between left and right pacemakers is interrupted by the surgery. br, brain; cb, central body; lam, lamina; med, medulla; ret, median eye retina. Data from Michel (1989).

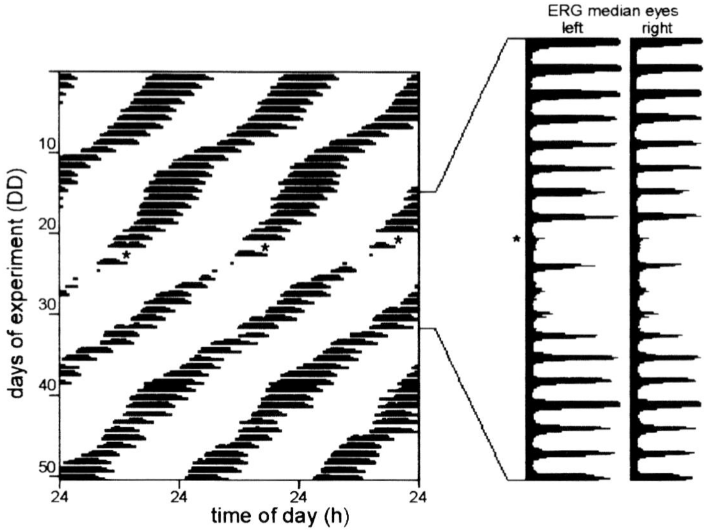

Figure 5.12 Spontaneous and repeated changes in the relation between subjective day and night under extended constant darkness. Sometimes (left panel) the relation between length of subjective night and subjective day can (under constant darkness, DD) spontaneously and periodically change, while free-running τ of the system as a whole may vary only little. This can be interpreted as changes of phase-angle difference between morning and evening oscillator which seem in this case to free-run on different τ's. A continuous plot of the ERG rhythms in the left and right median eye (right panel) shows how both median eyes undergo synchronous changes even when subjective night almost disappears (*). These data indicate a strong coupling between left- and right-side pacemakers and weaker feedback coupling between morning and evening oscillators. For reference, the same day is marked by an asterisk in both plots. Data for *Paruroctonus mesaensis,* from Hohmann and Fleissner, unpublished.

Second, the shortening of the night length may reach a critical value, leading to complete extinction of the night, though this may recover within the next cycle. This closely resembles the pattern of interference when two oscillators with different period lengths superimpose. Several conclusions can be drawn from this observation: coupling, which is invariably strong between left and right median eyes in scorpions, can vary in other parts of the system so that two components, probably morning and evening oscillation, change their phase relation continuously. In the case described above they may even free-run with different τ's. Considering that surgery sometimes produces patterns where morning and evening components seem to be separated, a new hypothesis develops to explain the functional significance of feedback between left and right. Either of the two pacemakers, left and right, may take over the function of the morning or evening oscillator, the phase relation of which determines the relation between the length of subjective day and night. Adjustment to seasonal variations of the photoperiod is thus accomplished despite an invariant circadian τ for the system as a whole. We have experimental evidence that the functional meaning of the left or right as morning or evening oscillator is mutually interchangeable.

Coupling between oscillators may thus have different qualities and can change independently. It may play its functional part fully when operating as a feedback loop.

CONCLUSION

This chapter has highlighted several research foci on the scorpion as a model system for neuroethological analysis of circadian systems. In particular, we considered five examples of feedback loops beginning with a generic view of circadian systems (fig. 5.1) and arrived at a more detailed understanding of its components (fig. 5.13). All these loops are well documented except loop 4 which, despite its likely existence, has not been adequately studied. The functional significance of all these loops deserves further investigation.

What, for example, are the possible functions of these loops? Do they serve

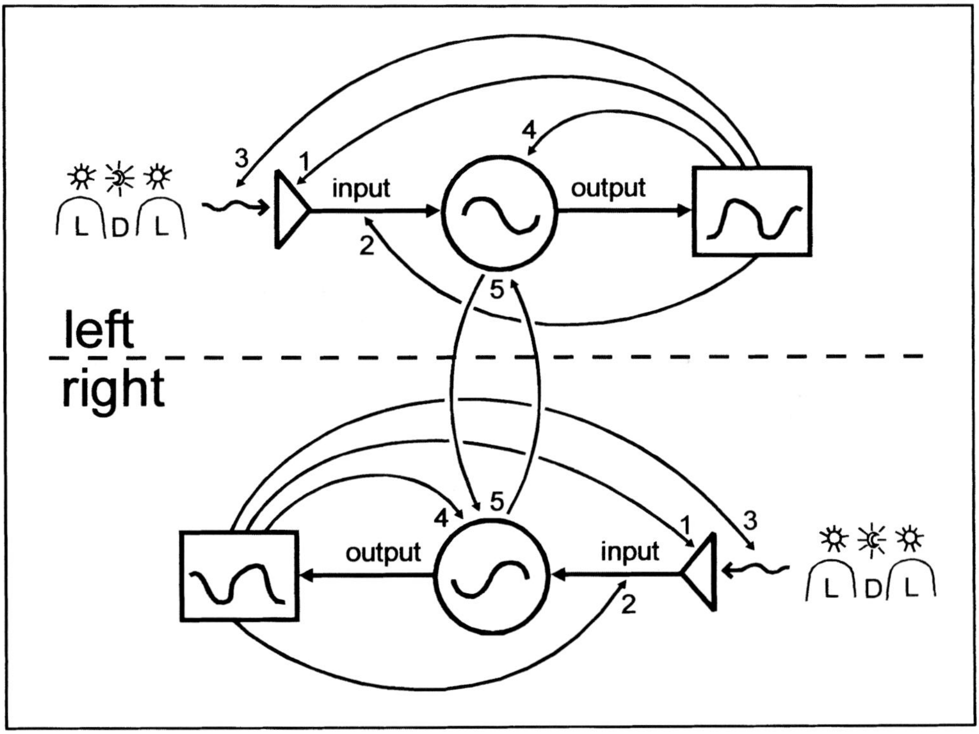

Figure 5.13 Summary of the described examples of feedback interactions regulating circadian behavior of scorpions: 1, 2, the feedback loops regulating the ERG, through which the central pacemaker controls its own photic Zeitgeber input (1) and its processing through optic neuropil (2) via the efferent neurosecretory fiber system; 3, the loop that controls the photic input to the pacemaker by way of clock-driven behavior; 4, the loop through which the pacemaker releases locomotor activity and the paradoxic photo-phase–response curve; 5, the loop coupling left and right central pacemakers. For basic components refer to fig. 5.1.

the processing of natural Zeitgeber, and if so, how? How do the different Zeitgeber inputs cooperate, and what are the mechanisms behind the paradoxic photo-PRC? How is locomotor activity involved as a causal factor in nonphotic events? How are the multiple pacemakers organized, and do they really share the function of morning and evening control?

In general, we may speculate that feedback loops will stabilize a complex circadian system, strengthen its internal coherency, and increase its flexibility in response to environmental challenges without compromising internal consistence. Loops may also improve information processing, for example, by filtering Zeitgeber information under the noisy conditions that exist in nature.

The particular advantages of the scorpion as a model for circadian research as discussed here, are worthy of emphasis. Scorpions are easy to keep in large numbers, and their size is small enough for convenience and large enough for manipulation. More important, the circadian rhythm of the eyes is, by magnitudes, one of the most distinct rhythms known. Much is already known about the neuronal organization of this circadian system, down to the single-cell level. The great variety of species inhabiting diverse photic environments, from deserts to caves, affords comparative study of clocks from different ecological settings. These are certainly strong indications that scorpions will continue to illuminate this field of research.

SUMMARY

Scorpions are excellent experimental animals for chronobiology and particularly for investigation of circadian rhythmicity (adaptation to the cycles of day and night) because their lifestyle is strongly controlled by circadian clocks and well adapted to constant long-term conditions without harm to their health. The possibility of simultaneously recording the clock-controlled photosensitivity and locomotor activity enables one to dissect the circadian system into its components. The scorpion thus became one of the first model systems where feedback loops have been elucidated at the level of neuronal networks. Five examples of those loops are described in this chapter, mostly dealing with processing of Zeitgeber stimuli. Paradoxic responses to light, the possible involvement of locomotor activity in stabilizing circuits, and ideas about the feedback interactions between bilaterally duplicate pacemakers have been highlighted.

Acknowledgments For the use of unpublished data, we are thankful to our students and earlier co-workers: Christian Dube, Wolfgang Hohmann, Michael Lüttgen, Stephan Michel, and Robert D. Stevenson. For technical assistance and animal care, we thank Elke Noering and Karl-Heinz Schmitt. Some of the scorpions were caught in Tunisia with the kind support of the Ministry of Agriculture, the General Director of Forests. Grants came from the Deutsche Forschungsgemeinschaft SFB 45 "Neurobiologie des Verhaltens."

References

Aschoff, J. (ed.) (1981) *Handbook of Behavioral Neurobiology* 4, *Biological Rhythms.* Plenum Press, New York.

Aschoff, J., Figala, J. and Poppel, E. (1973) Circadian rhythms of locomotor activity in the golden hamster (*Mesocricetus auratus*) measured with two different techniques. *J. Comp. Physiol. Psychol.* 85: 20–28.

Boulos, Z. and Rusak, B. (1982) Circadian phase response curves for dark pulses in the hamster. *J. Comp. Physiol.* A146: 411–417.

Cloudsley-Thompson, J.L. (1956) Studies in diurnal rhythms: VI Bioclimatic observations in Tunisia and their significance in relation to the physiology of the fauna, especially woodlice, centipedes, scorpions and beetles. *Ann. Mag. Nat. Hist.* 9: 305–329.

DeCoursey, P.J. (1986) Light-sampling behaviour in photoentrainment of a rodent circadian rhythm. *J. Comp. Physiol.* A159: 161–169.

DeCoursey, P.J. (1989) Photoentrainment of circadian rhythms: An ecologist's viewpoint. *In* Hiroshige, T. and Honma, K. (ed.) *Circadian Clocks and Ecology,* pp. 187–206. Hokkaido University Press, Sapporo.

Edgar, D., Martin, C.E. and Dement, W.C. (1991) Activity feedback to the mammalian circadian pacemaker: influence on observed measures of rhythm period length. *J. Biol. Rhythm* 6: 185–199.

Fleissner, G. (1971) Übr die Sehphysiologie von Skorpionen. Belichtungspotentiale in den Medianaugen von *Androctonus australis* L. und ihre tagesrhythmischen Veränderungen (Inaugural dissertation). JWG-University, Frankfurt/ Main, Germany.

Fleissner, G. (1972) Circadian sensitivity changes in the median eyes of the North African scorpion *Androctonus australis. In* Wehner, R. (ed.) *Information Processing in the Visual System of Arthropods,* pp. 133–139. Springer Verlag, Berlin, New York.

Fleissner, G. (1974) Circadiane Adaptation und Schirmpigmentverlagerung in den Sehzellen der Medianaugen von *Androctonus australis* L. (Buthidae, Scorpiones). *J. Comp. Physiol.* 91: 399–416.

Fleissner, G. (1975) A new biological function of the scorpion's lateral eyes as receptors of Zeitgeber stimuli. *Proc. 6th Int. Arachnol. Congr. 1974,* pp. 176–182.

Fleissner, G. (1977a) Entrainment of the scorpion's circadian rhythm via the median eyes. *J. Comp. Physiol.* A118: 93–99.

Fleissner, G. (1977b) Scorpion lateral eyes: extremely sensitive receptors of Zeitgeber stimuli. *J. Comp. Physiol.* A118: 101–108.

Fleissner, G. (1977c) Differences in the physiological properties of the median and the lateral eyes and their possible meaning for the entrainment. *J. Interdisc. Cycle Res.* 8: 15–26.

Fleissner, G. (1983) Efferent neurosecretory fibers as pathways for the circadian clock signals in the scorpion. *Naturwiss.* 70: 366–367.

Fleissner, G. (1985) Intracellular recordings of light responses from spiking and nonspiking cells in the median and lateral eyes of the scorpion. *Naturwiss.* 72: 46–48.

Fleissner, G. and Fleissner, G. (1978) The optic nerve mediates the circadian pigment migration in the median eyes of the scorpion. *Comp. Biochem. Physiol.* 61 A: 69–71.

Fleissner, G. and Fleissner, G. (1985) Neurobiology of a circadian clock in the visual system of scorpions. *In* F.G. Barth (ed.) *Neurobiology of Arachnids,* pp. 351–375. Springer Verlag, Berlin.

Fleissner, G. and Fleissner, G. (1986) Circadian rhythms in the compound eyes of the carabid beetle *Pachymorpha (Anthia) sexguttata* I. Sensitivity-rhythms and the bilateral circadian oscillator system. *In* den Boer, P.J., et al. (eds.) *Carabid Beetles: Their Adaptations and Dynamics,* pp. 3–17. Gustav Fischer, Stuttgart.

Fleissner, G. and Fleissner, G. (1988) Efferent control of visual sensitivity in arthropod eyes: with emphasis on circadian rhythms. *In* Lindauer, M. (ed.) *Information Processing in Animals.* 5, pp. 1–67. Gustav Fischer Verlag, Stuttgart.

Fleissner, G. and Fleissner, G. (1992) Feedback loops in the circadian system. *In* Zatz, M. (ed.) *Discussion in Neuroscience, FESN Study Group, VIII Circadian Rhythms,* pp. 79–84. Elsevier, Amsterdam.

Fleissner, G. and Fleissner, G. (1993) Seeing time. *In* Wiese, K., Gribakin, F., Popov, A.V. and Renninger, G. (eds.) *Sensory Systems of Arthropods,* pp. 288–306. Birkhaeuser, Basel.

Fleissner, G. and Fleissner, G. (1998) Natural photic Zeitgeber signals and underlying neuronal mechanisms in scorpions. *In* Y. Touitou (ed.) *Biological Clocks. Mechanisms and Applications,* pp. 171–180. Elsevier Science, Amsterdam.

Fleissner, G. and Heinrichs, S. (1982) Neurosecretory cells in the circadian clock system of the scorpion *Androctonus australis. Cell Tiss. Res.* 224: 233–238.

Fleissner, G. and Schliwa, M. (1977) Neurosecretory fibers in the median eyes of the scorpion *Androctonus australis* L. *Cell Tiss. Res.* 178: 189–198.

Hadley, N.F. (1990) Environmental physiology. *In* Polis, G.A. (ed.) *The Biology of Scorpions,* pp. 321–340. Stanford University Press, Palo Alto, CA.

Hohmann, W., Michel, S. and Fleissner, G. (1990) The scorpion as a model for feedback mechanisms in a circadian system. *Verh. Deutsch. Zool. Ges.* 1990: 647–648.

Kiesel, A. (1894) Untersuchungen des facettierten Auges. *S.B. Akad. Wiss.* 103: 97–139.

Koehler, W.K. and Fleissner, G. (1978) Internal desynchronisation of bilaterally organized circadian oscillators in the visual system of insects. *Nature* 274: 708–710.

Luettgen, M.A. (1993) Entrainment der circadianen Laufrhythmik durch Lichtzeitgeber: Untersuchung biologisch relevanter Lichtparameter am Beispiel der Lokomotionsrhythmik von *Androctonus australis* (Scorpiones, Buthidae). Inaugural Dissertation. JWG-Universitaet Frankfurt/Main, Germany.

Michel, S. (1989) Lauaktivitaet und Augenempfindlichkeit von Skorpionen: Untersuchungen der Kopplung in einem circadianen Multioszillatorsystem. Inaugural Dissertation. JWG-Universitaet Frankfurt/Main, Germany.

Moore-Ede, M.C. and Sulzman, F.M. (1981) Internal temporal order. *In* Aschoff, J. (ed.) *Handbook of Behavioural Neurobiology* 4, *Biological Rhythms,* pp. 215–241. Plenum Press, New York.

Mrosovsky, N., Reebs, S.G., Honrado, G.I. and Salmon, P.A. (1989) Behavioural entrainment of circadian rhythms. *Experientia* 45: 696–702.

Mrosovsky, N., Salmon, P.A., Menaker, M. and Ralph, M.R. (1992) Non-photic phase shifting in hamster clock mutants. *J. Biol. Rhythms* 7: 41–49.

Pittendrigh, C.S. (1974) Circadian oscillations in cells and the circadian organization of multicellular systems. *In* Schmitt, T.O. and Warden, F.G. (eds.) *The Neurosciences,* pp. 437–458. MIT Press, Cambridge, MA.

Pittendrigh, C.S. (1981) Circadian systems: entrainment. *In* Aschoff, J. (ed.) *Handbook of Behavioural Neurobiology* 4, *Biological Rhythms,* pp. 95–124. Plenum Press, New York.

Schmidt, K.P., Koehler, W.K., Fleissner, G. and Pflug, B. (1990) Locomotor activity accelerates the adjustment of the temperature rhythm in shift workers. *J. Interdisc. Cycle Res.*21: 243–245.

Terman, M., Reme, C.E. and Wirz-Justice, A. (1991) The visual input stage of mammalian circadian pacemaking system: II. The effect of light and drugs on retinal function. *J. Biol. Rhythms* 6: 31–48.

Turek, F.W. (1989) Effects of stimulated physical activity on the circadian pacemaker of vertebrates. *J. Biol. Rhythms* 4: 135–147.

von Holst, E. and Mittelstaedt, H. (1950) Das Reafferenzprinzip. *Naturwiss.* 37: 464–476.

6

Sensory Ecology and Orientational Behaviors

PHILIP BROWNELL

Sensory organs are among the most intricately structured features that animals possess. Beyond structure, the biochemical, physiological, and integrative capacities of sensory tissues provide some of biology's most exquisite examples of evolutionary adaptation to specialized environments. This intricacy of sensory structure and function has been shaped, through natural selection, for acquisition of information important to the animal's survival in a particular environment. A major concern of sensory ecology is to determine how this sensory information is generated, transmitted, and altered by passage through the environment. The term "ecology" is appropriately used here because the information of value passes through an environment having both physical and biotic character. Physical factors certainly limit the form and content of the signals produced by animals, but it is the ecological relationships between and among species—biotic factors—that give significance to the signals received. For sensory biology to advance beyond mechanistic understanding of sensory structure and function, a better appreciation of the physical and biological characteristics of an animal's habitat—its sensory *niche*—must be accurately and realistically assessed.

In this chapter I present the nocturnal sand scorpion, *Paruroctonus mesaensis,* as an advantageous model for sensory ecological studies (fig. 6.1). The primary attractions of this animal for investigation are the simplicity of their natural environment (the nocturnal sand dune ecosystem of the Mojave Desert) and the accessibility of the scorpion's primary sensory organs for anatomical and physiological analysis in the laboratory. Adding to this is the natural fluorescence of scorpion cuticle, an unusual attribute that provides a window for observation of this animal's behavior in a natural setting otherwise inaccessible to sensory biologists—the desert at night.

Like most other arachnids, scorpions are relatively secretive and inactive animals, with a conservative repertoire of behaviors related to their feeding, shelter, and reproductive needs (Alexander, 1957; Abushama, 1959; Stahnke, 1966; Shorthouse, 1971; Polis and Farley, 1979a,b; Shachak and Brand, 1983; Warburg and Polis, 1990). Although these arthropods are very sensitive to light (Fleissner and Fleissner, this volume), they are weakly visual animals (Locket, this volume) and rely more on mechanosensory and chemosensory signals to guide their behaviors (Brownell, 1977; Root, 1990). Scorpions inhabiting desert dune environments seem particularly insensitive to visual and auditory stimuli, leaving only the mechanical senses (touch and vibrational sensitivity) and chemosensory functions (taste or close-range olfaction) to fill their sensory ex-

Figure 6.1 The sand scorpion *Paruroctonus mesaensis* and its dune ecosystem in the Mojave Desert. (Left) Dark field image of *P. mesaensis* highlights the medial and lateral eyes and numerous mechanoreceptive hairs on the postabdomen, legs, and pedipalps. (Right) Large plants such as saltbush (*Atriplex*), mesquite (*Prosopis*), and creosote bush (*Larrea*) collect wind-blown sand and stabilize it in low mounds, leaving open and mildly sloped areas where *Paruroctonus* prefers to burrow (b) and hunt.

perience (Brownell and Farley, 1979a; Gaffin and Brownell, 1992; Gaffin et al., 1992). From an ecological standpoint, the sandy substrates into which sand scorpions burrow for shelter by day, and upon which they forage for food and mates at night, have comparatively simple physical and chemical features. Wind-blown sand is perhaps the most homogeneous and "quiet" of sensory environments, thus providing an optimal setting for observation and experimental manipulation of mechanical and chemical signals. The objectives of sensory ecological studies for sand scorpions are thus clearly defined: they begin with an assessment of the physical and biotic sources of sensory information important to the animal.

SENSORY ENVIRONMENTS

Temperature, Humidity, and Light

Extreme fluctuations of temperature and light intensity are hallmarks of the Mojave sand dune environment to which all its inhabitants show physiological and behavioral adaptations. As ectotherms, the active metabolic rate of sand scorpions is largely a function of environmental temperature (Hadley, 1990; Withers, 1992; Lighton et al., in press)—that is, of substrate temperature and the nearly still layer of air just above it (Geiger, 1957). Water is also a limiting resource in deserts, and what little scorpions assimilate comes mainly through their diet (Yokota, 1979). Relative humidity at the surface generally remains below 20%, and this, combined with elevated surface temperatures, creates a large vapor-pressure deficit and desiccation stress for actively respiring animals (Edney et al., 1974; Edney, 1977). Thus, *Paruroctonus* benefits metabolically from its nocturnal habits and generally minimizes the time spent outside the burrow to coincide with peak prey abundance (Hadley, 1968; Polis, 1980). Most of its 3- to 5-year life span is spent at rest in a semipermanent home burrow dug to a depth of 20–50 cm (Williams, 1966; Polis et al., 1986). Here, the seasonal extremes of surface temperature fluctuation (below freezing to above 60°C) are attenuated to tolerable levels (5–40°C), and the small but persistent amount of substrate moisture ($<$1% by weight) is sufficient to saturate the interstitial air below 20 cm depth. In this resting state the scorpion's basal metabolic rate is reduced to unusually low levels, approximately 30% of the lowest rate recorded for any comparably sized arthropod (Lighton et al., in press). These metabolic rates are very sensitive to temperature, however, with an average Q_{10} of about 3. It is not surprising that temperature is an important determinant of sand scorpion activity and behavior (Polis and Sissom, 1990).

From a sensory ecological perspective, the fluctuations in subsurface temperature may also have informational value as a timing signal for activation of feeding and other behaviors requiring the animal to surface. This potential is shown graphically in fig. 6.2 where light intensity at the surface and temperatures of the air and substrate are displayed over a 5-day period. Scorpions usually emerge from their burrows about an hour after sundown, as air and sand temperatures at the surface cool. During this period of surface cooling, burrow temperatures have been rising to their daily high, effectively warming the pre-emergent scorpion in anticipation of activity on the surface. The delay in subsurface heating relative to the surface is about 6 h (phase shift of 1/4 cycle) for burrows at 25 cm depth. This delayed conductance of a detectable thermal signal through sand to depths of about 30 cm creates an accurate and reliable indication of surface temperature and, therefore, of incident sunlight. The timing of this periodic signal as a function of burrow depth is predicted by the diffusion equation solved for one-dimensional flow (depth) of heat using a thermal diffusivity for sand of $\alpha = .000468\ \text{m}^2/\text{h}$ (Geiger, 1957). For surface heating that varies in amplitude (T_o) as a quasi-sinusoidal pattern over a 24-h period ($\omega = 0.262$ rad/h), temperature at depth x and time t is predicted by

$$T(x,t) = T_o \cdot e^{-x\sqrt{\frac{\omega}{2\alpha}}} \cdot Cos\left[\omega t - \left(x\sqrt{\frac{\omega}{2\alpha}}\right)\right].$$

Such thermal cycles may reinforce or substitute for photoperiodic Zeitgebers that scorpions cannot otherwise access during extended periods of inactivity (see Fleissner and Fleissner, chapter 5, this volume).

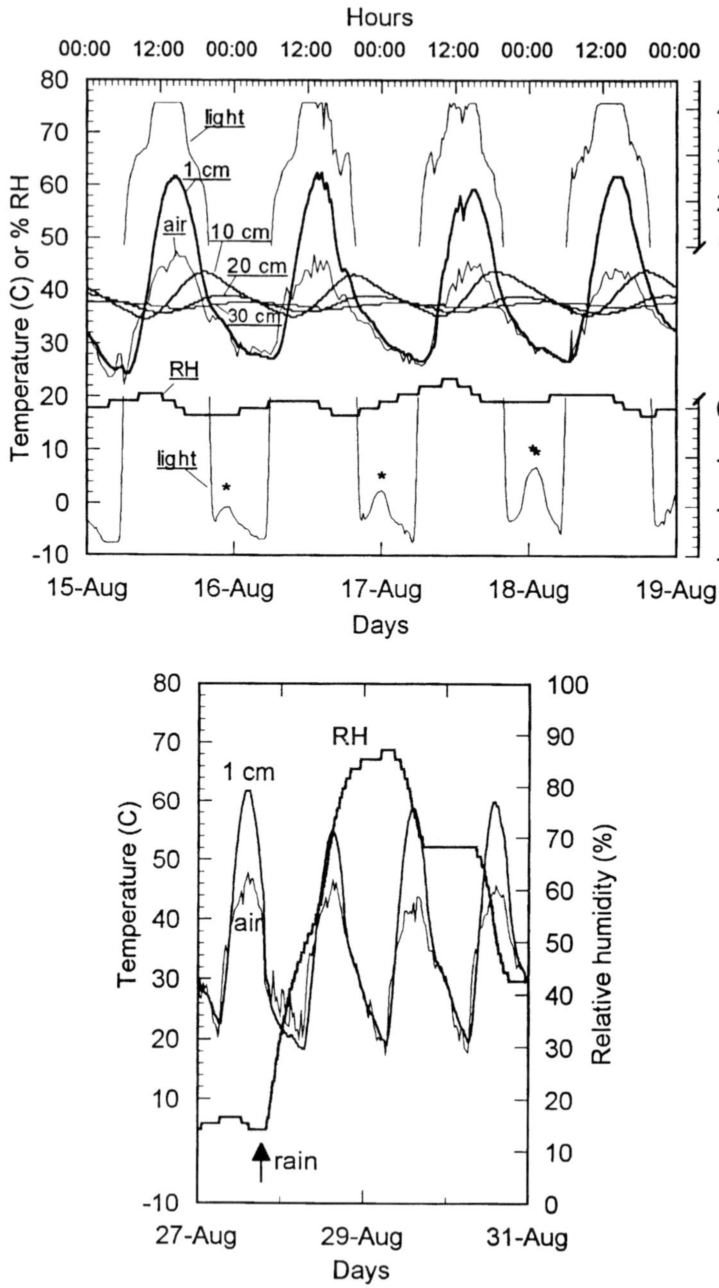

Figure 6.2 Fluctuations in temperature, light, and relative humidity for the Mojave sand dune environment. (A) Daily fluctuations in light intensity (log lumens/m^2; interrupted trace) vary more than seven orders of magnitude. Radiation absorbed at the surface warms the air (1 cm above surface) and substrate (1, 10, 20, 30 cm below surface) in a regular cycle. Average temperature at burrow depth (20–30 cm) is the same as the surface sand, but

Substrate

Perhaps the most dominating feature of the sand scorpion's natural environment is the substrate they live on—loosely packed sand. Natural populations of *Paruroctonus mesaensis* are found only on sandy substrates (Polis, 1980a; Polis and Farley, 1980; Fet et al., 1998), a good indication that their sensory adaptations are specialized for this peculiar and geologically recent (circa Pleistocene) habitat. The Mojave sands are wind-transported grains of spherical (subrounded to subangular) form composed of quartz (~70% of volume), feldspar (~25%), and platey micaceous minerals (3–5%) with no clay or fine silt particles to indicate organic content. The sands *Paruroctonus* prefers are well-sorted grains of uniform size, three-quarters of which are 0.1–0.25 mm in diameter. These dimensional properties of sand can be correlated with sensory and locomotory structures of *Paruroctonus* by representing structural size in the spatial frequency domain (fig. 6.3). Thus, tarsal hairs have a spatial frequency of about 3/mm, which corresponds well with spatial frequency (2–4/mm) for larger sand particles at the substrate surface (diameter 0.25–0.50 mm). This is consistent with the function of tarsal hairs as "snowshoes" for maintaining purchase on this substrate.

A correspondence between substrate particle size and sensory structure is evident in the pectinal sensory organs, two ventral appendages that scorpions apply to the substrate to sense its texture and chemical content (see Gaffin and Brownell, chapter 7, this volume). These comblike organs have sensilla-bearing teeth along their length at spatial frequencies between 5/mm (females) and 6/mm (male) in adult *Paruroctonus*. The individual sensilla on these teeth have much greater spatial frequencies of 100–200/mm, 10–100 times greater than sand particle frequencies. Such information is useful for accessing the sensory functions of the pectines. For example, the capacity of the pectines to sense substrate texture (Abushama, 1964; Boyden, 1978) is most likely mediated by flexure of the individual teeth along the pectine comb because these structures best fit substrates having particle frequencies of 6–7/mm (fig. 6.3). The spatial frequency of the sensilla on each tooth is better suited for detection of much more finely structured signals than substrate texture as speculated by Carthy (1958), and perhaps even enables detection of the specular chemical deposits residing on individual sand grains.

The chemical composition of dune sand is also remarkably homogeneous and featureless in its native state. Water content by weight is less than 0.1% for surface sand, increasing to about 1% at 10 cm depth and lower. Combustible organic matter is less than 1% at the surface, and what little there is may be transformed or degraded by high incident radiation and heat. The absence of organic matter also contributes to the low density and disaggregation of dune sands: as much as 50% of this substrate volume is air, a porosity that greatly modifies its

---

fluctuations are attenuated and delayed by 6 h (phase lag ≈ 90°). Night light intensity is significantly influenced (>2 log units) by rising and setting of the moon (*). (B) Relative humidity (RH) at 5 cm depth is consistently low (<20%) but abruptly rises toward saturation immediately after rain (arrow indicates 0.1″ of precipitation on 27 August). These fluctuations in light, temperature, and humidity may act as cues for optimal timing of foraging activities.

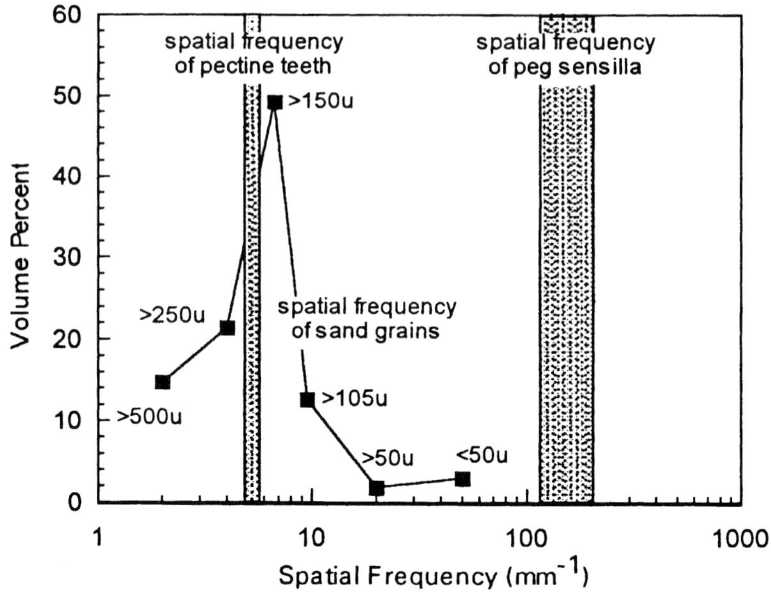

Figure 6.3 Spatial frequencies of dune sand in relation to prominent sensory and motor structures contacting the substrate (pectine teeth and peg sensilla). Sand grain diameter in micron (u) is indicated at each data point. Volume percent = percent by weight of total sand grains in a given volume of sand.

vibrational and sound-transmitting properties relative to other solids. Thus, the simple physical structure and chemical composition of sand, combined with its absorbent and constantly regenerating surface, make it a uniquely quiet sensory niche. This circumstance is important for detecting the subtle mechanical and chemosensory signals known to influence orientational behaviors of sand scorpions.

Biotic Factors

In addition to orienting their behaviors in time to accommodate the physical extremes of the dune environment, sand scorpions must orient accurately in space to obtain (or avoid) important biotic factors. The most prominent of these, food, refuge, mates, and predators, have been described for *Paruroctonus* by Polis (1979a,b; 1980a,b; 1990, chapter 12, this volume).

Spatial orientation occurs on both large (e.g., navigational) and small dimensions. During forays to feed, for example, *Paruroctonus* leaves its home burrow at dark to sit in ambush of insect prey. As hunting ends several hours later, these scorpions must relocate their burrows, a directional behavior they execute quickly and accurately from one to several meters distance (Polis et al., 1986). On the larger geographical scale, adult male *Paruroctonus* may travel hundreds of meters in one night during mating season while searching for females. Although it is highly unlikely these animals return to a home burrow, it is apparent that their walking is directionally consistent. On a local scale, the sand scorpion's success as a predator of small arthropods depends critically on its ability to localize suitable prey that move within a few decimeters of where it waits in ambush. These potential prey must be discriminated at a distance from larger scorpions and the vertebrates that readily prey upon *Paruroctonus* (Polis et al., 1981).

As a long-lived species residing in relatively stable populations, the biotic interactions sand scorpions experience outside the safety of their burrows must strongly influence their sensory evolution. In this regard special consideration must be given to the habit of mate cannibalism and the web of interspecific predatory interactions *Paruroctonus* encounters with sympatric species occupying the same sensory niche (Polis and Farley, 1980). In this ecological setting orientation of one species to conspecific mates of appropriate size (e.g., avoidance of heterospecific scorpions of larger size) strongly favors accurate and reliable mechanisms of species identification, preferably operating over a safe distance. Chemical signals are important factors for triggering reproductive behavior (see Gaffin and Brownell, chapter 7, this volume), thus creating a selective pressure for evolution of highly specific pheromonal signals. Alternatively, the benefits of intraguild predation for the larger species within the guild (e.g., *Hadrurus* and *Paruroctonus*) may favor the use of aggressive pheromone mimicry to attract males of smaller species as they search for mates. These ecological considerations help to direct laboratory investigations of scorpion chemical signaling systems, particularly of the pectines, which appear to be the primary organs for detection of semiochemicals.

SENSORY SYSTEMS AND ORIENTATIONAL BEHAVIORS

As strictly nocturnal animals, sand scorpions occupy a sensory niche unfamiliar to humans. Although scorpions have eyes with good lenses (Locket, chapter 3, this volume) and extremely sensitive photoreceptors (Fleissner, 1977a), their predatory and reproductive behaviors are undeterred by complete darkness or by covering their eyes. The eyes are sensitive enough to aid in navigation on moonless nights (Fleissner, 1977b), but this has not been demonstrated behaviorally. Auditory stimuli in the 0.1–10 kHz bandwidth also has little affect on scorpions except at the lowest frequencies and within a range of a few centimeters. Thus, in the absence of visual and auditory guidance, scorpions rely predominantly on tactile senses to detect disturbances in their local environment (Brownell and Farley, 1979a). Subtle motions of the substrate and air caused by prey movements distant (far field) and close (near field) to scorpions are detected by specialized mechanoreceptors on the legs (slit sensilla) and pedipalps (trichobothria). These mechanical cues are both necessary and sufficient to direct predatory orientational behavior (Brownell and Farley, 1979c; Brownell, 1984). Once objects of potential importance (food, potential mates, or predators) have been localized in space, chemosensory signals predominate in triggering subsequent consummatory or avoidance behaviors. The sensory channels of this information are the numerous pore-tipped chemosensory hairs on the legs and pedipalps and the dense fields of peg sensilla on their pectines. Thus, for sensory ecological analysis, nearly all of the sand scorpion's

Table 6.1 Primary sensory systems of scorpions and orientational behaviors they direct

Sensory channel	Sensory system	Stimulus mode	Orientational behavior
Mechanosensory-vibrational	Tarsal slit sensilla and hairs	Substrate vibrations	Predatory (far field)
	Trichobothria	Wind current	Predatory (near field)
Chemosensory	Pectines	Chemical deposits on substrate	Mate localization and identification
Vision	Principal eyes	Luminance pattern and polarization	Navigation

orientational behaviors can be accounted for by a few discrete sensory systems, each of definable structure and mode of stimulation. These are summarized in table 6.1.

Mechanosensory Orientation to Prey

Virtually all of the sand scorpion's prey are insects and other small arthropods which crawl or fly to the scorpion as it sits motionless in ambush outside the burrow (Polis, 1979). When a large insect such as a sand treader cricket or a burrowing cockroach (*Arenivaga investigata*) moves within 20–30 cm of a hunting *Paruroctonus,* it displays awareness of the prey by tensing its legs and opening its pedipalps. Further movements of the prey trigger a series of predatory orientation responses—distinct rotations of the body and pedipalps in the direction of the disturbance with forward movements that bring the scorpion 5–10 cm closer to its target (Brownell and Farley, 1979a). Each predatory orientation response is proceeded by a brief period of immobility when the prey must move on the sand again before the next orientation is elicited. As the tracking sequence progresses, the accuracy of turning increases until the pedipalps make contact and grasp the prey. Scorpions react readily to insects digging in or burrowing beneath the sand surface several decimeters away, but they are unresponsive to struggling insects held in air or on sand not contiguous with their substrate (Brownell and Farley, 1979a,c). These and other observations unequivocally demonstrate that prey location is perceived from mechanical waves conducted through dry sand. Specialized sensory organs on the tarsi (slit sensilla) have been identified as the sensory channels for detecting substrate vibrations, and their responses to these mechanical waves have been measured (Brownell and Farley, 1979b).

Vibrational signals in sand

Solid media conduct elastic energy in four different modes that are distinguished by their geometry of wavefront spreading and particle motions they cause within the substrate (Graff, 1975). Two of these waves, the compressional and shear waves, are body waves that propagate as spherically expanding wavefronts throughout the medium. These have particle motions aligned, respectively, parallel to and perpendicular to the direction of wave propagation. Two other waves in solid media, Rayleigh waves and Love waves, are surface waves that propagate along the air–solid interface. Their particle motions are elliptical and directed perpendicular (Rayleigh) or parallel (Love) to the surface plane. Each of the four waves travels at different speeds through the solid, with the compressional wave (also called sound or P-wave) propagating at about twice the velocity of the others.

For the purposes of understanding how psammophilic animals might make use of these signals as sensory cues, it is important to know how sand behaves as a conducting medium. For the signals of potential importance to animals (0.1–5 kHz bandwidth), dry sand is an elastic medium of homogenous, isotropic, and weakly stratified character. A distinctive attribute of surface sand as a solid is its high porosity, up to 50% of total volume being interstitial air. Because velocity is approximately equal to the square root of elasticity divided by density, this low density contributes to one of sand's most unusual characteristics relative to more consolidated natural media: its low velocity of signal conduction (Iida, 1938; Brownell, 1977; Markl, 1983; Aicher and Tautz, 1990). In sand the compressional wave travels at about 150 m/sec, whereas the Rayleigh wave conduction velocity is 40–50 m/sec. The frequency bandwidth for these vibrations is also comparatively low, ranging to 5 kHz near the source of disturbance and dropping sharply to 300–500 Hz at distances beyond 10 cm (Brownell, 1977; Aicher and Tautz, 1990). From these properties wavelength ($\lambda = v/f$) of surface waves traveling at 45 m/sec (300–500 Hz frequency content) can be calculated as 9–15 cm. Such signal wavelengths are sufficiently near the leg span of scorpions (~5 cm in adults) that time-of-arrival cues (or phase delay) across the spatially separated and vibration-sensitive tips of the legs can be used to estimate wave source direction (Michelsen and Larsen, 1985; Dusenbery, 1992; Gerstner et al., 1996).

Another favorable property of sand as a medium for information transfer is its elasticity for vibrations at biologically important frequencies (Brownell, 1977). Signal energy dissipates with distance of transmission as a result of geometrical expansion of the wave front (G_r) and frictional losses due to frequency-dependent absorption and scattering by the medium (α). For wave-induced particle accelerations of intensity, A_o, conducting through a homogeneous medium, signal attenuation with distance (r) is a product of geometric spreading and absorbency losses:

$$A_r = A_o\, G_r\, e^{-\alpha r}.$$

Loss of signal strength for spherically spreading body waves (i.e., compressional wave) is more sensitive to distance ($G_r = r^{-2}$) than the circular wave fronts of surface waves ($G_r = r^{-1}$). Thus, even in a perfectly elastic medium, the surface wave energy has a substantial advantage for conduction into the far field. This applies especially to disturbances of the substrate surface because most of this energy enters and propagates as a surface wave (White, 1965). Attenuation due to frictional losses (absorption and scattering) is very small for dry sand (α = 0.23 db/cm), as measured most recently by Aicher and Tautz (1990). Such small loss of signal should preserve biologically detectable signals for several decimeters while otherwise ensuring a quiet vibrational background for scorpions to detect signal above background noise.

Thus, by its wave-conducting and elastic properties, sand is an unexpectedly good conductor of information over distances and bandwidths of significance to sand scorpions. The extent to which surface waves are used to detect and localize vibrational sources, such as moving prey, can only be determined through physiological and behavioral experiments. However, it is evident from the physical properties of surface waves that localization of substrate disturbances will most likely involve measurement of time (or phase) delays between stimulation of sensors at different points on the substrate (Barth, 1985). Distance-dependent changes in amplitude or spectral content of the conducted signal are too insignificant to be of reliable value for localizing targets beyond a few centimeters.

Sensory detection of substrate vibration

As described above, contact between the scorpion's tarsal leg segments and the body of sand being disturbed is both necessary and sufficient to evoke accurate orientational responses. Of the several mechanosensory systems located on the tarsi (fig. 6.4; Brownell and Farley, 1979b; Foelix and Schabronath, 1983), the basitarsal slit sensilla are most important for vibration source localization. This dependence was demonstrated in the laboratory by selectively ablating these organs and observing the effect on turning accuracy. Agitated *Paruroctonus* defend themselves by rotating the pedipalps and body to face, with high accuracy, substrate disturbances presented 8–10 cm away (fig. 6.5). When the slit sensilla on one side are ablated, accuracy of turning to either side (left, right) is systematically affected, especially for targets presented in the direction of the ablated sensors. Experimental variation of the pattern of sensory ablation further demonstrates that input from all eight tarsal slit sensilla are used to calculate target direction, with those nearest the source having greater impact. Moreover, by controlling the patterns and timing of tarsal movements, it has been shown that the target angle is determined most effectively by the time delay between stimulation of opposing legs and not the relative intensity of tarsal displacement (Brownell and Farley, 1979c).

The tarsal slit sensilla function as seismometers in detecting the elastic motions of sand produced by local disturbances. Of the two types of energies that propagate from the source into the far field, the particle displacements for surface (Rayleigh) waves are vertical to the plane of the substrate and most appropriate for stimulation of the slit sensilla. The form and amplitude of surface wave movements can be reconstructed by an electromechanical transducer and applied directly to a tarsus to measure its electrophysiological responses (Barth and Geethabali, 1982). By varying the amplitude and frequency of such stimuli it is possible to show that the slit sensilla are highly sensitive to oscillations in the 300–500 Hz range, the frequency bandwidth that carries most surface wave energy beyond a few centimeters (fig. 6.6). With recent advancements in methods for intracellular recording from slit sensillar neurons (Seyfarth et al., 1995), it may be possible to analyze the mechanisms of transduction that tune these cells to particular sensory environments.

The final stage of this modeling process is to predict the integrated sensory response for tarsal slit sensilla on each of the scorpion's eight legs. This can be accomplished by calculating the entire stimulus field (temporal pattern and amplitude of substrate movements at each point of tarsal contact) from the equation for frequency-dependent attenuation (above) and conduction velocity for surface waves (Brownell and Farley, 1979c; Stürzl et al., 2000). The sensory re-

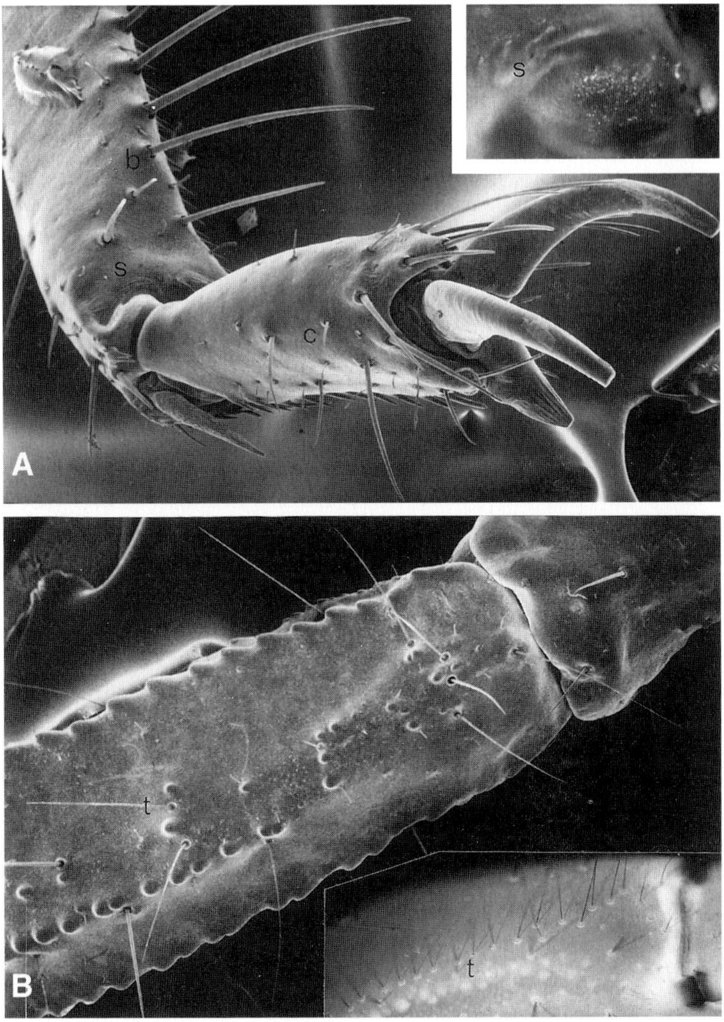

Figure 6.4 Mechanosensory sensilla of sand scorpions. (A) Scanning electron micrograph of the tarsal and pretarsal leg segments showing the vibration-sensitive slit sensilla (s), numerous bristles (b), and tactile and chemosensory (c) hairs. (B) Filamentous hairs, called trichobothria (t), are distributed in rows (inset) along the outer and ventral surfaces of the pedipalps. These articulate freely in one plane and may encode the direction of prey movements nearby.

sponse is then estimated from the stimulus-response characteristics of tarsal slit sensilla. Such comprehensive modeling of information input to a sensory system is only possible where the pattern of stimulation is of simple and predictable form and the structure and function of the sensory receptive field is accurately quantifiable. These attributes of substrate movements in sand and sensory response of tarsal vibration detectors of sand scorpions permit highly

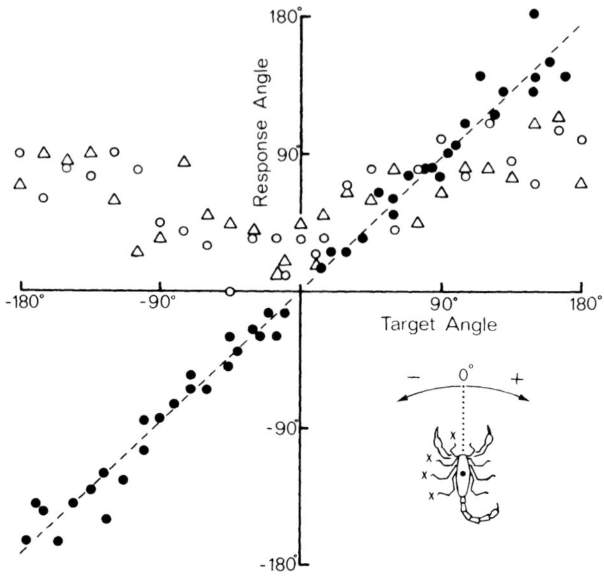

Figure 6.5 Directional orientation of *Paruroctonus mesaensis* to nearby disturbances of the sand. With eyes covered, *P. mesaensis* accurately turns to face the source of a substrate disturbance 8–10 cm away (filled circles). When the basitarsal slit sensilla are ablated unilaterally on all four legs, the angle of turning (open symbols) becomes biased in the direction of legs with sensors remaining intact. (+,) Indicate right and left directions, respectively.

accurate, predictive modeling of orientational behavior, for comparison of theory to experimental results (Stürzl et al., 1998).

Trichobothria and response to air currents

The primary organ of prey capture for all scorpions is the pedipalps, the pincerlike anterior appendages that first contact and seize objects. Once orientation to prey in the far field (~3–30 cm) has succeeded in bringing the sand scorpion close to its prey, the final stages of prey capture require precise positioning of the pedipalps using sensory cues from the near field (<5 cm distance). One of the best candidates for mediating this sensitivity are the filiform hairs, or trichobothria, found exclusively on the pedipalps (fig. 6.4B).

The trichobothria are highly elongated, low-mass hairs (500–1500 μm long, 4–8 μm diameter) with extreme sensitivity to air currents or low-frequency sound (threshold velocities as low as 1 mm/sec) (Ignat'yev et al., 1976; Reissland and Gorner, 1985). Trichobothria of scorpions have sensitivities consistent with air movement detection (Hoffman, 1967), and Linsenmair (1968) attributes the scorpion's ability to walk directionally in wind (anemotaxis) to these structures. Studies of spiders (Barth et al., 1995) and insects (Tautz, 1978; Miller et al., 1991) give convincing evidence of a role for trichobothria in prey detection and predator avoidance, but these aspects of their behavioral functions in scorpions have not been examined.

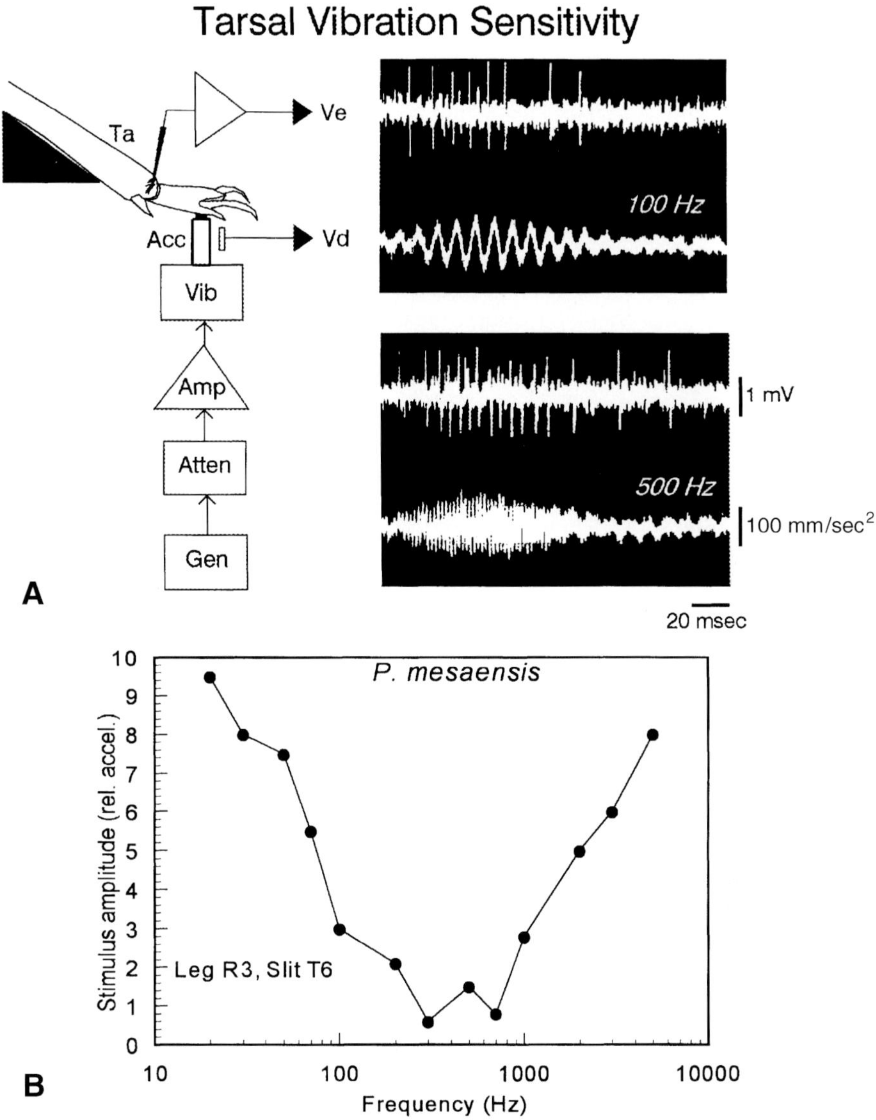

Figure 6.6 Vibration sensitivity of tarsal slit sensilla. (A) Apparatus for measuring the vibration sensitivity of tarsal slit sensilla (left). Panels to the right show the sensory response (Ve) of a single slit sensillum to monofrequency movements of a tarsus at 100 and 500 Hz, as monitored (Vd) by an accelerometer (Acc) at the tip of the vibrator (Vib). (B) Threshold curve showing relative amplitude of stimulation required to evoke a detectable response from a single slit at various frequencies of tarsal vibration. The greatest sensitivity (i.e., lowest stimulus amplitude to evoke a threshold response) occurs within the range 300–700 Hz, which correlates well with the natural frequencies of signals conducted by dune sand.

Compared to spiders, which have thousands of trichobothria distributed over their legs and body surfaces, the trichobothria of *Paruroctonus mesaensis* are few in number and restricted to the pedipalps. About 70 trichobothria extend from the outer lateral surfaces of the patellar segment and dorsoventrally along femur, patellar, and tibio-tarsal segments (fixed and moveable fingers of the pincer). In these locations air stream velocity from convectional flow over the scorpion is expected to be greatest. Flexural movements of these trichobothria are directional, with trichobothria along outer lateral surfaces of the pedipalps generally articulating in a plane parallel to appendage axis and those along dorsal and ventral surfaces and ridges flexing perpendicular to this axis. By inspection under magnification these trichobothria are exquisitely responsive to movements of prey centimeters away and thus may encode this information for directing the pedipalps to their target. Another stimulus of potential interest is signals produced by movements of other scorpions nearby: juddering behavior, tail wagging, and the defensive lurching movements of scorpions in stilted postures are particularly good candidates for trichobothrial detection.

Air flow around the pedipalps and trichobothria is a complex stimulus field to model quantitatively when compared to the vibrational inputs to tarsal slit sensilla. However, progress in describing air fluid dynamics on submillimeter scales (Humphrey et al., 1993; Barth et al., 1995) is encouraging. Once the response properties of scorpion trichobothria are better known, it should be possible to model this aspect of the sensory input and its role in directing predatory orientational behavior (see Theunissen and Miller, 1991).

Chemosensory Systems and Orientation to Food, Water, and Mates

When successful, the orientational responses to mechanical disturbances of the substrate and air bring the scorpion into physical contact with the target. Thereafter, behavioral responses will be influenced strongly by chemical properties of the target as sensed by chemotactic hairs at the tips of the pedipalps. Chemosensory hairs of scorpions are of distinctive form and smaller size relative to other setaform sensilla, the most evident of which are mechanosensory. More than 1000 of these smaller hairs extend from all cuticular surfaces of *Paruroctonus,* but most are concentrated ventrally where they contact the substrate. The tarsal leg segments and tips of the pedipalps are particularly well endowed with chemosensory hairs, each with a curved tip and single pore opening characteristic of contact chemoreceptors in other arthropods (Foelix and Mueller-Vorholt, 1983). Behavioral observations of *Paruroctonus* indicate most of the chemical signals of importance to scorpions (food, water, and potential mates) can be identified through this sensory channel alone. Gaffin et al. (1992) have used electrophysiological procedures to demonstrate sensitivity of these hairs to water and salts, but more comprehensive studies of other natural stimulants are needed.

The most remarkable condensation of chemotactic sensilla on scorpions, or perhaps on any arthropod, occurs on the pectines, unique ventral appendages of scorpions that are common to all species (Schroeder, 1908; Ivanov and Balashov, 1979; Brownell, 1989). Several hundreds to thousands of minute (<10 μ length) sensilla (called "pegs") are arranged in a flat field on the ventral sur-

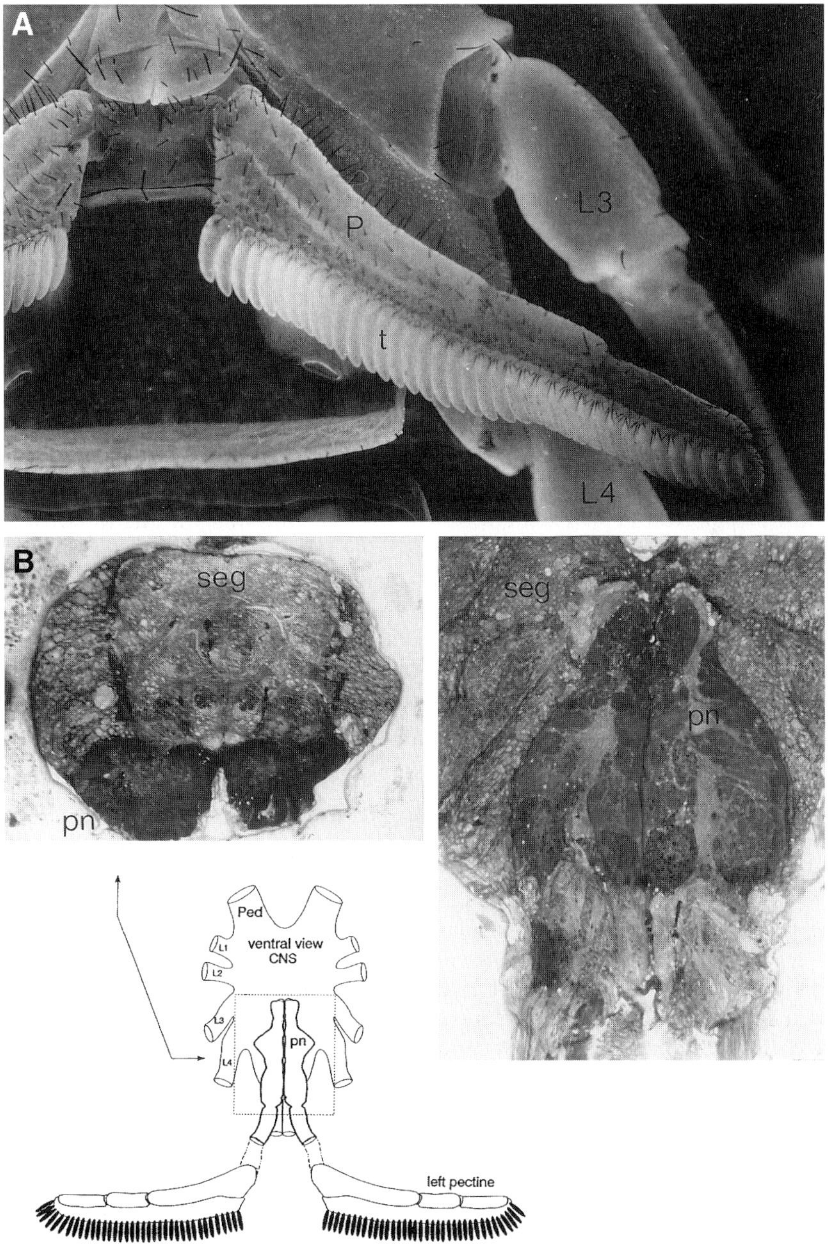

Figure 6.7 The pectinal sensory system of scorpions. (A) Ventral view of an adult male *Paruroctonus* (blacklight illumination) showing the left pectine and its 38 teeth (t). L3, L4, third and fourth legs. (B) Histological sections through the subesophageal ganglion (seg) taken in cross-section (left) and frontal section (right) through pectinal neuropil. Arrow and box in diagram show the level and placement of these sections. The pectine neuropil is large and darkly staining, with numerous glomerular substructures.

faces of each pectine tooth. Together, the aligned teeth on each pectine form a two-dimensional array of chemoreceptors that is applied to substrate surface each time the pectines are extended (fig. 6.7). In adult male sand scorpions the pectinal sensory field may contain more than 1 million sensory neurons, each receiving chemical cues from one point (one peg sensillum) within the entire two-dimensional array of more than 10,000–100,000 sensilla. In this way the pectines appear to function like a chemosensory "retina" capable of reporting the chemical content, position, and shape of chemical deposits on the sand. The electrophysiological response properties of these peg sensilla are described in chapter 7 of this volume and in Gaffin and Brownell (1997a,b). The spatial organization of this input can be assessed histologically by mapping the connections of sensory cells in the pectine to their terminal endings in the central nervous system.

Pectinal sensory projection

The pectinal nerves enter the subesophageal ganglion posteriorly and terminate in large, darkly staining (chromophilic) neuropil (Brownell, 1998). The cytoarchitecture of pectinal "lobes," or neuropil, is reminiscent of the olfactory lobes of insects (Homberg et al., 1989), especially in frontal sections where numerous glomeruli are evident (fig 6.7B). In cross-sections (fig. 6.7B) the pectinal neuropil is located medially and on the ventral surface of the brain. Enlarged cross-sections through the pectinal neuropil show distinctive layers composed of axonal bundles along the outer surface and synaptic terminals within the inner layers (fig. 6.8).

By selectively ablating portions of the pectinal sensory field (e.g., removal of pectinal teeth in distinct patterns), it is possible to correlate, or map, the topography of connections between the sense organ and its connections in the central nervous system. Figure 6.8 shows the results of an experiment where sensilla-bearing teeth (1–6 and 21–25) on the left pectine were removed along with teeth (11–14 and 32–36) on the right pectine. About 8 days after ablation, axonal degeneration appeared in the central nervous system as a discrete pattern easily related to the position of removed teeth. Similar patterns of connectivity were observed when these peripheral sensory structures were labeled with dyes that migrate along axons into the central nervous system. Thus, by its neuroanatomical arrangement, the input from each pectine has topographic order (e.g., point-to-point mapping), and all of its chemosensory input arrives in the central nervous system at coordinate locations within a spatially ordered array.

Topographical organization and layered cytoarchitectures are rarely seen in chemosensory pathways of other animals, so their functional significance for perception is unknown (Hildebrand and Shepherd, 1997). Some indication of a possible role comes from behavioral and electrophysiological observations (Hoffman, 1964; Krapf, 1986; Gaffin and Brownell, 1992) and from consideration of the scorpion's sensory ecology. In *Paruroctonus,* for example, the pectines appear to have both tactile (texture) and chemosensory functions. When the pectines are applied directly to the substrate, several thousand peg sensilla make momentary contact with grains of sand at the surface. Because the spatial frequencies for peg sensilla (>100/mm) are much greater than those of individual sand grains (~5/mm) (see fig. 6.3), each grain may contact many sensilla. This raises the possibility that chemical deposits having spatial dimen-

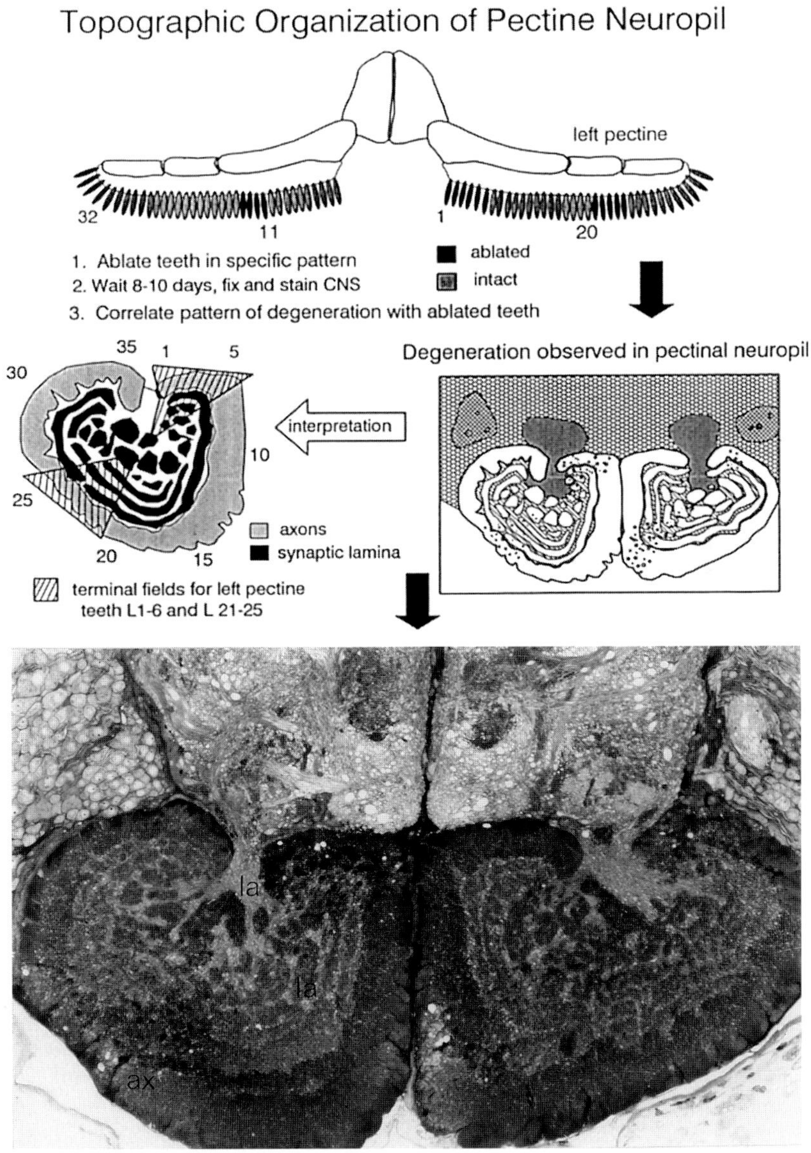

Figure 6.8 Topographic organization of chemosensory input to the pectinal neuropil. Selective ablations of teeth 11–14 and 32–36 on the right pectine and of teeth 1–6 and 21–25 on the left pectine induced a predictable pattern of axonal degeneration (ax, light spots) and nerve terminal degeneration (dark spots) in the central nervous system. Drawings illustrate the topographical pattern of sensory neuron projection indicated by these results. Note that microglomeruli are arranged in layers (la) within the pectinal neuropil.

sions less than the width of a grain of sand may be detected by this unique sensory organ.

Vision and Navigation

Except for some troglodyte species (Franke, 1978), all scorpions have conspicuous eyes. As discussed by Locket and Fleissner and Fleissner in this volume, the principal, or median, eyes have exceptional light-gathering ability, and the sensitivity of the lateral eyes to light is among the highest described for any animal (Fleissner, 1977b; Land, 1979). Nevertheless, *Paruroctonus* is a strictly nocturnal animal, rarely appearing on the surface within an hour of dusk and preferring moonless nights for surface activities of feeding and mating (Hadley, 1968; Polis, 1980a). Under these circumstances starlight is probably sufficient for visual navigation relative to large objects and the horizon (Fleissner, 1977b). Two potential uses of vision for *Paruroctonus* have been suggested by quantitative measurements of directed locomotion under low luminance and under polarized light.

Orientation to polarized light

Scorpions are relatively slow and persistent walkers, an attribute that makes them good subjects for experimental analysis of navigational cues guiding locomotory behavior. A particularly useful instrument for quantifying the direction of free walking is the locomotion compensator, or "Krammersphere," a large ball that rotates under servo-control to continuously reposition a walking animal to its apex (fig. 6.9A). Brownell and Weber (unpublished data) have used a Krammersphere to examine the sensitivity of scorpions to visual stimuli of low intensity. In the absence of all other stimuli, sand scorpions readily pursue small, detectable sources of illumination. When the source of light is diffuse and of uniform intensity across the visual field, walking is randomly directed and frequently interrupted by long pauses. However, when the illumination has directional cues, such as a gradient in intensity of the superimposed light, walking direction becomes fixed relative to the orientation cue and walking velocity increases. In the experiment shown in fig. 6.9, this acute sensitivity has been used to reveal an unexpected sensitivity of scorpions to polarized light. The sensors for e-vector orientation are the median eyes. Reversible covering of these, but not the lateral eyes, invokes randomly directed but continuous walking, as though sensitivity to light remains but without its directional cues.

Although polarization vision has well-established navigational importance for many diurnal arthropods (Rossel, 1989; Wehner, 1989), its potential significance for exclusively nocturnal animals is enigmatic. Within 1 h of sundown the night sky should lack polarized light of direct solar origin (Stair and Johnston, 1953). If polarized light has significance as a navigational cue for sand scorpions, then other potential sources of directed illumination, such as moonlight or the Milky Way, must be considered.

Fluorescence and vision

It is remarkable that the most striking and unique feature of many scorpions, their cuticular fluorescence under short wavelength illumination, has no known

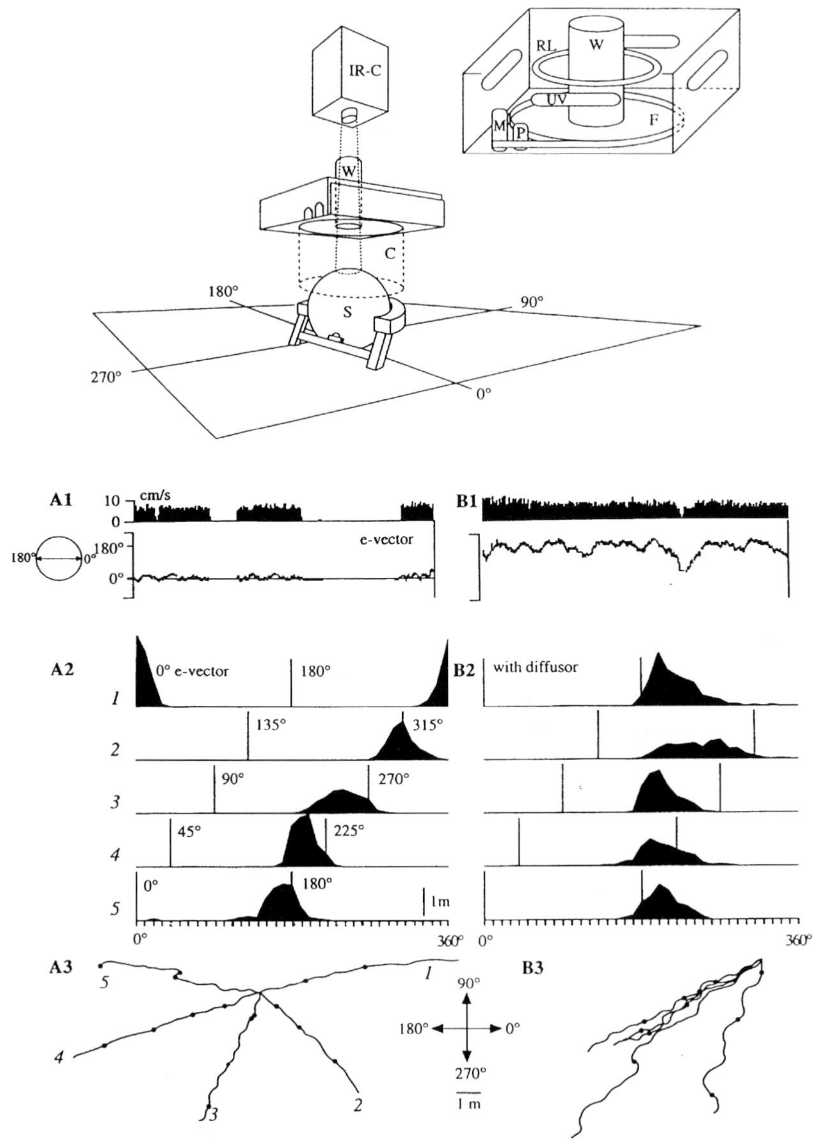

Figure 6.9 Sensitivity to polarized light and oriented walking. A locomotion compensator (Krammersphere) apparatus records the speed and direction of walking under illumination of variable (0–360°) orientation of e-vector polarity. A scorpion is placed at the apex of the sphere (S) is continuously monitored and repositioned by an infrared camera (IR) and rotation of the sphere under servocontrol. Direction and speed of walking are calculated from compensatory movements of the sphere. A low-intensity light source (RL) and rotatable polarization filter (F) are the only source of illumination. (A1) Plots showing velocity (upper trace) and direction (lower trace) of walking when overhead illumination is polarized at 0/180°. (A2, A3) As direction of light polarization varies from 0° to 180° in 45° increments, direction of walking varies accordingly to the e-vector (vertical lines in graphs). (B1, B2, B3) When the polarity of illuminating light was disrupted by a diffusion filter, walking became continuous and direction was unaffected by changes in the direction of the polarizer filter.

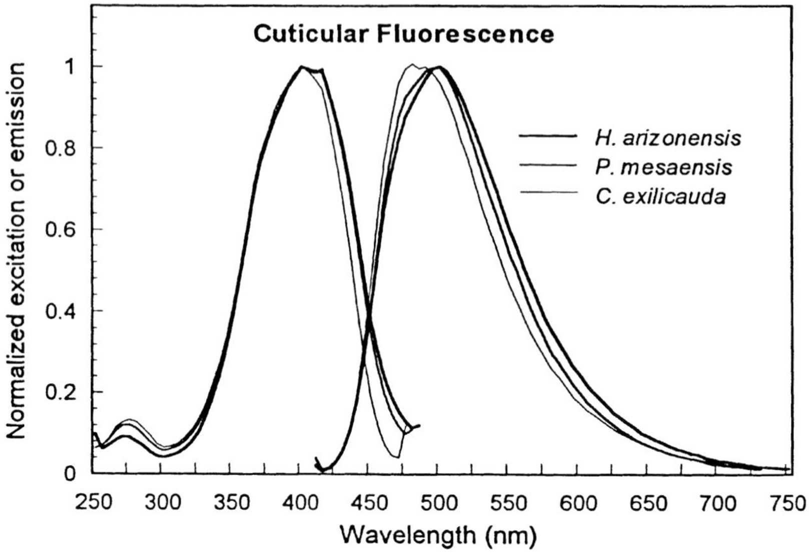

Figure 6.10 Excitation and emission spectra for cuticular fluorescence in three species of scorpion. Scorpions from three different families (*Hadrurus arizonensis* [Iuridae], *Paruroctonus mesaensis* [Vejovidae], and *Centruroides exilicauda* [Buthidae]) have similar excitation-emission spectra of cuticular fluorescence. Multiple peaks of both excitation (400, 415 nm) and emission (480, 490, 500 nm) indicate several chromophores contribute to fluorescence and to the slight variance in color between species (*C. exilicauda* is slightly bluer than the other two species).

function. Analysis of the physical characteristics of this property and some careful studies of scorpion behavior in its natural setting may soon resolve this interesting puzzle. Figure 6.10 shows the excitation and emission spectra for *Paruroctonus* and two other species of relatively close (*Hadrurus*) and distant (*Centruroides*) taxonomic relationship. The cuticle of each species shows the same pattern of excitation (peak absorbance near 400 nm) and emission (peaks at 495–505 nm), indicating that a family of related fluorochromes present in the epicuticle are responsible for the phenomenon. Close inspection of these spectra (irregularities in spectra at peaks) confirm that more than one molecule is responsible as also concluded from biochemical analysis of cuticle fluorochromes (Stachel et al., 1999). The emission spectra are noteworthy for their complimentarity to visual sensitivity of scorpion eyes (Fleissner, 1968; Machan, 1968) and those of other arachnids (Yamashita, 1985; Walla et al., 1996), suggesting that the absorbance (darkness) and emission (lightness) are seen by other scorpions as a contrast-enhanced image at these wavelengths. The eye lens of scorpions is one of the few cuticular structures that do not fluoresce, and Locket's analysis (chapter 3 this volume) of the visual field and optical properties of the median eye lens are consistent with a role in detection of weak light sources to the side or above the animal. Fluorescence may also function as an attractant for nocturnal insects, although efforts to demonstrate this in the field have not been successful.

COMPREHENSIVE ASSESSMENTS OF SENSORY EXPERIENCE

As our understanding of the sand scorpion's sensory biology and ecological milieu become ever more comprehensive, it is tempting to speculate about its evolved strategy for survival as a nocturnal predator of small arthropods. Underlying the entire strategy is the animal's ability to sense and avoid the physical extremes of its environment. This it accomplished through daytime refuge in a burrow and long periods of metabolic inactivation when conditions for feeding are not optimal. A precise mechanism for timing the exit and reentry into the burrow is of paramount importance, and this seems well accommodated by one of the most precise photoperiodic clocks yet described for any animal (Fleissner and Fleissner, 1985). The extent to which diurnal fluctuations in substrate temperature aid in time keeping during the lightless periods of inactivity in the burrow is unknown but easily tested.

For orientation to important biotic factors, the sand scorpion's effectiveness as a predator of insects seems well accounted for by the extraordinary capabilities of its mechanosensory organs. The vibration-sensitive slit sensilla and hairs of the tarsal leg segments are highly effective devices for localizing insect prey as they move on the surface in darkness. At close quarters the trichobothria hairs on the pedipalps detect air currents induced by movements of the prey and guide the prey-capturing appendages to the target. Identification and seduction of potentially cannibalistic mates is a paradoxical problem the scorpion appears to solve with highly specific chemicals, pheromones, released onto the substrate. This sensitivity is mediated to a large extent by the chemosensory pectines. The topographic organization of inputs to the central nervous system from the pectines suggest that spatial structure of the pheromonal signal, as well as its chemical composition, may be information-bearing qualities of the stimulus. For males wandering in pursuit of reproductive females, it is possible that the pattern of celestial brightness or of polarized light may be navigational cues.

Finally, the extent to which cuticular fluorescence may play a role in feeding behavior as a low-intensity visual lure for nocturnal flying insects or as an attractant to other scorpions for purposes of mate finding or territorial spacing is unresolved. The answers to these and other questions raised in this chapter may eventually lead to a comprehensive understanding of the sand scorpion's sensory ecology.

Acknowledgments I thank Dr. Victor Fet for critically reading the manuscript and for helpful suggestions for improving its organization and clarity. Lorin Davis provided information for modeling of heat flow into sand, and Drs. Tom Plant and David Morton aided in measurements of cuticular fluorescence properties. The behavioral studies of orientation to polarized light were done in collaboration with Dr. Theo Weber of the Max Planck Institute at Seewiesen, and measurements of displacement sensitivity were done with the help of Prof. Freiderich Barth and his laboratory in Vienna. All of these studies were supported by the National Science Foundation with help from a DAAD Fellowship and the DFG of Germany.

References

Abushama, F.T. (1959) Courtship and mating in the buthid scorpions. *Proc. Zool. Soc.* (London) 133: 145–169.

Abushama, F.T. (1964) On the behavior and sensory physiology of the scorpion *Leiurus quinquestriatus. Anim. Behav.* 12: 140–153.
Aicher, B. and Tautz, J. (1990) Vibrational communication in the fiddler crab, Uca puilator. 1. Signal transmission through the substratum. *J. Comp. Physiol.* 166A: 345–353.
Alexander, A.J. (1957) The courtship and mating of *Opisthophthalmus latimanus. Proc. Zool. Soc* (London) 128: 529–544.
Barth, F.G. (1985) Neuroethology of the spider vibration sense. *In* Barth, F.G. (ed.) *Arachnid Neurobiology,* pp. 3–19. Springer-Verlag, Berlin.
Barth, F.G. and Geethabali. (1982) Spider vibration receptors; threshold curves of individual slits in the metatasal lyriform organ. *J. Comp. Physiol.* 148A: 175–185.
Barth, F.G., Humphrey, J.A.C., Wastl, U., Halbritter, J. and Brittinger, W. (1995) Dynamics of arthropod filiform hairs. III: Flow patterns related to air movement detection in a spider (Cupiennius salei). *Phil. Trans. R. Soc. Lond.* B 347: 397–412.
Boyden, B.H. (1978) *Substrate selection by Paruroctous boreus (Girard) (Scorpionida: vejovidae)* (masters thesis). Idaho State University, Boise.
Brownell, P.H. (1977) Compressional and surface waves in sand used by desert scorpions to locate prey. *Science* 197: 479–482.
Brownell, P.H. (1998) Glomerular cytoarchitectures in chemosensory systems of Arachnids. *Ann. N.Y. Acad. Sci.* 855: 502–507.
Brownell, P.H. and Farley, R.D. (1979a) Prey localizing behaviour of the nocturnal desert scorpion, *Paruroctonus mesaensis: orientation to substrate vibrations. Anim. Behav.* 27: 185–193.
Brownell, P.H. and Farley, R.D. (1979b) Detection of vibrations in sand by tarsal sense organs of the nocturnal scorpion, *Paruroctonus mesaensis. J. Comp. Physiol.* A 131: 23–30.
Brownell, P.H. and Farley, R.D. (1979c) Orientation to vibrations in sand by the nocturnal scorpion *Paruroctonus mesaensis:* mechanism of target localization. *J. Comp. Physiol.* A 131: 31–38.
Brownell, P.H. (1984) Prey detection by the sand scorpion. *Sci. Am.* 251: 86–97.
Brownell, P.H. (1989) Neuronal organization and functions of the pectine chemosensory system of scorpions. *Chemical Senses* 10: 557.
Carthy, J.D. (1968) The pectines of scorpions. *Symp. Zool. Soc. Lond.* 23: 251–261.
Dusenbery, D.B. (1992) *Sensory Ecology.* W.H. Freeman, San Francisco.
Edney, E.B. (1977) *Water Balance in Land Arthropods.* Springer-Verlag, Berlin.
Edney, E.B., Haynes, S. and Gibo, D. (1974) Distribution and activity of the desert cockroach *Arenivaga investigata* in relation to microclimate. *Ecology* 55: 420–427.
Fet, V., Polis, G.A., and Sissom, W.D. (1998). Life in sandy deserts: the scorpion model *J. Arid Environ.* 39: 609–622.
Fleissner, G. (1968) Untersuchungen zur Sehphysiologie der skorpione. Verhandlungen der Deutschen Zoologischen Gesellschaft (Innsbruck) 61: 375–380.
Fleissner, G. 1977a. Scorpion lateral eyes: extremely sensitive receptors of Zeitgeber stimuli. *J. Comp. Physiol.* 118: 101–108.
Fleissner, G. 1977b. The absolute sensitivity of the median and lateral eyes of the scorpion *Androctonus australis* L. *J. Comp. Physiol.* 118: 109–120.
Fleissner, G. and Fleissner, G. 1985. Neurobiology of a circadian clock in the visual systems of scorpions. *In* Barth, F.G. (ed.) *Neurobiology of Arachnids,* pp. 351–375. Springer-Verlag, Berlin.
Foelix, R.F. and Müller-Vorholt, G. (1983) The fine structure of scorpion sensory organs. II. Pecten sensilla. *Bull. Br. Arachnol. Soc.* 6: 68–74.

Foelix, R.F. and Schabronath, J. (1983) The fine structure of scorpion sensory organs. I. Tarsal sensilla. *Bull. Br. Arachnol. Soc.* 6: 53–67.

Francke, O.F. 1978. New troglobite scorpion of genus *Diplocentrus. Entomol. News* 89: 39–45.

Gaffin, D.D. and Brownell, P.H. (1992) Evidence of chemical signaling in the sand scorpion, *Paruroctonus mesaensis* (Scorpionida: Vaejovidae). *Ethology* 91: 59–69.

Gaffin, D.D. and Brownell, P.H. (1997a) Response properties of chemosensory peg sensilla on the pectines of scorpions. *J. Comp. Physiol.* 181A: 291–300.

Gaffin, D.D. and Brownell, P.H. (1997b) Electrophysiological evidence of synaptic interactions within chemosensory sensilla of scorpion pectines. *J. Comp. Physiol.*181A: 301–307.

Gaffin, D.D., Wennstrom, K.L. and Brownell, P.H. (1992) Water detection in the desert sand scorpion *Paruroctonus mesaensis* (Scorpionida, Vejovidae). *J. Comp. Physiol.* 170A: 623–629.

Geiger, R. (1957) *The Climate Near the Ground.* Harvard University Press, Cambridge, Mass.

Gerstner, W., Kempter, R., van Hemmen, J.L. and Wagner, H. (1996) A neuronal learning rule for sub-millisecond temporal coding. *Nature* 383: 76–78.

Graff, K.F. (1975) *Wave Motion in Elastic Solids.* Clarendon Press, Oxford.

Hadley, N.F. (1990) Environmental physiology. *In:* Polis, G.A. (ed.) *The Biology of Scorpions,* pp. 321–340. Stanford University Press, Palo Alto, CA.

Hadley, N.F. (1968) Surface activities of some North American scorpions in relation to feeding. *Ecology* 49: 726–734.

Hildebrand, J.G. and Shepherd, G.M. (1997) Mechanisms of olfactory discirmination: converging evidence for common principles across phyla. *Annu. Rev. Neurosci.* 20: 595–631.

Hoffmann, C. (1964) Zur Funktion der Kammformigen Organ von Skorpionen. *Naturwiss.* 7: 172.

Hoffmann, C. (1967) Bau und Funktion der Trichobothrien von *Euscorpius carpathicus* L. *Z. Vergl. Physiol.* 5: 290–352.

Homberg, U., Christensen, T. and Hildebrand, J.G. (1989) Structure and function of the deutocerebrum in insects. *Annu. Rev. Entomol.* 34: 477–501.

Humphrey, J.A.C., Devarakonda, R., Iglesias, J. and Barth, F.G. (1993) Dynamics of arthropod filiform hairs. I. Mathematical modelling of the hair and air motions. *Phil. Trans. R. Soc. Lond.* B 340: 423–444.

Ignat'yev, A.M., Ivanov, V.P. and Balashov, Y.S. (1976) The fine structure and function of the trichobothria in the scorpion *Buthus eupeus,* Scorpiones, Buthidae. *Entomol. Rev.* 55(3): 12–18.

Iida, K. (1938) The velocity of elastic waves in sand. *Bull Earthq. Res. Inst. Tokyo* 16: 131–145.

Ivanov, V.P. and Balashov, Y.S. (1979) The structural and functional organization of the pectine in a scorpion *Buthus eupeus* Koch (Scorpiones, Buthidae) studied by electron microscopy. *Tr. Zool. Inst. Leningrad* 85: 73–81.

Krapf, D. (1986) Contact chemoreception of prey in hunting scorpions (Arachnida: Scorpiones). *Zool. Anz.* 217: 11–29.

Land, M. (1979) Comparative physiology and evolution of vision in invertebrates. *In Handbook of Sensory Physiology,* vol. 7, pt. 6A, p. 486. Springer-Verlag, Berlin.

Linsenmair, K.E. 1968. Anemomenotaktische Orientierung bei Skorpionen (Chelicerata, Scorpiones). *Z. Vergl. Physiol.* 6: 445–449.

Lighton, J.R.B., Brownell, P.H., Joos, B., and Turner, R.J. (In press) Low metabolic rate, high biomass, and cannibalism among scorpions.

Machan, L. 1968. Spectral sensitivity of scorpion eyes and the possible role of shielding pigment effect. *J. Exp. Biol.* 4: 95–105.
Markl, H. (1983) Vibrational communication. *In* Huber, F. and Markl, H. (eds.) *Neuroethology and Behavioral Physiology.* Springer, Berlin.
Michelsen, A. and Larsen, O.N. (1985) Hearing and sound. *In* Kerkut, G.A. and Gilbert, L.I. (eds.) *Comprehensive Insect Physiology, Biochemistry and Pharmacology,* vol. 6, pp. 495–556. Pergamon Press, Oxford.
Miller, J.P., Jacobs, G.A. and Theunissen, F.E. (1991) Representation of sensory information in the cricket cercal sensory system. I. Response properties of the primary interneurons. *J. Neurophysiol.* 66: 1680–1689.
Polis, G.A. (1979) Prey and feeding phenology of the desert sand scorpion *Paruroctonus mesaensis* (Scorpionida: Vaejovidae). *J. Zool.* 188: 333–346.
Polis, G.A. (1980a) Seasonal patterns and age-specific variation in the surface activity of a population of desert scorpions in relation to environmental factors. *J. Anim. Ecol.* 49: 1–18.
Polis, G.A. (1980b) The effect of cannibalism on the demography and activity of a natural population of desert scorpion. *Behav. Ecol. Sociobiol.* 7: 25–35.
Polis, G.A. (1990) Ecology. *In* Polis, G.A., (ed.) *The Biology of Scorpions,* pp. 247–293. Stanford University Press, Palo Alto, CA.
Polis, G.A. and Farley, R.D. (1979a) Behavior and ecology of mating in the cannibalistic scorpion, *Paruroctonus mesaensis* Stahnke (Scorpionida: Vaejovidae). *J. Arachnol.* 7: 33–46.
Polis, G.A. and Farley, R.D. (1979b) Characteristics and environmental determinants of natality, growth and maturity in a natural population of a desert scorpion, *Paruroctonus mesaensis* (Scorpionida: Vaejovidae). *J. Zool.* 187: 517–542.
Polis, G.A. and Farley, R.D. (1980) Population biology of a desert scorpion: survivorship, microhabitat and the evolution of life history strategy. *Ecology* 61: 620–629.
Polis, G.A., Myers, C.A. and Quinlan, M.A. (1986) Burrowing biology and spatial distribution of desert scropions. *J. Arid Environ.* 10: 137–146.
Polis, G.A. and Sissom, W.D. (1990) Life history. *In* Polis, G.A. (ed.) *The Biology of Scorpions,* pp. 161–223. Stanford University Press, Palo Alto, CA.
Polis, G.A., Sissom, W.D. and McCormick, S.J. (1981) Predators of scorpions: field data and a review. *J. Arid Environ.* 4: 309–326.
Reissland, A. and Gorner, P. (1985) Trichobothria. *In* Barth, F.G. (ed.) *Arachnid Neurobiology,* pp. 3–19. Springer-Verlag. Berlin.
Root, T. M. (1990) Neurobiology. *In* Polis, G.A. (ed.) *The Biology of Scorpions,* pp. 341–413. Stanford University Press, Palo Alto, CA.
Rossel, S. (1989) Polarization sensitivity in compound eyes. *In* Stavenga, D.G. and Hardie, R.C. (eds.) *Facets of Vision,* chapt. 13. Springer-Verlag, Berlin.
Schröder, O. (1908) Die sinnesorgane der skorpionskamme. *Z. Wiss. Zool.* 9: 436–444.
Seyforth, E.A., Sanders, E.J., and French, A.S. (1995) Sodium channel distribution in a spider mechanosensory organ. *Brain Research* 683(1): 93–104.
Shachak, M. and Brand, S. 1983. The relationship between sit-and-wait foraging strategy and dispersal in the desert scorpion *Scorpio maurus palmatus. Oecologia* 6: 371–377.
Shorthouse, D.J. (1971) Studies on the biology and energetics of the scorpion, *Urodacus yaschenkoi* (Birula, 1904) (Ph.D. dissertation). Australian National University, Canberra.
Stachel, S.J., Stockwell, S.A., and Van Vranken, D.L. (1999). The fluorescence of scorpions and cataractogenesis. *Chem. and Biol.* 6(8): 531–539.
Stahnke, H.L. (1966) Some aspects of scorpion behavior. *Bull. Calif. Acad. Sci.* 65: 65–80.

Stair, R. and Johnston, R. (1953) Ultraviolet spectral radiant energy reflected from the moon. *J. Res. Natl. Bureau Stand.* 51: 81–84.

Stürzl, W. (1998) Ein neuronales Modell zur Beutelokalisation de Wüstenskorpions Paruroctonus mesaensis. Ph.D thesis. Inst. für Theoretische Physik, Tech. U. München.

Stürzl, W., Kempter, R., and Van Hemmen, J.L. (2000) Theory of arachnid prey localization. *Phys. Rev. Let.* 88: 5668–5671.

Tautz, J. (1978) Reception of medium vibration by thoracal hairs of caterpillars of Barathra brassicae L. (Lepidoptera, Noctuidae). II. Response characteristics of the sensory cell. *J. Comp. Physiol.* 125A: 67–77.

Theunissen, F.E. and Miller, J.P. (1991) Representation of sensory information in the cricket cercal sensory system. II. Information theoretic calculation of system accuracy and optimal tuning-curve widths of four primary interneurons. *J. Neurophysiol.* 66: 1690–1703.

Walla, P., Barth, F.G. and Eguchi, E. (1996) Spectral sensitivity of single photoreceptor cells in the eyes of the ctenid spider *Cupiennius salei* Keys. *Zool. Sci.* 13: 199–202.

Warburg, M.R. and Polis, G.A. (1990) Behavioral responses, rhythms and activity patterns. *In* Polis, G.A. (ed.) *The Biology of Scorpions,* pp. 224–246. Stanford University Press, Palo Alto, CA.

Wehner, R. (1989) Neurobiology of polarization vision. *Trends Neurosci.* 12: 353–359.

White, J.E. (1965) *Seismic waves—radiation, transmission and attenuation.* McGraw-Hill, New York.

Williams, S.C. (1966) Burrowing activities of the scorpion *Anuroctonus phaiodactylus* (Wood) (Scorpionida: Vaejovidae) *Proc. Calif. Acad. Sci.* 3: 419–420.

Withers, P.C. (1992) *Comparative Animal Physiology.* Saunders Publishing, New York.

Yamashita, S. (1985) Photoreceptor cells in the spider eye: spectral sensitivity and efferent control. *In* Barth, F.G. (ed.) *Arachnid Neurobiology,* pp. 3–19. Springer-Verlag, Berlin.

Yokota, S.D. (1979) *Water, Energy and Nitrogen Metabolism in the Desert Scorpion Paruroctonus mesaensis* (Ph.D dissertation). University of California, Riverside.

7

Chemosensory Behavior and Physiology

DOUGLAS GAFFIN
PHILIP BROWNELL

Of all the sensory capacities animals possess, the ability to access chemical content of the surrounding environment is most primitive and influential: visual, auditory, and mechanosensory systems often guide animals to food and prospective mates, but it is the chemistry of contact that most often dominates the release of adaptive behaviors. It is paradoxical, therefore, that of all the senses used by animals, the chemical senses remain least understood at the physiological and biochemical levels. This is perhaps attributable to the enormous diversity of chemical substances in natural environments and to the complexity of peripheral and central nervous systems that process chemosensory information. In this chapter we describe several aspects of scorpion chemosensory physiology and behavior that make these animals favorable models for future research in this area.

Among the metazoa, mandibulate arthropods have already contributed inordinately to our understanding of the chemicals sensed by animals and to the physiological and behavioral processing of this information (Dethier, 1976; Kaissling, 1987; Derby and Atema, 1988). The success of these investigations is due largely to the comparative simplicity of chemical signals used for communication between individuals and the accessibility of their externalized organs for detection, the antennae. In the most familiar examples from insects, female Lepidoptera commonly advertise their reproductive readiness using volatile, species-specific compounds—pheromones—that powerfully activate stereotyped searching behaviors in conspecific males. Many aspects of the sensory anatomy and physiology mediating the male's behavior are known, even to the level of central neuronal circuits conveying sensory input to premotor pathways of locomotion. As we penetrate ever more deeply into studies of the insect-antennal model of chemosensory biology, it should be instructive to study other members of the phylum, particularly those using entirely different strategies and mechanisms for chemical signaling. By comparing these different but related organisms we gain insights into the evolutionary processes that shape the structure and function of chemosensory systems.

Chelicerate arthropods, particularly the several orders of terrestrial arachnids, offer many excellent candidates for comparative study. Aside from the obvious antennalike functions of nonambulatory anterior legs of uropygids and amblypygids, the malleolar organs of solpugids and the pectines of scorpions are now known to be specialized appendages of chemosensory function (Brownell and Farley, 1974; Ivanov and Balashov, 1979; Gaffin and Brownell, 1997a). Although all of these appendages are used for sensing properties of the substrate, the malleoli and pectines are especially important because, unlike any other arthropod chemosensory organ, they extend ventrally from posterior appendages. Both organs (pectines and malleoli) make intimate contact with the substrate and their sensillar microanatomy is appropriate for detection of nonvolatile substances. Thus, by gross anatomical criteria, we can consider the two great successes in terrestrial arthropod evolution (i.e., insects, arachnids) as having followed two lines of chemosensory specialization: a "chelicerate" mode using specialized ventral organs for detecting nonvolatile stimuli on substrates, and a "mandibulate" (insect, decapod crustacean) mode using anterior and dorsal appendages (e.g., antennae) more specialized for detecting olfactants in air or water.

The pectines seem especially advantageous for study of substrate-mediated chemical signaling. Although several properties of these organs suggest they are similar in function to the antennae of insects, they possess unique anatomical and physiological qualities that ensure they will become an important new preparation for basic research.

CHEMOSENSORY BEHAVIOR

In selecting appropriate model systems for study of chemosensory biology, special consideration must be given to the potential for ethological analysis. Natural behaviors are often incompatible with the laboratory environment, even when animal survivorship in captivity is good. We find in sand scorpions an excellent model for combined field and laboratory studies.

Our preferred subject for study is *Paruroctonus mesaensis* from the Mojave Desert of Southern California. This animal is comparatively innocuous, abundant, and easily maintained in the laboratory. Many aspects of its sensory biology, physiology, and ecology are well described because it is easily observed in the field and it withstands physiological manipulations in the laboratory (Stahnke 1966; Hadley 1974; Brownell, 1977; Polis, 1979, 1980). After the first molt, juvenile and adult *P. mesaensis* live singly in burrows, emerging for a few hours at night to forage for insect prey within a meter of the burrow entrance (Polis, 1980; Brownell, 1984). As ambush predators they assume motionless hunting postures and respond to vibrational stimuli created by arthropod prey (including members of their own species) moving nearby (Brownell, 1977). During summer months mature males leave the relative safety of their home burrows and wander over the dune surface in search of potential mates (Polis, 1980). Once a sexually receptive female is located, both sexes display stereotyped reproductive behaviors, including the elaborate *promenade aux deux,* a mating behavior in which the male grasps the female by the pedipalps and leads her to an appropriate substrate where a spermatophore is deposited (see Benton, this volume). Fertilization occurs when the male positions the female over the spermatophore and she takes it into her genital aperture.

Given the nocturnal habits of *P. mesaensis* and the limited visual and auditory abilities of scorpions in general, there is much uncertainty about the sensory information that guides the male scorpion to reproductively mature females. Several anecdotal observations indicate that chemical cues may direct the mate-finding behavior. For example, researchers commonly find clusters of two or more males near burrows of conspecific females, suggesting that species-

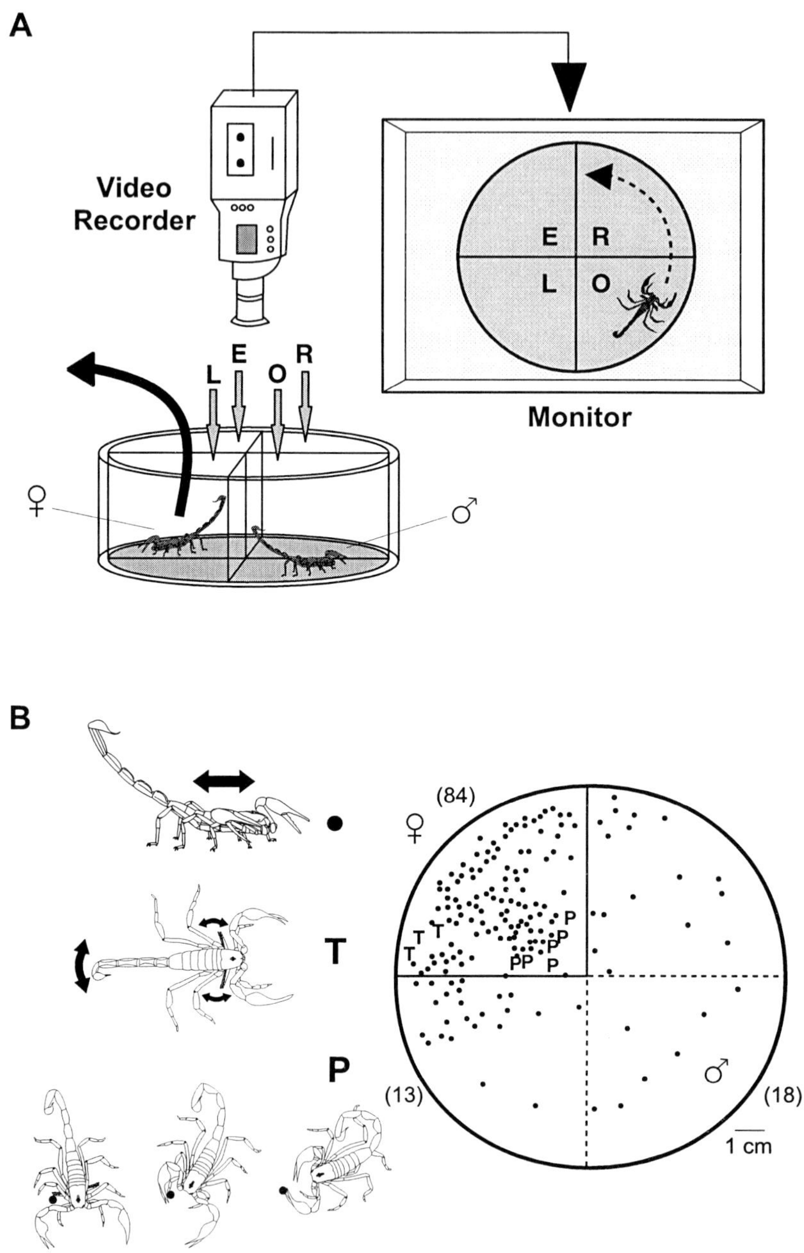

Figure 7.1 Behavioral evidence of chemosensitivity in the sand scorpion *P. mesaensis.* (A) Video monitoring of scorpion movements through quadrants in a round arena. The test animal is restrained behind a fence in the quadrant opposite the test stimuli. A trial is initiated by lifting the partition, allowing the responder access to the remainder of the arena. Stimuli include natural deposits from conspecific females and organic washes from sympatric and conspecific scorpions. Illumination is via a fluorescent black light bulb.

specific chemical signals may be the male attractant (G.A. Polis, personal communication). Also, before males make direct contact with females, they commonly display stereotyped courtship behaviors, such as "juddering," a vibrational movement suspected to be a mate-calling behavior at close range. Other workers have observed that males of *Pandinus imperator* and *Bothriurus bonariensis* judder or show other pre-courtship behavior after contact with objects or substrates previously exposed to female conspecifics (Krapf, 1986; Peretti, 1995), further suggesting that chemical signals may stimulate mating behavior. The communicative importance of juddering is unknown, but it may alert the female to an approaching male and subsequently suppress her reflexive predatory behaviors. The risk of predation or cannibalization during mate search (Polis and Farley, 1979a) underscores the importance of accurate and effective communication between these animals.

The importance of chemical signals in guiding mating behavior of *P. mesaensis* can be investigated in the laboratory. One method of study is shown in fig. 7.1A. When males encounter sand substrates previously occupied by conspecific females, their forward steps shorten and their tendency to turn increases (Gaffin and Brownell, 1992). Juddering and other precourtship behaviors (e.g., tail wagging, pedipalp reaching) are released as the response gains intensity near the center of the treated substrate. These behaviors also are elicited in response to organic extracts of female cuticle dried onto sand or filter paper substrates (fig. 7.1B). Moreover, changes in male behavior generally occur only after contact with the treated substrate, a further indication that the chemical signal is of low volatility and requires gustation or near-range olfaction for detection (Gaffin and Brownell, 1992).

CHEMOSENSORY ORGANS

If mate-finding behavior involves chemical signals associated with the substrate, which of the several sensory structures contacting the substrate are important? Numerous pore-tipped sensilla on the pedipalps and tarsal leg segments contact the substrate, where they might mediate chemosensitivity (Foelix and Schabronath, 1983). Behavioral and physiological experiments have already indicated that setaform hairs on the tarsi detect substrate-associated moisture (Gaffin et al., 1992), but their role in detection of intraspecific chemical signals has not been examined. In spiders, a porous cuticular depression on the dorsal surface of each tarsus, called the tarsal organ (Foelix and Schabronath, 1983), may detect sex-specific signals because they respond differentially to extracts derived from male and female cuticle (Dumpert, 1978). In scorpions the

(B) Response of male *P. mesaensis* to substrates containing cuticular extracts (chloroform:methanol, 2:1, dried onto 1 g sand) from conspecific female. Occurrence of three stereotyped behaviors, juddering (·), tail wagging (T), and pedipalp reaching (P), are indicated for the first 5 min of animal movement. The frequency of pectine taps on the substrate per minute is indicated by numbers in parentheses for responding male as it first moved from the initial quadrant (O) through quadrant L and into the treated quadrant (E). (Adapted from Gaffin and Brownell, 1992).

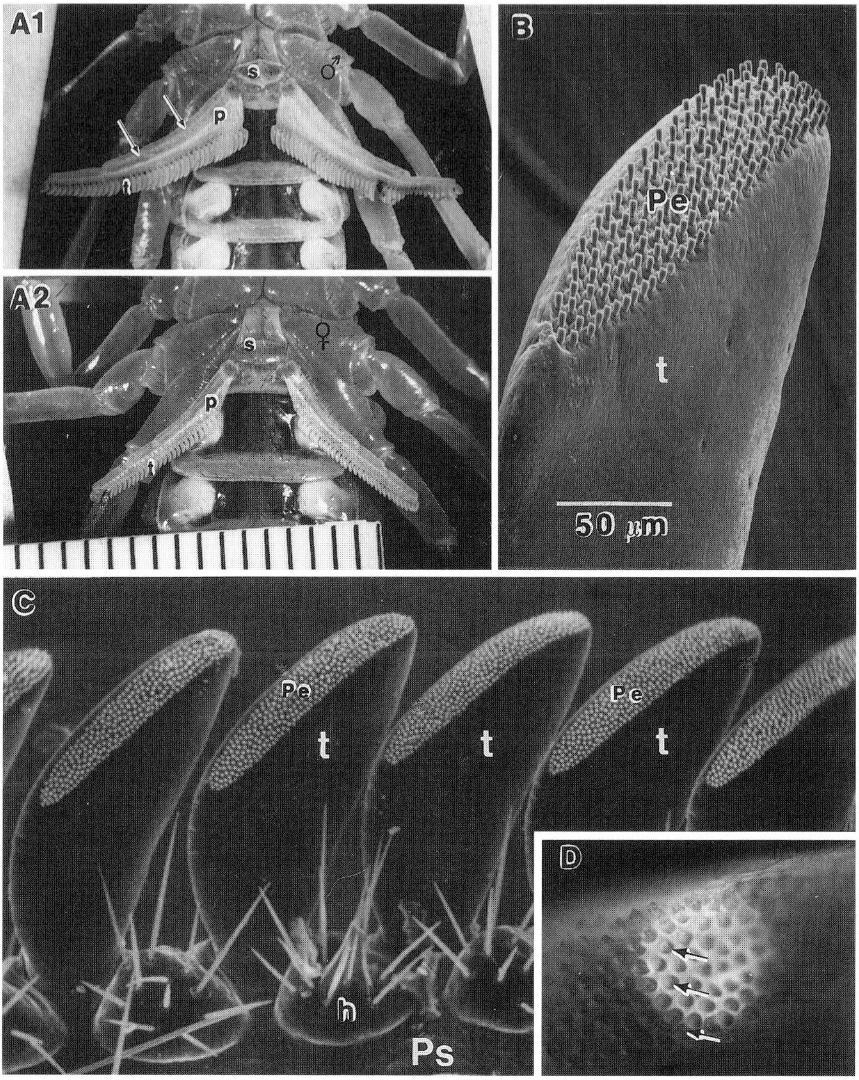

Figure 7.2 External morphology of the pectines. (A) Ventral view of *P. mesaensis* showing the larger size of pectines (p) and greater numbers of sensilla-bearing teeth (t) in males (A1) compared to females (A2) (males = 36.0 ± 1.6; females = 25.5 ± 1.4). (B) Scanning electron micrograph of distal face of female tooth showing dense arrays of peg-shaped sensilla (Pe). (C) Peg fields of adjacent teeth overlap in the horizontal axis; only peg sensilla and tactile hairs contact substrate when the pectines are swept forward during locomotion; h, tuft of mechanosensory hairs; Ps, spine of pectine. (D) Epifluorescent illumination of peg sensilla in a live animal. Arrows point to individual pegs.

tarsal organs respond to humid air stimuli (Gaffin et al., 1992), but their sensitivity to pheromonal and other chemical cues remains unexamined.

The largest and most elaborate sensory appendages of scorpions are the pectines (fig. 7.2), two ventromedial appendages that actively sweep over the substrate as the animal walks. Early morphologists considered these comblike extensions to be manipulative or respiratory appendages (see Cloudsley-Thompson, 1955, for summary of early work) until behavioral studies showed they were used as tactile organs to select substrates for placement of the spermatophore (Alexander, 1957, 1959). Neuroanatomically, the pectines were known to rival the innervation density of the antennae of insects and crustaceans (Schröder, 1908) suggesting a more important sensory role than mechanoreception. However, additional behavioral studies of the pectines only confirmed their role as mechanoreceptors (Abushama, 1964). Most important, the initial studies of pectine ultrastructure by Carthy (1966, 1968) failed to reveal pores at the tips of their numerous peg sensilla (a criterion of chemosensitivity), and Hoffmann's (1964) electrophysiological studies found only mechanosensory responses to various modalities of pectine stimulation, including direct application of chemostimulants. These findings dampened interest in further study of the organ, and their exclusive function as mechanoreceptors for substrate texture went unchallenged for almost two decades. In retrospect, these findings contributed strongly to the misconception that scorpions, and perhaps arachnids in general, lacked chemosensitivity comparable to antenna-bearing mandibulate arthropods.

The rebirth of interest in chemosensory biology of scorpions can be traced to the work of Ivanov and Balashov (1979) and Foelix and Müller-Vorholt (1983), who independently reexamined pectine ultrastructure using improved fixation techniques for scanning and transmission electron microscopy. These studies clearly showed an external opening at the terminal end of each peg sensillum in *Mesobuthus eupeus* (Buthidae) *and Euscorpius italicus* (Chactidae), respectively. Pore openings have since been found in representatives of several families, including *Paruroctonus mesaensis* (Vaejovidae), *Hadrurus arizonensis* (Iuridae), and *Pandinus gregoryi* (Scorpionidae) (Brownell, 1988, 1989). In each case the morphology of sensory and accessory cells within the pegs were typical of arthropod contact chemoreceptors, thus inviting reevaluation of their physiological and behavioral functions.

An important advance in understanding pectine function came with the realization that male scorpions may use them to detect pheromones deposited on the substrate by conspecific females. An example of this behavior is illustrated in fig. 7.1B where a male scorpion strongly increased the frequency of pectine "tapping" as it crossed substrate labeled only with the wash from the cuticle of a conspecific female. Similarly, in the example shown in fig. 7.3A, an adult male *H. arizonensis* increased the frequency of its pectine tapping as it entered the vacant home container of male or female conspecifics, but much more dramatically when the previous occupant was a mature female (Brownell, 1988). In another experiment (fig. 7.3B), the precourtship behavior of pectine-ablated males was significantly depressed relative to intact males when both were exposed to substrates treated with cuticular washes from conspecific females (Gaffin, 1994). The simplest explanation of these results is that the pectines are important organs for chemosensory identification of prospective mates.

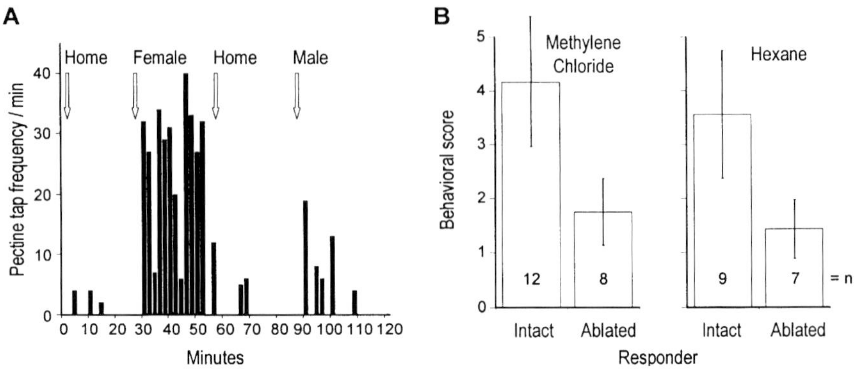

Figure 7.3 Behavioral evidence of pectine involvement in pheromone detection. (A) Frequency of substrate tapping by the pectines of male *Hadrurus arizonensis* as it is moved from its home container to a container previously occupied by a conspecific female. Placement of the male in a container used by a conspecific male failed to evoke intense tapping. (B) Effect of pectine ablation on male precourtship behavioral responses to extracts from conspecific females. Histograms show mean behavioral scores (±SE) of intact and pectine-ablated male *P. mesaensis* to methylene chloride and hexane extracts from conspecific female cuticle. Pooled cuticular extracts from 10 females were dried onto a small volume of sand; see Gaffin and Brownell, 1992, for description of scoring system. Pectines were surgically removed from animals at least 1 week before testing to allow recovery. Locomotory behavior and activity of pectineless animals were otherwise normal at time of testing.

STRUCTURE AND PHYSIOLOGY OF PECTINE SENSILLA

Sensory Morphology

The importance of the male's pectines for mate localization, courtship, and reproduction is consistent with the high degree of sexual dimorphism evident in most species. Pectines are mobile appendages with three flexible segments constituting the "spine," or marginal lamella of the pectine comb (see chapter 2). To the marginal lamellae, as few as five (*Euscorpius*), or as many as 40 (*Parabuthus*), "teeth" are attached. These teeth support dense, two-dimensional arrays of chemoreceptive setae, called "pegs" (fig. 7.2). The pectines of males are longer and typically contain more sensilla-bearing teeth than those of females (table 7.1). Although the number of teeth per pectine does not change from birth, the rate of pectine elongation in adult males may exceed that of females. For example, in *P. mesaensis,* pectine elongation in sexually mature males is greater than in immature males or females at any stage of development (Polis and Farley, 1979b). Most important, in *Paruroctonus* (Vaejovidae) and *Parabuthus* (Buthidae), the number of chemosensory pegs in males increases dramatically relative to females as they become sexually mature (Brownell, 1989) and differs in adult animals even in the few species where males and females have similar numbers of teeth (e.g., *Superstitionia donensis*) (Swoveland, 1978). Such sex-

Table 7.1 Sexual Dimorphism of Scorpion Pectines

Family	Species	Sex	No. teeth/ 2 pectines	No. pegs/ tooth	Total no. pegs	No. pegs (♂)/ no. pegs (♀)	Reference
Vaejovidae	*Anuroctonuis phaiodactylus*	♂	19	201	3,819	2.55	1
		♀	13	115	1,495		
	Nullibrotheas allenii	♂	24	160	3,840	2.75	1
		♀	17	82	1,394		
	Paruroctonus mesaensis	♂	75	1,600	120,000	13.89	1
		♀	48	180	8,640		
	Uroctonus mordax	♂	27	1,100	29,700	9.71	1
		♀	20	153	3,060		
	Vaejovis confusus	♂	34	1,000	34,000	4.31	1
		♀	27	292	7,884		
	Vaejovis spinigerus	♂	46	1,000	46,000	4.01	1
		♀	37	310	11,470		
Buthidae	*Centruroides exilicauda*	♂	50	800	40,000	1.86	1
		♀	44	490	21,560		
	Parabuthus pallidus	♂	72	1,650	118,800	1.82	2
		♀	62	1,050	65,100		
Chactidae	*Superstitionia donensis*	♂	12	800	9,600	1.63	1
		♀	12	490	5,880		
Diplocentridae	*Didymocentrus comondae*	♂	17	1,050	17,850	7.83	1
		♀	15	152	2,280		
Scorpionidae	*Pandinus gregoryi*	♂	36	1,400	50,400	1.65	2
		♀	34	900	30,600		

Refercences: (1)Swoveland, 1978; (2) Brownell, unpublished data.

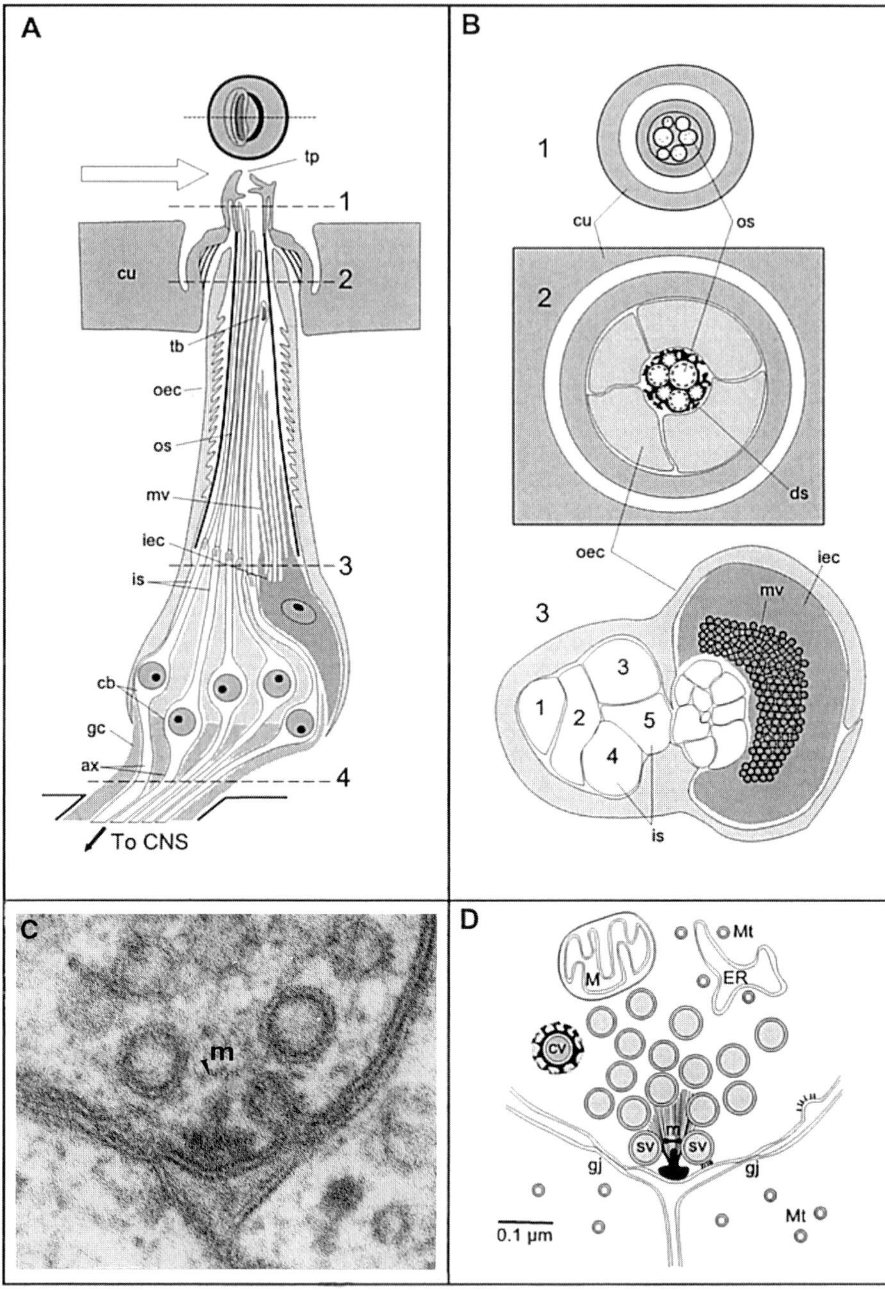

Figure 7.4 Morphology of peg sensilla. (A) Diagram of a longitudinal section through sensillum showing arrangement of dendrites, cell bodies, inner and outer enveloping cells, and structural elements (after Ivanov and Balashov, 1979; Foelix and Müller-Vorholt, 1983). (B) Cross-sections taken from levels 1–3 of panel A showing (1) double-walled shaft containing several dendritic outer segments (os) inside receptor lymph; (2) cuticular base of

ual and developmental differentiation is strongly suggestive of a function related to reproduction.

Blanchard (1853) was first to show that the pectines are richly innervated, and Gaubert (1889) and Schröder (1908) showed that most of the neural elements extend through the teeth and into the peg-shaped sensilla. By comparison to other chemosensory organs of arthropods, Schröder (1908) concluded that the pectines were both mechanosensory and chemosensory in function and were perhaps important in mediating sex recognition in scorpions. Nearly a century later these insightful speculations from histological observations are largely confirmed.

The peg sensilla of scorpions have many structural features in common with chemosensory sensilla of insects (Foelix and Müller-Vorholt, 1983). In external appearance the pegs look like truncated hairs extending from circular sockets spaced 2–10 μm apart. In *P. mesaensis* the pegs are about 2 μm in diameter and 8 μm long, with double-walled shafts that gradually flatten toward the peg tip (Fet and Brownell, unpublished data). Here, a slit-shaped pore opens to a fluid-filled chamber inside the sensillum (fig. 7.4). The reported number of dendrites varies from 10 to 18 per peg (Ivanov and Balashov, 1979; Foelix and Müller-Vorholt, 1983; Brownell and Adams, unpublished data). Most of the bipolar sensory neurons extend through the sensillar chamber to within a few microns of the slit opening. Foelix and Müller-Vorholt (1983) described one neuron as a mechanoreceptor having a characteristic tubular body in a dendrite terminating near the peg base. The dendritic outer segments are surrounded by a dendritic sheath and extensions of outer enveloping cells that appear to isolate the outer segments from hemolymph. At the level of the ciliary regions of the dendritic inner segments are numerous microvillar projections from an inner enveloping cell. The electron-dense inclusions of this microvillar cell are perhaps secretory vesicles containing substances unique to fluids of the sensillar chamber, including putative odorant-binding proteins (Slipher, 1970; Foelix and Müller-Vorholt, 1983; Bulseco and Brownell, 1989). The inner segments of the dendrites are short (<10 μm) compared to the outer segments (20–40 μm). The inner segments of five of the sensory neurons, or about one-third of the sensory cells of each peg, show a highly regular array opposite the microvillar projections of the inner enveloping cell (Brownell, 1989). This reveals a polarity in the arrangement of dendrites within each peg and may provide a means of identifying individual sensory neurons within each sensillum for correlation with physiological properties (see below). The cell bodies of sensory cells are spindle shaped and form a layer or "plate" of somata approximately 50 μm below the cuticular surface.

An interesting feature of peg sensillum morphology is the appearance of chemical synaptic profiles between sensory cell axons. These synapses are lo-

---

sensilla, extensions of outer enveloping cells (oec), dendritic sheath (ds), and dendritic outer segments with microtubules arranged at periphery; and (3) regular arrangement of five dendritic inner segments (is) relative to microvillar (mv) projections of inner enveloping cell (iec). (C) Typical dyad synapse and axo-axonic profiles as seen just below sensory cell body layer in pectine (level 4 in panel A) of *Androctonus australis* (×82,500; courtesy of R. Foelix). (D) Diagram of a dyad synapse; m, capping membrane; sv, synaptic vesicles; M, mitochondrion; ER, smooth endoplasmic reticulum; Mt, microtubules; cv, coated vesicles; gj, gap junctions. (From Foelix, 1985 with permission).

cated just below the somata region where axonal fibers fasciculate into bundles wrapped by glial cell extensions (Foelix and Müller-Vorholt, 1983). The most common pattern of contact is the dyad type, where one fiber is presynaptic to two others (fig. 7.4C,D). Presynaptic cells are distinguished by the presence of synaptic vesicles and bars, capping membranes and presynaptic organelles such as mitochondria and endoplasmic reticulum (Foelix, 1975, 1985). The axons vary markedly in size, and it is not known if all of these sensory axons continue to the central nervous system. It is conceivable that only a subset of these neurons project axons to the scorpion brain, as commonly observed for retinal sensory systems. This question can be resolved by counting the number of neural elements at several levels along the neural projection to the central nervous system.

Morphological evidence of synapses between peripheral sensory neurons is rare among insects, but such synapses may be a common feature for arachnids, having been also described in the antennaform first-leg pair of whip spiders (amblypygids) and harvestmen (Opiliones) (Foelix, 1975). Only recently have there been physiological studies to corroborate this morphology. Below, we describe methods for recording neural activity from pectinal peg sensilla as well as physiological evidence of interactions between active neural elements within these structures.

Electrophysiological Response

Electrical recordings from individual sensory receptors indicate their mode of response and sensitivity to natural stimuli. For peg sensilla on the pectines, the impulse activities of individual sensory neurons can be used to assay the organ's response to potential food and pheromonal stimuli. As a rule, electrophysiological recordings are most useful when applied to sensilla containing few sensory cells, as in pheromone-sensitive hairs of saturniid moth antennae (Kaissling, 1987). Peg sensilla of the pectine contain as many as 18 sensory neurons, but fewer than one-third of these are spontaneously active. Furthermore, these units discharge at low rates (<5 Hz) and can be identified individually by their impulse waveform. These features make it possible to access the most effective stimuli for several of the peg sensillar neurons.

In a typical electrophysiological preparation of scorpion pectine, the animal is anesthetized (by cooling or CO_2) and fixed ventral-side up to a rigid stage. The pectines are further immobilized on a small platform using double-sided tape to arrange the teeth for maximal exposure of the peg fields. A recording electrode with a tip diameter of <0.1 μm is inserted through the flexible cuticle at the peg base (fig. 7.5A), and a reference recording electrode is placed in contact with hemolymph at some distance from the recording site (e.g., soft cuticle at the distal end of the pectinal spine). Conventional electronic instrumentation is used to amplify, visualize, and record the electrical signals for later analysis. In this recording configuration, chemical stimuli can be applied to the sensillar pore directly (gustatory stimulation) or blown across the preparation as a volatile gas (olfactory stimulation). Such recordings are stable for several days without appreciable degeneration of signal form (Gaffin and Brownell, 1997a). The spontaneously active sensory cells produce extracellular action potentials of approximately 1–2 mV, typically with a signal-to-noise ratio >5. Because hundreds of peg sensilla are accessible within small regions of a single tooth, a

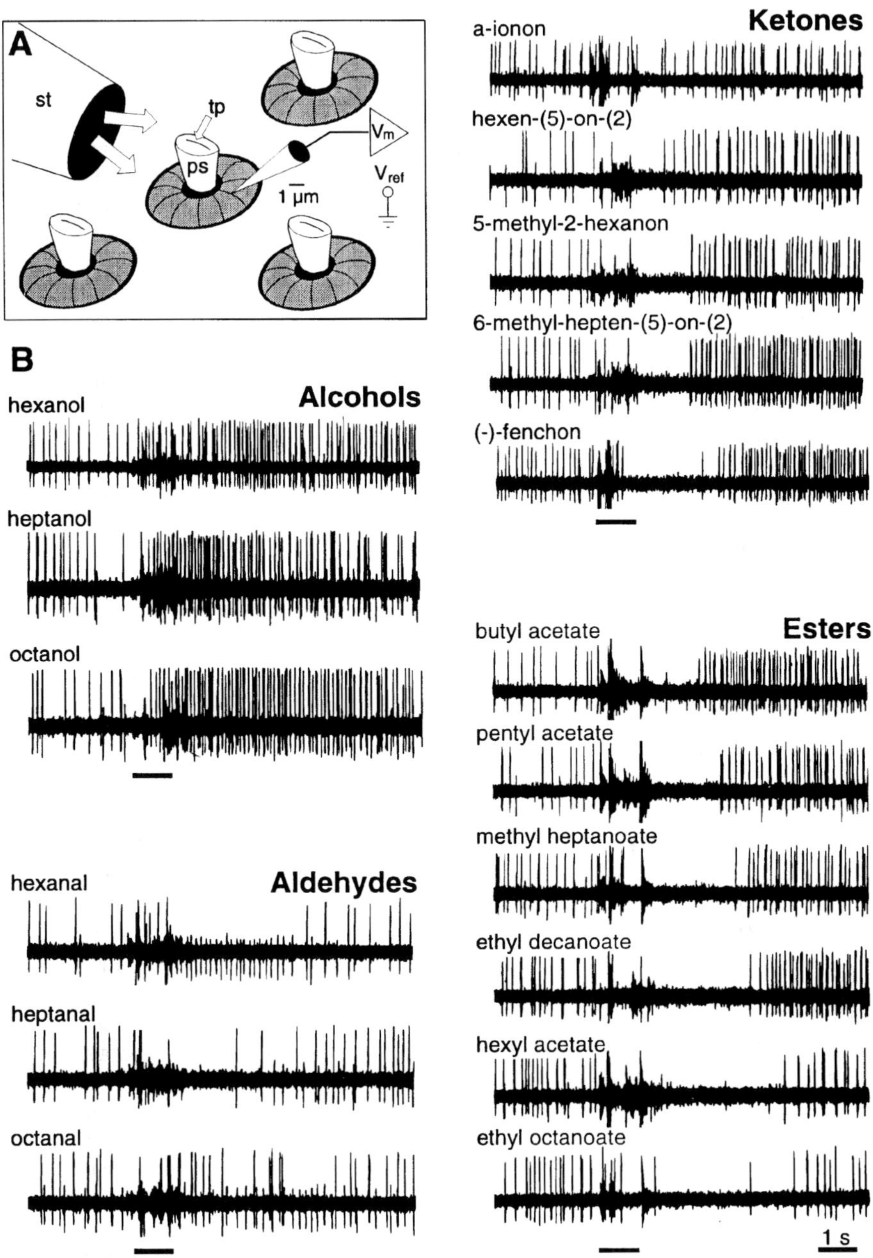

Figure 7.5 Electrophysiological response of peg sensillum to general chemical stimuli. (A) Extracellular recording configuration from peg sensillum. Electrolytically etched tungsten electrode (V_m) is inserted through the flexible cuticle at the base of peg sensillum (ps), allowing stimuli (st) access to terminal pore (tp); reference electrode (V_{ref}) is placed in contact with hemolymph at some distance from recording electrode. (B) Unfiltered, multiunit responses of a single peg sensillum to stimulation by pure substances. Volatile alcohols, aldehydes, ketones, and esters were blown in 1-sec pulses across the sensillum pore. Solid bars at bottom of traces indicate the stimulus duration.

single animal preparation can quickly generate a profile of sensillar responses to many different substances (Gaffin and Brownell, 1997a).

Individual peg sensilla on *P. mesaensis* respond strongly to stimulation with short-chained organic compounds blown across the sensillar pore (Gaffin and Brownell, 1997a). Representative responses of a single peg sensillum to stimuli from four molecular families (alcohols, aldehydes, ketones, and esters) are shown in figure 7.5B. The high threshold for olfactory stimulation suggests that the peg sensillum functions best as a contact chemoreceptor. In some sensilla, simple, straight-chained alkyl-alcohols (C_6–C_8) excite large triphasic units in a dose-dependent manner, whereas aldehydes (C_6–C_8) excite a small, biphasic unit, which appears to inhibit firing of the larger units. Stimulation with ketones and esters of the C_6–C_{11} range initially inhibit, then tonically excite, activity of the large units. In most cases the time course and intensity of a response depends on the concentration and type of stimulus. Thus, although scorpion peg sensilla have morphology typical of contact chemoreceptors and appear to be used for gustation of the substrate, they are capable of responding vigorously to olfactory stimuli. This convenient method of stimulation allows for efficient investigation of their chemosensory response properties.

SYNAPTIC INTERACTION BETWEEN CHEMOSENSORY NEURONS

Arthropod chemoreceptor cells typically fire at rates that vary smoothly over time as a function of stimulus concentration. As clearly evident in fig. 7.5, some neurons within the peg sensilla respond unevenly to stimulation. There is reason to believe that this interrupted pattern of activity is the product of synaptic interactions between these sensory cells. For example, in fig. 7.6, the spontaneous firing pattern of a large unit (labeled A in the figure) is interrupted when a smaller unit (labeled B) begins to discharge. The effect of B on A appears to be inhibitory because the larger spike tends not to fire immediately after the occurrence of the smaller spike. Synaptic interactions between chemosensory neurons of a single sensillum are virtually undescribed for insect olfactory systems where afferents generally make their first synapses only after entering the central nervous system (Homberg et al., 1989).

Cross-Correlation Analysis of Sensory Neuron Activity

Putative synaptic interactions between two neurons, either excitatory or inhibitory, can be examined physiologically by showing dependence of one cell's

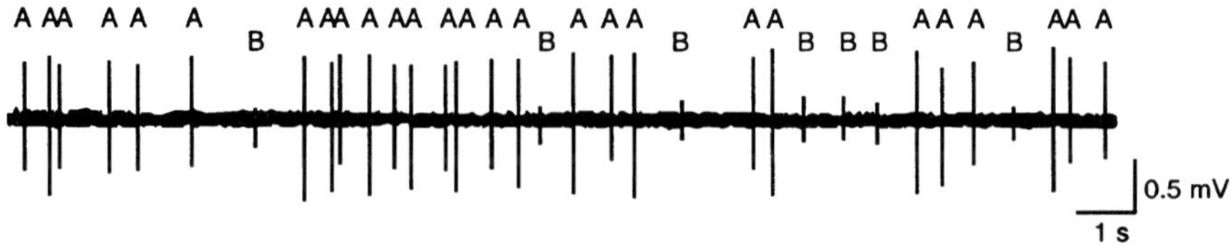

Figure 7.6 Baseline activity from peg sensillum showing interruptions of spiking activity of large unit A by small unit B. From Gaffin and Brownell (1997b), with permission.

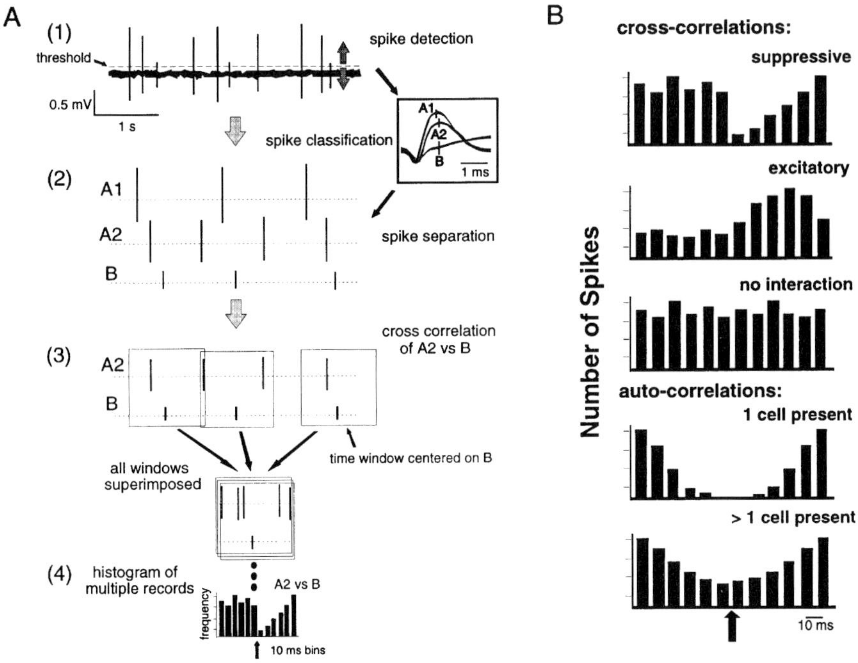

Figure 7.7 Automated spike sorting methods. (A) Flow chart of computer algorithm used in spike sorting and interspike interaction analysis (see text for details). (B) Hypothetical correlation histograms representing various types of cell–cell interactions. Top three profiles are cross-correlations, bottom two profiles are autocorrelations (see text for details). From Gaffin and Brownell (1997b), with permission.

activity on another within a brief (<50 msec) period. To identify single units within the complex multiunit recordings obtained from peg sensilla (fig. 7.7A), spike waveforms for all active units are categorized and parsed by computer so that a template, or averaged waveform, of each cell's electrical output can be identified. Once the integrated activity of the cells in one sensillum is sorted by shape, the activity of individual neurons can be segregated from the multiunit record and displayed as individual traces (fig. 7.7A). Cross-correlation analysis of these traces is then used to reveal interactions between the cells.

For correlation analysis, all spikes of one type, the "reference cell," are detected, recorded, and redisplayed as superimposed impulses in the center of a time window. Spikes from a second class of neuron ("target cell") are then displayed relative to the reference cell's activity, and the number of target cell spikes falling within discrete bins of time before and after the reference spike are counted and displayed in histogram form (fig. 7.7A). In this way, the timing of one sensory neuron's activity can be seen relative to the activity of any other. Possible synaptic interactions are revealed as an increase (excitation) or decrease (inhibition) in spike activity of the target cell relative to the reference cell.

In fig. 7.7B, several hypothetical examples of correlation histograms are shown for possible types of interactions between sensory cells. The top three histograms show cross-correlation profiles expected for synaptically coupled cells that suppress (inhibit), excite, or have no interaction. The bottom two profiles are correlations of the activity of a spike class referenced against itself (autocorrelation). Autocorrelations provide evidence that all spikes assigned to a particular spike class originate from a single neuron. The first autocorrelation profile shown in fig. 7.7B predicts activity that would be expected if only one cell were active. Autocorrelations for a class containing more than one cell would appear as shown in the bottom profile (see Eggermont, 1990, for further description of correlation analysis). Through such comparisons, the characteristics of synaptic connectivity within a discrete circuit of neurons (e.g., peg sensilla) can be defined.

Evidence of Synaptic Interactions within Peg Sensilla

When correlation analysis is applied to electrical recordings from a single peg sensillum, a clear indication of synaptic interaction between its sensory neurons emerges. In the example of fig. 7.8, the occurrence of a small, biphasic spike of distinctive waveform appears to inhibit the activities of two other sensory cells for a period of about 100 msec. Figure 7.8B shows all possible cross- and autocorrelations of the three classes of spikes for 30 min of unstimulated activity. The small biphasic unit (spike class B in fig. 7.8) clearly suppresses the activity of the two larger triphasic units (A1 and A2), while A1 and A2 have no apparent mutual effect. The immediate suppression of A1 and A2 following activity in B and the appearance of a rebound effect in the activity of the suppressed spike (e.g., see A1 versus B histogram about 200 msec after reference spike B in fig. 7.8B) are further indications of synaptic, as opposed to receptor, level of interaction. Because these records were obtained from a surgically isolated pectine, the site of interaction between these sensory neurons must be local and not involve feedback from the central nervous system (Gaffin and Brownell, 1997b). Although the generality of these sensory interactions in scorpions is unknown, they have been observed in each of five species (representing four families) sampled to date (Vaejovidae, *P. mesaensis;* Iuridae, *Hadrurus arizonensis;* Buthidae, *Parabuthus gregoryi, Androctonus australis;* Scorpionidae, *Pandinus gregoryi;* Gaffin and Brownell, unpublished data).

CONCLUSIONS

It is evident from these initial investigations that scorpions, and probably arachnids as a group, have been neglected as experimental systems for the study of chemosensory biology. As in other arthropods, numerous chemotactic hairs are distributed across the body surface and appendages of scorpions, but these channels pale in comparison to the pectines, which, by numbers of primary sensory neurons, rank among the most densely innervated sensory organs in arthropods. The accumulated findings of anatomical, behavioral, and physiological research indicate that these unique structures mediate at least two crucial roles in the reproductive biology of scorpions: chemosensory identification and localization of potential mates, and tactile discrimination of suitable substrates for spermatophore deposition. A strong sexual dimorphism of the pectines in most species reflects this role in mating. But other behaviors, such as feeding,

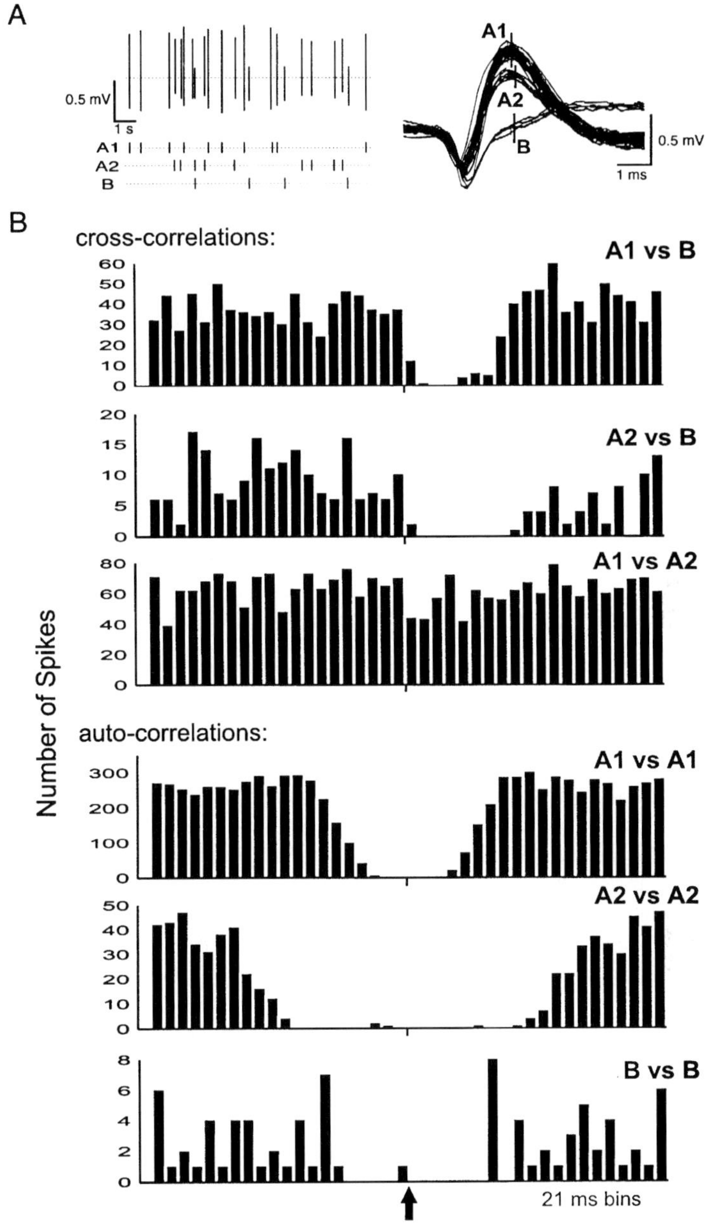

Figure 7.8 Physiological evidence of interaction between spiking units in a single sensillum. (A) Sample of baseline recording from peg sensillum from a detached pectine of *P. mesaensis.* This record has been filtered and categorized by computer algorithm into three spike classes (A1, A2, and B). Sixty of these 10-sec records, sampled from 30 min of continuous baseline activity, were used to create the histograms shown in panel B (see fig. 7.7 for procedure). (B) Histogram profiles of inter- and intraclass interactions grouped by 21-msec bins. The nomenclature "A1 vs. B" indicates that B is the centered (reference) spike. Note the period of suppression of activity of spike types A1 and A2 in reference to the activity of B (top two profiles). Spike types A1 and A2 show no apparent effect in relation to each other (third profile). Bottom three profiles are autocorrelations, the result of referencing each spike type against itself. Adapted from Gaffin and Brownell (1997b).

predator avoidance, and navigation, are also influenced by chemical cues, raising the possibility that future studies will expand our knowledge of the organ's greater utility to the animal. Finally, in thinking about their function, it is noteworthy that pectinal appendages have retained similar form in all scorpions since their first appearance on Silurian aquatic species (Kjellesvig-Waering, 1986). Whatever the pectines do for modern terrestrial scorpions, it is possible that they worked as well on water-submerged substrates with little or no alteration of form.

Because of their clear specialization for use on substrates, the pectinal organs are an important addition to the arsenal of experimental systems used for research in chemical ecology. Compared to odorants transported by air or water, we know little about the nature of chemical signals deposited on substrates or their mode of detection. For example, nonvolatile chemical stimuli attached to solid surfaces could have physical shape that conveys useful information about the signal source. Such spatial information may be detectable by a two-dimensional array of sensors, such as the fields of peg sensilla on the scorpion's pectines, which project to ordered glomeruli in the scorpion brain (Brownell, 1998; Brownell, this volume). Chemical signals transported in fluid media—air or water—have unstable spatial structure, and this may account for the general absence of topography in the olfactory pathways of most animals. By developing a model system for investigating substrate-mediated chemical signaling, we may begin to understand how physical structure of chemical deposits is encoded by nervous systems.

Apart from its value as an arachnid model for comparative chemosensory studies, the pectinal sensory system has utility for basic studies of information processing. Peripheral synaptic connections have been recognized morphologically and/or physiologically in several arthropods including Xiphosuran chelicerates (Griffin and Fahrenbach, 1977; Hayes and Barber, 1982), various arachnids (Foelix, 1985), and insects (Moulins and Noirot, 1972; Steinbrecht, 1989; Gödde and Haug, 1990; White et al., 1990), but in scorpion pectines these connections are particularly prevalent and accessible for physiological analysis. Each peg sensillum contains 10–18 sensory cells, a rather large number compared to other chemotactic sensilla in arthropods (Slipher, 1970; Foelix and Müller-Vorholt, 1983). The potential complexity of response from each peg is increased by the apparent synaptic interaction between several of these cells. Yet the electrical signature of individual cells is sufficiently different that the activity of any one unit can be resolved within a multiunit extracellular record (Gaffin and Brownell, 1997b). This makes possible a physiological dissection of the integrated response from complex chemosensory circuits outside the central nervous system.

The morphological and physiological evidence of synaptic connectivity within individual peg sensilla suggests new lines of research on the neurobiology of processing of chemical and mechanical information. Among the important questions to ask are: What is the role of sensory cell interaction in processing sensory input from natural chemical signals (e.g., pheromones, potential food, or predators)? Do the synaptic interactions between neurons of the same sensillum extend to neurons of neighboring sensilla? Are mechanoreceptors within the sensillum gating or otherwise affecting the response of chemosensory cells? Development of methods to directly stimulate individual pegs (chemically and mechanically) will be useful for characterizing and quantifying response patterns to more natural modes of stimulation. Of paramount importance will be the identification and purification of the species-specific pheromonal substances detected by the pectines. Apart from their obvious importance to the coordination of reproductive behavior of these animals, pheromonal stimulants would provide a powerful tool for analyzing circuit-level response and the encoding of information of great behavioral significance to the animal.

Acknowledgments We thank J. Boeckh, J. Gödde, and R. Loftus of the University of Regensburg, Germany, for early assistance in morphological and electrophysiological studies. A grant from the National Science Foundation (BNS-8709890) and DAAD Fellowships made this research possible.

References

Abushama, F.T. (1964) On the behavior and sensory physiology of the scorpion *Leiurus quinquestriatus*. *Anim. Behav.* 12: 140–153.

Alexander, A.J. (1957) The courtship and mating of the scorpion, *Opisthophthalmus latimanus*. *Proc. Zool. Soc.* (London) 128: 529–544.

Alexander, A.J. (1959) Courtship and mating in the buthid scorpions. *Proc. Zool. Soc.* (London) 133: 145–169.

Blanchard, A. (1853) L'organisation du règne animal. *Arachnides* 46.

Brownell, P.H. (1977) Compressional and surface waves in sand: used by desert scorpions to locate prey. *Science* 197: 479–482.

Brownell, P.H. (1984) Prey detection by the sand scorpion. *Sci. Am.* 251: 86–97.

Brownell, P.H. (1988) Properties and functions of the pectine chemosensory system of scorpions. *Neurosci. Abstr.* 15: 1289.

Brownell, P.H. (1989) Neuronal organization and functions of the pectine chemosensory system of scorpions. *Chem. Senses* 10: 557 (Abstract).

Brownell, P.H. (1998) Glomerular cytoarchitectures in chemosensory systems of arachnids. *Annals of the New York Academy of Sciences* 855: 502–507.

Brownell, P.H. and Farley, R.D. (1974) The organization of the malleolar sensory system in the solpugid, *Chanbria sp. Tissue Cell* 6: 471–485.

Bulseco, D.A. and Brownell, P.H. (1989) Biochemical properties of sensilla-specific proteins in sand scorpions. *Neurosci. Abstr.* 15: 750.

Carthy, J.D. (1966) Fine structure and function of the sensory pegs on the scorpion pecten. *Experientia* 22: 89.

Carthy, J.D. (1968) The pectines of scorpions. *Proc. Zool. Soc.* (London) 23: 251–261.

Cloudsley-Thompson, J.L. (1955) On the function of the pectines of scorpions. *Ann. Mag. Nat. Hist.* 8: 556–560.

Derby, C.D. and Atema, J. (1988) Chemoreceptor cells in aquatic invertebrates: peripheral mechanisms of chemical signal processing in decapod Crustacea. *In* Atema, J., Popper, A.N., Fay, R.R. and Tavolga, W.N. (eds.) *Sensory Biology of Aquatic Animals,* pp. 365–385. Springer, New York.

Dethier, V.G. (1976) *The Hungry Fly.* Harvard University Press, Cambridge, MA.

Dumpert, K. (1978) Spider odor receptor: electrophysiological proof. *Experientia* 34: 754–755.

Eggermont, J.J. (1990) *The Correlative Brain; Theory and Experiment in Neural Interaction.* Springer-Verlag, Berlin.

Foelix, R.F. (1975) Occurrence of synapses in peripheral sensory nerves of arachnids. *Nature* 254: 146–149.

Foelix, R.F. (1985) Sensory nerves and peripheral synapses. *In* Barth, F.G. (ed.) *Neurobiology of Arachnids,* pp. 189–200. Springer-Verlag, Berlin.

Foelix, R.F. and Müller-Vorholt, G. (1983) The fine structure of scorpion sensory organs. II. Pecten sensilla. *Bull. Br. Arachnol. Soc.* 6: 68–74.

Foelix, R.F. and Schabronath, J. (1983) The fine structure of scorpion sensory organs. I. Tarsal sensilla. *Bull. Br. Arachnol. Soc.* 6: 53–67.

Gaffin D.D. (1994) Chemosensory physiology and behavior of the desert sand scorpion, *Paruroctonus mesaensis* (Ph.D. thesis). Oregon State University, Corvallis.

Gaffin, D.D. and Brownell, P.H. (1992) Evidence of chemical signaling in the sand scorpion, *Paruroctonus mesaensis* (Scorpionida: Vaejovidae). *Ethology* 91: 59–69.

Gaffin, D.D. and Brownell, P.H. (1997a) Response properties of chemosensory peg sensilla on the pectines of scorpions. *J. Comp. Physiol.* A 180: 291–300.

Gaffin, D.D. and Brownell, P.H. (1997b) Electrophysiological evidence of synaptic interactions within chemosensory sensilla of scorpion pectines. *J. Comp. Physiol.* A 180: 301–307.

Gaffin, D.D., Wennstrom, K.L. and Brownell, P.H. (1992) Water detection in the desert sand scorpion *Paruroctonus mesaensis* (Scorpionida, Vaejovidae). *J. Comp. Phys.* A 170: 623–629.

Gaubert, M. (1889) Notes sur la structure anatomique du peigne des scorpions et des raquettes coxales des Galeodes. *Bull. Soc. Philomaire* (Paris) 2: 57–58.

Gödde, J. and Haug, T. (1990) Analysis of the electrical responses of antennal thermo- and hygroreceptors of *Antheraea* (Saturniidae, Lepidoptera) to thermal, mechanical and electrical stimuli. *J. Comp. Physiol.* A 167: 391–401.

Griffin, A.J. and Fahrenbach, W.H. (1977) Gill receptor arrays in the horsehoe crab (*Limulus polyphemus*). *Tissue Cell* 9: 745–750.

Hadley, N.F. (1974) Adaptational biology of desert scorpions. *J. Arachnol.* 2: 11–23.

Hayes, W.F. and Barber, S.B. (1982) Peripheral synapses in *Limulus* chemoreceptors. *Comp. Biochem. Physiol.* 72: 287–293.

Hoffmann, C. (1964) Zur Funktion der Kammformigen Organ von Skorpionen. *Naturwiss.* 7: 172.

Homberg, U., Christensen, T. and Hildebrand, J.G. (1989) Structure and function of the deutocerebrum in insects. *Annu. Rev. Entomol.* 34: 477–501.

Ivanov, V.P. and Balashov, Y.S. (1979) The structural and functional organization of the pectine in a scorpion *Buthus eupeus* Koch (Scorpiones, Buthidae) studied by electron microscopy. *Tr. Zool. Inst. Leningrad* 85: 73–81.

Kaissling, K-E. (1987) *R.H. Wright Lectures on Insect Olfaction.* Simon Fraser University, Burnaby, B.C.

Kjellesvig-Waering, E.N. (1986) A restudy of the fossil Scorpionida of the world. *Paleont. Am.* 55. Ithaca, N.Y. Paleontological Research Institute.

Krapf, D. (1986) Contact chemoreception of prey in hunting scorpions (Arachnida: Scorpiones). *Zool. Anz.* 217: 119–129.

Moulins, M. and Noirot, C. (1972) Morphological features bearing on transduction and peripheral integration in insect gustatory organs. *In* Schneider, D. (ed.) *Olfaction and Taste IV*, pp. 49–55. Wissensch Verlagsges, Stuttgart.

Peretti, A.V. (1995) Análisis de la etapa inicial del cortejo de *Bothriurus bonariensis* (C.L. Koch) (Scorpiones, Bothriuridae) y su relación con el reconocimiento sexual. Revue Arachnologique 11: 35–45.

Polis, G.A. (1979) Prey and feeding phenology of the desert sand scorpion *Paruroctonus mesaensis* (Scorpionida: Vaejovidae). *J. Zool.* (London) 188: 333–346.

Polis, G.A. (1980) Seasonal patterns and age-specific variation in the surface activity of a population of desert scorpions in relation to environmental factors. *J. Anim. Ecol.* 49: 1–18.

Polis, G.A. and Farley, R.D. (1979a) Behavior and ecology of mating in the cannibalistic scorpion, *Paruroctonus mesaensis* Stahnke (Scorpionida: Vaejovidae). *J. Arachnol.* 7: 33–46.

Polis, G.A. and Farley, R.D. (1979b) Characteristics and environmental determinants of natality, growth and maturity in a natural population of a desert scorpion, *Paruroctonus mesaensis* (Scorpionida: Vaejovidae). *J. Zool.* (London) 187: 517–542.

Schröder, O. (1908) Die Sinnesorgane der Skorpionskamme. *Z. Wiss. Zool.* 9: 436–444.

Slipher, E.H. (1970) The structure of arthropod chemoreceptors. *Annu. Rev. Entomol.* 15: 121–142.

Stahnke, H.L. (1966) Some aspects of scorpion behavior. *Bull. Calif. Acad. Sci.* 65: 65–80.

Steinbrecht, R.A. (1989) The fine structure of thermo-/hygrosensitive sensilla in the silkmoth *Bombyx mori:* receptor membrane substructure and sensory cell contacts. *Cell Tiss. Res.* 255: 49–57.

Swoveland, M.C. (1978) External morphology of scorpion pectines (Master's thesis). California State University, San Francisco.

White, P.R., Chapman, R.F. and Ascoli-Christensen, A. (1990) Interactions between two neurons in contact chemosensilla of the grasshopper, *Schistocerca americana. J. Comp. Physiol.* A 167: 431–436.

8

Structure and Neurotoxicity of Venoms

ERWANN LORET
BRUCE HAMMOCK

The toxicity of scorpion venom derives mainly from small proteins synthesized in the venom gland of the terminal tail segment, or telson (Miranda and Lissitzki, 1961). As for many of the most powerful animal venoms, the molecular targets of these proteins are the different ionic channels localized in excitable membranes of nerve and muscle cells (Catterall, 1980; Couraud and Jover, 1984; Miller et al., 1985). Venoms may also contain mixtures of proteolytic and lipolytic enzymes that increase tissue penetration and action of the neurotoxic components (El Asmar et al., 1984) or modify exocrine secretion from the pancreas (Possani et al., 1991). Neurotoxic proteins bind with high specificity for their molecular targets, a characteristic that made them critically important tools for biochemical purification of ion channels and for high throughput drug screening. In this chapter we examine how scorpion venom proteins have been applied recently to studies of the structure–function relationship between neurotoxins and their well-defined receptors. The chapter begins with a description of essential structural motifs within neurotoxins that give these proteins their specificity of action in insects and mammals. A closing section demonstrates how such information can be applied to the design of a new class of insecticides—toxinlike recombinant proteins expressed in insect-specific viruses.

Scorpions are an ancient and biogeographically diverse group. As such, their venoms present a rich resource of proteins with specialized but related functions. Scorpions of the family Buthidae, the primary subjects of this chapter, are found throughout the world in an area roughly limited by the 50th parallels north and south (fig. 8.1). The separation of South America from Gondwana to join North America during the Mesozoic era possibly divided buthids into the New and Old World groups and set the stage for divergent evolution of their venoms. Neurotoxins of New and Old World buthids now differ in amino acid sequence, pharmacological action, and immunological properties (Rochat et al., 1979; Couraud and Jover, 1984; Watt and Simard, 1984; Granier et al., 1989) but still share characteristics indicative of their common ancestry and conserved function. More recently, a toxin having the properties expected of an ancestral type for New and Old World toxins was discovered as a minor component in the venom of a North African Buthid (Loret et al., 1991).

The most potent toxins in Buthidae have evolved specialized actions limited to either vertebrate or insect targets (Zlotkin et al., 1971). For example, some insect-specific toxins can be 2500 times more toxic to insects than insecticides such as DDT, yet they have no effect on mammals (Loret et al., 1992). Specificity

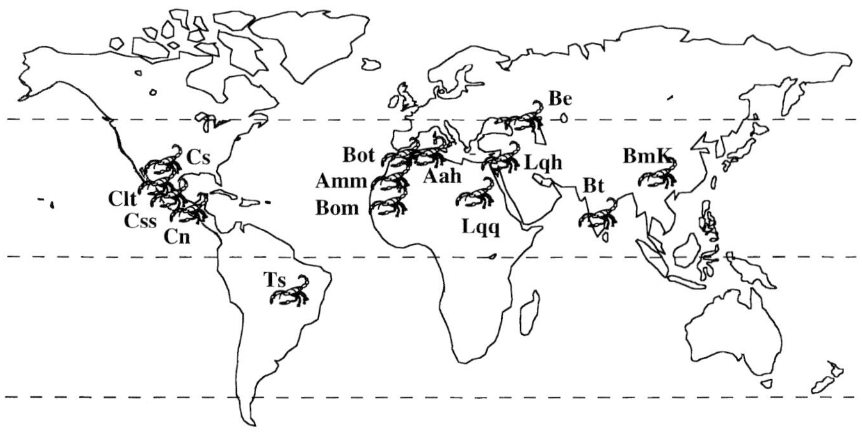

Figure 8.1 Geographic distribution of buthid scorpions most often used in venom research. Dashed lines indicate equator and 50° latitudes north and south. Aah, *Androctonus australis hector;* Amm, *Androctonus mauretanicus mauretanicus;* Me, *Mesobuthus epeus;* Bom, *Buthus occitanus mardochei;* Bot, *Buthus occitanus tunetanus;* Ce, *Centruroides exilicauda* (=sculpturatus); Clt, *Centruroides limpidus;* Css, *Centruroides suffusus suffusus;* Cn, *Centruroides noxius;* Lqq, *Leirus quinquestriatus quinquestriatus;* Lqh, *Leiurus quinquestriatus hebraeus;* MmK, *Mesobuthus martensii* Karsh; Mt, *Mesobuthus tamulus;* Ts, *Tityus serrulatus.*

of toxins for particular targets evolved by variation of length and composition of amino acid sequences in toxin proteins (Possani, 1984), and this characteristic is especially useful for classification, localization, and purification of biological receptors of subtly different types. This is especially true of toxins acting on ion channels because channel activity is difficult to measure by methods other than molecular binding assays. Thus, as research improves our understanding of toxin structure in diverse species, we can expect to find common motifs of molecular structure that are essential for toxin action and related functions.

The first studies of scorpion venoms were aimed at developing antisera against their action in humans (Hassan, 1984; El Ayeb and Delori, 1984). Although the number of lethal envenomations continues to decline, hundreds of scorpion stings still cause deaths worldwide, especially among children (Simard and Watt, 1990). Scorpion venom intoxication is a health problem both in the North African Maghreb and in the Americas. Health information campaigns have decreased the number of stings in Tunisia from 3000 per year in 1967 to about 1000 per year in the 1980s and 1990s. Recently, serotherapy as reduced mortality to 0.35%, with most deaths occurring in underweight children. However, in light of the large number of countries in which there is a risk of scorpion envenomation during summer months, its prevention and treatment remain a major problem. The advent of serotherapy using monoclonal antibodies against scorpion toxins holds promise as a cure (Bahraoui et al., 1988; Granier et al., 1989) but has not been developed. A synthetic peptide mimicking a North African scorpion toxin and produced by chemical solid-phase synthesis seems to be a very efficient vaccine for mice (Zenaouaki et al., 1997) and could be an economical alternative to monoclonal antibodies.

Aside from serotherapy, the drugs chosen for treatment of scorpion venom intoxication vary according to symptoms. The efficacy of drugs typically used as antidotes (e.g., atropine, propranolol) has been compared to that of less conventional treatments (e.g., dipyridamole, doxapram, quinine formate, lysine-acetyl-salicylate, valproate and verapamil) using preventive and therapeutic effects of the drugs on mice as an experimental assay. Although most of these drugs prolonged survival, the administration of verapamil, aspirin or quinine formate gave 50 to 100% reduction of venom toxicity, depending on the drug and the origin of the venom. Propranolol, doxapram, atropine, dipyridamole and valproate had little or no protective effect. If these results are confirmed in humans, the systematic use of these drugs could become a simple means for treating scorpion stings (Guieu et al., 1993).

MOLECULAR TARGETS OF SCORPION NEUROTOXINS

All scorpions produce toxins, but only members of the family Buthidae produce neurotoxic venoms dangerous to humans. Taking into account the small amount injected, the lethal capacity of these venoms is impressive: the deadly North African scorpion *Androctonus australis hector,* for example, injects only 100–200 μg dry weight of whole venom, only 5% of which is the active neurotoxin. Such high toxicity is possible because these small proteins bind with a very high affinity to ion channels in nerve and muscle tissue (Rochat et al., 1979). Initial studies of scorpion venoms focused on neurotoxic components affecting voltage-dependent sodium channels, the integral plasma membrane proteins responsible for the rapid increase in sodium ion permeability during the rising phase of the action potential in nerve and muscle (Catterall, 1977; Hille, 1991). More recently the neurotoxic components acting on potassium channels (Miller et al., 1985), calcium channels (Valdivia et al., 1991), and chloride channels (DeBin and Strichartz, 1991) have been identified, thus greatly increasing our knowledge about the potential complexity and specificity of scorpion venom actions in vertebrates and insects. The following sections discuss the classification of buthid neurotoxins according to their pharmacological targets.

Sodium Channel

The first neurotoxins isolated from scorpion venom came from *Androctonus australis* (Rochat et al., 1967; Miranda et al., 1970). Several of these toxins acted on the voltage-gated sodium channels from mammals (Catterall, 1977), but *A. australis* venom also contained toxins that specifically affected the sodium channel of insects (Zlotkin et al., 1971; Gordon et al., 1984). This distinction gave an early indication of the subtle variations that exist between venom proteins of nearly identical function found in the same scorpion species.

Vertebrate sodium channels

Toxins acting on vertebrate sodium channels are of two types, α and β, that are generally distinguishable by their pharmacological and electrophysiological effects (Couraud and Jover, 1984). The binding of α toxins prolongs the opening phase of ion channel conductance to sodium ions; β toxins trigger repetitive bursts of channel opening and closing (Catterall, 1980). These effects on channel gating behavior can vary between tissues, for example, in cardiac muscle where β toxins slow activation of the gating mechanism (Yatani et al., 1988). The α and β toxins do not compete for the same sites on the sodium channel, and immunological experiments show no cross-reaction among antibodies specific for α and β toxins (Granier, 1989). The independence of these binding sites is further supported by the observation that sea anemone toxins compete with α toxins, but not with β toxins, for binding sites on vertebrate sodium channels (Schweitz et al., 1985). Scorpion toxins also have greater toxicity toward mammals than toward amphibians (Duval et al., 1989); thus their affinity may vary among classes of vertebrates.

A biogeographical distinction in buthid venom phenotypes is also evident: α toxins are found mainly in the venoms of Old World species, whereas β toxins are prevalent in New World species. The rare exceptions to this distinction are a β toxin found in the venom of Old World spieces, *Leiurus quinquestriatus* (Rack et al., 1987), and α toxins in the venoms of two New World species, *Centruroides exilicauda,* or *C. sculpturatus* (Possani, 1984), and *Tityus serrulatus* (Kirsch et al., 1989). More recently, another toxin isolated from the venom of *Androctonus australis* was shown to compete for both α and β toxin binding sites in both vertebrates and insects, indicating a possible ancestral structure for this toxin molecule (Loret et al., 1991).

With the resolution of the molecular structure of voltage-gated sodium channel in mammals (Numa, 1989) came a detailed understanding of where scorpion toxins bind to their targets. In rats the ion channel is composed of three glycoprotein subunits, an α subunit of 260-kDa molecular weight, which is covalently linked to a β2 subunit (33 kDa) and a noncovalently bound β1 subunit (36 kDa) (Catterall, 1988). Molecular cloning of the α subunit subsequently demonstrated different subtypes within and between tissues (Numa, 1989), all having a common structural domain of 300–400 amino acids repeated four times within the 1800 amino acid residues that constitute the subunit (Hille, 1991). Each of these domains contains six transmembrane segments (S1 through S6), with the receptor site for scorpion toxin located within domains I and IV, in two loops (between the segments 5 and 6, see fig. 8.4) that protrude to the outside surface of the membrane (Thomsen and Catterall, 1989; Tejedor and Catterall, 1990). The receptor site for α scorpion toxin is located on the α subunit of the ion channel; the receptor site for β scorpion toxin is located on the β1 subunit (Jover et al., 1988).

Insect sodium channels

The venoms of the Old World Buthidae contain some toxins that act only on insect sodium channels (Loret et al., 1992). These insect-specific neurotoxins can be also classified into α and β subtypes like their vertebrate homologues. The α toxin specific for the insect sodium channel is characterized by electrophysiological properties similar to those observed with α toxins specific to the vertebrate sodium channel (Eitan et al., 1990). Another site on the sodium channel of insects appears to bind two different types of toxins, a contraction-inducing toxin that causes rapid excitatory paralysis when assayed in larvae of the fly *Sarcophaga argyrostoma,* and depressant toxins that induce a slow, flaccid paralysis (Zlotkin et al., 1985). Each of these toxins show distinct electrophysiological effects in voltage-clamped axonal preparation of cockroaches (*Periplaneta americana*). Only the contraction-inducing toxins show electrophysiological effects similar to β toxins acting on vertebrates (Couraud and Jover, 1984).

Insect-specific β toxins purified from Old World buthid venoms do not cross-react with vertebrate-specific β toxins from New World species. However, some New World β toxins are toxic to insects (Watt and Simard, 1984) and bind to both vertebrate and insect sodium channels (Lima et al., 1989). Old World β toxins are strictly specific for insects and have no effect on mammals (Zlotkin et al., 1971; de Dianous et al., 1987). Because the sodium channels of insects appear to be composed of a single, 190-kDa subunit, homologous to the α subunit of the vertebrate sodium channel (Lima et al., 1988) or 260 kDa (Gordon et al., 1988), the receptor site for β toxin on the insect sodium channel should be located on this subunit (Lima et al., 1988). In the context of developing insecticides having little or no side effects on vertebrates, it is essential to know the difference between the sodium channels of insects and vertebrates.

Potassium Channels

Of the many different types of potassium channels now described, scorpion toxins act selectively on at least three types: the "delayed rectifier" or axonal potassium channels (K), the transient or A-type potassium channels (K_A), and the calcium-dependent potassium channels (K_{Ca}) commonly present on cell bodies and dendritic processes (Hille, 1991). Toxins from the venom of *Centruroides noxius,* such as noxiustoxin (NTX), *Tityus serrulatus* and *Leiurus quinquestriatus hebraeus* are most selective for K channels (Carbone et al., 1982; Possani, 1984; Blaustein et al., 1991; Garcia et al., 1994), but they also block calcium-activated potassium channels (K_{Ca}) at 10-fold higher concentrations (Valdivia et al., 1988). The unfractionated venom from *Pandinus imperator* blocks the K_A channel, as do other venoms from Old World scorpions (Pappone et al., 1988), whereas venoms from the New world species selectively block the K channels (Blaustein et al., 1991).

Scorpion toxins specific to the K_{Ca} channels have been found mainly in Old World scorpions. The charybdotoxin (CTX), from *Leiurus quinquestriatus,* blocks the BK or apamine-insensitive K_{Ca} channel (Miller et al., 1985), and another toxin from this venom, leiurotoxin (LTX, or scillatoxin), acts on the SK or apamine-sensitive K_{Ca} channel (Chichi et al., 1988; Auguste et al., 1990). The iberiotoxin (IbTX) from *Mesobuthus tamulus* acts on K_{Ca} channels by decreasing the probability of channel opening as well as decreasing average open time of the channel. Iberiotoxin interacts at a distinct site on the channel and modulates CTX binding by an allosteric mechanism (Galvez et al., 1990). The kaliotoxin (KTX) from *Androctonus mauretanicus* induces a blockade of the K_{Ca} channel (voltage independent), in contrast to CTX (Crest et al., 1992). All potassium channels appear to be composed of a single domain repeated four times in a tetrameric structure (MacKinnon, 1991), a structure quite different from the continuous amino acid chain making up the sodium channels of vertebrates, but the toxin receptor site of CTX is still located in the loop between transmembrane domains S5 and S6 of each subunit (MacKinnon et al., 1990) (fig. 8.2).

The integral membrane proteins constituting many voltage-dependent potassium channels are associated with an additional protein known as the beta sub-

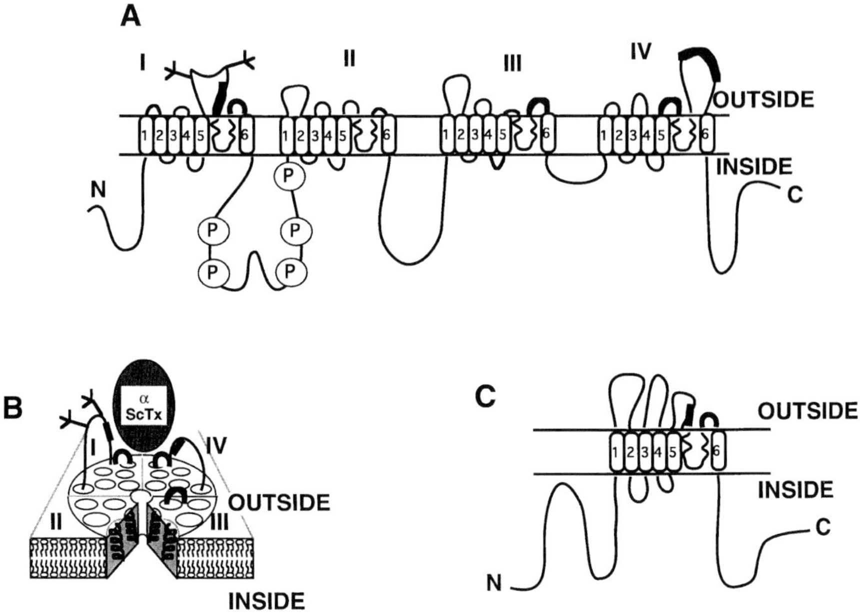

Figure 8.2 Ion channel structure and sites of toxin action. (A) Two-dimensional representation proposed for the transmembrane folding of the sodium channel α subunit from rat brain depicting in bold the sites of interaction with α scorpion toxins (Thomsen and Catterall, 1990). (B) Three-dimensional view of the sodium channel α subunit showing apposition of homologous domains I and IV and formation of a transmembrane pore. The α scorpion toxin (ScTx) binds to the channel through interaction with extracellular sequences between segments S5 and S6 of domains I and IV (Thomsen and Catterall, 1990). (C) Two-dimensional view of the calcium-activated potassium channel from *Drosophila* showing binding sites for the CTX toxin as bold lines (Mac Kinnon et al., 1990; modified from Catterall, 1988).

unit. One function of beta subunits is to modify K+ channel gating. The structure of the conserved core of mammalian beta subunits was determined by X-ray crystallography at 2.8 Å resolution (Gulbis et al., 1999). Like the integral membrane component of K^+ channels, beta subunits form a four-fold symmetric structure. Each subunit is an oxidoreductase enzyme combined with a nicotinamide co-factor in its active site. Several structural features of the enzyme's active site, including its location with respect to the four-fold axis of symmetry for the channel, imply that it may interact with the K+ channel's voltage sensor. This arrangement suggests a mechanism for coupling biochemical status of the cell directly to its electrical excitability.

Calcium and Chloride Channels

Pharmacological and electrophysiological experiments show scorpion toxins are also blockers of calcium and chloride channels, although the molecular sites of action on these ion channels are not well characterized (Hille, 1991). Toxin

II.6 from the Mexican scorpion *Centruroides limpidus* strongly depresses the activation of calcium channels (Alagon et al., 1988), and the venom from *Hottentotta hottentotta* selectively increases binding of the ligand ryanodine to calcium channels (involved in the mechanism of neurotransmitter release from nerve terminals) (Valvidia et al., 1991). The venom of *Leiurus quinquestriatus* produces a significant and reversible inhibition of reconstituted chloride channels of the small conductance type, found in rat colonic epithelial cells (DeBin and Strichartz, 1991).

Given their broad spectrum of actions within excitable tissues, scorpion venoms have become a valuable source of highly specific molecular probes for localization and study of ionic channels in nerve and muscle tissues. As our understanding of the complexity and diversity of these ionic channels continues to grow, the utility of these toxins and basic research concerning them will intensify. Beyond these practical uses as molecular tools, toxin proteins and ion channels they target present an interesting problem to evolutionary competition, where small changes in molecular structure, even single amino acids within a large protein, can profoundly alter the effectiveness of a venom or defense against it.

STRUCTURE OF SCORPION TOXINS

The scorpion neurotoxins studied so far are short peptides of 30–70 residues, cross-linked by 3 or 4 disulfide bridges, and having net positive charge. Structural studies of all scorpion toxins show rather conserved secondary structures consisting of an α-helix adjacent to a double-stranded antiparallel β-sheet. This core structure is fixed by two disulfide bridges connecting the α-helix with one strand, while the third conserved disulfide bridge connects the other strand to a residue preceding the N-terminal side of the α-helix (see diagram below).

Comparison of toxin amino acid sequences (most of which act on sodium channels) shows there are homology groupings but no consensus sequence. When the six cysteines residues of the well-conserved disulfide bridges are aligned (fig. 8.3; Dufton and Rochat, 1984), the common features of primary and secondary structure for all toxins peptides can be represented as follows:

NH2-[...]-C-[...]-C-X-X-X-C-X-(K or R)-[.....]-G-X-C-[.....]-C-X-C-[..........]-COOH

α-helix β-strand β-strand

where C = cysteine, G = glycine, K = lysine, R = arginine, and X = other amino acids. The brackets between secondary structures (double-headed arrows) are segments of variable length, depending on the structural group to which each toxin belongs. A basic amino residue (Lys or Arg) is present at the C-terminus of the α-helix in most of the known sequences, the one exception being Be-M9 (fig. 8.3), which lacks this basic residue. The motif G-X-C is well conserved among scorpion toxins except for the toxin NTX (= G-X-X-C). The conserved glycine residue minimizes the distance between the α-helix and β-sheet, thus eliminating space for insertion of other amino acid sides chains (Bontems et al., 1991).

```
                       10        20        30        40        50        60        70        80
                        |         |         |         |         |         |         |         |
A.a.h. I    (1)  -KRDGYIVYPN-NCVYHCVPP-----CDGLCKKN-GGSS-GSCSFLVPSGLACWC-KDLP-DNVPIKDT--SRK-CT-
A.a.h. I'   (1)  -KRDGYIVYPN-NCVYHCIPP-----CDGLCKKN-GGSS-GSCSFLVPSGLACWC-KDLP-DNVPIKDT--SRK-CT-
A.a.h. I''  (2)  -KRDGYIVYPN-NCVYHCVPP-----CDGLCKKN-GGSS-GSCSFLVPSGLACWC-KDLP-DNVPIKDT--SRK-CTR
A.a.h. III  (3)  -VRDGYIVNSK-NCVYHCVPP-----CDGLCKKN-GAKS-GSCGFLIPSGLACWCVA-LP-DNVPIKDP--SYK-CHS
A.a.h. IV  (35)  -GRDGYIVDSK-NCVYHCYPP-----CDGLCKKN-GAKS-GSCGFLVPSGLACWC-NDLP-ENVPIKDP--SDD-CHK
B.o.m. III (13)  -GRDGYIAQPE-NCVYHCFP-G-SSGCDTLCKEK-GATS-GHCGFLPGSGVACWC-DNLP-NKVPIVVG--GEK-CH-

A.a.h. II   (4)  -VKDGYIVDDV-NCTYFCGR---NAYCNEECTKL-KGES-GYCQWASPYGNACYCYK-LP-DHVRTKGP--GR--CH--
B.o.t. III  (5)  -VKDGYIVDDR-NCTYFCGR---NAYCNEECTKL-KGES-GYCQWASPYGNACYCYK-VP-DHVRTKGP--GR--CN--
L.q.q. V    (7)  -LKDGYIVDDK-NCTFFCGR---NAYCNDECKKK-GGES-GYCQWASPYGNACWCYK-LP-DRVSIKEK--GR--CN--
A.m.m. V    (8)  -LKDGYIIDDL-NCTFFCGR---NAYCDDECKKK-GGES-GYCQWASPYGNACWCYK-LP-DRVSIKEK--GR--CN--
B.o.t. XI   (6)  -LKDGYIVDDR-NCTYFCGT---NAYCNEECVKL-KGES-GYCQWVGRYGNACWCYK-LP-DHVRTVQA--GR--CR--
B.e. M10    (9)  -VRDGYIADDK-DCAYFCGR---NAYCDEECKK--GAES-GKCWYAGQYGNACWCYK-LP-DWVPIKQKVSG-K-CN--
B.e. M 9   (10)  -ARDAYIAKPH-NCVYECYNPK-GSYCNDLCTEN-GAES-GYCQILGKYGNACWCIQ-LP-DNVPIR-IP-G-K-CH--
B.m. I     (34)  -VRDAYIAKPH-NCVYECAR---NEYCNDLCTKD-GAKS-GYCQWVGKYGNGCWCIE-LP-DNVPIR-VP-G-K-CH
B.m. II    (34)  -VRDAYIAKPH-NCVYECAR---NEYCNDLCTKD-GAKS-GYCQWVGKYGNGCWCIE-LP-DNVPIR-IP-G--NCH
B.o.t. I   (12)  -GRDAYIAQPE-NCVYECAQ---NSYCNDLCTKN-GATS-GYCQWLGKYGNACWC-KDLP-DNVPIR-IP-G-K-CHF
B.o.t. II  (12)  -GRDAYIAQPE-NCVYECAK---NSYCNDLCTKN-GAKS-GYCQWLGRWGNACYCI-DLP-DKVPIR-IE-G-K-CHF
L.q.h. αIT (28)  -VRDAYIAKNY-NCVYECFR---DAYCNELCTKN-GASS-GYCQWAGKYGNACWCYA-LP-DNVPIR-VP-G-K-CR
B.e. M14   (10)  -ARDAYIADDR-NCVYTCALN-P--YCDSECKKN-GAD-GSYCQWLGRFGNACWC-KNLP-DDVPIRKIP-GEE-CR--
L.q.q. IV  (14)  GVRDAYIADDK-NCVYTCGS---NSYCNTECTKN-GAES-GYCQWLGKYGNACWCIK-LP-DKVPIR-IP-G-K-CR
C.s. V     (11)  -KKDGYPVDSG-NCKYECLK---DDYCNDLCLER-KADK-GYCYW-GKV--SCYC--GLP-DNSPTKT--SG-K-CNPA


C.l.t. I   (23)  --KEGYLVNHSTGCKYECFKLGDNDYCLRECRQQYGKGAGGYCYAF-----GCWC-THLY-EQAVVWPLP--NKTCS-
C.s.s. IV  (33)  --KEGYLVNSYTGCKFECFKLGDNDYCLRECRQQYGKGSGGYCYAF-----GCWC-THLY-EQAVVWPLP--NKTCN-
C.s.s. VI  (33)  --KEGYLVNSYTGCKFECFKLGDNDYCKRECKQQYGKSSGGYCYAF-----GCWC-THLY-EQAVVWPLP--NKTCN-
C.s.s. II  (18)  --KEGYLVSKSTGCKYECLKLGDNDYCLRECKQQYGKSSGGYCYAF-----ACWC-THLY-EQAVVWPLP--NKTCN-
C.s. v I   (16)  --KEGYLVKKSDGCKYDCFWLGKNEHCDTECKAKNQGGSYGYCYAF-----ACWC-EGLP-ESTPTYPLP--NK-CS-
C.s. v II  (16)  --KEGYLVNKSTGCKYGCLKLGENEGCDKECKAKNQGGSYGYCYAF-----ACWC-EGLP-ESTPTYPLP--NK-CSS
C.s. v III (16)  --KEGYLVKKSDGCKYGCLKLGENEGCDTECKAKNQGGSYGYCYAF-----ACWC-EGLP-ESTPTYPLP--NKSC--
C.s. I     (17)  --KDGYLVEK-TGCKKTCYKLGENDFCNRECKWKHIGGSYGYCYGF-----GCYC-EGLP-DSTQTWPLP--NK-CT-
C.n. II-14 (14)  --KDGYLVDAK-GCKKNCYKLGKNDYCNRECRMKHRGGSYGYCYGF-----GCYC-EGLS-DSTPTWPLT--NKTC--

T.s. VII   (19)  --KEGYLMDHE-GCKLSCF-IRPSGYCGRECGIKKG-SS-GYCAW-P----ACYCY-GLPNWVKVWDRAT--NK-C--

A.a.h. IT4 (24)  --EHGYLLNKYTGCKVWCVI--NNEECGYLCN-KRRGGYYGYCYF---WKLACYCQGARK-SE-LWNYKT--NK-CDL

L.q.h. IT2 (32)  ---DGYIKRRD-GCKVACLI-G-NEGCDKEC--KAYGGSYGYCWT---WGLACWC-EGLP-DDKTWKSET--NT-CG
L.q.q. IT2 (27)  ---DGYIRKRD-GCKLSCLF-G-NEGCNKEC--KSYGGSYGYCWT---WGLACWC-EGLP-DEKTWKSET--NT-CG
B.j.   IT2 (32)  ---DGYIRKKD-GCKVSCII-G-NEGCRKEC--VAHGGSFGYCWT---WGLACWC-ENLP-DAVTWKSST--NT-CG

A.a.h. IT  (20)  -KKNGYAVDSS-GKAPECLL---SNYCNNQCTK-VHYADKGYCCLL-----SCYCF-GLNDDKKVLEISDTRKSYCDTTIIN
A.a.h. IT1 (25)  -KKNGYAVDSS-GKAPECLL---SNYCNNECTK-VHYADKGYCCLL-----SCYCF-GLNDDKKVLEISDTRKSYCDTTIIN
A.a.h. IT2 (25)  -KKDGYAVDSS-GKAPECLL---SNYCYNECTK-VHYADKGYCCLL-----SCYCF-GLNDDKKVLEISDTRKSYCDTPIIN
A.a.h. IT3 (26)  -KKNGYAVDSS-GKAPECLL---SNYCYNECTK-VHYADKGYCCLL-----SCYCF-GLNDDKKVLEISDTRKSYCDTPIIN
L.q.q. IT1 (27)  -KKNGYAVDSS-GKAPECLL---SNYCYNECTK-VHYADKGYCCLL-----SCYCV-GLSDDKKVLEISDARKKYCDFVTIN

B.e.I5a    (15)              MCMP-CFTTDPNMA--KKCRDCCGG-N-GKC-----FGPQCLCNR

C.n.   NTX (21)            TI-INVKCTS---PKQCSKPCKELYGSSAGAKCMNG-----KCKCYNN
L.q.h. CTX (22)            EFT-NVSCTT---SKECWSVCQRLHNTSR-GKCMNK-----KCRCYS
Bt    IbTX (29)            EFT-DVDC---SVSKECWSVCKDL-FGVDRGKCMGK-----KCRCYQ
A.m.m. KTX (30)            GVEINVKC---SGSPQCLKPCKDA--GMRFGKCMNR-----KCHCTP
L.q.h. LTX (31)            AF-----CNL----RMCQLSCRSL--GLL-GKCIG-D----KCECVKH

                        |         |         |         |         |         |         |         |
                       10        20        30        40        50        60        70        80
```

Figure 8.3 Primary structure of buthid neurotoxins. Amino acid sequences are aligned with the half-cystines to maximize sequence homology according to Rochat et al. (1979). Most of these toxins act on sodium channels, except CTX, IbTX, LTX, NTX, and KTX (bottom) which act on the potassium channels. (1) Rochat et al. (1970); (2) Martin and Rochat (1984); (3) Kopeyan et al. (1979); (4) Rochat, H. et al. (1972); (5) El Ayeb et al. (1983); (6) Sampieri et al. (1987); (7) Kopeyan et al. (1978); (8) Rosso and Rochat (1985); (9) Grishin et al. (1981); (10) Volkova et al. (1985); (11) Watt and Simard (1984); (12) Gregoire and Rochat (1983); (13) Vargas et al. (1987); (14) Kopeyan et al. (1985); (15) Arseniev et al. (1984); (16) Babin et al. (1974); (17) Babin et al. (1975); (18) Possani et al. (1985); (19) Bechis et al. (1984); (20) Darbon et al. (1982); (21) Possani (1984); (22) Gimenez-Gallego et al. (1988); (23) Martin et al. (1988); (24) Loret et al. (1991), (25) Loret et al. (1990a); (26) Bougis et al. (1989); (27) Kopeyan et al. (1990); (28) Eitan et al. (1990); (29) Galvez et al. (1990); (30) Crest et al. (1992); (31) Chichi et al. (1988); (32) Zlotkin et al. (1991); (33) Sampieri et al. (personal communication); (34) Ji et al. (personal communication).

Structure of Scorpion Toxin Acting on Sodium Channels

Primary structure

There are several consistent features of primary and secondary structure for scorpion venoms. All scorpion toxins specific for sodium channels are single chains of 60–70 amino acid residues (Rochat et al., 1970) cross-linked by four disulfide bridges (Kopeyan et al., 1974). The positions of the disulfide bridges are conserved among toxins acting on mammals (Rochat et al., 1979), but in insect-specific β-toxins they are shifted by one half-cystine residue (Darbon et al., 1982). As a group these toxins have relatively high content of aromatic residues and proline, with methionine present only in toxins of *Mesobuthus eupeus*.

Scorpion toxins specific for sodium channels can be classified into eight groups (A–G; fig. 8.4) by comparison of their amino acid sequences. This classification generally confirms pharmacological and immunological specificity of

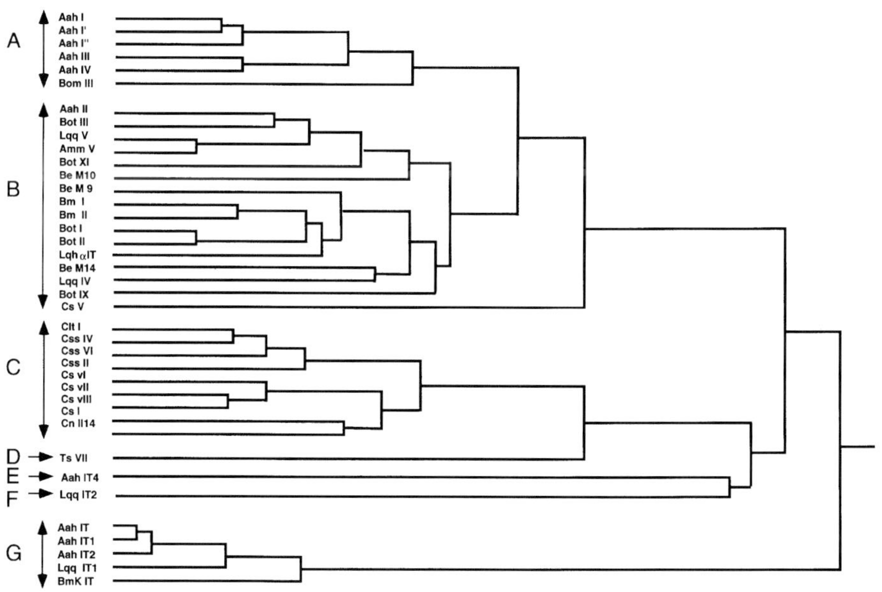

Figure 8.4 Dendrogram of sodium-channel–blocking toxins constructed from sequence homology and gap events. Groups A and B are α toxins binding to mammalian sodium channels, they are purified from Old World buthids except Ce V which comes from the venom of a New World species (Centruroides). These two groups have different immunological properties. Group C is β toxins with affinity for vertebrate sodium channels which are mostly purified from the venoms of New World scorpions. Group D is represented by a single toxin Ts VII, which binds to a β site on mammal and insect sodium channel. Group E is represented by Aah IT4, a possible "ancestral toxin" that binds to α and β sites on mammalian sodium channels and β sites on insect sodium channels. Group F is represented by Lqq IT2, which binds only to β insect sodium channels and paralyzes fly larvae. Group G is toxins binding only to β toxin site on insect sodium channels (they induce contractions of fly larvae), and they are found only in venoms from Old World buthids. Modified from Loret et al., (1991).

the toxins, except for the placement of an insect-specific α-toxin (Lqh-IT in group B, which otherwise contains vertebrate-specific α-toxins; the insect-specific β-toxins form group G, clearly distinct with specific immunological properties) (Granier et al., 1989). The structural similarities between the insect-specific α-toxins is striking considering geographical separation of *Buthus martensis* (Karsh) scorpions collected in East Asia and *A. australis* scorpions collected in North Africa (fig. 8.1). Sequence homology indicates that the major division of scorpion toxins acting on sodium channels is due to differences between the α and β types rather than to geographical proximity of scorpion species or toxin specificity for vertebrates or insects.

Toxins acting on mammals were further classified by the distance analysis (Possani, 1984), a method for estimating genetic distance between two amino acid sequences based on their possible codon sets. These data suggest the following scheme of evolutionary relationship among buthid toxins acting on mammals, where toxins from North African species are nearest the ancestral condition:

North African (GroupA) → North African (Group B) → New World (Group C)

North African (Group B) ↓ Central Asian (Group B)

Secondary and tertiary structure

Circular dichroism (CD) spectroscopy is a useful method for assessing secondary structures of proteins suspendended in solution (Johnson, 1985). Comparative CD studies of 10 scorpion toxins representing groups A, B, C, and G show that toxins within each grouping have overall structural homology with minor differences. Comparison further suggests that the toxins are related conformational variants having the following order of group similarity: B > A > C > G (Dufton et al., 1986). This conformational variation between groups is likely related to the percentage of secondary structure in α-helix and β-sheet configuration (or extended structures) in each group. For example, group B toxins (represented by Aah II in fig. 8.5) show CD spectra with high percentage in β-sheet structure, which is characterized by a positive CD band at 200 nm and negative CD band near 215 nm; group G toxins show a higher content of α-helix characterized by an intense positive band at 190 nm and a negative band at 205 nm. Table 8.1 shows that the number of amino acid residues in α-helix configuration shifts from 9 to 10 for the group B and C toxins, and to 15 residues in group G toxins. Similarly, the number of residues in β-sheet configuration or extended structure shifts from 36 residues for group B toxins to 22–23 residues in the group C and G peptides. Group E is characterized by a high content of non-determinate structures. These results suggest that secondary structure varies within an otherwise conservative structural scheme for the toxin proteins. Such variation may play an important role in the pharmacological specificity of scorpion toxins.

X-ray crystallographic studies and two-dimensional nuclear magnetic resonance measurements (Pashkov et al., 1988; Fontecilla-Camps et al., 1988; Darbon et al., 1991; Oren et al., 1998; Polykarpov et al., 1999) of scorpion tox-

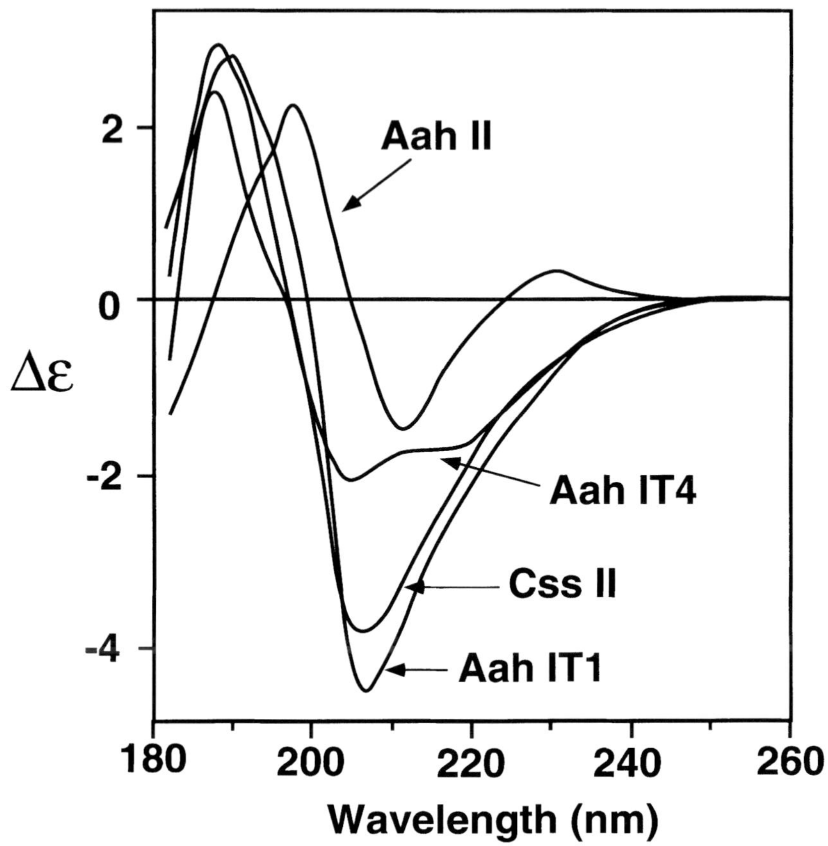

Figure 8.5 Circular dichroism spectra of four scorpion toxins showing differences in secondary structures between toxin groups B, C, E, and G (see fig. 8.4). Aah II representative of group B, Css II representative of group C, Aah IT4 representative of group E, and Aah IT1 representative of group G.

Table 8.1 Analyses of circular dichroism spectra of scorpion toxins[a]

Toxin[b]	α-Helix	Antiparallel β sheet	Parallel β sheet	β-Turn	Other structure	Total
Aah II (group B)	0.14 (9)[b]	0.42 (27)	0.13 (9)	0.12 (8)	0.18 (11)	1.00 (64)
Css II (group C)	0.16 (10)	0.35 (23)	0.00 (0)	0.26 (17)	0.24 (16)	1.01 (66)
Aah IT4 (group E)	0.13 (9)	0.23 (15)	0.06 (4)	0.20 (13)	0.35 (23)	0.98 (65)
Aah IT2 (group G)	0.22 (15)	0.31 (22)	0.01 (0)	0.23 (15)	0.26 (18)	1.00 (70)

[a]According to the method of Manavalan and Johnson (1987).

[b]Numbers in parentheses correspond to the number of amino acid residues.

ins confirm that they have a common structural core with supplementary secondary structures and differences in protruding loops that may confer their pharmacological specificity (fig. 8.6). The size of the loops protruding from this structural core differs between the toxins from group B and group C (Fontecilla-Camps et al., 1988). For example, in comparison to the group B and C toxins, the toxins from group G (fig. 8.6) possess a supplementary turn of α-helix and a fourth antiparallel β-sheet strand (Darbon et al., 1991). Likewise, stacked side chains from aromatic residues compose a hydrophobic surface unique to the Aah II and Aah IT1 toxins (Fontecilla-Camp et al., 1988; Darbon et al., 1991).

Structure of Toxins Acting on Potassium Channels

Toxins that target potassium channels are markedly different from sodium channel toxins. The five potassium channel-specific toxins sequenced to date include CTX (Gimenez-Galego et al., 1988) and LTX (Chichi et al., 1988) toxins from *Leiurus quinquestriatus hebraeus,* NTX toxin from *Centruroides noxius* (Possani, 1984), IbTX toxin from *Mesobuthus tamulus* (Galvez et al., 1990), and KTX toxin from *Androctonus mancetanius mancetanius* (Crest et al., 1992). These toxins are of smaller size (31–39 residues) and have only three disulfide bridges. They have few or no aromatic residues, but at least one methionine residue is present in each toxin. The secondary structure of CTX toxin, as determined by nuclear magnetic resonance imaging, shows a triple-stranded β-sheet of 13 amino acid residues, an α-helix of 11 amino acid residues, and an extended fragment of 5 amino acid residues. This structure is similar to the core of toxins acting on sodium channels (fig. 8.6), except for the lack of a hydrophobic surface (Bontems et al., 1991). These differences in structure between large (sodium channel specific) and small (potassium channel specific) toxins lead to some consideration of the structure-function relationships between scorpion toxins and their ion channel targets.

Maurotoxin (MTX), purified from the scorpion *Scorpio maurus,* is a potent ligand for different types of potassium ion channels. It shows broad reactivity with voltage-gated potassium channels, as well as on small-conductance calcium-activated potassium channels. The solution structure of MTX shows a bended helix from residues 6 to 16 connected by a loop to a two-stranded antiparallel beta sheet (residues 23 to 26 and 28 to 31). Its structure is similar to other short scorpion toxins despite its peculiar disulfide pairing (Blanc et al., 1997). The interaction of MTX with the pore region of the Kv1.2 potassium channel has been modeled according to charge anisotropy of the ligand and its receptor.

STRUCTURE–FUNCTION RELATIONSHIPS

The comparison of primary, secondary, and tertiary structures of scorpion toxins with different pharmacological characteristics helps us to understand their interaction with receptor sites on targeted proteins. A good understanding of the structure–function relationship in scorpion toxins is a key factor in elucidating information about their targets and the potential use of these toxins for insect control.

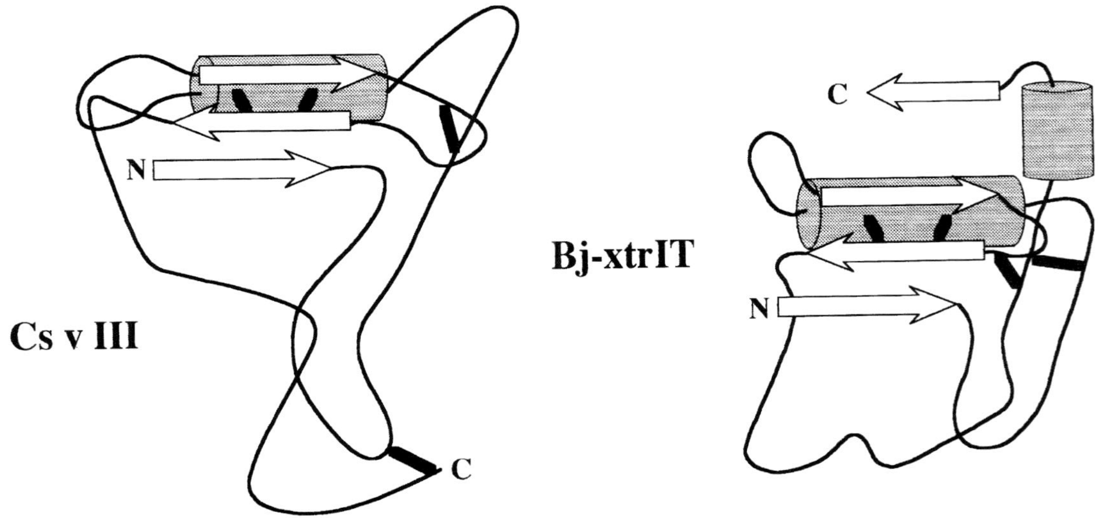

Figure 8.6 Secondary structure and folding of toxin Cs v III and Aah IT1. Arrows are β-sheet structures, cylinder is an α-helix. N and C are the amino and carboxy terminals of the toxin peptide chain, respectively. The four disulfide bridges are shown in bold. Structure of Cs v III redrawn from Almassy et al., (1983) and of Bj-xtrIT redrawn from Oren et al. (1998).

Binding of Toxins to Ionic Channels

As illustrated in fig. 8.2, the binding sites for scorpion toxins on sodium and potassium channels are located on loops extending above the outer membrane surface. The toxin binding relationships for various ionic channels is likely similar, and so the elegant study made by MacKinnon et al. (1989, 1991), using point mutations both on the K_{Ca} channel of *Drosophila* and the CTX toxin, is exemplary. Chemical esterification of carboxyl groups on external acidic amino acid residues of the K_{Ca} channel affects ion conductance and gating, and it reduces the affinity for binding of CTX toxin. Conversely, CTX bound to the channel dur-

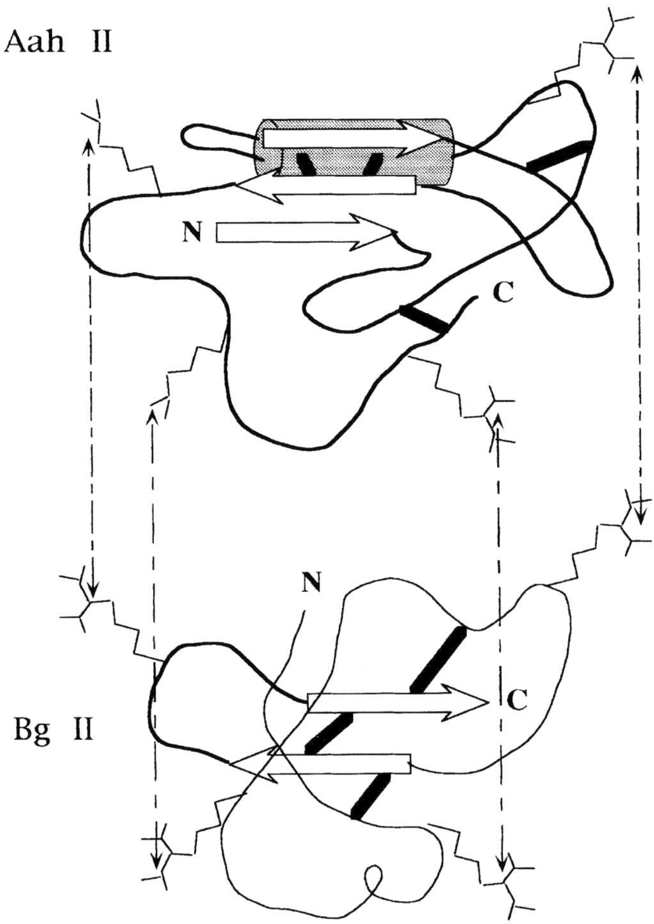

Figure 8.7 Comparison of structures for scorpion toxin Aah II and a sea anemone toxin Bg II. Although different in overall structure, the two toxins compete for the same binding site on the sodium channel. Superimposition of the two structures reveals four basic amino acid residues (dashed double-headed arrows) located in the same position within the two proteins (Arg 62, Lys 58, Lys 50 and Arg 18 in Aah II and Arg 5, Arg 14, Arg 27 and Arg 36 in Bg II (Loret et al., 1994).

ing chemical agent exposure prevents the reduction of channel conductance. Genetic manipulation of the CTX binding site shows that negative charges on its surface are necessary for a high toxin affinity and that substitution of positively charged amino acid residues further reduces binding affinity. These experiments suggest that CTX interacts with the negatively charged carboxyl group present on the surface of the binding site.

Scorpion toxins are basic proteins with a positive charge that appears to be critically important for toxin binding. For example, the toxin Bot XI is structurally homologous to Aah II but is far less potent, possibly because of a valine residue that replaces the lysine at position 58 in the Aah II toxin (Sampieri et al., 1987). Likewise, chemical modification of lysine residues in scorpion toxins acting on mammalian sodium channels dramatically decrease their toxicity (Sampieri and Habersetzer, 1978; Darbon, 1990), whereas chemical modification of the conserved basic residue between the α-helix and β-sheet (see above) only decreases the toxicity of Aah IT1 without markedly changing toxin binding affinity (Loret et al., 1990a).

The important role of basic residues was further confirmed by comparing the 3D structures of Aah II toxin with that of the sea anemone toxin, Bg II (Loret et al., 1994). Bg II has the characteristic structure of other sea anemone toxins which are typically short proteins of 4000 to 5000 DA cross-linked by four disulfide bridges (Kem, 1988). Although BgII has no α helix and the folding of its peptidic skeleton is distinctly different from Aah II (Widmer et al., 1989), it has similar pharmacological properties to the scorpion toxin and competes for the same binding sites. Importantly, the superimposition of 3D structures for Bg II and Aah II reveal four basic amino acid residues located in similar position and orientation, structural features that would constitute two positively charged poles on the surfaces of both toxins (Fig. 8.7).

The importance of these basic residues for β toxin action was confirmed when the crystal structure of neurotoxin Ts1, a major component of the venom of the Brazilian scorpion *Tityus serrulatus,* was resolved. The folding of the polypeptide chain of Ts1 is similar to another scorpion toxin, Cs vIII (fig. 8.6), but Ts1 is much more toxic to mammals than is Cs vIII. Superimposition of 3D structures for the two toxins revealed several basic amino acid residues on one surface that may account for the greater toxicity of Ts1 (Polikarpov et al., 1999). The role of basic residues in recognizing both the α and β receptor sites was further outlined by mutational experiments and chemical modifications (Zilberberg et al., 1997; Hassani et al., 1999).

Toxicity and Binding Affinity

The first pharmacological studies of scorpion toxins attempted to relate potency of toxin action to its affinity for the receptor (Couraud and Jover, 1984). Studies of insect sodium channels have since shown that toxins can bind to receptors with similar affinities and still have different electrophysiological effects (Zlotkin et al., 1985; Lima et al., 1989) and behavioral toxicity (Lima et al., 1986, 1989). Selective chemical modifications of scorpion toxins now show it is possible to decrease toxicity without altering toxin binding affinity (Loret et al., 1990a). Likewise, toxin binding affinity does not determine toxicity: the "ancestral" scorpion toxin from Aah described earlier in this chapter binds to the sodium channel of insects with the same affinity as insect-specific β-toxins, but it has lower toxicity (Loret et al., 1991). Therefore, the ability of a scorpion toxin to bind to a receptor is not correlated to its effect once it has bound.

Toxin Size and Target Diversity

Although structurally related, scorpion toxins differ in their amino acid composition, in the length of their polypeptide chains, and in content of secondary structures. With few exceptions, scorpion toxins are specific for particular receptors, and their structural diversity complements the variety of different pharmacological targets they affect. However, every toxin appears to have common structural features—namely, an α-helix and three β-sheets connected by three disulfide bridges. This structural uniformity among toxins probably relates to common structure in the ionic channels they target, especially the external loops of channel proteins where the toxin binding sites appear to be localized (fig. 8.2). Beyond their common features, scorpion toxins have fine structures that are conserved only among toxins acting on the same pharmacological receptor. This is reflected in the different sizes of scorpion toxins acting on particular receptors (fig. 8.3). For example, the toxins acting on potassium channels are shorter peptides (31–37 amino acid residues) compared to those acting on sodium channels (61–70 residues). Even among the toxins acting on sodium channels, those belonging to group G are always larger (70 residues) than those from other toxin groups (61–66 amino acid residues). This correlation between size and pharmacological specificity suggests that toxin structures modify, sometimes markedly, to act on diverse targets.

Toxin Secondary Structures and Target Diversity

Although toxins from groups D, E, F, and G bind to the same site in the insect sodium channel (fig. 8.4), comparison of their covalent structure shows they have differences in amino acid composition and chain length. There is no consensus sequence to indicate what is common to these toxins that enables them to have the same pharmacological target. Similar secondary structures are a possibility. A circular dichroism study of 10 scorpion toxins showed that within discrete groups the content in secondary structure is similar, but that between groups this content varied (Dufton et al., 1986). This study suggests that changes in secondary structures play an important role in target recognition. This is supported by another study using three toxins belonging to groups C, D, and G. The D toxin binds to β sites on mammal and insect sodium channels, the C toxin binds only to β sites on mammal sodium channels, and the G toxin binds only to β sites on insect sodium channels (fig. 8.4). The D toxin undergoes changes in content of secondary structures that are not observed in the two other toxins. It was suggested that this flexibility could modulate the structure and function of the binding site, which allowed the D toxin to bind to both mammal and insect sodium channels (Loret et al., 1990b). This flexibility was observed in circular dichroism studies of an ancestral scorpion toxin, one that binds on all known toxin binding sites on mammalian and insect sodium channels. This toxin is unusual in having no proline residues in its sequence (Loret et al., 1991), whereas prolines are commonly located in conserved positions within scorpion toxins with high target specificity (fig. 8.3). Proline residues interrupt secondary structures such as α-helixes and β-sheets (Mathews and Van Holde, 1990). Thus, toxins may evolve greater activity on specific classes and types of ion channels through modification of secondary structures subsequently stabilized by the addition of proline residues.

The insect-specific α toxin Bj-xtrIT is one example where additional secondary structures may define target specificity (fig. 8.6). The X-ray crystal structure of this protein (Oren et al., 1998) shows an α helix in the C-terminal region that is present in two other insect-specific α toxins, Aah IT1 and Aah IT2 (Loret et al., 1990a), but not in mammal-specific α toxin (fig. 8.6). This α helix constitutes part of the bioactive surface of insect-specific α toxins as revealed by mutational experiments (Froy et al., 1999).

Evolution of Scorpion Toxins

Venoms containing highly toxic proteins have obvious advantages for scorpion predation and defense, so their specificity for critical targets is expected to increase over evolutionary time. By this reasoning the Aah IT4 toxin may be a relatively primitive or ancestral protein because it is relatively nonspecific and has poor toxicity (Loret et al., 1991). In this regard, it is noteworthy that this toxin is from a North African scorpion venom (*Androctonus australis*). If we take into account the differentiation of scorpion toxins proposed in fig. 8.4, the two oldest toxins, Aah IT4 and Lqq IT2, belong to North African scorpions (fig. 8.1). Similarly, distance analysis of the sequences for 18 toxins active on mammals shows that North American, South American, and Central Asian toxins may be derivatives of the North African toxins (Possani, 1984). The analysis of Buthidae scorpion venom might indicate a North African origin for this family.

There is a stronger sequence homology between insect-specific β toxins than between toxins active on mammals. The latter often differ markedly, even for toxins from the same venom. For example, the insect contraction-inducing toxin, Aah IT1, purified from the venom of a North African scorpion, has 77% sequence homology with BmK IT toxin purified from the Chinese scorpion, *Mesobuthus martensii*. Conversely, in *A. australis* venom there is only 50 to 60% sequence homology between Aah IT1 and other venom toxins acting on mammalian targets. In this connection, it is interesting that a family of closely related insect toxins are present in the venom of *A. australis* together with a family of more distantly related mammal-specific toxins. Scorpions venoms are mainly directed against arthropods, especially soft-bodied insects that are potential prey (Zlotkin, 1983), and the masterwork of scorpion venom evolution may be the insect contraction-inducing toxins. These are large peptides (fig. 8.3), and the high degree of sequence homology observed among them probably indicates that mutations generally decrease the venom's toxicity and are not retained. From this perspective the use of these toxins for insect control has the great advantage of combining high toxicity and specificity for insects with a very low probability that they will revert to toxins having activity in mammals (Gordon, 1997; Froy et al., 1999).

Scorpion Toxins and New Strategies for Drug Design

While scorpion toxins will continue to be valuable tools in the study of ion channels, they are also contributing importantly to the study of ligand-receptor interactions. Toxic proteins are produced by a diversity of venomous animals from various phyla. Typically these are small molecules with a high density of disulfide bonds, and they exert multiple effects on a variety of molecular targets, including enzymes and ion channels. We saw in the comparison of scorpion and sea anemone toxins that proteins with unrelated folding structure can exert similar biological functions. Conversely, toxin proteins having similar folded structure may have distinctly different biological actions. Moreover, it has recently been shown that part of the active site of a foreign protein can be grafted onto an animal toxin "scaffold," opening new perspectives within the field of protein engineering (Menez, 1998).

The first attempts to manufacture drugs for specific molecular targets of action used rigid molecules selected for perfect fit by X-ray analysis of crystalline structures. This approach failed to give the expected results since, under physiological conditions, most target proteins are flexible molecules with labile structure. Studies of scorpion toxin proteins show that high specificity of binding to target sites can be achieved even though the active chemical functions (e.g., the guanidinium and amine groups) are connected by a flexible aliphatic chains (arginine and lysine amino acids, respectively) to the rigid peptidic backbone of the toxin molecule. New drug design strategies use this concept to connect chemical functions, called probes, to a rigid molecular scaffold through flexible aliphatic chains. For most drugs the scaffold is usually an aromatic structure and the attached probes may be chemical functions with ionic charge or proton donor/acceptor capability. The SP3 hybridization of carbon bonds in the aliphatic chains connecting the probes to the scaffold allows subtle variation in positions of probes for facilitation of target binding. In scorpion toxins, the different probes (basic residues) connected to the scaffold provide for multipoint interaction with the target, and there is a correlation between the number of probes and target affinity. The structure of the scaffold and the length of the aliphatic chains provide the specificity for target binding. This evolutionary strategy for improving the effectiveness of the scorpion's toxins now finds application in the design of better drugs.

USE OF SCORPION TOXINS IN INSECT CONTROL

Previous studies of scorpion toxins were largely motivated by the desire to find antidotes and antisera for humans and domestic animals stung by dangerous species. The recent demonstration that insect-specific natural toxins can be used to construct recombinant viral insecticides gives new incentive to apply scorpion toxins to another urgent problem: control of agricultural pest insects (Mc Cutchen et al., 1991; Stewart et al., 1991). The dramatic effects obtained when a gene coding for the Aah IT toxin from *Androctonus australis* was inserted into the genome of an insect-specific baculovirus vector provides ample justification to search for other venoms with still greater toxicity. Scorpions that are innocuous to human health may yield potent toxins against key insect groups, so a greater appreciation of scorpion bionomics is warranted.

The advantages of using insect-specific viruses as vectors for peptide toxin administration are evident. Oral administration through spraying or direct expression of peptide toxins by the plant tissues in need of protection is impractical because most peptides do not penetrate insect cuticle and they are digested rapidly in the gut. Some blood-feeding Diptera absorb large proteins across the gut lining, making them reasonable targets for ingested peptide toxins, but this is an unusual circumstance. Chemical derivatives of peptide toxins may enhance penetration, as they do for some classical carbamate pesticides (Fukuto and Fahmy, 1981), and gut permeability may be increased by agents that disrupt the gut membrane or the peritrophic membrane (e.g., toxins of the bacterium *Bacillus thuringiensis*). Although each of these approaches is feasible, none has been demonstrated to be effective in the field.

Recombinant DNA technology provides a new approach to application of biological pesticides, one which may avoid many of the problems inherent in the administration of classical insecticides (Williams, 1967; Hammock, 1985; 1998). By recombining genes for potent neurotoxins into viral vectors with high specificity for pest insects, these toxins can be "injected" into viral hosts in much the same way that scorpions paralyze their prey. Of many potential vectors, baculoviruses are preferred for initial studies because they are double-stranded, circular DNA viruses found only in arthropods. The best known candidates in this group are multinuclear polyhedrosis viruses (mNPVs), so named because many virons are enclosed in a protective mass of polyhedron protein. The mNPV viruses are often genus or species-specific for a particular insect host, and significant oral transmission of the virus seldom extends beyond one family of insect, even for more broadly contagious viruses. These viruses are used commercially in the United States and other countries as biological pesticides, but their primary importance derives from their excellent properties as cloning systems from recombinant genes. The polyhedron and p10 promoters of baculovirus expression systems are the most powerful promoters known, making them a popular choice for studies of eukaryotic genes and ensuring that additional baculovirus expression systems will continue to develop. Several expression systems are now commonly in use (see reviews by Miller, 1988; Maeda, 1989a; Luckow, 1991).

Baculoviruses can be extremely effective control agents, but they have the disadvantage of being slow acting. Natural selection favors viruses that program their hosts to make as many viral progeny as possible, thus inducing unnaturally high rates of feeding and growth in hosts prior to death. Although this behavior is advantageous to natural biological control, it can be disastrous when baculoviruses are used as biological insecticides on standing crops (Benz, 1986; Hammock and Soderlund, 1986). The first unsuccessful efforts to engineer viruses with faster rates of host killing involved the cloning of a buthid toxin from *Mesobuthus* into the nuclear polyhedrosis virus of the caterpillar larva, *Autographa californica* (AcmNPV) (Carbonell et al., 1988). Maeda (1989b) subsequently increased the speed of kill using a gene coding for diuretic hormone-like peptide expressed in the nuclear polyhedrosis virus of the silkworm, *Bombyx mori* (BmmNPV). Subsequent attempts to use a variety of neurohormones to regulate growth and reproduction have been disappointing, possibly because insects have compensatory mechanisms for regulation of endogenous hormone actions. O'Reilly and Miller (1989) discovered that the virus may have "captured" an insect enzyme to delay molting and that removal of this gene slightly increased the speed of kill of AcmNPV. In 1990, Hammock et al. reported that expression of natural juvenile hormone esterase from the virus AcmNPV in the cotton bollworm *Heliothis virescens* reduced the amount of food consumed by the pest insect. A subsequent study, using a modified enzyme, gave greatly increased speed of kill (Bonning and Hammock, 1990; Bonning et al., 1997).

The most dramatic improvements in lethality of engineered baculoviruses have come from recombinations of genes for natural neurotoxins. Tomalski and Miller (1991) cloned a toxin from the mite *Pyemotes tritici* into a polyhedron-deficient AcmNPV vector, and three studies have used the insect-specific toxin from the scorpion *Androctonus australis hector* (Aah IT) in a similar fashion. The AcmNPV vector is commercially important because it infects larvae of noctuid moths, the pest species accounting for more than half of the insecticide use worldwide. To avoid problems of unintentional release into the environment, Maeda et al. (1991) first cloned genes into the polyhedron-deficient strain of mNPV for *Bombyx mori,* ensuring that it was orally inactive and rapidly killed by light. Because *B. mori* is not indigenous to the United States, initial trials with this nocturnal moth provided additional protection against accidental release. Administration of the virus dramatically reduced feeding of host larvae 40 h after infection, and most of the experimental animals died within 60 h, compared to control animals, which died 96 h after treatment with vector (minus toxin) alone. The moribund behavior of virus-infected larvae was indistinguishable from larvae injected with the AahIT toxin purified from scorpion venom (both exhibited continuous rotation of anterior body segments, dorsal arching, and tremors). Subsequently, Weyer et al. (1990) developed an AcmNPV vector capable of expressing AahIT toxin in polyhedron-positive strains, thus making them stable and orally active in the field. In these studies feeding was dramatically reduced, and the speed of kill by the recombinant virus was increased over that of control viruses (McCutchen et al., 1991; Stewart et al., 1991). The current status of this work was recently reviewed (Hammock et al., 1998; 1999).

In retrospect, it is surprising that viral expression of AahIT constructs were so effective in moths because this toxin has marginal activity when injected into lepidopteran larvae (Herrmann et al., 1990). Pharmacokinetic studies suggest that Lepidoptera, unlike most other insect orders, have a relatively impermeable sheath surrounding the nervous system that could prevent toxin entrance. Also, the level of toxin expression and export by AcmNPV are poor in comparison to other peptides. Nevertheless, viruses expressing AahIT kill insects at lower levels of toxin in the blood than the levels needed to kill by injection. This raises the possibility that the virus has greater access to the nervous system of the moth than other circulating substances or that chronic exposure to low levels of toxin is more effective than acute exposure to high concentration. Whatever the mechanism of action, the unexpectedly strong effect of the AahIT toxin on lepidopterans suggests that even higher levels of control in other orders of pest insects (e.g., flies, beetles) can be achieved as vector systems for these groups are discovered.

Another consideration is to improve the pharmacokinetic properties of the toxin. Because the scorpion depends on rapid immobilization of its prey by its venom, long-term stability of the toxin may not have been optimized through natural selection. Expression systems initially produce low levels of toxin which increase only gradually, making it important to block the major routes of metabolism of the toxin in pest insects. Improved expression of the toxin could be facilitated through use of natural toxin cDNAs or synthetic genes having proper export sequences. A variety of approaches may lead to better toxin expression, but it is most important to isolate genes for scorpion toxins that are highly specific in targeting insects without potential side effects on vertebrates. Recombinant viruses have been shown to act synergistically with several classical pesticides and may be useful in the management of insects that have developed pesticide resistance. Scorpions, like many venomous animals, commonly use synergistic mixtures of toxins in their venoms. The same approach can be used with recombinant viruses to enhance potency of action and delay development of resistance (Hammock, 1998).

Synthetic chemical pesticides set a high standard of efficacy for new agents of control (Hammock and Soderlund, 1986). However, there are many applications in integrated pest management where toxin-expressing viruses already offer significant advantages over natural viruses, and they are competitive in cost and speed of kill with classical insecticides. However, there will be many regulatory questions regarding the safety of these viruses. Several laboratories are taking cautious and well-planned steps toward field evaluation of recombinant viruses (Hammock, 1988). After more than four years of field tests in Europe, North America, and Asia, recombinant viruses are showing a high level of efficacy, a rapid rate of kill even against highly resistant pest species, and a high commercial potential. Importantly, they appear to have fewer collateral disruptive effects on natural enemies than classical pesticides or even sprays of *Bacillus thuringiensis.* The ultimate success of these materials will depend upon how well they integrate with new integrated pest management strategies, new variants of conventional pesticides, and other recombinant technologies (Hammock, 1998). The most important problem concerning use of engineered viruses are the collateral effects they may have on nontarget (desirable) lepidopteran species. Although engineered viruses probably present far less risk to nontarget species than chemical pesticides, this risk is being reduced through continued improvements in methods (Williamson, 1991; Hammock, 1992; 1998).

The public and regulatory reaction to use of engineered viruses coding for the expression of scorpion toxins has been very positive in North America. Although it is difficult to imagine a safer toxin to mammals than AahIT (de Dianous et al., 1987; Loret et al., 1992), or a more selective vector than baculoviruses, it is important that researchers take the responsible and conservative approach of attempting to disprove the safety of this new technology. The successful control of serious insect pests with recombinant baculoviruses expressing scorpion toxins is a wonderful example of the practical and very positive value of research on scorpions for the world community. Similar recombinant technology may revolutionize vaccine production. However, analogous approaches could utilize vertebrate expression vectors and the highly potent mammalian toxins from some scorpion species to engineer agents for biological warfare. Thus, expanded work on these interesting chemicals is needed not only for medical, agricultural, and environmental benefit, but also to insure that the knowledge is not used for inappropriate purposes. Further studies of scorpion venoms will certainly play a key role in risk evaluation as the molecular basis of toxin selectivity for mammalian versus insect targets is being resolved. In knowing these distinctions and by demonstrating the improbability of converting an insect-selective toxin to one with mammalian activity, the safety and acceptability of these agents is substantially increased. This integration of structural biochemistry with the tools of molecular biology is demonstrating a usefulness for scorpion toxins that we could not have previously imagined.

Acknowledgments We thank Drs. Ibrahim Abalis, Andrew Baver, and Rafi Hermann for fruitful discussions. E.P.L. gratefully acknowledges Professor Hervé Rochat for support of his research. The authors are greatly indebted to Dr. François Sampieri for providing unpublished results and Jeannine Lawrence for review of the manuscript. E.P.L. was supported by a grant of the IBRO-MacArthur Foundation. B.D.H. was supported by U.S.D.A. grant 91–37302–6185 and 23–696, NSF DCB-9119332 and a BARD agricultural research grant.

References

Alagon, A.C., Guzman, H.S., Martin, B.M., Ramirez, A.N., Carbone, E. and Possani, L.D. (1988) Isolation and characterization of two toxins from the Mexican scorpion *Centruroides limpidus limpidus* Karsch. *Comp. Biochem. Physiol.* B89: 153–161.

Almassy, R.J., Fontecilla-Camps, J.C., Suddath, F.L. and Bugg, C.E. (1983) Structure of variant-3 scorpion neurotoxin from *Centruroides sculpturatus* Ewing, refined at 1.8 Å resolution. *J. Mol. Biol.* 170: 497–527.

Auguste, P., Hugues, M., Gravé, B., Gesquiére, J.-C., Maes, P., Tartar, A., Romey, G., Schweitz, H. and Lazdunski, M. (1990) Leiurotoxin I (Scyllatoxin), a peptide ligand for Ca^{2+}-activated K^+ channels. Chemical synthesis, radiolabeling and receptor characterization. *J. Biol. Chem.* 265: 4753–4759.

Arseniev, A.S., Kondakov, V.I., Mairov, V.N. and Bystrov, V.F. (1984) NMR solution spatial structure of short scorpion insectotoxin I5A. *FEBS Lett.* 165: 57–62.

Babin, D.R., Watt, D.D., Goos, S.M. and Mlejnek, R.V. (1974) Amino acid sequences of neurotoxic protein variants from the venom of *Centruroides sculpturatus* Ewing. *Arch. Biochem. Biophys.* 164: 694–706.

Babin, D.R., Watt, D.D., Goos, S.M. and Mlejnek, R.V. (1975) Amino acid sequence of neurotoxin I from *Centruroides sculpturatus* Ewing. *Arch. Biochem. Biophys.* 166: 125–134.

Baharaoui, E.M., Pichon, J., Muller, J.M., Darbon, H., El Ayeb, M., Granier, C., Marvaldi, J. and Rochat, H. (1988) Monoclonal antibodies to scorpion toxins. *J. Immunol.* 141: 214–220.

Bechis, G., Sampieri, F., Yuan, P.-M., Brando, T., Martin, M.-F., Diniz, C.R. and Rochat, H. (1984) Amino acid sequence of toxin VII, a β-toxin from the venom of the scorpion *Tityus serrulatus*. *Biochem. Biophys. Res. Commun.* 122: 1146–1153.

Benz, G.A. (1986) Introduction: Historical perspectives. *In* Granados, R.R. and Federici, B. (eds.) *Biology of Baculoviruses,* vol. I, pp. 1–35. CRC Press, Boca Raton, FL.

Blanc, E., Sabatier, J.M., Kharrat, R., Meunier, S., el Ayeb, M., Van Rietschoten, J. and Darbon, H. (1997) Solution structure of maurotoxin, a scorpion toxin from *Scorpio maurus,* with high affinity for voltage-gated potassium channels. *Proteins* 29(3): 321–333.

Blaustein, M.P., Rogowski, R.S., Schneider, M.J. and Krueger, B.K. (1991) Polypeptide toxins from the venoms of Old Word and New World scorpions preferentially block different potassium channels. *Mol. Pharmacol.* 40: 932–942.

Bonning, B.C. and Hammock, B.D. (1996) Development of recombinant baculovirus for insect control. *Ann. Rev. Entomol.* 41:191–210.

Bonning, B.C., Merryweather, A.T. and Possee, R.D. (1991) Genetically engineered baculovirus insecticides. *Ag. Biotech. News Inform.* 3: 29–31.

Bonning, B.C., Ward, V.K., van Meer, M.M.M., Booth, T.F., and Hammock, B.D. (1997) Disruption of lysosomal targeting associated with insecticidal potency of juvenile hormone esterase. *Proc. Natl. Acad. Sci. USA.* 94: 6007–6012.

Bontems, F., Roumestand, C., Gilquin, B., Ménez, A. and Toma, F. (1991) Refined structure of charybdotoxin: common motifs in scorpion toxins and insect defensins. *Science* 254: 1521–1523.

Bougis, P.E., Rochat, H. and Smith, L.A. (1989) Precursors of *Androctonus australis* scorpion neurotoxins. Structures of precursors, processing outcomes and expression of a functional recombinant toxin II. *J. Biol. Chem.* 264: 19259–19265.

Bürchel, W. (1971) Classification, biology and venom extraction of scorpions. *In* Bürchel, W. and Buckley, E. (eds.) *Venomous Animals and their Venoms,* vol. III, pp. 317–3447. Academic Press, New York.

Carbone, E., Wanke, E., Prestipino, G., Possani, L.D. and Maelicke, A. (1982) Selective blockage of voltage-dependent K^+ channel by a novel scorpion toxin. *Nature* 296: 90–91.

Carbonell, L.F., Hodge, M.R., Tomalski, M.D. and Miller, L.K (1988) Synthesis of a gene coding for an insect-specific scorpion neurotoxin and attempts to express it using baculovirus vectors. *Gene* 73: 409–418.

Castle, N.A. and Strong, P.N. (1986) Identification of two toxins from scorpion (*Leiurus quinquestriatus*) venom which block distinct classes of calcium-activated potassium channels. *FEBS Lett.* 209: 117.

Catterall, W.A. (1977) Membrane potential dependent binding of scorpion toxin to the action potential Na^+ ionophore. Studies with a toxin derivative prepared by lactoperoxydase-catalysed iodination. *J. Biol. Chem.* 252: 8660–8668.

Catterall, W.A. (1980) Neurotoxins that act on voltage-sensitive sodium channels in excitable membranes. *Annu. Rev. Pharmacol. Toxicol.* 20: 15–43.

Catterall, W.A. (1988) Structure and function of voltage-sensitive ion channels. *Science* 242: 50–61.

Chichi, G.G., Gimenez-Gallego, G., Ber, E., Garcia, M.L., Winkist, R. and Cascieri, M.A. (1988) Purification and characterization of a unique, potent inhibitor of apamin binding from *Leiurus quinquestriatus hebraeus* venom. *J. Biol. Chem.* 263: 10192–10197.

Couraud, F. and Jover, E. (1984) Mechanism of action of scorpion toxins. *In* Tu, A.T. (ed.) *Handbook of Natural Toxins,* vol. 2, pp. 659–678. Marcel Dekker, New York.

Crest, M., Jacquet, G., Gola, M., Zerrouk, H., Benslimane, A., Rochat, H., Mensuelle, P. and Martin-Eauclaire, M.F. (1992) Kaliotoxin, a novel peptidyl inhibitor of neuronal BK-type Ca^{2+}-activated K^+ channels characterized from *Androctonus mauretanicus mauretanicus* venom. *J. Biol. Chem.* 267: 1640–1647.

Darbon, H. (1990) Scorpion neurotoxin: a molecular model of interaction with the voltage-dependent sodium channel. *In*Zaidi, Abbasi, Smith (eds.) *Protein Structure-Function,* pp. 169–181. TWEL Publishers.

Darbon, H., Weber, C. and Braun, W. (1991) Two-dimensional 1H nuclear magnetic resonance study of Aah IT1, an anti-insect toxin from the scorpion *Androctonus australis Hector.* Sequential resonance assignments and folding of the polypeptide chain. *Biochemistry* 30: 1836–1845.

Darbon, H., Zlotkin, E., Kopeyan, C., Van Rietschoten, J. and Rochat, H. (1982) Covalent structure of the insect toxin of the North African scorpion *Androctonus australis Hector. Intl. J. Pep. Prot. Res.* 20: 320–330.

DeBin, J. A. and Strichartz, J. (1991) Chloride channel inhibition by the venom of the scorpion *Leirus quinquestriatus. Toxicon* 29: 1403–1408.

de Dianous, S., Hoarau, F. and Rochat, H. (1987) Re-examination of the specificity of the scorpion *Androctonus australis hector* insect toxin towards arthropods. *Toxicon* 25: 411–417.

Dufton, M.J., Drake, A.F. and Rochat, H. (1986) Conformational variations among scorpion toxins. *Biochim. Biophys. Acta* 869: 16–22.

Dufton, M.J. and Rochat, H. (1984) Classification of scorpion toxins according to amino acid composition and sequence. *J. Mol. Evol.* 20: 120–127.

Duval, A., Malécot, C.O., Pelhate, M. and Rochat, H. (1989) Changes in Na channel properties of frog and rat skeletal muscles induced by the Aah II toxin from the scorpion *Androctonus australis. Eur. J. Physiol.* 415: 361–371.

Eitan, M., Fowler, E., Herrmann, R., Duval, A., Pelhate, M. and Zlotkin, E. (1990) A scorpion venom neurotoxin paralytic to insects that affects sodium current inactivation: purification, primary structure and mode of action. *Biochemistry* 29: 5941–5947.

El Amin, E.O. (1992) Issues in management of scorpion sting in children. *Toxicon* 30: 111–115.

El Asmar, M.F. (1984) Metabolic effect of scorpion venom. *In* Tu, A.T. (ed.) *Handbook of Natural Toxin,* vol. 2, pp. 551–575. Marcel Dekker, New York.

El Ayeb, M. and Delori, P. (1984) Immunlogy and immunochemistry of scorpion neurotoxins. *In* Tu, A.T. (ed.) *Handbook of Natural Toxin,* vol. 2, pp. 607–638. Marcel Dekker, New York.

El Ayeb, M., Delori, P. and Rochat, H. (1983) Immunochemistry of scorpion α-toxins: antigenic homologies checked with radioimmunoassays (RIA). *Toxicon* 21: 709–716.

Fontecilla-Camps, J.L., Habersetzer-Rochat, C. and Rochat, H. (1988) Orthorhombic crystals and three-dimensional structure of the potent toxin II from the scorpion *Androctonus australis* Hector. *Proc. Natl. Acad. Sci. USA* 85: 7443–7447.

Froy, O., Zilberberg, N., Gordon, D., Turkov, M., Gilles, N., Stankiewicz, M., Pelhate, M., Loret, E.P., Oren, D.A., Shaanan, B. and Gurevitz, M. (1999) The putative bioactive surface of insect-selective scorpion excitatory neurotoxins. *J. Biol. Chem.* 274: 5769–5776.

Fukuto, T.R. and Fahmy, M.A.H. (1981) Sulfur and propesticide action. *In* Rosen, J.D., McGee, P.S. and Casida, J.E. (eds.) Sulfur and pesticide metabolism and action. *American Chemical Society Symposium Series 158,* pp. 35–49. American Chemical Society Publications, Washington, DC.

Garcia, M.L., Garcia-Calvo, M., Hidalgo, P., Lee, A. and MacKinnon, R. (1994) Purification and characterization of three inhibitors of voltage-dependent K+ channels from *Leiurus quinquestriatus hebraeus* venom. *Biochemistry* 33(22): 6834–6839.

Gordon, D. (1997) A new approach to insect-pest control: combination of neurotoxins interacting with voltage sensitive sodium channels to increase selectivity and specificity. *Invert Neurosci.* 3: 103–116.

Galvez, A., Gimenez-Gallego, G., Reuben, J.P., Roy-Contancin, L., Feigenbaum, P., Kaczorowski, G.J. and Garcia, M.L. (1990) Purification and characterisation of a unique, potent, peptidyl probe for the high conductance calcium-activated potassium channel from venom of the scorpion *Buthus tamulus. J. Biol. Chem.* 266: 11083–11090.

Gordon, D., Jover, E., Couraud, F. and Zlotkin, E. (1984) The binding of the insect selective neurotoxin (AaIT) from scorpion venom to locust synaptosomal membranes. *Biochim. Biophys. Acta* 778: 349–358.

Gordon, D., Merrick, D., Wollner, D.A. and Catterall, W.A. (1988) Biochemical properties of sodium channels in a wide range of excitable tissues studied with site-directed antibodies. *Biochemistry* 27: 7032–7038.

Gimenez-Galego, G., Navia, M., Reuben, J.P., Katz, M.G., Kaczorowski, G.J. and Garcia, M.L. (1988) Purification, sequence and model structure of charybdotoxin, a potent selective inhibitor of calcium-activated potassium channels. *Proc. Natl. Acad. Sci. USA* 85: 3329–3333.

Granier, C., Novotni, J., Fontecilla-Camps, J.C., Fourquet, P., El Ayeb, M. and Bahraoui, E. (1989) The antigenic structure of scorpion toxin. *Mol. Immunol.* 26: 503–513.

Gregoire, J. and Rochat, H. (1983) Covalent structure of toxins I and II from the scorpion *Buthus occitanus tunetanus. Toxicon* 21: 153–162.

Grishin, E.V., Soldatova, L.N., Shakhparonov, M.I. and Kazarov, V.K. (1981) The amino-acid sequence of neurotoxin M 10 from the venom of the Central Asian scorpion *Buthus eupeus. Bioorg. Khim.* 6: 390–398.

Gueron, M. and Ovsyshcher, I. (1987) What is the treatment for the cardiovascular manifestations of scorpion envenomation? *Toxicon* 25: 121–124.

Guieu, R., Kopeyan, C., Rochat, H. (1993) Utilization of aspirin, quinine and verapamil in the prevention and treatment of scorpion venom intoxication. *Life Sci.* 53(26): 1935–1946.

Gulbis, J.M., Mann, S. and MacKinnon, R. (1997) Structure of a voltage-dependent K+ channel beta subunit. *Cell* 25; 97(7): 943–952.

Habersetzer-Rochat, C. and Sampieri, F. (1976) Structure-function relationships of scorpion neurotoxins. *Biochemistry* 15: 2254–2261.

Hammock, B.D. (1985) Regulation of juvenile hormone titer: degradation. *In* Kerkut, G.A. and Gilbert, L.I. (eds.) *Comprehensive Insect Physiology, Biochemistry and Pharmacology,* pp. 431–472. Pergamon Press, New York.

Hammock, B.D. (1992) Virus release evaluation. *Nature* 355: 119.

Hammock, B.D. and Soderlund, D.M. (1986) Chemical strategies for resistance management. *In* Glass, E. (ed.) *Pesticide Resistance, Strategies and Tactics for Management,* pp. 111–129. National Academy of Sciences Press, Washington, DC.

Hammock, B.D., Wroblewski, V., Harshman, L., Hanzlik, T., Maeda, S., Philpott, M., Bonning, B. and Possee, R. (1990) Cloning, expression and biological activity of the juvenile hormone esterase from *Heliothis virescens. In* Hagedorn, H., Hildebrand, J., Kidwell, M. and Law, J. (eds.) *Molecular Insect Science,* pp. 49–56. Plenum Press, New York.

Hammock, B.D. (1998) Status of recombinant baculoviruses in insect pest control. Reviews in Toxicology. *In* Pesticides and the Future: Minimizing Chronic Exposure of Humans and the Environment, (Kuhr, R.J., and N. Motoyama, eds.), pp. 205–221. IOS Press, Amsterdam, The Netherlands.

Hammock, B.D. (1999) The quest for a recombinant viral pesticide. A tribute to Susumu Maeda. *RIKEN Rev.* 22: 9–16.

Hammock, B.D, McCutchen, B.F., Beetham, J., Choudary, P., Ichinose, R., Ward, V.K., Vickers, J., Bonning, B.C., Harshman, L.G., Grant, D., Uematsu, T., Lanzrein, B. and Maeda, S. (1992) Development of recombinant viral insecticides by expression of an insect specific enzyme in nuclear polyhedrosis viruses. *Arch. Biochem. Biophys,* 22: 315–344.

Hassan, F. (1984) Production of scorpion antivenin. *In* Tu, A.T. (ed.) *Handbook of Natural Toxin,* vol. 2, pp. 577–605. Marcel Dekker, New York.

Hassani, O., Mansuelle, P., Cestele, S., Bourdeaux, M., Rochat, H., Sampieri, F. (1999) Role of lysine and tryptophan residues in the biological activity of toxin VII (Ts gamma) from the scorpion *Tityus serrulatus. Eur. J. Biochem.* 260(1): 76–86.

Herrman, R., Fishman, L. and Zlotkin, E. (1990) The tolerance of Lepidopterous larvae to an insect-selective neurotoxin. *Insect Biochem.* 20: 625–637.

Hille, B. (1991) *Ionic Channels of Excitable Membranes.* Sinauer Associates, Sunderland, MA.

Johnson, W. C. Jr. (1985) Circular dichroism and its emperical application to biopolymers. *Methods Biochem. Anal.* 31: 61–163.

Jover, E., Massacrier, A., Cau, P., Martin, M.-F. and Couraud, F. (1988) The correlation between Na^+ channel subunits and scorpion toxin-binding sites. A study in rat brain synaptosomes and in brain neurons developing *in vitro*. *J. Biol. Chem.* 263: 1542–1548.

Kem, W.R. (1988) *In* Hessinger, D., and Lenhoff, H., eds.) *The Biology of Nematocysts*, pp. 375–405, Academic Press, New York.

Kirsch, G.E., Skattebol, A., Possani, L.D. and Brown, A.M. (1989) Modification of Na channel gating by an α scorpion toxin from *Tityus serrulatus*. *J. Gen. Physiol.* 93: 67–83.

Kirshbaum, J.B. (1985) Potential implication of genetic engineering and other biotechnologies to insect control. *Annu. Rev. Entomol.* 30: 51–70.

Kopeyan, C., Mansuelle, P., Sampieri, F., Brando, T., Bahraoui, E.M., Rochat, H. and Granier, C. (1990) Primary structure of scorpion anti-insect toxins isolated from the venom of *Leiurus quinquestriatus quinquestriatus FEBS Lett.* 261: 423–426.

Kopeyan, C., Martinez, G., Lissitzky, S., Miranda, F. and Rochat, H. (1974) Disulfide bonds of toxin II of the scorpion *Androctonus australis hector*. *Eur. J. Biochem.* 47: 483–489.

Kopeyan, C., Martinez, G. and Rochat, H. (1978) Amino acid sequence of neurotoxin V from the scorpion *Leiurus quinquestriatus quinquestriatus*. *FEBS Lett.* 89: 54–58.

Kopeyan, C., Martinez, G. and Rochat, H. (1979) Amino acid sequence of neurotoxin III of the scorpion *Androctonus australis Hector*. *Eur. J. Biochem.* 94: 609–615.

Kopeyan, C., Martinez, G. and Rochat, H. (1985) Primary structure of toxin IV of *Leiurus Quinquestriatus Quinquestriatus*. Characterization of a new group of scorpion toxins. *FEBS Lett.* 181: 211–217.

Lima, M.E., Couraud, F., Lapied, B., Pelhate, M., Diniz, C.R. and Rochat, H. (1988) Photoaffinity labeling of Scorpion toxin receptor associated with insect synaptosomal Na^+ channels. *Biochem. Biophys. Res. Commun.* 151: 187–192.

Lima, M.E., Martin, M.-F., Diniz, C.R. and Rochat, H. (1986) *Tityus serrulatus* toxin VII bears pharmacological properties of both β-toxins and insect toxin from scorpion venoms. *Biochem. Biophys. Res. Commun.* 139: 296–302.

Lima, M.E., Martin, M.F., Hue, B., Loret, E.P., Dinitz, C.R. and Rochat, H. (1989) On the binding of two scorpion toxins to the central nervous system of the cockroach *Periplaneta americana*. *Insect Biochem.* 19: 413–422.

Loret, E.P., Mansuelle, P., Rochat, H. and Granier, C. (1990a) Neurotoxins active on insects: Amino acid sequences, chemical modifications and secondary structures estimation by circular dichroism of toxins from the scorpion *Androctonus australis Hector*. *Biochemistry* 29: 1492–1501.

Loret, E.P., Martin-Eauclaire, M.F., Sampieri, F., Granier, C. and Rochat, H. (1991) An anti-insect toxin purified from the scorpion *Androctonus australis Hector* also acts on the α- and β-sites of the mammalian sodium channel: sequence and circular dichroism study. *Biochemistry* 30: 630–640.

Loret, E.P., Sampieri, F., Miranda, F., Granier, C. and Rochat, H. (1992) Scorpion toxins active on insects. *Methods Neurosci.* 8: 381–395.

Loret, E.P., Sampieri, F., Roussel, A., Granier, C. and Rochat, H. (1990b) Conformational flexibility of a scorpion toxin active on mammals and insects: a circular dichroism study. *Proteins* 8: 164–172.

Luckow, V.A. (1991) Cloning and expression of heterologous genes in insect cells with baculovirus vectors. *In* Prokop, A., Bajpai, R.K. and Ho, C. (eds.) *Recombinant DNA Technology and Applications*, pp. 97–152. McGraw-Hill, New York.

MacKinnon, R., Heginbotham, L. and Abramson, T. (1990) Mapping the receptor site for charybdotoxin, a pore-blocking potassium channel inhibitor. *Neuron* 5: 767–771.

MacKinnon, R. (1991) Determination of the subunit stoichiometry of a voltage-activated potassium channel. *Nature* 350: 232–235.

MacKinnon, R., Latorre, R. and Miller, C. (1989) Role of surface electrostatics in the operation of a high-conductance Ca^{2+}-activated K^+ channel. *Biochemistry* 28: 8092–8099.

Maeda, S. (1989a) Expression of foreign genes in insects using baculovirus vectors. *Annu. Rev. Entomol.* 34: 351–372.

Maeda, S. (1989b) Increased insecticidal effect by a recombinant baculovirus carrying a synthetic diuretic hormone gene. *Biochem. Biophys. Res. Commun.* 165: 1177–1183.

Manavalan, P. and Johnson, W.C. (1987) Variable selection method improves the prediction of proteins secondary structure from circular dichroism spectra. *Ann. Biochem.* 167: 76–85.

Martin, B.M., Carbone, E., Yatani, A., Brown, A.M., Ramirez, A.N., Gurrola, G.B. and Possani, L. D. (1988) Amino acid sequence and physiological characterization of toxins from the venom of the scorpion *Centruroides limpidus tecomanus* Hoffmann. *Toxicon* 26: 785–794.

Martin, M.-F. and Rochat, H. (1984) Purification and amino acid sequence of toxin I" from the venom of the North-African scorpion *Androctonus australis Hector. Toxicon* 22: 695–703.

Mathews, C.K. and van Holde, K.E. (1990) *Biochemistry.* Benjamin/Cummings Inc., Menlo Park, CA.

McCutchen, B.F., Choudary, P.V., Crenshaw, R., Maddox, D., Kamita, S.G., Palekar, N., Volrath, S., Fowler, E., Hammock, B.D. and Maeda, S. (1991) Development of a recombinant baculovirus expressing an insect-selective neurotoxin: potential for pest control. *Biol. Technol.* 9: 848–952.

Menez, A. (1988) Functional architectures of animal toxins: a clue to drug design? *Toxicon* 36(11): 1557–1572.

Miller, C. (1990) Diffusion-controlled binding of a peptide neurotoxin to its K^+ channel receptor. *Biochemistry* 29: 5320–5325.

Miller, C., Moczydlowski, E., Latorre, R. and Phillips, M. (1985) Charybdotoxin, a protein inhibitor of single Ca^{2+} activated K^+ channnels from mammalian skeletal muscle. *Nature* 213: 316–318.

Miller, L.K., Lingg, A.J. and Bulla Jr., L.A. (1983) Bacterial, viral and fungal insecticides. *Science* 219: 715–721.

Miller, L.K. (1988) Baculoviruses as gene expression vectors. *Annu. Rev. Microbiol.* 42: 177–199.

Miranda, F., Kupeyan, C., Rochat, H., Rochat, C. and Lissitzky, S. (1970) Purification of animal neurotoxins: isolation and characterization of eleven neurotoxins from the venoms of the scorpions *Androctonus australis Hector, Buthus occitanus tunetanus* and *Leiurus quinquestriatus quinquestriatus. Eur. J. Biochem.* 16: 514–523.

Miranda, F. and Lissitzki, S. (1961) Scorpamins: the toxic proteins of scorpion venoms. *Nature* 190: 443–445.

Numa, S. (1989) A molecular view of neurotransmitter receptors and ionnic channels. *Harvey Lect.* 83: 121–165.

O'Reilly, D.R. and Miller, L.K. (1989) A baculovirus blocks insect molting by producing ecdysteroid UDP-glucosyl transferase. *Science* 245: 1110–1112.

Oren, D.A., Froy, O., Amit, E., Kleinberger-Doron, N., Gurevitz, M., Shaanan, B. (1998) An excitatory scorpion toxin with a distinctive feature: an additional alpha helix at the C terminus and its implications for interaction with insect sodium channels. *Structure* 6(9): 1095–1103.

Pappone, P.A. and Lucero, M.T. (1988) *Pandinus imperator* scorpion venom blocks voltage-gated potassium channels in GH_3 cells. *J. Gen. Physiol.* 91: 817–833.

Park, C.-S., Hausdorff, S.F. and Miller, C. (1991) Design, synthesis and functional expression of a gene for charybdotoxin, a peptide blocker of K^+ channels. *Proc. Natl. Acad. Sci. USA* 88: 2046–2050.

Pashkov, V.S., Maiorov, V.N., Bystrov, V.F., Hoang, A.N., Volkova, T.M. and Grishin, E.V. (1988) Solution spatial structure of 'long' neurotoxin M_9 from the scorpion *Buthus eupeus* by ^{1}H-NMR spectroscopy. *Biophys. Chem.* 31: 121–113.

Polikarpov, I., Junior, M.S., Marangoni, S., Toyama, M.H., Teplyakov, A. (1999) Crystal structure of neurotoxin Ts1 from *Tityus serrulatus* provides insights into the specificity and toxicity of scorpion toxins. *J. Mol. Biol.* 290(1): 175–184.

Polis, G.A. (1990) *In* Polis, G.A. (ed.) *The Biology of Scorpions,* pp. 247–293. Stanford University Press, Palo Alto, CA.

Possani, L.D. (1984) Structure of scorpion toxins. *In* Tu, A.T. (ed.) *Handbook of Natural Toxin,* vol. 2, pp. 513–550. Marcel Dekker, New York.

Possani, L.D., Martin, B.M., Fletcher, M.D. and Fletcher, P.L. (1991) Discharge effect on pancreatic exocrine secretion produced by toxins purified from *Tityus serrulatus* scorpion venom. *J. Biol. Chem.* 266: 3178–3185.

Possani, L.D., Martin, B.M., Svendsen, I., Rode, G.S. and Erickson, B.W. (1985) Scorpion toxins from *Centruroides noxius* and *Tityus serrulatus.* Primary structures and sequence comparison by metric analysis. *Biochem. J.* 229: 739–750.

Rack, M., Richter, D. and Rubly, N. (1987) Purification and characterization of a β-toxin from the venom of the African scorpion *Leiurus quinquestriatus. FEBS Lett.* 214: 163–166.

Rochat, H., Bernard, P. and Couraud, F. (1979) Scorpion toxins: chemistry and mode of action. *In* Cecarelli, B. and Clementi, F. (ed.) *Advances in Cytopharmacology,* vol. 3, pp. 325–334. Raven Press, New York.

Rochat, C., Rochat, H., Miranda, F. and Lissitzky, S. (1967) Purification and some properties of the neurotoxins of *Androctonus australis Hector. Biochemistry* 6: 578–585.

Rochat, H., Rochat, C., Miranda, F., Lissitzky, S. and Edman, P. (1970) The amino acid sequence of neurotoxin I of *Androctonus australis hector. Eur. J. Biochem.* 17: 262–266.

Rochat, H., Rochat, C., Sampieri, F., Miranda, F. and Lissitzky, S. (1972) The amino acid sequence of neurotoxin II of *Androctonus australis Hector. Eur. J. Biochem.* 28: 381–388.

Rosso, J.-P. and Rochat, H. (1985) Characterization of ten proteins from the venom of the Moroccan scorpion *Androctonus mauretanicus mauretanicus,* six of which are toxic to the mouse. *Toxicon* 23: 113–125.

Sampieri, F. and Habersetzer-Rochat, C. (1978) Structure-function relationships in scorpion neurotoxins: Identification of the superreactive lysine residue in toxin I of *Androctonus australis hector. Biochim. Biophys. Acta* 535: 100–109.

Sampieri, F., Habersetzer-Rochat, C., Martin, M.-F., Kopeyan, C. and Rochat. (1987) Amino acid sequence of toxin XI of the scorpion *Buthus occitanus tunetanus.* Evidence of a mutation having an important effect upon neurotoxic activity. *Intl. J. Pept. Prot. Res.* 29: 231–237.

Schweitz, H., Bidard, J.N., Frelin, C., Pauron, D., Vijverberg, H.P.M., Mahasneh, D.M. and Lazdunski, M. (1985) Purification, sequence and pharmacological properties of sea anemone toxins acting on the sodium channel. *Biochemistry* 24: 3554–3561.

Schweitz, H., Bidard, J.-N., Maes, P. and Lazdunski, M. (1989) Charybdotoxin is a new member of the K^+ channel toxin family that includes dendrotoxin I and mast cell degranulating peptide. *Biochemistry* 28: 9708–9714.

Simard, M. and Watt, D.D. (1990) Venoms and toxins. *In* Polis, G.A. (ed.) *The Biology of Scorpions*, pp. 415–444. Palo Alto, CA.

Sissom, W.D. (1990) Systematics, biogeography and paleontology. *In* Polis, G.A. (ed.) *The Biology of Scorpions*, pp. 64–161. Stanford University Press, Palo Alto, CA.

Stahnke, H.L. (1966) Some aspects of scorpion behaviour. *Bull. 5th Calif. Acad. Sci.* 65: 65–80.

Stewart, L.M.D., Hirst, M., Ferber, M.L., Merryweather, A.T., Cayley, P.J. and Possee, R.D. (1991) Construction of an improved baculovirus insecticide containing an insect-specific toxin gene. *Nature* 352: 85–88.

Summers, M.D. and Smith, G.E. (1987) *Texas Agric. Exp. Stat. Bull.* 1555: 1–56.

Tejedor, F.J. and Catterall, W.A. (1990) Photoaffinity labeling of the receptor site for α-scorpion toxins on purified and reconstituted sodium channels by a new toxin derivative. *Cell. Mol. Neurobiol.* 10: 257–265.

Thomsen, W.J. and Catterall, W.A. (1989) Localization of the receptor site for α-scorpion toxins by antibody mapping: implications for sodium channel topology. *Proc. Natl. Acad. Sci. USA* 86: 10161–10165.

Tomalski, M.D. and Miller, L.K. (1991) Insect paralysis by baculovirus-mediated expression of a mite Neurotoxin Gene. *Nature* 352: 82–85.

Valdivia, H.H., Fuentes, O., El-Hayek, R., Morrissette, J. and Coronado, R. (1991) Activation of the ryodine receptor Ca^{2+} release channel of sarcoplasmic reticulum by a novel scorpion venom. *J. Biol. Chem.* 266: 19135–19138.

Vargas, O., Martin, M.-F. and Rochat, H. (1987) Characterization of six toxins from the venom of the Moroccan scorpion *Buthus occitanus mardochei*. *Eur. J. Biochem.* 162: 589–599.

Volkova, T.M., Garsia, A.F., Telezhinskaya, I.N., Potapenko, N.A. and Grishin, E.V. (1985) Study of neurotoxins from the venom of Central Asian scorpion *Buthus eupeus*. *Bioorg. Khim.* 11: 1445–1456.

Watt, D.D. and Simard, J.M. (1984) Neurotoxic proteins in scorpion venom. *J. Toxicol.-Toxin Rev.* 3: 181–221.

Weyer, U., Knight, S., Possee, R.D. (1990) Analysis of very late gene expression by *Autographa* california nuclear polyhedrosis virus and the further development of multiple expression vectors. *J. Gen. Virol.* 71: 1525–1534.

Widmer, H., Billeter, M., and Wuthrich, K. (1989) Three-dimensional structure of the neurotoxin ATX Ia from *Anemonia sulcata* in aqueous solution determined by nuclear magnetic resonance spectroscopy. *Proteins* 6(4): 357–371.

Williams, C.M. (1967) Third-generation pesticides. *Sci. Am.* 217: 13–17.

Williamson, M. (1991) Biocontrol risks. *Nature* 353: 394.

Yatani, A., Kirsch, G.E., Possani, L.D. and Brown, A.M. (1988) Effects of New World scorpion toxins on single-channel and whole cell cardiac sodium currents. *Am. J. Physiol.* 254: H443–H451.

Zenouaki, I., Kharrat, R., Sabatier, J.M., Devaux, C., Karoui, H., Van Rietschoten, J. el Ayeb, M., Rochat H. (1997) *In vivo* protection against *Androctonus australis hector* socrpion toxin and venom by immunization with a synthetic analog of toxin II. *Vaccine* 15(2): 187–194.

Zilberberg, N., Gordon, D., Zlotkin, E., Pelhate, M., Gafni, A., Schauerte, J., Loret, E.P. and Gurevitz, M. (1997) Genetic Alteration of the anti-insect/mammal toxicity ratio of an alpha scorpion neurotoxin expressed in *Escherichia Coli*. *J. Biol. Chem.* 272: 14810–14816.

Zlotkin, E. (1983) Insect selective toxins derived from scorpion venoms: an approach to insect neuropharmacology. *Insect Biochem.* 13: 219–236.

Zlotkin, E., Eitan, M., Bindokas, V.P., Adams, M.E., Moyer, M., Burkhart, W. and

Fowler, E. (1991) Functional duality and structural uniqueness of depressant insect-selective neurotoxins. *Biochemistry* 30: 4814–4821.

Zlotkin, E., Kadouri, D., Gordon, D., Pelhate, M., Martin, M.-F. and Rochat, H. (1985) An excitatory and a depressant insect toxin from scorpion venom both affect sodium conductance and possess a common binding site. *Arch. Biochem. Biophys* 240: 877–887.

Zlotkin, E., Miranda, F., Kupeyan, C. and Lissitzky, S. (1971) A new toxic protein in the venom of the scorpion *Androctonus australis hector. Toxicon* 9: 9–13.

9

Metals in Cuticular Structures

ROBERT M.S. SCHOFIELD

Inorganic building materials have shaped the evolution of groups such as mollusks, vertebrates, and radiolarians. The functional and evolutionary significance of the calcium and silicon minerals filling the shells, skeletons, and teeth of these and many other organisms is well explored. In contrast, the significance of a distinct inorganic structural material that is also widely employed has not yet been explored. This alternative structural material is distinguished from space-filling calcium and silicon biomineralizations by relatively low tissue concentrations of less common elements such as manganese, copper, and zinc. Relatively small quantities of these heavy metals are precisely localized in the contact regions of jaws, claws and other "tools" for manipulating the environment. Often, zinc and manganese are both present, each segregated into a different region of a structure. The study of the properties and occurrences of these heavy-metal biomaterials will increase our understanding of the functional and evolutionary importance of inorganic structural modification. At the same time, the study of the form of these materials promises to reveal novel metal biochemistry: the metal concentrations are usually too low for space filling biominerals and often too high (up to 25% of dry mass) for the metals to be held in place by metalloproteins.

Of the organisms known to produce Mn, Cu, or Zn-enriched structures (table 9.1), scorpions show the highest metal concentrations and the greatest variety of metal enrichment in their cuticle. This chapter describes the distribution of metal-enriched cuticular structures in several families of scorpions, using high-energy ion microscopy, and discusses possible forms and functions of this biomaterial. As relatively large arthropods, scorpions are likely to become preferred experimental preparations for future studies of mechanical and chemical properties of heavy-metal–enriched materials.

HISTORY AND METHODOLOGY

Manganese, copper, and zinc-enriched structures of animals were not recognized until recently, probably because materials containing these metals lack visible distinctions from surrounding material. Bryan and Gibbs (1979) were the first to describe accumulations of these metals in structural features when they attempted to relate the levels of toxic metals in the bodies of polychaete worms to those of the environment. Zinc was found in the jaws of nereid worms, and Cu in the jaws of glycerids, both at concentrations that were independent of en-

Table 9.1 The prevalence of heavy metals in mechanical structures

Phylum / Subphylum / Class / Order	Metals (enriched species /species examined)[a] Mn	Fe	Cu	Zn		Structure(s)[b]	Reference
Annelida							
Polychaeta							
Eunicida	0		0	0	/4		Bryan and Gibbs (1979)
Phyllodocida	1		5	9	/15	jaws	Bryan and Gibbs (1979, 1980); Gibbs and Bryan (1980a,b)
Phyllodocida	3	1	2	3	/5	jaws, Fe:paragnaths	Schofield (1990, unpublished data)
Arthropoda							
Mandibulata							
Insecta							
Orthoptera	0			7	/11	mand.	Hillerton and Vincent (1982)
Lepidoptera	0			6	/6	mand.	Hillerton and Vincent (1982)
Dermaptera	0			0	/1		Hillerton and Vincent (1982)
Coleoptera	57			53	/61	mand.	Hillerton and Vincent (1982); Hillerton et al. (1984)
Coleoptera	2	0	0	0	/2	mand.	Schofield (unpublished data)
Hymenoptera	5			5	/5	mand.	Hillerton and Vincent (1982)
Hymenoptera	3	0	0	6	/10	mand., t. claw	Schofield (1988, 1990, unpublished data)
Diptera	0	0	0	0	/2		Schofield (unpublished data)
Hemiptera	1	0	0	1	/2	t. claw	Schofield (unpublished data)
Diplopoda							
Callipodida	0	0	0	0	/1		Schofield (unpublished data)
Glomerida	0	0	0	0	/1		Schofield (unpublished data)
Spirostreptida	0	0	0	0	/1		Schofield (unpublished data)
?	0	0	0	0	/2		Schofield (1990)

continued

Table 9.1 Continued

Phylum / Subphylum / Class / Order	Metals (enriched species /species examined)[a]					Structure(s)[b]	Reference
	Mn	Fe	Cu	Zn			
Chilopoda							
Geophilomorpha	0	0	0	3	/3	forcipule, t. claw	Schofield (unpublished data)
Scutigeromorpha	0	0	0	0	/1		Schofield (unpublished data)
Scolopendromorpha	0	0	0	1	/1	forcipule, t. claw, mand.	Schofield (unpublished data)
Lithobiomorpha	0	0	0	0	/1		Schofield (unpublished data)
Crustacea							
Copepoda							
Calanoida	0	0	0	1[c]	/1	mand.	Perry et al. (1988)
Malacostraca							
Isopoda	0	0	0	0	/3		Schofield (1990)
Decapoda	0	0	0	0	/1		Schofield (1990)
Amphipoda	0	0	0	0	/1		Schofield (unpublished data)
Chelicerata							
Merostomata	0	0	0	0	/1		Schofield (1990)
Pycnogonida	0	0	0	0	/2		Schofield (unpublished data)
Arachnida							
Scorpiones	16	4	0	16	/16	chel., pedip., t. claw, sting	Schofield (1989, 1990, this study)
Araneae	6	0	0	9	/9	chel., pedip., t. claw	Schofield (1989, 1990); Schofield and Lefere (1989)
Acari	0	0	0	0	/2		Schofield (1990)

Amblypygi	1	0	0	1	/1	chel., pedip.	Schofield (unpublished data)
Opiliones	2	0	0	0	/3	chel., pedip., t. claw	Schofield (unpublished data)
Pseudoscorpiones	1	4	0	2	/4	chel., pedip., t. claw	Schofield (1990, unpublished data)
Solifugae	2	0	0	1	/2	chel., pedip.	Schofield (1990, unpublished data)
Uropygi	0	0	0	1	/1	chel., pedip., t. claw.	Schofield (unpublished data)
Chaetognatha							
Aphragmophora	0	0	0	1	/1	teeth, spines, cut. pl.	Bone et al. (1993)
Aphragmophora	0	0	0	1[d]	/1	spines	Schofield (unpublished data)
Chordata							
Myxini							
Myxiniformes	0	0	0	1[e]	/1	tips of teeth	Schofield (1990)
Nemertia							
Enopla							
Hoplonemertia	0	0	0	1	/1	stylet (with Ca & P)	Schofield (unpublished data)

[a]Nonzero entries indicate detected presence for reports prior to 1988 and, for later reports, enrichment relative to surrounding material. Concentrations were on the order of 1% dry mass or greater unless noted.

[b]mand., mandibles; t. claw., tarsal claw; chel., chelicera; pedip., pedipalp; cut. pl., cuticular plate.

[c]Concentration was not reported but probably reached only about 0.05% of dry mass (Schofield, 1990).

[d]Concentration reached only 0.2% of dry mass. The head was enriched with Zn to about 0.1%.

[e]Concentration reached only 0.1% of dry mass.

vironmental levels (Bryan and Gibbs, 1979, 1980; Gibbs and Bryan, 1980a,b). They suggested that metal accumulations in the jaws might improve their mechanical properties. Hillerton and Vincent (1982) and Hillerton et al. (1982, 1984) concurred with this view when they found Zn and Mn in the hard cuticle at the cutting edge of the mandibles of herbivorous insects. A third study detected Zn in chaetognath teeth and spines (Bone et al., 1983). In each instance the metals were quantified using chemical techniques that could only approximate metal concentrations in localized regions because of poor spatial resolution. The advent of high-energy ion microscopy has greatly improved methods for surveying and measuring localized element concentrations, revealing the uniquely high metal concentrations in these structures, the patterns of multiple-metal deposition, the diversity of organisms in which these accumulations occur, and their occurrence in structures other than mouthparts.

High-Energy Ion Microscopy

High-energy ion microprobes operate by focusing a beam of accelerated protons or other ions onto a specimen in a manner analogous to scanning electron microprobes. They have two major advantages over the latter—namely, in limits of detection (for most elements this can be 2 or 3 orders of magnitude greater; Legge et al., 1988), and, more importantly for this work, in greater tissue penetration (about 1 mm versus a few microns for electron probes). Specimens can thus be surveyed for a diversity of elements at fair depth with little or no sectioning (Schofield and Lefevre, 1993). Two high-energy ion microscopy techniques, STIM (scanning transmission ion microscopy; Lefevre et al., 1991) and PIXE (particle induced X-ray emission; Johansson et al., 1995), were used in the present study to simultaneously measure the concentrations of many elements in tissue specimens.

Figure 9.1 shows an example of images produced by STIM and PIXE using a scanned beam of 4 MeV protons. Figure 9.1A is a STIM image of the fang and smaller marginal teeth on the chelicera of a garden spider, *Araneus diadematus*. This image represents the kinetic energy lost by individual protons that pass through the specimen, with lighter shadings indicating regions where protons lost more energy. The energy-loss measurements can be used to calculate the quantity of material penetrated with only an approximate knowledge of sample composition (Lefevre et al., 1991). The STIM technique thus provides a measure of mass for the sampled volume.

PIXE provides information on the quantity of individual elements in the sampled volume. As protons pass through the specimen they eject electrons from inner orbitals, which, in turn, emit X-rays of element-specific energies as they repopulate these orbitals. Each of the smaller images (Fig. 9.1B–E) shows the locations of various elements based on their characteristic X-ray emissions. Zinc X-rays, for example, are induced in an infinitesimal volume of the specimen at a rate proportional to the quantity of Zn in that volume. The quantity of Zn in the sampled volume is determined by counting these X-rays, but for thicker specimens this determination is complicated by two effects: first, not all of the emitted X-rays can be detected because some are reabsorbed by the specimen. Second, the rate at which X-rays are emitted depends on the energy of the protons, which is not constant but decreases as they pass through the specimen. Fortunately, data obtained from STIM can be used to directly compensate for

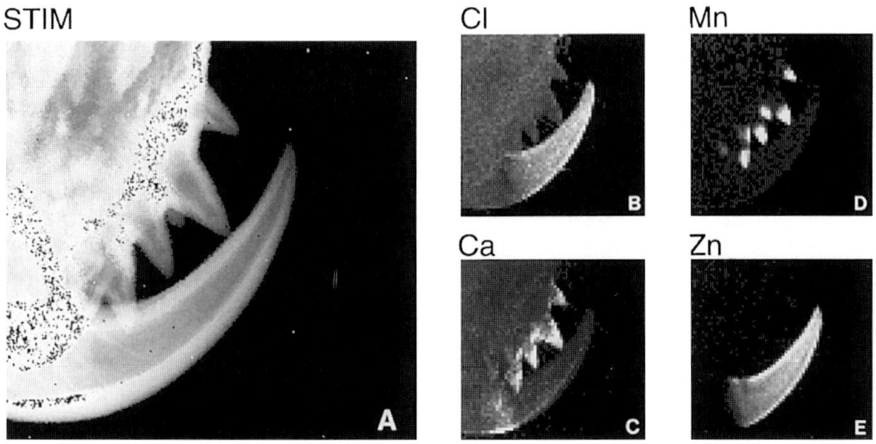

Figure 9.1 Images produced by high-energy ion microscopy showing a cheliceral fang and marginal teeth from a garden spider, *Araneus diadematus.* (A) Scanning transmission ion microscopy image in which lighter shades indicate larger projected densities. (B, C, D, E) Proton-induced X-ray emission images of the same field as panel A showing the origin of, respectively, Cl, Ca, Mn, and Zn X-rays. Lighter shades indicate greater numbers of detected X-rays. The fang is enriched with Zn and Cl. The tips of the marginal teeth contain Mn. Ca is enriched in the teeth, surrounding regions, and the portion of the fang not enriched with Zn. Backgrounds have not been subtracted. Frame size: 1 mm $\times$ 1 mm.

these thick-specimen effects or to set limits on the possible errors they introduce (Schofield and Lefevre, 1993). The larger measurement uncertainties reported here for thicker specimens and lower-energy characteristic X-rays reflect these thick-specimen effects.

Together then, STIM and PIXE enable measurement of element concentrations in microscopic volumes of unsectioned specimens: the quantity of specific elements is given by PIXE, the total quantity of material by STIM. The data plotted in fig. 9.1 can be used to calculate a concentration value for each of the 4096 (64×64) pixels, or picture elements, in figs. 9.1B–E. Near the tip of the spider fang, the measured Zn concentrations reached $15\pm3\%$ of dry mass (ranges throughout this chapter indicate a 68% confidence interval). At this location the chlorine concentration was found to be $3\pm1\%$. In the marginal teeth, Mn concentrations reached $3.7\pm0.7\%$ (Schofield, 1990). Each of these values is derived from a volume represented by a single pixel in fig. 9.1. Because these volumes extend along the beam path through the specimen and may thus include unenriched material, the concentration values are lower limits to the peak local values. To measure concentrations in local volumes that may be completely internal, we developed tomography using PIXE and STIM (Schofield, 1995). For this procedure, data are collected at a number of sample orientations, and the internal element distribution is reconstructed using a computer.

We are using the PIXE-STIM techniques and PIXE alone to survey a variety of organisms for structural accumulations of elements between 11 and 36 in atomic number (the most sensitive range of our probe). Our findings to date of heavy metals are summarized in table 9.1, along with the findings of others using other methods of determination. Although heavy-metal–rich structural materials adorn a wide range of organisms in several phyla, it is difficult to predict which organisms will have them. For instance, these accumulations are not associated solely with herbivorous feeding behaviors as earlier investigators had suspected. No such consistent association with behavioral or environmental factors is evident. Moreover, these accumulations are not strictly associated with organisms that do not synthesize materials containing Ca, Si, or Fe; thus Mn, Cu, and Zn are not used as substitutes for the more common elements. Nereid worms, for example, have Fe-containing teeth (paragnaths) surrounding their Zn-rich jaws (Schofield, 1990). Even the phylogeny of the higher taxa does not reliably predict the presence of heavy metal features. Within the insect order Hymenoptera, high concentrations of metals have been found in the jaws and tarsal claws of ants, but not in the closely related wasp and bee families. Some more distantly related hymenopterans (gall wasps) have metals, others (saw flies) do not (Schofield, 1990). On the other hand, phylogenetic relationship is predictive at a family level and below. We have not found major differences in metal distributions among the examined species within any arthropod family.

METALS IN SCORPION CUTICLE

Distribution

High concentrations of heavy metals were found in the teeth (cuticular protuberances) lining the chelicera and pedipalps and in the tarsal claws and sting of every scorpion species for which these structures were examined. In each structure, metals were localized in regions susceptible to abrasion and mechanical force through contact with the environment. Three heavy metals, Mn, Fe, and Zn, were found in two very different enrichment patterns, one characteristic of the family Buthidae and the other characteristic of all other scorpion families. Figure 9.2 is a diagram of these two patterns and the species in which they were found. Pattern A occurs in nonbuthid scorpions, pattern B in buthids. The buthoids are thought to be a primitive sister group of the other hypothesized superfamilies, the Diplocentroidea and the Chaeriloidea (Lamoral, 1980; Sissom, 1990), both of which are represented in the nonbuthid families sampled here.

Figures 9.3 and 9.4 show typical examples of cuticular metals in scorpions. Figure 9.3 shows cheliceral fixed fingers from two scorpion species representing the two major accumulation patterns. The finger on the left is from the buthid *Centruroides exilicauda,* and the one on the right is from the vaejovid *Paruroctonus mesaensis.* The teeth of both scorpions are enriched with Zn and, to a differing degree in each scorpion, with Mn. Large quantities of Fe, however, are found only in the buthid. The regions of deeply tanned cuticle that appear dark in visible light correspond roughly to the Mn-enriched region of the buthid teeth and to the Zn-enriched region in the nonbuthid. However, visible features do not distinguish the smaller Zn- and Fe-enriched regions in the buthid. Structural iron accumulations are unusual among the arthropods that we have examined. Accumulations of Fe similar to those in buthid teeth have only been found in some pseudoscorpions (Schofield, 1990, unpublished observations).

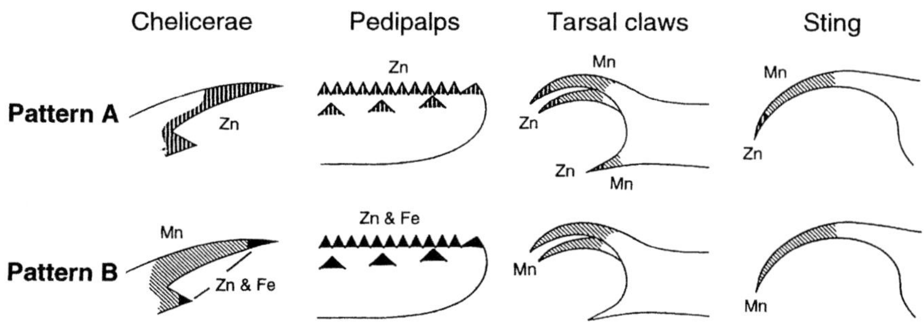

Species	Family	No. of individuals contributing specimen(s) of: Chelicerae	Pedipalps	Tarsal claws	Sting
Pattern A					
Paruroctonus mesaensis	Vaejovidae	2	3	5	2
Paruroctonus boreus	Vaejovidae	0	0	3	2
Vaejovis confusus	Vaejovidae	3	1	4	2
Vaejovis spinigerus	Vaejovidae	1	0	0	1
Anuroctonus phaiodactylus	Vaejovidae	3	3	3	2
Hadrurus arizonensis	Iuridae	1	2	3	4
Hadrurus spadix	Iuridae	0	0	0	2
Pandinus gregoryi	Scorpionidae	3	3	3	3
Urodacus yaschenkoi	Scorpionidae	2	2	2	2
*Diplocentrus whitei**	Diplocentridae	1	1	1	1
*Chaerilus variegatus**	Chaerilidae	1**	1	1**	1
Pattern B					
Centruroides exilicauda	Buthidae	3	6	4	6
Centruroides sp.	Buthidae	1	1	1	1
Androctonus australis	Buthidae	3	4	3	4
*Leiurus quinquestriatus**	Buthidae	1	1	1	1

* : Specimen preserved in liquid. All others were either dry or freeze substituted with acetone.

**: The smaller teeth (serrations) on the movable finger of the chelicera were not enriched. The single examined specimen of the chaerilid family conformed to pattern A with the exception that the median tarsal claw was not enriched.

Figure 9.2 A diagrammatic summary of the metal accumulations found in the chelicerae, pedipalps, tarsal claws, and stings of representative scorpions. Except where Fe and Zn overlap, only the prevalent metal in each region is shown. The buthid family was characterized by pattern B, members of other families conformed to pattern A. Similar regional specializations of Zn and Mn were found in the tarsal claws and sting of pattern A and the cheliceral teeth of pattern B.

Such unusual accumulations may be helpful in determining phylogenetic relationships.

Figure 9.4 shows tarsal (leg) claws, demonstrating metal accumulations in structures that are not normally used in feeding. On the left is a leg tip from the buthid *Centruroides exilicauda* and on the right one from the vaejovid *Vaejovis confusus.* Metals in the vaejovid claws are found in an arrangement reminiscent of buthid cheliceral teeth and nonbuthid stings, a Mn-fortified shaft capped with Zn-enriched material. The buthid claws are not enriched with Zn, but the two lateral claws contain considerably more Mn than the nonbuthid claws.

In the scorpions and other arthropods we have examined, darkly tanned cuticle is not a reliable indicator of metal deposition. The buthid median claw is not enriched with heavy metals, but it does not differ from the lateral claws in color or darkness of tanning. Even when metals are present, they often do not precisely overlap the darkly tanned regions; usually the metal-enriched regions

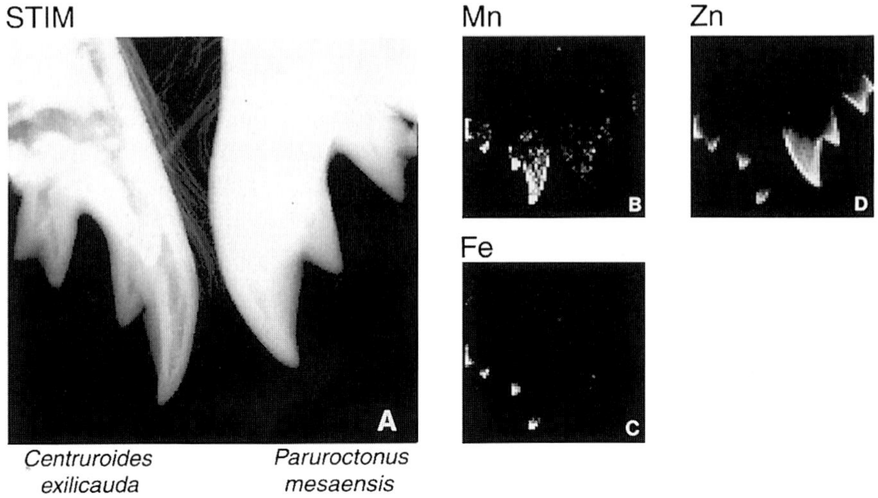

Figure 9.3 Family-specific patterns of structural metal accumulations in the fixed cheliceral fingers of two scorpions. The buthid *Centruroides exilicauda* (left) and vaejovid *Paruroctonus mesaensis* (right) are compared. (A) Scanning transmission ion microscopy image. (B, C, D) Proton-induced X-ray emission images in Mn, Fe, and Zn X-rays, respectively, showing the same field as panel A. Mn, Fe, and Zn are found in *Centruroides.* Zn is prevalent and more broadly distributed in *Paruroctonus.* Frame size: 1.2 mm × 1.2 mm.

are smaller and have more abrupt boundaries. Metals have not been found in light cuticle.

The only distinction between non-buthid families was enrichment of the pedal spurs, tarsal structures that help maintain purchase during locomotion (not diagrammed in fig. 9.2). One or two pedal spurs protrude from a joint close to the leg tip and contact the substrate when it is relatively flat (Couzijn, 1976). In some non-buthids the spurs are enriched with Zn. Figure 9.5 shows the prolateral pedal spur on a specimen of *Pandinus gregoryii.* Metals in the spur are distributed in the familiar pattern found also in the tarsal claws and sting: Zn is located at the tip, Mn along the shaft. This scorpion is a member of the diplocentroid line of families that lack retrolateral spurs (Sissom, 1990). The single spur on leg specimens representing this group of families had Zn-enriched tips. Zinc was not found in either spur of specimens representing the families that have both prolateral and retrolateral spurs. (Zn was found in the single pedal spur of specimens of: *Pandinus gregoryii, Urodacus yaschenkoi,* and *Diplocentrus whitei.* Zn was not found in either pedal spur of specimens of: *Anuroctonus phaiodactylus, Paruroctonus mesaensis, Hadrurus arizonensis, Chaerilus variegatus, Centruroides exilicauda* and *Leiurus quinquestriatus.* Tibial spurs, when present, also did not contain Zn.) Perhaps the Zn-enriched single spurs serve a more active role during locomotion than do the unenriched paired spurs.

The spines in figure 9.5 do not contain Zn but are enriched with small quantities of Mn (this is more obvious if the Fe and Mn maps are compared (i.e., figs.

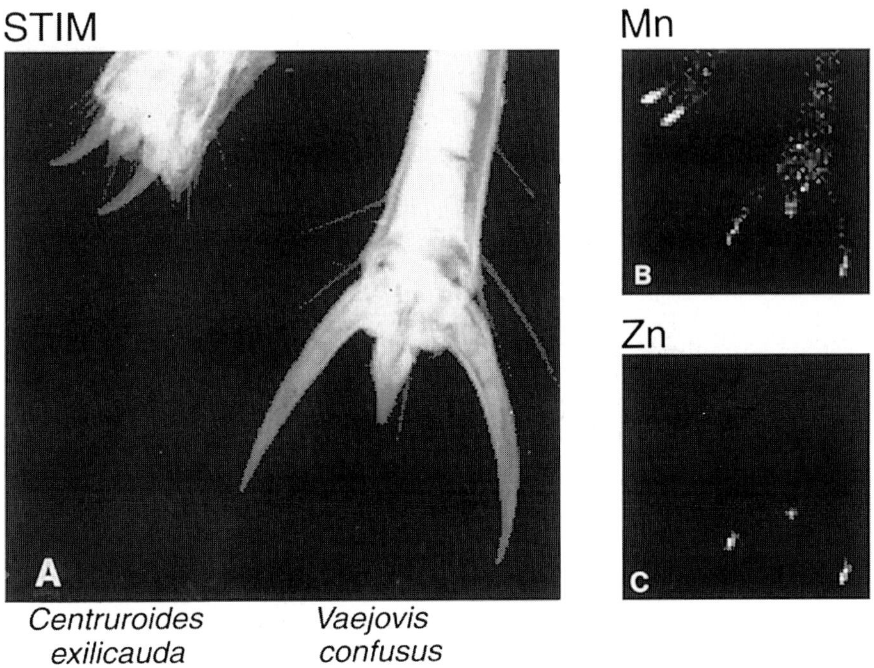

Figure 9.4 Family-specific metal accumulations in the tarsal (leg) claws of two scorpions. The buthid *Centruroides exilicauda* (upper left), and vaejovid, *Vaejovis confusus* (lower right) are compared. (A) Scanning transmission ion microscopy image showing lateral claws (ungues) and the smaller median claw (dactyl) in each specimen. (B, C) Proton-induced X-ray emission maps showing the location of Mn and Zn in the field of panel A. The lateral claws, but not the median claw, of *C. exilicauda* are highly enriched with Mn. Comparably less Mn is found in all three claws of *V. confusus*. The claw tips of *V. confusus* are fortified with Zn. Frame size: 1.5 mm × 1.5 mm.

5c,b). The joint is somewhat enriched with Mn and Fe. The gray scales for the Mn and Fe maps have been set to emphasize lesser quantities of metals than in previous figures; the Fe concentration in this joint is about an order of magnitude lower than that in the buthid teeth. Spine and joint accumulations like these were found in most scorpions.

To search for other enriched regions, the entire cuticular surface of a scorpion, (*Pandinus* sp.) was systematically surveyed. Selected regions from a variety of other species were also examined including coxal mouthparts, pedipalpal gnathobases, legs, regions of tanned cuticle, and the pectines. Metal concentrations greater than 1% were found only in the structures shown in figure 9.2 and in the pedal spurs described above.

In summary, high concentrations of cuticular metals were found exclusively in protrusive structures that contact surfaces or objects in the environment. In contrast to Ca biomineralizations, these metal accumulations involve several cations distributed in intricate regional patterns. The precise localization of these cations suggests much greater control of cuticular metal chemistry during development than previously known.

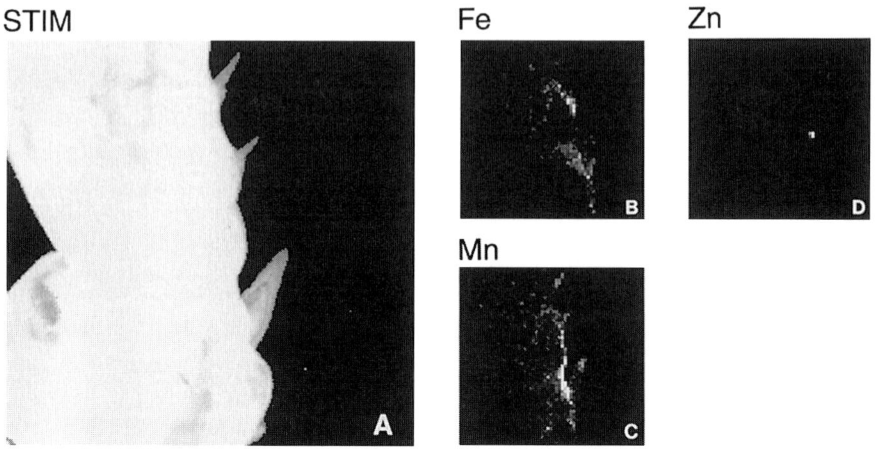

Figure 9.5 Metals in the spur and joint region between tarsomeres I and II on a leg specimen of *Pandinus gregoryii* (Scorpionidae). (A) Scanning transmission ion microscopy image. (B, C, D) Proton-induced X-ray emission element maps showing the distribution of Fe, Mn, and Zn, respectively. Gray scales for the Mn and Fe images have been set to emphasize smaller quantities of metals than in figs. 9.3 and 9.4. The lower and largest of the three pointed protrusions visible in panel A is the pedal spur; the other two are spines. At the tip, the pedal spur is highly enriched with Zn. Different regions of the joint are enriched with Mn and Fe. The pedal spur and spines are enriched with Mn but not Fe. Frame size: 2 mm × 2 mm.

Concentrations

Table 9.2 lists the peak metal concentrations measured in the cuticular structures of several scorpions. Most of the values in table 9.2 were derived from data similar to that displayed in figs. 9.1, 9.3, and 9.4. However, concentration values obtained using PIXE and STIM tomography and values obtained with an electron microprobe for a sectioned chelicera are also included in the table. Tomography and electron microscopy indicated that the predominant metals were localized in surface layers of the integument (Schofield et al., 1989; Schofield and Lefevre, 1992). Near edges in the images the two-dimensional STIM-PIXE technique samples only the surface layer, so the values in table 9.2 are good indications of the peak local concentrations even though they are averaged over depth. The results obtained from all three methods were in satisfactory agreement (Schofield et al., 1989, Schofield and Lefevre, 1993).

When Zn and Fe were present in the same structure, the tabulated values are for a single location. This highlights an apparent correlation between Zn and Fe contents of between two to four Zn atoms per Fe atom, with a ratio of 3:1 falling within the 68% confidence interval in all but one sample. This ratio was not found near the edges of some accumulations because Zn- and Fe-rich regions did not always overlap precisely. Thus the Zn and Fe cannot be bound together in a stoichiometric or invariant complex.

The Mn and Zn concentration values in table 9.2 are unusually high for elements that are not abundant in nature. These Zn values are higher than any re-

Table 9.2 Peak metal concentrations in cuticular features.

Specimen No.[b]	Feature	Percent of dry mass[a]		
		Manganese	Iron	Zinc
Centruroides sp.				
1	Cheliceral tooth	0.24 ± 0.05[c]	8 ± 2[c]	24 ± 5[c]
	Pedipalp tooth		2.8 ± 0.7	7 ± 2
	Tarsal claw	0.5 ± 0.2		
	Sting	1.5 ± 0.2		
Centruroides exilicauda				
1	Cheliceral tooth	0.6 ± 0.2	2.9 ± 0.5	13 ± 2
	Pedipalp tooth		5 ± 2	15 ± 3
	Sting	2.4 ± 0.4		
2	Pedipalp tooth		7 ± 1	20 ± 4
	Sting	0.8 ± 0.2		
3	Pedipalp tooth		2.4 ± 0.6	6 ± 1
	Sting	3.9 ± 0.7		
4	Pedipalp tooth		9 ± 2	19 ± 4
Paruroctonus mesaensis				
1	Cheliceral tooth	0.07 ± 0.02		10 ± 2
	Pedipalp tooth			20 ± 4
	Sting	0.9 ± 0.3		7 ± 2
Hadrurus arizonensis				
1	Sting	3.8 ± 0.5[d]		25 ± 3[d]
Pandinus gregoryi				
1	Pedal spur	0.1 ± 0.05		3.1 ± 0.8

Empty cell indicates that element was either not detected, not enriched relative to surrounding cuticle, or (apparently) a dust constituent.

[a]Ranges (±) indicate 68% confidence intervals, computed by numerically convoluting simplified error distributions for each source of systematic and random error (Schofield and Lefevre, 1993). Unless indicated, each concentration value is the peak value for a 64x64 array of cylindrical volumes traversed by the approximately 15μm diameter beam. The uncertainty is for a single measurement and is not related to the spread among the 4096 measured values.

[b]Specimens were either dry or freeze substituted with acetone.

[c]Higher resolution electron microprobe measurement on a sectioned chelicera (Schofield et al., 1989).

[d]PIXE-STIM microtomography measurement: peak local values in the interior of the specimen (Schofield and Lefevre, 1992).

ported in normal living tissue (Underwood, 1977). In table 9.2, peak Zn values range between 3% and 25% of dry mass, Mn between 0.1% and 4%, and Fe between 2% and 9%. The variation between individuals of the same species was often significant. For example, in *C. exilicauda* the peak Zn values in the pedipalp teeth of four individuals ranged from 6±1 to 20±4% of dry mass. Heavy metal concentrations measured in other body cuticle were generally on the order of 0.01% or less.

The high levels of metal accumulation are further indications that cuticular metal chemistry is precisely controlled, with metal concentrations varying by three orders of magnitude on either side of abrupt boundaries. These high values also indicate that the accumulations are not spurious and suggest that their deposition may entail a significant energetic cost to the organism.

Associations between elements

Associations between Zn, Mn, Cl, and Ca were found repeatedly in the cuticle of scorpions and in the other organisms that we examined (listed in table 9.1). These associations are demonstrated for scorpions in fig. 9.6, a series of PIXE element maps for Cl, Ca, Mn, and Zn showing a buthid sting at the top and a nonbuthid sting at the bottom. Chlorine is enriched along with Zn in the nonbuthid sting. This Zn-Cl association was first noted by Hillerton and Vincent (1982) in insects. Chlorine enrichment accompanies Zn enrichment in most of the arthropods that we have examined, but not in the buthid scorpions or in the centipedes. In the examined polychaete worms, Cl-enrichment occurs in the Zn-enriched portion of nereid jaws and in the Cu-enriched portion of glycerid jaws. The Zn-Cl association does not appear to be a simple chemical relationship. The atomic ratios are not consistent with $ZnCl_2$, and the observed ranges in the atomic ratio (Zn to Cl: 7±2 in an ant; 1.6±0.5 in a spider) indicate that a single chemical complex could not account for the major portion of Zn present in every organism (Schofield, 1990). Chlorine and other halogens may well influence cuticular properties on their own. In a few arthropods Cl has been found without Zn, and in most aquatic arthropods that we have examined (horseshoe crabs, pycnogonids, crustaceans [even terrestrial isopods]), cuticular structures were enriched with Br instead of heavy metals (Schofield, 1990, unpublished observations). Nevertheless, when Cl and Zn were found together, as in nonbuthid scorpions, the enriched regions matched precisely, suggesting that although these accumulations are not likely to be stoichiometrically related, they are closely coupled.

A correlation between Ca and Mn enrichment was also apparent, but Ca enrichment was often found over a broader region than Mn (this is more evident in the spider chelicera of fig. 9.1 than in fig. 9.6), and it was found in most darkened cuticular structures. Thus Mn and Ca enrichments are not as clearly coupled as Zn and Cl.

Negative correlations were also observed. Zinc and halogen deposition seem to interfere with Ca deposition, as seen in figs. 9.1 and 9.6 where Ca content is depressed in regions of Zn abundance. This depression has been observed in structures solely enriched with Zn, Cl, or Br, respectively, in certain centipedes, millipedes, and crustaceans (Schofield, 1990, unpublished observations). Thus it seems that each of these elements interferes with Ca accumulation and not just where they are in combination. Zinc accumulation may interfere with Mn

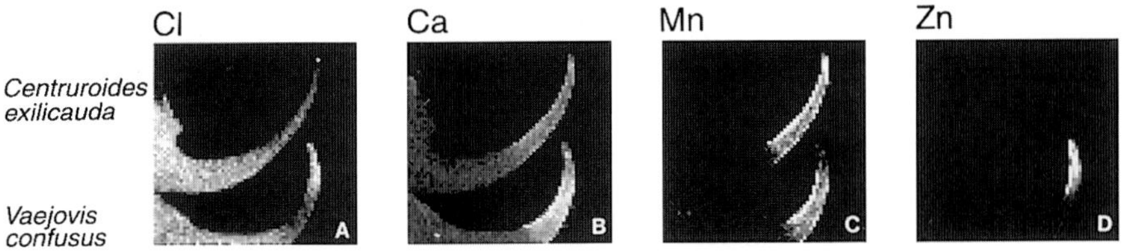

Figure 9.6 Related accumulations of Cl, Ca, Mn, and Zn in the stings of two scorpions. The buthid *Centruroides exilicauda* (top) and the vaejovid *Vaejovis confusus* (bottom) are compared. Proton-induced X-ray emission element maps showing the distribution of (A) Cl, (B) Ca, (C) Mn, and (D) Zn. Chlorine enrichment is correlated with Zn enrichment, whereas Ca and Mn enrichment are negatively correlated with Zn enrichment. Unlike Mn and Zn, Ca and Cl are abundant in arthropod cuticle, although a portion of the displayed data is background. Frame size: 2.3 mm × 2.3 mm.

accumulation, as in fig. 9.6 where two Mn-enriched stings show comparable distribution of this metal except in the Zn-enriched region of the lower sting.

These relationships between Cl, Ca, Mn, and Zn distribution in the cuticle suggest that they accumulate interactively. There may be two distinct processes, one associated with Zn and halogen accumulation, and a second associated with Ca and Mn. The negative correlation between these two groups of accumulations may indicate that part of the process shared by Mn and Ca conflicts with, or is modified for, Zn and halogen accumulation.

The absence of correlation between Zn and some other elements is also chemically significant. The concentration of Zn and that of Al, Si, P, or S was not correlated, indicating that most Zn deposits are not bound as minerals containing these other elements. Of the common Zn minerals, only $ZnCO_3$ and ZnO remain as candidates. Similarly, the absence of a detectable association between Zn and S is strong evidence that most Zn is not bound to a metallothionein. The atomic binding ratio of Zn to S in metallothionein is about 1:3 (Taylor and Simkiss, 1989), and the association would have been detectable if it existed. Furthermore, the cysteine enrichment characteristic of metallothioneins was not detected in amino acid analyses of entire pedipalps and stings (D. Cox, University of Oregon, unpublished results).

Table 9.3 summarizes the observed characteristics of the Zn and Mn accumulations found in scorpions, other arthropods, and annelids. The consistency of the correlations among Cl, Ca, Mn, and Zn in scorpions, other arthropods, and even annelids is compelling evidence of an undescribed and generic form of metal chemistry. These correlations also suggest that functional properties associated with heavy metal accumulations are similar throughout the variety of structures and organisms in which they have been found.

Possible Molecular Forms

Biological metals are known to exist in animals in biominerals, such as the magnetite in the radular teeth of certain mollusks, and in organic molecules, such as metallothionein proteins (Kagi and Schaffer, 1989; Mann et al., 1989). The concentrations of heavy metals observed in cuticular structures, particularly those in scorpions (table 9.2), are difficult to reconcile with either of these molecular forms.

Table 9.3 Characteristics of zinc and manganese enrichment in scorpions and the other organisms

	Zn	Mn
Sites	Usually pointed tips or sharp edges	Often shafts and rounded regions; more broadly distributed than Zn
Concentrations	Up to 25% of dry mass	Up to 4% of dry mass
Correlated with	Cl, but not in buthids and centipedes; P and Fe in buthids	Ca, but not closely associated like Cl and Zn (see fig. 9.1)
Negatively correlated with	Ca, Mn	Zn, mostly segregated into separate regions of a structure or layers of the cuticle[a,b]
Location within the cuticle	Surface layer[a]	Surface layer, subsurface layer (below Zn) or distributed throughout the cuticle[a,b]
Visible characteristics	Usually not detectable, but enriched material sometimes orange, vitreous and shatter prone; Zn-rich tips sometimes severed at enrichment boundary	Not detectable

[a]Schofield et al. (1989).

[b]Schofield and Lefevre (1992).

Biominerals

Cuticular accumulations of common Mn-, Fe-, or Zn-containing minerals could not modify mechanical properties in the same way that Ca and Si structural biominerals do (Vincent, 1982) because the minerals would fill only a small fraction of the cuticle volume. This volume fraction can be calculated from the measured metal concentration using published mineral densities and assuming that the density of organic cuticular components is 1.5 g/cm^3. If Zn is incorporated as $ZnCO_3$ or ZnO (other common inorganic salts are excluded because Al, Si, P, and S were not associated with Zn), then the mineral would occupy only about 0.25 ($ZnCO_3$) or 0.1 (ZnO) of the volume in the most metal-rich cuticle we have found (25% of dry mass). In contrast, the volume fraction of $CaCO_3$ in decapod (Crustacea) cuticle is typically about 0.6 (calculated from Vincent, 1982; Horst and Freeman, 1993).

Of course, it is possible that Zn accumulates in biomineral form as a byproduct of a process that affects cuticular properties but that, by itself, the Zn biomineral does not substantially influence cuticular properties. It seems unlikely, however, that cuticular metals play a passive role because the regional segregation of Mn, Fe, and Zn implies that different processes are associated with each metal. A passive role would imply that three different processes in different regions happened to have specific heavy metal byproducts.

There are some similarities between the accumulations in scorpions and previously studied cases of biomineralization. For example, the Fe biomineral in chiton (mollusk) teeth is associated with Zn as small amounts (about 0.1%) of Zn accumulate during mineralization (Webb et al., 1989). There are other similarities as well. Examinations of Mn- and Zn-enriched cuticle using X-ray and electron diffractometry do not indicate cuticular metals bound in crystalline form (Schofield, 1990; Nesson and Schofield, unpublished data), although they could be bound as amorphous biominerals. Although there are no known structural examples of Zn or Mn biominerals for comparison with the scorpion accumulations (Lowenstam, 1986; Mann et al., 1989), amorphous Fe biominerals have been reported. The Fe accumulations in the sternal shields and tube linings of some polychaete annelids (Lowenstam, 1972) may be similar to the Fe accumulations in buthid scorpions. The accumulation in annelids, whose functional significance is uncertain, are thought to be amorphous hydrous ferric oxide gels and hydrous ferric-phosphates. The Fe found in the buthid scorpions was also associated with phosphorus, but here too the cation concentrations were much lower (less than 10% in the scorpion) compared to polychaetes (about 30%).

Organically bound metals

At the same time that metal concentrations seem too low to affect mechanical properties in biomineral form, the 25% concentration of Zn in scorpion cuticle is too high to be accommodated by organic binding. Hillerton and Vincent (1982) presented a hypothesis that Zn in the mandibles of insects might increase the number of secondary bonds in the cuticle and thereby increase the density and fracture toughness of these structures. While the lowest concentrations of cuticular metals might be explained by this configuration, the 25% Zn concentrations are inordinately high, especially since the concentration in the binding proteins would have to be considerably higher to account for cuticle components that do not bind Zn. For comparison, less than 10% of the mass of metallothioneins, which are considered to be extremely metal rich, consists of Zn, Cu, or Cd (calculated from Kagi and Schaffer, 1989).

The difficulty in reconciling the high Zn concentrations with organic binding can be illustrated by estimating the maximum possible Zn concentrations for two postulated binding mechanisms: binding to cuticular proteins or binding during cuticle sclerotization. The amount of Zn that cuticular proteins could bind was estimated using an analysis of the amino acid composition of proteins from the wing cuticle of *Locusta migratoria* (Andersen, 1988). It was assumed that tyrosine or its metabolites and histidine would bind the Zn (Breslow, 1973; Freeman, 1973; Parraga et al., 1988). Cystine, another possible binding site, was not found in the locust samples or in the scorpion samples. Each tyrosine residue or derivative was assumed to bind one Zn atom and each pair of histidine residues one Zn atom (Avdeef et al., 1978; Kagi and Schaffer, 1989). According to this scheme, the most enriched of the 13 locust proteins would contain only 7% Zn by dry mass. This scheme also yielded a maximum of 7% for an amino acid analysis of a nereid worm jaw (Schofield, 1990).

A similarly low value is obtained if it is assumed that the metals are bound by catecholate ligands during cuticle sclerotization. (Catechols and their oxidation products are involved in cuticle sclerotization and are known to avidly form complexes with transition metals [Avdeef et al., 1978; deLearie and Pierpont, 1986; Kramer and Hopkins, 1987].) To estimate the maximum possible Zn concentration in cuticle, it was assumed that all of the mass added during tanning consists of catechols and that a Zn atom is bound by each catecholate ligand. The largest known mass increment during cuticle sclerotization and tanning is 25% (Vincent, 1982). Using this value and assuming the molecular weight of the catechols to be that of *N*-acetyldopamine, a cuticular Zn concentration of only 8% of dry mass was obtained (Schofield, 1990), less than one-third of that found in scorpion cuticle.

In conclusion, no mechanism is known that could account for such extreme quantities of metal being bound as individual atoms to the organic cuticular matrix. On the other hand, these quantities of metal could be accommodated as constituents of biomineral particles bound to proteins, analogous to biomineral particles within ferritin molecules. The Fe concentration for ferritin molecules, which contain up to 4500 Fe atoms in oxide form, can reach 50% by mass (calculated from Kaim and Schwederski, 1994). Another possibility is binding in organic salts, which, in contrast with common inorganic salts, could fill the cuticle enough to alter mechanical properties. The volume fraction of zinc oxyacetate [$(CH_3-CO_2)_6\ Zn_4O$], for example, would be about 50% where Zn concentrations reached 25% of dry mass. However, where Zn concentrations reached only 3%, the volume fraction would reach only about 5%. Because it is likely that cuticular metals affect mechanical properties over the wide range of concentrations encountered, it seems more likely that they are bound in ferritin-scale accumulations that alter cuticular properties in conjunction with proteins, rather than in bulk accumulations similar to those of well-known structural biominerals. In any case, whether the metals are incorporated in unusual biominerals, organic salts, proteins, or in some combination of these, the solution of the molecular form of the heavy metal biomaterials promises to lead to the identification of novel modes of metal incorporation into biological material.

Preliminary biochemical investigations

The question of whether Zn is associated with the protein or chitin (polysaccaride) fraction of the cuticle has been investigated. Whole and ground scorpion stings were incubated with a protein-digesting enzyme (papain), a chitin-digesting enzyme (chitinase), or, as control samples, boiled enzyme or no enzyme, and then centrifuged. Zinc and Mn were detected (using PIXE) in the supernatant of only the papain incubate (Cox and Schofield, unpublished data). This suggests that Zn and Mn are associated with protein rather than chitin in the cuticle.

Recently, EXAFS (extended X-ray absorption fine structure) analyses of Zn-rich material have been obtained for three insect and one nereid worm species. All analyses yielded similar results: O and/or N donors, coordination numbers between 3.9 and 4.5, and first-shell radii between 1.96 and 1.99 Å (J.D. Fawke, J.G. McClements, and P. Wyeth, personal communication). These data are not consistent with Zn binding in smithsonite ($ZnCO_3$) but are consistent with Zn binding in oxides (Inorganic Crystal Structure Database, FIZ Karl Sruhe, Berlin, Germany). The EXAFS data are also consistent with binding in numerous organic salts, such as zinc oxyacetate (Cambridge Structural Database: Bowen et al., 1958; Allen and Kennard, 1993), and the values are typical for Zn binding by water, imidazole, or carboxyl ligands (Bertini and Luchinat, 1994).

Possible Function

Few explanations can account for all the metal accumulation sites we observe in scorpions because the sites are so varied. Among the many enriched features, the sting is perhaps the most distinctive because it is the only enriched structure on the scorpions or any of the other arthropods listed in table 9.1 that is not at the distal tip of a leg homologue. An explanation that could account for the specific sites of each of the metal accumulations must depend on similarities that link the sting and the structures on leg homologues. The only apparent similarities between these structures are their protrusive shapes and mechanical functions.

High cuticular concentrations of Zn have only been found in protrusive structures. This association with shape might arise from special conditions needed to produce sharp or protruding structures. For example, Zn accumulations might be associated with pre-ecdysial tanning. During the molting of many insects, structures that protrude, including the tarsal claws and mandibles, are hardened before the old cuticle is shed (Cottrell, 1964). Recent data, however, indicate that Zn is deposited after ecdysis in mandibles of the ant *Atta sexdens* (Schofield and Wyeth, unpublished results) and after ecdysis in all enriched features of second instars of the scorpion *Vaejovis spinigerus* (Schofield and Nesson, unpublished results). Zinc is thus not involved in pre-ecdysial tanning in these and probably other cases and it is apparently not needed to produce protrusive structures.

All known Zn-enriched structures are associated by function: they are effectively the "tools," or working ends of structures that the organism uses to manipulate its environment. The uses of these Zn-enriched structures include cutting, piercing, digging, grasping, clamping, and traction. Even the pedal spur contacts the environment and may be important in traction (Couzijn, 1976). Within the enriched structures, Zn is found in the regions that seem most susceptible to mechanical forces and abrasion: the teeth of the pedipalp claw, the tip and serrated edges of the tarantula fang (Schofield and Lefevre, 1989). The performance and durability of each of the Zn-containing structures would seem to be enhanced by hardening and especially by increased resistance to wear. However, there are other structures that might benefit from mechanical enhancement; high concentrations of Zn have not been found in leg joints or in proximal mouthparts. But the enriched structures differ from the unenriched structures in that they routinely come in direct contact with environmental features that may be much harder than ordinary cuticle.

Hardness Measurements

Each of the cuticular structures enriched with Zn would benefit by increased hardness relative to normal cuticle. One method of quantifying hardness is to press a small diamond pyramid into the specimen and measure the size of the resulting indentation; harder materials indent less. Hillerton and Vincent (1982) discovered Zn in the mandibles of a locust only after they observed that the cutting edge was harder than surrounding cuticle. The Zn was localized on the side of the mandibles yielding the highest hardness values. Nevertheless, it is premature to attribute the high hardness values to Zn-enrichment because equally high values were measured in unenriched mesothorax (Hillerton et al., 1982), and because dessication may differentially alter the hardness of adjacent regions.

Hardness measurements were attempted on fresh spider fangs and worm jaws using a Vickers micro-hardness indenter with limited success (Schofield, 1990) because of curvature and texture differences between enriched and control regions. More consistent results obtained along the relatively smooth stings of scorpions suggested that the Zn-enriched region was hardened (Schofield, unpublished results), but these measurements may have been compromised by the large size of the indentations and the decreasing radius of curvature near the enriched tip of the sting. The group at Southampton has avoided surface topography by making measurements on polished sections. Zinc-enriched material in sections of dried insect mandibles and worm jaws was found to be harder than surrounding material (Edwards et al., 1993; McClements et al., 1993). Nevertheless, measurements on wet and unsectioned specimens are desirable and can now be made by using new atomic force microscopy (AFM)-based instruments to make indentations that are small compared to surface features. We have recently begun to use an AFM-based device (Hysitron, Inc.; Minneapolis, Minnesota, USA), to make indents that are hundreds of nanometers in width (1 or 2 orders of magnitude smaller than indents made using the Vickers device). Preliminary measurements on scorpion stings, chelicerae, and an ant mandible suggest that both Zn- and Mn-enriched cuticle are harder than unenriched cuticle. The AFM-based technique may be ideally suited for unraveling the complexities of the mechanical properties associated with heavy-metal enrichment because of its small scale and because it yields quantitative measurements of modulus of elasticity and coefficient of friction in addition to hardness values.

Future Directions

The unparalleled evolutionary success of arthropods is largely attributable to the enormous diversity of forms and functions made possible by a cuticular exoskeleton. Structured deposition of metal is yet another example of the remarkable properties of these cuticular integuments. The epithelial cells that excrete cuticle form setae, knobs, slit sensors, localized regions with and without resilin, iridescent regions, and a variety of other variations in structure and composition. The selective and precise distribution of heavy metals in the cuticle may also be controlled by the epithelial cells and may be a useful system for studying the extraordinary regional specialization of these cells.

Along with the ability to sequester metals in the cuticle, these organisms must assimilate large quantities of trace metals from the environment, store them, and transport them to fixation sites. High-energy ion microscopy may be useful in studies of the deposition process because it resolves the spatial locations of metals at some depth within the tissue. This procedure was used to reveal Zn accumulations near the stomach wall of an intact ant, a potential uptake site (Schofield et al., 1988). Similar accumulations of Zn have been found in the Malpighian tubules of fruit fly larvae; these stores are apparently reabsorbed in the pupal stage (Schofield et al., 1997). Scorpions may also store Zn internally so that large quantities are available during molting.

The metal-enriched materials seem to represent a subtle alternative to biomineralization with large quantities of minerals. Each cation may be associated with different mechanical properties. Zinc is found in edges or tips, where hardness and resistance to wear are important. Manganese is usually found in the shafts or rounded regions of structures, suggesting an association with rigidity. Enrichment may have drawbacks such as increased brittleness. The Zn-rich tips of tarsal claws were broken off on nearly all legs of some specimens. If Mn-rich material mitigates this tendency to fracture, the commonly observed pattern of Mn and Zn enrichment would have functional significance (Schofield and Lefevre, 1992).

The remarkable chemistry associated with cuticular metals is apparently not unique to arthropods. The jaw of the worm *Nereis virens* (Polychaeta), for example, resembles the sting of nonbuthid scorpions in that it contains large quantities of Zn and Cl in the distal portion and Mn and Ca segregated into the base (Schofield, unpublished observations). These similarities suggest that the unexplored biochemistry of the scorpion accumulations is important and widely employed. The molecular associations involved may blur the boundary between organic and inorganic binding. Metal concentrations lie between those typical of structural biomineralization and of organic binding and are difficult to accommodate within either model. Perhaps the biological metals are bound in combined modes, possibly as nanometer-scale oxide particles bound to proteins. Because of their size and the richness of their metal features, scorpions are a model of choice for further studies.

SUMMARY

High-energy ion microscopy was used to survey the scorpion integument for structural metal accumulations similar to those found recently in a variety of other organisms. High concentrations of Mn, Fe, or Zn were found in the teeth (cuticular protuberances) of the chelicerae and pedipalps, in tarsal claws, stings (aculei), and in some pedal spurs. Accumulation patterns in buthid scorpions differed dramatically from those in other sampled families. In all cases, the metals were localized in parts of structures that experience abrasional and deformational forces. The metal concentrations in cuticle were thought to be too low for the metals to affect mechanical properties as constituents of common minerals, and often too high (up to 25% Zn by dry mass) for the metals to be bound to the organic matrix as individual atoms.

Acknowledgments A major portion of this work was undertaken in collaboration with Dr. Harlan W. Lefevre, using the microprobe facility at the University of Oregon, with additional studies completed in collaboration with Dr. Klas Malmqvist, University of Lund, using the Lund microprobe. I am especially grateful to Dr. David Cox, who was kind enough to provide unpublished biochemical results for this chapter and to tutor me in biochemistry. Dr. Philip Brownell and members of his laboratory at Oregon State University participated in helpful discussions, provided many specimens, and initially prompted me to examine scorpion cuticular structures. Drs. Brownell, David Cox, Paul Wyeth, Julian Vincent, and Roger Farley made helpful comments on the manuscript. Many others have made suggestions and provided specimens. Among these are Drs. G. A. Polis, W. D. Sissom, R. C. Anderson, V. F. Lee, R. Roeloffs, W. R. Lourenço, V. Fet, S. C. Williams, K. Barricklow, M. Nesson, K. Hogan, L. Cederholm and the Lund University Museum of Zoology, and H. Enghoff and the Zoological Museum of Copenhagen University. This work was supported in part by fellowships from the Department of Physics, University of Oregon, from the Swedish Institute, the Department of Nuclear Physics, Lund University, and NSF grant IBN 9817206.

References

Allen, F.H. and Kennard, O. (1993) 3D search and research using the Cambridge Structural Database. *Chem. Design Automat. News.* 8: 31–37.

Andersen, S.O. (1988) Characterization of proteins from pharate adult wing cuticle of the migratory locust *Locusta migratoria. Insect Biochem.* 18: 415–421.

Avdeef, A., Sofen, S.R., Bregante, T.L. and Raymond, K.N. (1978) Coordination chemistry of microbial iron transport compounds. 9. Stability constants for catechol models of enterobactin. *J. Am. Chem. Soc.* 100: 5362–5370.

Bertini, I. and Luchinat, C. (1994) The reaction pathways of zinc enzymes and related biological catalysts. *In* Bertini, I., Gray, H.B., Lippard, S.J. and Valentine, J.S. (eds.) *Bioinorganic Chemistry,* pp. 37–106. University Science Books, Mill Valley, CA.

Bone, Q., Ryan, K.P. and Pulsford, A.L. (1983) The structure and composition of the teeth and grasping spines of chaetognaths. *J. Mar. Biol. Assoc. U.K.* 63: 929–939.

Bowen, H.J.M., Donohue, J., Jenkin, D.G., Kennard, O., Wheatley, P.J. and Whiffen, D.H. (1958) *Tables of Interatomic Distances and Configuration in Molecules and Ions.* The Chemical Society, London.

Breslow, E. (1973) Metal-protein complexes. *In* Eichhorn, G.L. (ed.) *Inorganic Biochemistry,* pp. 227–247. Elsevier, Amsterdam.

Bryan, G.W. and Gibbs, P.E. (1979) Zinc—a major inorganic component of nereid polychaete jaws. *J. Mar. Biol. Assoc. U.K.* 59: 969–973.

Bryan, G. W. and Gibbs, P.E. (1980) Metals in nereid polychaetes: the contribution of metals in the jaws to the total body burden. *J. Mar. Biol. Assoc. U. K.* 60: 641–654.

Cottrell, C.B. (1964) Insect ecdysis with particular emphasis on cuticular hardening and darkening. *In* Beament, J.W.L., Treherne, J.E. and Wigglesworth, V.B. (eds.) *Advances in Insect Physiology,* Vol. 2, pp. 175–212. Academic Press, London.

Couzijn, H.W.C. (1976) Functional anatomy of the walking-legs of Scorpionida with remarks on terminology and homologization of leg segments. *Neth. J. Zool.* 26: 453–501.

deLearie, L.A. and Pierpont, C.G. (1986) Catecholate complexes of high oxidation state metal ions, synthesis and characterization of tris (3, 5-di-*tert*-butylcatecholato)—rhenium (VI). *J. Am. Chem. Soc.* 108: 6393–6394.

Edwards, A.J., Fawke, J.D., McClements, J.G., Smith, S.A. and Wyeth, P. (1993) Correlation of zinc distribution and enhanced hardness in the mandibular cuticle of the leaf-cutting ant *Atta sexdens rubropilosa. Cell Biol. Intl.* 17: 697–698.

Freeman, H.C. (1973) Metal complexes of amino acids and peptides. *In* Eichhorn, G.L (ed.) *Inorganic Biochemistry,* pp. 121–162. Elsevier, Amsterdam.

Gibbs, P.E. and Bryan, G.W. (1980a) Copper—the major metal component of glycerid polychaete jaws. *J. Mar. Biol. Assoc. U. K.* 60: 205–214.

Gibbs, P.E. and Bryan, G.W. (1980b) A note on the elemental composition of the jaws of *Goniada maculata* (Polychaeta: Goniadidae). *J. Mar. Biol. Assoc. U. K.* 60: 541–542.

Hillerton, J. E., Reynolds, S.E. and Vincent, J.F.V. (1982) On the indentation hardness of insect cuticle. *J. Exp. Biol.* 96: 45–52.

Hillerton, J.E., Robertson, B. and Vincent, J.F.V. (1984) The presence of Zn or Mn as the predominant metal in the mandibles of adult stored-product beetles. *J. Stored Prod. Res.* 20: 133–137.

Hillerton, J.E. and Vincent, J.F.V. (1982) The specific location of Zn in insect mandibles. *J. Exp. Biol.* 101: 333–336.

Horst, M.N. and Freeman, J.A.F. (1993) *The Crustacean Integument.* CRC Press, Boca Raton, FL.

Johansson, S.A.E., Campbell, J.L. and Malmqvist, K.G. (eds.) (1995) *Particle-Induced X-Ray Emission Spectrometry (PIXE).* John Wiley & Sons, New York.

Kagi, J.H.R. and Schaffer, A. (1989) Biochemistry of Metallothionein. *In* Neurath, H. (ed.) *Perspectives in Biochemistry,* pp. 46–52.American Chemical Society, Washington, DC.

Kaim, W. and Schwederski, B. (1994) *Bioinorganic Chemistry: Inorganic Elements in the Chemistry of Life.* John Wiley & Sons, New York.

Kramer, K.J. and Hopkins, T.L. (1987) Tyrosine metabolism for insect cuticle tanning. *Arch. Insect Biochem. Physiol.* 6: 279–301.

Lamoral, B.H. (1980) A reappraisal of suprageneric classification of recent scorpions and of their zoogeography. *Proc. 8th Intl. Congr. Arachnol., Vienna* 439–444.

Lefevre, H.W., Schofield, R.M.S., Bench, G.S. and Legge, G.J.F. (1991) STIM with energy loss contrast: an imaging modality unique to MeV ions. *Nucl. Instrum. Methods Phys. Res.* B54: 363–370.

Legge, G.J.F., O'Brien, P.M., Kirby, B.J. and Allan, G.L. (1988) Proton microscopy and microanalysis—biological applications. *Ultramicroscopy* 24: 283–298.

Lowenstam, H.A. (1986) Mineralization processes in monerans and protoctists. *In* Leadbeater, B.S.C. and Riding, R. (eds.) *Biomineralization in Lower Plants and Animals,* pp. 1–18. Clarendon Press, Oxford.

Lowenstam, H.A. (1972) Phosphatic hard tissues of marine invertebrates: their nature and mechanical function and some fossil implications. *Chem. Geol.* 9: 153–166.

Mann, S., Webb, J. and Williams, R.J.P. (1989) Preface. *In* Mann, S., Webb, J. and Williams, R.J.P. (eds.) *Biomineralization,* pp. v–viii. V.C.H. Publishers, New York.

McClements, J.G., Smith, S.A. and Wyeth, P. (1993) Chemical and physical characterisation of zinc-replete biocomposites. *In* Welch, A.J. and Chapman, S.K. (eds.) *The Chemistry of the Copper and Zinc Triads.* The Royal Society of Cambridge, Cambridge, England.

Parraga, G., Horvath, S.J., Eisen, A., Taylor, W.E., Hood, L., Young, E.T. and Klevit, R.E. (1988) Zinc-dependent structure of a single-finger domain of yeast ADR1. *Science* 241: 1489–1492.

Perry, C.C., Grime, G.W. and Watt, F. (1988) An X-ray analytical study of mandibles from *Calanus pacificus. Nucl. Instrum. Methods Phys. Res.* B30: 367–371.

Schofield, R.M.S. (1995) Application of ion-beam tomographic element microanalysis (ITEM). *Nucl. Instrum. Methods Phys. Res.* B 104: 212–221.

Schofield, R.M.S. (1990) X-ray microanalytic concentration measurements in unsectioned specimens: a technique and its application to Zn, Mn and Fe enriched mechanical structures of organisms from three phyla (Ph.D. dissertation). University of Oregon, Eugene.

Schofield, R.M.S. and Lefevre, H.W. (1993) Analysis of unsectioned specimens: 2D and tomographic PIXE with STIM. *Nucl. Instrum. Methods Phys. Res.* B 77: 217–224.

Schofield, R.M.S. and Lefevre, H.W. (1992) PIXE - STIM microtomography: zinc and manganese concentrations in a scorpion stinger. *Nucl. Instrum. Methods. Phys. Res.* B 72: 104–110.

Schofield, R. and Lefevre, H. (1989) High concentrations of Zn in the fangs and Mn in the teeth of spiders. *J. Exp. Biol.* 144: 577–581.

Schofield, R.M.S., Lefevre, H.W., Overley, J.C. and MacDonald, J.D. (1988) X-ray microanalytical surveys of minor element concentrations in unsectioned biological samples. *Nucl. Instrum. Methods Phys. Res.* B 30: 398–403.

Schofield, R., Lefevre, H. and Shaffer, M. (1989) Complementary microanalysis of Zn, Mn and Fe in the chelicera of spiders and scorpions using scanning MeV-ion and electron microprobes. *Nucl. Instrum. Methods Phys. Res.* B 40/41, 698–701.

Schofield, R.M.S., Postlethwait, J.H. and Lefevre, H.W. (1997) MeV-Ion microprobe analyses of whole *Drosophila* suggest that zinc and copper accumulation is regulated storage not deposit excretion. *J. Exp. Biol.* 200: 3235–3243.

Sissom, W.D. (1990) Systematics, biogeography and paleontology. *In* Polis, G.A. (ed.) *The Biology of Scorpions,* Stanford University Press, Palo Alto, CA.

Taylor, M.G. and Simkiss, K. (1989) Structural and analytical studies on metal ion-containing granules. *In* Mann, S., Webb, J. and Williams, R.J.P. (eds.) *Biomineralization,* pp. 427–453. V. C. H. Publishers, New York.

Underwood, E.J. (1977) *Trace Elements in Human and Animal Nutrition.* Academic Press, New York.

Vincent, J.F.V. (1982) *Structural Biomaterials.* John Wiley & Sons, New York.

Webb, J., Macey, D.J. and Mann, S. (1989) Biomineralization of iron in molluscan teeth. *In* Mann, S., Webb, J. and Williams, R.J.P. (eds.) *Biomineralization,* pp. 345–385. V. C. H. Publishers, New York.

10

Brood Care and Social Behavior

DIETER MAHSBERG

Brood care is generally considered to be "any form of parental behavior that appears likely to increase the fitness of a parent's offspring" (Clutton-Brock, 1991, p. 8). Sociality is defined as the temporary or permanent cooperation in a group of communicating conspecifics (Wilson, 1971). An extension of parent–offspring interactions beyond the age when young can feed themselves may lead to a social system. Such a scenario, with its roots in parental care, occurs along the subsocial route in the evolution of social behavior.

Scorpions are predators, all are venomous, and many are cannibals par excellence—a seemingly inappropriate setting for evolution of brood care and sociality. But within the chelicerates, scorpions show a predisposition toward sociality. All 1259 described species care for their newborn. After emergence from the mother's gonopore, young scorpions climb onto her back to pass their first instar stage. For less than 1% of all scorpion species, prolonged care and even advanced sociality is reported (table 10.1; Polis and Lourenço, 1986).

It is generally accepted that parental care is most highly developed in animals where broods or newborn face adverse abiotic conditions, strong interspecific competition, and heavy predation or parasitism (Clutton-Brock, 1991). An effective way to increase offspring survival is to bear offspring live, a trait characteristic of all scorpions (see Farley, this volume). Viviparity was likely an adaptation to terrestrial life (Little, 1990) and appears to have fixed early within the modern lineage of scorpions, predisposing them to develop brood care. Viviparity is rare in other arachnids. Only some mites show this form of prenatal care. In the straw itch mite, *Pyemotes tritici,* progeny develop to the adult stage within the mother's body, with males assisting the birth (Wrensch and Bruce, 1991).

The extent of parental care, tolerance, and cooperation are each affected by environmental conditions; the environment influences the balance between costs and benefits for brood-caring parents and group-living conspecifics. However, our knowledge of the costs and benefits of parental care is limited (Clutton-Brock, 1991). Scorpions are particularly amenable to the study of these types of problems. They occur in a variety of habitats, from temperate regions to tropical forests, and they are most diverse in harsh subtropical deserts (Polis, 1990), a habitat offering unique physiological challenges to survival. Where congeneric species (e.g., in *Heterometrus, Opistophtalmus, Pandinus, Urodacus, Vaejovis)* or subspecies (e.g., in *Scorpio maurus*) are distributed along gradients of changing habitat quality, a comparative study of reproductive strategies (see Benton, this volume) and parental care may be especially accessible.

Table 10.1 Scorpion species with extended brood care and cooperation[a]

Taxon	Mother–young association	Reference
Diplocentridae		
Didymocentrus caboensis	2–4 months (?)	Polis and Lourenço (1986)
Didymocentrus lesuerii	?	Polis and Lourenço (1986)
Scorpionidae		
Scorpio maurus palmatus	3–4 months	Shachak and Brand (1983)
Heterometrus longimanus	end of third instar (6–9 months)[b]	Schultze (1927)
Heterometrus fulvipes	1–2 years	T. Shivashankar (personal communication, 1991)
Pandinus imperator	1–3 years	Mahsberg (unpublished data)
Ischnuridae		
Liocheles spp.	?	J. König, (personal communication, 1990)
Opisthacanthus asper	?	Lourenço (1991)
Opisthacanthus cayaporum[a]	?	Lourenço (1991)
Bothriuridae		
Urophonius iheringi	5.5–6.5 months	Maury (1969, 1973 in Polis and Lourenço, 1986)
Urophonius brachycentrus	5.5–6.5 months	Maury (1969, 1973 in Polis and Lourenço, 1986)

[a]All species are obligate burrowers. Except the apoikogenic Bothriuridae, all other species belong to katoikogenic families (see Farley, this volume).

[b]Estimated value according to data of captive-bred H. fulvipes at the end of instar 3 (Mahsberg, unpublished data).

In contrast to the annual life span of most spiders (the exceptions are orthognaths; Foelix, 1996), scorpions are among the longest-lived arthropods, with life spans from 1 to 25 years (Polis and Sissom, 1990; Warburg and Elias, 1998; Mahsberg and Warburg, 2000). Further, iteroparity is widespread in the order. As scorpions lengthen their reproductive life span (e.g., to >10 years), and reproduce repeatedly (see table 10.2; Benton, this volume), parental effort may vary with age. In this way, iteroparous scorpions provide an opportunity for quantitative laboratory studies. For example, one might ask how pre- and postnatal care varies throughout a female's reproductive life span, or how the breeding success of females is controlled by nutritional factors. Such influences of food quality and quantity should be measurable in the laboratory because growth to maturity in scorpions is determinate.

Scorpions such as *Pandinus imperator* (Koch) or *Heterometrus fulvipes* (Koch) facilitate research on brood care and sociality. These tropical arachnids are easier to rear in captivity than desert scorpions and reproduce successfully (25 of 30 pregnant females gave birth in the last 6 years with the first laboratory-reared individuals now being parents themselves; see also table 10.2). This contrasts with the overall low success of rearing most other scorpion species (Sissom et al., 1990). Our "senior *Pandinus*" joined the laboratory 11 years ago, and a *Heterometrus indus* died after 15 years of captivity. Such life-history char-

Table 10.2 Life history parameters of *Pandinus imperator* and *Heterometrus fulvipes*

	Pandinus imperator (*n*)[a]	*Heterometrus fulvipes* (*n*)[a]
Time to birth from insemination[1]	370 days (6) [298/455][b] 753 days (3)[c] [734/1063]	436 days (5) [413/465]
Iteroparity	Yes	Yes
Litter size, mean ± SD	19.2 ± 5.7 (24)	32.2 ± 6.9 (5)
Newborn weight[d]	0.339 g (26) [0.224/0.423]	0.162 (5) [0.140/0.216]
Duration of first instar, median	14 days (24) [9/26]	12 days (5) [9/14]
Period on mother's dorsum, median	17 days (24) [10/36]	14 days (3) [12/14]
Maternal tolerance	Great	Great
Intragroup tolerance	$\gg$ 1 year	About 1 year
Age to maturity	2.5–? years	2.5–? years
Longevity	>10 years	>10 years
Postpartum body weight (g), mean ± SD	26.7 ± 6.4 g (25)	21.2 ± 5.4 g (4)
Postpartum width of prosoma, mean ± SD	20.5 ± 1.4 mm (22)	17.2 ± 1.3 mm (4)

[a]All data are from scorpions kept under laboratory conditions. *n* refer to the number of captive breedings or the number of individuals.

[b]Numbers in brackets are minimum/maximum values.

[c]Multiple broods from a single insemination.

[d]Median of mean individual weight in *n* litters.

acteristics (see table 10.2 for data from captive scorpions) allow long-term observations both in the laboratory and in the field.

A newborn *Pandinus* is relatively large, easy to observe, and easy to manipulate. With a mean birth weight of more than 0.3 g (see table 10.2), at 1 year of age it nearly equals the size of adult *Paruroctonus mesaensis,* a relatively large sand scorpion from temperate deserts of North America (Polis and Farley, 1979). Captive *Pandinus* can be individually marked with color dots or reflecting adhesive labels, which is easily accomplished in such large animals. This marking technique is reliable under tropical field conditions as well. The loss of marks at ecdysis of immature individuals is unavoidable, but due to variability in body weight, a reconstruction of an individual's code is possible in most cases even when kept in groups. Individual long-term identification of adult scorpions can also be accomplished using preprogrammed microchips attached to the scorpion's tail and radio frequency transponders for remote tracking (Mahsberg, unpublished observations). The encoded animal's number is read by a portable scanner. This technique is especially useful in burrow-inhabiting nocturnal scorpions: animals can be identified when not visible from outside their shelter. For observations of scorpions at night, portable ultraviolet lights are important tools (Sissom et al., 1990). On the other hand, UV light also attracts insects.

To avoid biased field data due to artificially high insect abundances, we use UV lights only to detect scorpions. For longer lasting observations of stationary burrowing species such as *Pandinus,* white light of low intensity was a good alternative and did not disturb the scorpions. Even video recording is practical under these conditions.

Field and laboratory measurements of growth rates allow us to trace an individual's life history and to estimate the effects of intraspecific interactions or reproduction. Scorpions can be bred in captivity, with food resources under qualitative and quantitative control. Such attributes help to test hypotheses on natal dispersal and factors affecting group size (Bekoff, 1987). A comparison of laboratory-reared *Pandinus* with marked sib groups from the field yielded similar growth rates (Mahsberg, unpublished data), which reaffirms the value of life-history studies conducted under laboratory conditions (see Sissom et al., 1990, for the possible problems that exist with laboratory-reared scorpions).

Despite some scorpions' special suitability for general studies into brood care and sociality, much remains to be learned. This chapter begins with a description of brood care as it has been observed in scorpions to date and how it affects offspring survival. I also consider infanticide and its possible evolutionary significance for brood care. This discussion is followed by a summary of evidence for sociality in scorpions, gathered from the literature and from field and laboratory observations of cooperative and conflict behavior in scorpion groups. The final section relates the life habits of social scorpions to environmental factors that may be selective forces for establishing sociality in these arachnids.

BROOD CARE IN SCORPIONS

Maternal versus Paternal Roles

The differential allocation of resources for reproduction by males and females (Trivers, 1972) predisposes female scorpions to care for their young after birth. In laboratory matings, males of *P. imperator* and *H. fulvipes* are capable of depositing a new spermatophore about 1 week after their last mating. Thus, in theory, one male could fertilize more than 50 females per year (in reality mating probably occurs only 1–5 times annually). Such male investment stands in sharp contrast to the energy and time females must invest per offspring (see litter size and newborn weight, table 10.2). The cohabitation of male and female for days, known from *Euscorpius flavicaudis* (Benton, 1992) and *Pandinus* (Mahsberg, personal observations) suggests a last-male advantage (Birkhead and Møller, 1992; see Benton, this volume). As far as we know, male scorpions do not assist in any care of their progeny. This may allow them additional opportunities for mating during the reproductive season. Paternal care occurs rarely in other terrestrial arthropods for reasons not easily explained by a single hypothesis (Zeh and Smith, 1985). In arachnids, paternal care is reported in some tropical harvestmen (Opiliones), where males guard eggs for extended periods (Mora, 1990; Martens, 1993; Machado and Oliveira, 1998). Eggs without guardian males suffer high mortality from fungal infections, cannibalism, and interspecific predation. In two subsocial spider mite species (Acari, Tetranychidae), females and males cooperate in brood defense against predators (Saito, 1997).

Adaptive Value of Maternal Care

Birth behavior

Most embryonic growth in scorpions probably occurs during the periods of peak prey abundance (Polis and Farley, 1979). Before birth, gravid females retreat to natural shelters or self-dug subterranean burrows (also see Benton, this volume). The construction of special burrows for parturition (e.g., in *Urophonius* spp. *Parabuthus planicauda:* Polis, 1990) is costly in terms of time and energy and contributes to maternal expenditures on brood care.

Postnatal care of newborns is obligate for all scorpion species and starts with a specific birth behavior (reviewed by Francke, 1982; Polis and Sissom, 1990). At birth, a female assists her emerging young in several ways: she assumes a stilting posture, forms a "birth basket" to catch her newborn with one or two pairs of legs, removes residuals of embryonic membranes, and aids the offspring to ascend her back. Newborn scorpions (first instars) are unable to feed and depend on nutrient reserves stored during the prenatal period. They are defenseless, with a soft cuticle and a nonfunctional sting. Sensory organs such as the trichobothria are still undeveloped. An important exception, however, are the pectines, which support a small number of chemosensory peg sensilla at birth (see Gaffin and Brownell, this volume). These may be especially important in newborns for recognizing their mother (Mahsberg, 1990). Elevation onto the mother's back protects young in the event their burrow floods during heavy rains, and it reduces the impact of potential predators like ants or other scorpions.

Maternal protection

Benton's (1991, this volume) experiments with *E. flavicaudis* demonstrate that maternal care functions mainly to prevent predation on young by cannibalistic conspecifics. The increased aggressiveness seen in pregnant and young-carrying females (Vannini et al., 1978, Mahsberg, 1990) supports this argument. *Pandinus* mothers with offspring (second instars or older) show a high readiness to attack adult males (25% aggressive encounters in 32 observations) compared to the tolerance shown to other females carrying young (no attacks in 69 observations; Eichler, 1991). Such behavior suggests that male conspecifics present a high risk to infants. Similarly, maternal aggression occurs in many mammals; in rodents, for example, unfamiliar male mice are vigorously attacked by nursing females (Elwood, 1992).

In enclosure experiments with two unrelated *Pandinus* family groups, females were equally tolerant of first and second instars from either brood or "adopted" foreign young (Mahsberg, 1990; Eichler, 1991). In contrast, the young scorpions preferred to be near their birth mother. Here, young within a social group may use familiarity with sibs as a guide to kin discrimination (Barnard, 1991), thereby avoiding contact with a potential cannibal (males, nongravid females, older siblings). Chemical factors from the cuticle probably mediate discrimination (Torres and Heatwole, 1967; Vannini and Ugolini, 1980), and it is likely that the pectine organs are the detectors of this signal (see Gaffin and Brownell, this volume). In solitary scorpions, avoiding contact with conspecifics by hiding as an outsider should be the safer strategy. Therefore we should expect in species that terminate brood care after the dependent stage no ability to discriminate between related or unrelated conspecifics. Tolerance in *Pandinus* may also point to communal breeding of close kin, where cannibalistic females would risk killing their own offspring. Males are more vagrant and thus often unrelated to females, whereas females live in related sister groups. Thus kin selection may act to minimize cannibalistic tendencies of females toward related females and their young from the same group, as is shown in *Polistes* wasps (Crespi, 1992) and amphibians (Blaustein and O'Hara, 1982).

Maternal care also reduces the potential impact of predators, such as ants, which easily prey on infant scorpions. In West Africa, the large stink ant, *Paltothyreus tarsatus* (Ponerinae), often foraged near *Pandinus* burrows. Third and fourth instar *Pandinus* were observed to escape after vigorous attacks of *Paltothyreus,* which may severely hurt these juveniles (Mahsberg, unpublished observations). Driver ants are probably the most dangerous enemies of these scorpions: patrolling *Dorylus nigricans* dispatched several adult *Pandinus,* which roamed outside their burrows (Mahsberg, personal observation). On the other hand, sit-and-wait predation on driver ants by adult *Pandinus* occasionally occurs (A. Nourney, personal observation, 1992, in the Ivory Coast, where *Dorylus* colonies are especially abundant; Leroux, 1977, in Hölldobler and Wilson, 1990). We have no evidence that maternal defense and a burrowing existence actually protect *Pandinus* against these group raiders, but site fidelity of *P. imperator* suggests that burrow sites are especially safe, even after several years of close encounters with foraging driver ants.

Mother scorpions normally reduce surface activity, remaining within their retreat (beneath stones and bark or in a subterranean burrow), where microclimatic conditions are moderate and the risk of predation is reduced (also see Benton, this volume). Even scorpions that normally live in scrapes construct deep burrows where they care for their progeny (Polis, 1990). However, exceptions contradict the general appearance of a reclusive life for brooding scorpions: the sympatric buthids *Hottentotta hottentotta* (Fabricius) and *Babycurus kirki* (Pocock) were observed to forage for termites in the sunny savanna, the former with first instars on her dorsum, the latter even with second instars (Mahsberg, personal observations). *Centruroides exilicauda* (Wood) from Mexico are frequently observed foraging while carrying first and second instar young (G.A. Polis, personal communication). During pregnancy and postpartum care of young the non-burrowing *Centruroides vittatus* avoids running or decreases sprint speed. This has been suggested to reduce potential costs of parental care (Shaffer and Formanowicz, 1996). One may ask what are the costs of such activity compared to brood care in a defended burrow. Do maternal benefits balance the costs of reduced offspring survival in these cases? These questions require experimental studies to determine how specific ecological conditions affect brood care behavior.

Mothers as a source of water?

Vannini et al. (1985) concluded from experiments with tritium-labeled water that the close contact between the mother and first instars would favor transfer of maternal water to newborn scorpions and protect against dehydration. However, an exchange of water seems to be of minor importance for offspring survival, at least in species that prefer to stay in moist microhabitats during the postnatal period when young are present (there is no information available for

Table 10.3 Birth weight (mean ± SD) of first instar *Pandinus imperator* compared with their weight after successful molt to second instar stage (age 18 days)[a]

	Weight at birth(g)	Weight after first molt (g)	Relative weight loss (%)
On mother's back	0.443 ± 0.034 (9)	0.381 ± 0.012 (9)	−13.9 %
Isolated offspring	0.424 ± 0.007 (8)	0.385 ± 0.018 (8)	−9.1 %
p	ns	ns	

[a]One group of juveniles (controls) resided on the female's back while their siblings were separated from the mother in a wire cage nearby; *n* (in parentheses), number of individuals, *p*, Student's *t* test.

species where mothers migrate with their babies). Isolation of first instar *Pandinus* from their mothers did not negatively affect survival, molting success, or growth of young compared to siblings settled on their mother's back (table 10.3). Both groups were kept under the microclimatic conditions of their natural burrows (27°C, >90% relative humidity; Mahsberg, unpublished data). The difference in relative weight loss of both experimental and control group during the first instar stage was insignificant. Such decrease in body weight is common in newborn scorpions (they rely on stored yolk reserves during this instar). The results agree with Benton (1991) on similar experiments with *E. flavicaudis*.

Maternal Care in Scorpions: Cannibalism versus Protection

Occasionally, scorpion mothers devour some or all their young. In the buthid *Tityus trivittatus*, maternal cannibalism increased under stress conditions (Peretti, 1994). In some cases this is likely a laboratory artifact (McCormick and Polis, 1990), but our data on housed *P. imperator* suggest cannibalism of young may be a normal part of maternal postpartum behavior. For example, during the period when newborn are aided by their mother in the ascent to her dorsum, the function of such help might be to sort healthy and strong young, who may pass to the back, from weaker young who are then eaten. Table 10.4 summarizes the occurrence of maternal infanticide in 10 *Pandinus* parturitions. Newborn were classified "intact" if they were not hindered by remains of birth envelopes or by deformations or "handicapped" if they had problems with separation from birth membranes or exhibited distorted appendages or other anomalies. Although infanticide did not eliminate all handicapped newborn, it was significantly more frequent against these individuals compared to their intact brothers and sisters (Mahsberg, unpublished observations). This elimination hypothesis set forth in Hrdy's (1979) review on cannibalism is supported by observations of other authors who found that first instar *Pandinus* that dropped from their mother's back during the first molt were frequently eaten by the female (Larrouy et al., 1973). A high level of anomalies in *Opisthacanthus asper* (Peters) was a probable cause for cannibalism of entire clutches (Lourenço, 1991). There is evidence that a female's decision to eat her young is influenced by the temporal parameters of the newborn's ascent (length of time and number of trials; Mahsberg, unpublished observations) and by mechanochemical properties of newborn (young are often intensively palpated by the maternal chelicerae). Discrimination of sick young

Table 10.4 Selective maternal infanticide in *Pandinus imperator*[a]

	Intact	Handicapped	*p*
Maternal cannibalism	3	32	
			<.0001
Accepted offspring	93	30	
Total	96	62	

[a]Intact first instars had no visible defects; handicapped newborns were malformed or entangled in birth membranes; accepted offspring were those making successful ascent to the mother's back; *p*, chi-square test ($\chi^2 = 51.3$, df = 1, $n = 158$). Data from 10 litters.

and subsequent cannibalism is observed in birds and rodents (see references in Elwood, 1992; Stanback and Koenig, 1992). In our laboratory population the overall rate of observed infanticide in a total of 344 newborn *Pandinus* (20 parturitions from 18 females) was 24.2%. The victims were exclusively first instars, with maternal cannibalism generally occurring shortly after birth (see also Peretti, 1994).

But why do parents eat their offspring (see FitzGerald, 1992)? How might these traits—brood care and cannibalism—interact from a sociobiological perspective? Such early brood reduction is consistent with theoretical considerations of the optimal energy allocation in young of superior reproductive value (Clutton-Brock, 1991). Infanticide is favored if parental care decreases the amount of resources, time, and energy available for future reproduction, and thus the parent's residual reproductive value (Pianka and Parker, 1975). Early brood reduction could be advantageous for an iteroparous scorpion if the survival probability of cannibalized offspring is low. Handicapped offspring, which survive the first molt, have little chance to reach maturity: growth is often decreased, ecdysis to later stages is inhibited, cannibalism by sibs is frequent, and mortality is generally high, even under optimal environmental conditions in the laboratory (fig. 10.1).

The potential benefits to mothers that eat their offspring can be estimated as energy gained. The postnatal body weight of *Pandinus* ($n = 15$) was 18.5–42.9% (median: 27.5%) lower than the weight of mothers shortly before parturition (for further data on changes in maternal mass, see Polis and Sissom, 1990; Benton, 1991; Brown and Formanowicz, 1996). This weight loss suggests a high cost of reproduction. In the field, *Pandinus* mothers arrested their postpartum weight level, while their offspring grew rapidly (Mahsberg, unpublished data). Such energetic costs could be reduced by selective cannibalism (Polis, 1981). It is important to emphasize that under laboratory conditions, the postnatal weight of female *Pandinus* did not increase for the months that offspring were present: although mothers captured prey for their progeny, they often did not participate in meals (Mahsberg, 1990). Six of 17 *Pandinus* mothers that lived in terraria with their offspring showed a progressive decrease in body weight for up to 3 years and finally starved to death although they had access to food. This could suggest a natural end to life for aged females with low or no probability of future reproduction (note that the females remained alive during the early, critical years of their last brood). It would be intriguing to compare the morphology of the ovariuterus of females who survived prolonged brood care phases with

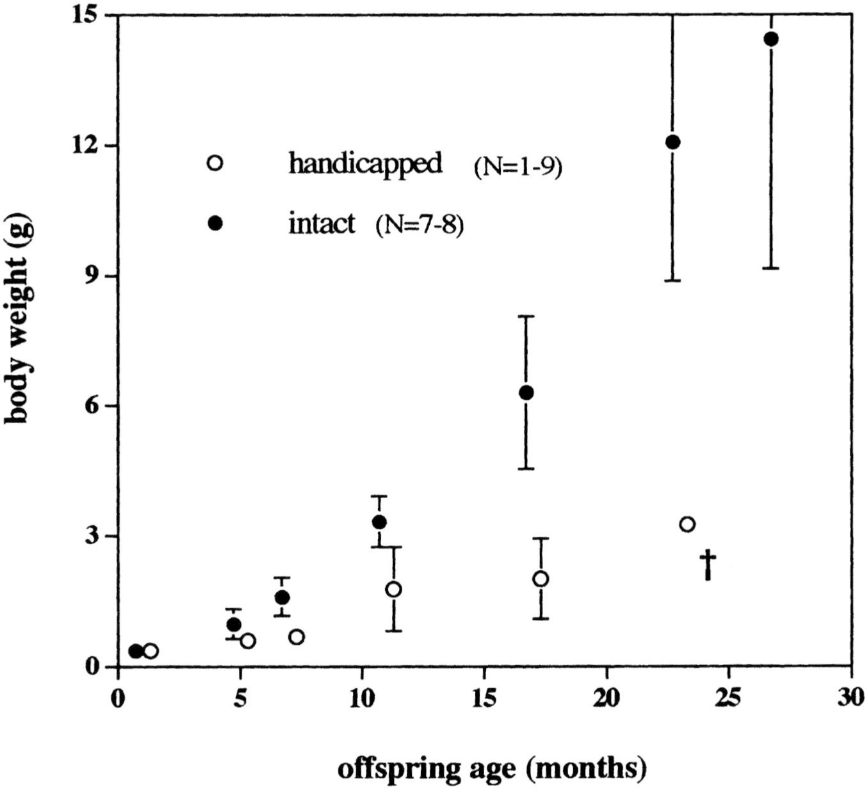

Figure 10.1 Growth of *Pandinus imperator* siblings (mean body weight ± SD) born intact (closed circle) or handicapped (open circle) by malformed extremities, etc. Cross indicates death of last handicapped offspring.

those who died. The number of degenerated diverticula, which may correlate to the lifetime reproductive effort, should be higher in older females (Warburg and Rosenberg, 1992, 1993; Warburg and Elias, 1998; Mahsberg and Warburg, 2000). However, phenotypic variation may confound measurements of reproductive cost (Clutton-Brock, 1991; Stearns, 1992). Breeding experiments with reared *Pandinus* siblings, where conditions are controlled and individual life histories are known, could shed more light on the costs of maternal care in scorpions.

SOCIALITY

Living in a Group: Mutual Tolerance and Siblicide

It is impressive to observe a crowd of 10–20 juvenile *P. imperator* peacefully coexisting where cannibalism might otherwise prevail. Even as adults they can be kept together for many years (Mahsberg et al., 1999). In the field, young *Pandinus* congregate for about 1 year, and even 2-year-old offspring fail to disperse from the natal group (Mahsberg, unpublished data). Within groups, *Pandinus* exemplify mutual tolerance, the most important prerequisite for the evolution of sociality. Even in the highly vulnerable state of molting, susceptible individuals appear to be safe among their conspecifics, indicating that harmonious cooperation is not merely a consequence of equal defense capabilities among the offspring.

As in social spiders (Uetz and Hieber, 1997), tolerance among individual scorpions bypasses the major impediment for development of social behavior in predatory and potentially cannibalistic taxa. In *Pandinus,* familiarity with conspecifics (independent of relatedness) controls the switch between tolerance and occasional agonistic encounters in groups (Eichler, 1991). *P. imperator* which were permanently isolated from each other reacted aggressively in 47.6% of 63 contacts with a previously unfamiliar intruding conspecific; familiar animals were always tolerant of each other (30 observations). What cues, then, mediate sib cannibalism in these scorpions? Siblicide was the cause of death of 28.5% of 137 second through eighth (= adult) instar *Pandinus* that were raised in the laboratory (Mahsberg, unpublished data). Most of the preyed upon individuals were handicapped, growth-retarded, or they failed to molt properly: in 15 of these 39 siblicidal cases, the victim was killed (and collectively eaten) during, or shortly after, ecdysis.

An analysis of growth rates indicates an enormous increase in body weight of cannibalizing young compared to noncannibalizing siblings (Mahsberg, unpublished data). Cannibalism of siblings increases group variance in body size, leading to shorter instar duration, younger age at maturity, and probably a larger adult size of cannibalistic group members. Corresponding trends are reported from several scorpion species (Polis and Sissom, 1990). However, it has not yet been demonstrated in scorpions whether a diet of conspecifics provides a better balance of nutrients than do heterospecific prey of the same dry weight or energy content (for amphibians, see review in Crump, 1992).

If harsh environmental conditions, low or season-dependent prey availability, and high predation pressure act against early dispersal of young, it may be advantageous for those scorpions that delay dispersal occasionally to eat smaller or deformed sibs as "intragroup food caches." Sibling cannibalism on "runt" siblings is known among egrets, where sibling rivalry maximizes the parental breeding success under conditions of food scarcity (Mock, summarized in Alcock, 1993).

Siblings serve as "transitory food caches" (Alexander, 1974; Polis, 1981) in the desert scorpion *Paruroctonus mesaensis* or "as packages of live meat for their kin" (Polis, 1980). These scorpions are intolerant to each other and cannibalism of conspecifics constitutes more than 25% of total diet biomass. The comparison of *Pandinus* and *Paruroctonus* shows that food caches may also be a component of a strategy, where fewer, better provisioned young are produced (Polis, 1981).

Both *P. imperator* and *H. fulvipes* (Shivashankar et al., 1990) have comparatively elaborate social systems. Each species is relatively large, has a tropical distribution, and belongs to the family Scorpionidae; each exhibits extended care of offspring, and each lives in burrows. Individuals show mutual tolerance to conspecifics, but cannibalism may occasionally occur. Although few data exist on the postnatal development of *H. fulvipes,* it appears that intragroup cannibalism among juveniles is probably more frequent than in *Pandinus.* The rate of cannibalism among adults (Shivashankar et al., 1990) seems to be as low as in *Pandinus* and stands in contrast to the aggressive behavior of many other scorpion species (Polis, 1981), where risks of being killed by the potential victim or eating relatives may be lower than in communal scorpions.

Cooperation versus Individualism in Foraging Scorpions

The example of *P. imperator* shows that brood care and group living may lead to conflicts between mother and offspring or among siblings. But social groups should be organized in a cooperative manner (Wilson, 1971), which balances the trade-off between costs and benefits. There are only a few observations of cooperation in scorpions (see table 10.1; Polis, 1990); they describe communal actions in the establishment of burrows *(Opisthacanthus cayaporum)* and in foraging *(Didymocentrus caboensis)*. More recent data on *H. fulvipes* (Shivashankar et al., 1990) and *P. imperator* (Mahsberg, unpublished) permit studies on the cooperation and communication among scorpions in regard to those ecological conditions that favor the establishment of social groups.

Foraging behavior in group-living *H. fulvipes* and *P. imperator* looks quite similar. Both scorpionids are typical sit-and-wait predators that forage from the entrance of their burrow and restrict their activity to areas near their shelter. In *P. imperator*, the presence of a strong and aggressive mother in the communal burrow enables offspring access to large or dangerous prey, which they otherwise could not subdue. I have observed a group of nine second instar *Pandinus* that were unable to immobilize an adult cockroach (*Blaptica dubia*) because their telsons were ineffective to penetrate cockroach's cuticle. Sib groups isolated from their mother at the second instar stage starved to death if they were offered only such difficult live prey (Mahsberg, 1990).

A laboratory study of capture success in single versus group-foraging scorpions has been done by Ehrlich (1993) using three instars (stage two to five) of *P. imperator* and *H. fulvipes*. Scorpions were offered live arthropods ranging from easily killed crickets to large millipedes that are difficult to subdue. As a mea-

Table 10.5 Weight ratio between various prey items and scorpions[a] attacking solitarily or in groups

Prey[b]	Killed by individuals	Killed by groups	Fisher Exact Probability[c]
Orthoptera	89.5% (24)	145% (13)	<0.00001
Beetle larva	76.5% (12)	133% (8)	<0.0099
Cockroach	86% (10)	131.5% (6)	<0.0594
Millipede	72% (8)	298.5% (4)	< 0.0455
Sum for all prey	85.0% (54)	139.0% (31)	<0.00001

[a]Scorpions: instar 2, 3, and 5 of *Heterometrus fulvipes* or *Pandinus imperator*. Scorpions in a group were brothers and sisters.

[b]Prey: *Acheta domesticus*, *Gryllus* spp., *Locusta migratoria* (Orthoptera); *Zophobas morio*, Tenebrionidae (beetle larva); *Blaptica dubia*, *Leucophaea madeirae* (cockroach); *Telodeinopus assiniensis*, *Pelmatojulus tigrinus* (millipede).

[c]Statistic; median test; number of trials in parentheses. The median percentage of *n* trials is reported (after Ehrlich 1993). Percent values indicate reflect weight of prey killed to scorpion body weight. For the group, the mean individual weight of the three group members was used.

sure of success, the ratio between live weight of the heaviest subdued prey item and weight of the individual predator was calculated (table 10.5). One scorpion alone could subdue only prey somewhat lighter than its own body weight (prey-to-predator ratio between 0.72 and 0.89). In cooperative attacks by three scorpions, much heavier arthropods (prey-to-predator ratio between 1.31 and 2.98) were killed and communally eaten. A comparison of growth rates over a period of 100 days showed faster development of group members, with shorter intermolt periods compared to isolated individuals. Group living in early instars of *Opisthacanthus cayaporum* produces similar results: the variance in mean body weight was significantly lower in siblings that lived in groups than among those that were kept isolated (Lourenço, 1991).

Group foraging may be especially beneficial for juvenile scorpions in habitats where easy prey are less frequent than difficult ones and additional constraints of the environment (low probability to find a suitable place to live) tip the balance in favor of delayed dispersal (Emlen, 1982). On a few nights in the rainy season in tropical West Africa, swarming termites were locally abundant and easily captured by *Pandinus* of all age classes. However, for most of the year, prey were sparse and mainly consisted of large, sclerotized arthropods, many of which were capable of injuring young scorpions (e.g., capricorn beetles, millipedes, ants). In pitfall-traps placed in a *Pandinus* habitat during a 2-week period, 78.5% of the total biomass (11.5 g) of arthropods captured were stink ants *(Paltothyreus tarsatus)* and millipedes (Mahsberg, unpublished data). Under such conditions, young scorpions attacking in groups should capture some food, whereas individuals foraging alone may starve. In other taxa, the tropical pseudoscorpion *Paratemnus elongatus* is capable of dispatching large prey (mainly ants) only by cooperation (Zeh and Zeh, 1990). Rypstra and Tirey (1991) argue that ephemeral large food resources may reinforce the social system of *Anelosimus eximius,* a theridiid spider of tropical forests of South America. Taken together, these observations suggest that constraints of the tropical environment may lead to retention of siblings within the natal group, an important initial condition favoring sociality in scorpions, pseudoscorpions, and spiders alike (Wilson, 1971; Buskirk, 1981).

Nocturnal observations of *P. imperator* activity in the Comoé National Park in the Ivory Coast/West Africa (Mahsberg, unpublished) offer anecdotal insight to the complex interactions that occur among predator and prey and among conspecifics. The fluttering of swarming termites (*Macrotermes bellicosus*) alerted a group of 11 third instar *Pandinus,* which began foraging in front of their communal burrow until a large mantid landed, causing a retreat of all young scorpions to their shelter. The mother of this brood then emerged from the burrow and captured the mantid. Mutual intraguild predation (Polis et al., 1989) among these scorpions and mantids, carabids, ants, solpugids, spiders, and centipedes is likely. But as a consequence of the large size of adult *Pandinus* and *Heterometrus* and the protective role of parental care, mutual intraguild predation and ontogenetic diet shifts should be less important to the dynamics of these scorpions than for solitary scorpion species (Polis, 1981). It would be intriguing to extend such comparisons to desert scorpions with prolonged brood care, such as the scorpionid *Scorpio maurus* (Shachak and Brand, 1983). Aside from their contrasting habitats, foraging behavior of *Scorpio* and *Pandinus* is similar.

The participation of different age classes in prey capture as a function of prey type and prey size suggests that scorpions are good models to test hypotheses concerning the adaptiveness of cooperative feeding. In spiders, "local enhancement" (direction of an individual's attention to a food location by group members) and the "ricochet effect" (successful prey capture after bouncing off several webs in succession) may increase the prey capture efficiency in group foragers compared to solitary individuals (Rypstra, 1989; Uetz, 1989). Communal hunting on large noctuid moths (more than twice as heavy as juvenile scorpions) by small groups of young *Pandinus* was also successful due to local enhancement: the more a captured moth struggled, the more scorpions emerged to the surface and joined the attack. Information about potential prey is likely encoded in vibratory signals because young *Pandinus* are attracted to fluttering termites and moths but are repulsed by large buzzing beetles (Mahsberg, unpublished field observations). As in spiders (Barth, 1982), substrate vibrations are of great significance in scorpions and may be used to discriminate predator from prey (Brownell and Farley, 1979; Krapf, 1986). Airborne vibrations created by tail-wagging behavior of juvenile *Pandinus* may facilitate regular spacing of conspecifics in a feeding group (Ehrlich, 1993). *Pandinus* possesses nearly twice as many wind-sensitive trichobothria on the pedipalps than *Heterometrus* and other scorpion species (Vachon, 1973). These sense organs could be useful for detecting prey by near-field sound or air currents they produce or for mediating communication between conspecifics during feeding.

We have no evidence that a true division of labor occurs among foraging scorpions as described by Shivashankar et al. (1990) for *H. fulvipes.* In *Heterometrus,* as in *Pandinus,* only some members of a family group are active during the same night. In both species, adults forage from the main entrance of the communal burrow with juveniles stationed in smaller side entrances. All entrances are connected to the central burrow chamber; the number of small entrances is greater in burrow systems without adults than those with adults. Burrow construction requires considerable energy expenditure and risk: the excavation of hard soil, exposure to potential predators, and defense of a territory. The observations of *Heterometrus* suggest that one benefit of long-lasting brood care is the time and energy saved in burrow construction and foraging. It would be of interest to know if sand-dwelling scorpions ever show group living, as burrows are easily constructed in sandy habitats.

Social Foraging: Trade-offs and Risk Sensitivity

The many benefits accruing to juvenile scorpions that stay with their mother and siblings are balanced by costs: individuals foraging as part of a group may decrease the temporal variance in feeding frequency (individuals feed more regularly and often), but mean size of the ingested meal decreases (they share their food). Furthermore, young animals may be cannibalized during times of severe food shortage while, at other times, survivorship is enhanced by association with a group relative to solitary existence. The theory of risk-sensitivity describes such trade-offs (Caraco, 1980; Real and Caraco, 1986). Risk-sensitive models are based on the starvation probability of an organism independent of the stochasticity of environmental resources. This probability is not only affected by the average quantity of food obtained but also by its variance. There is no difference in the mean reward of two alternative strategies, which, however, differ in variances. Risk aversion is the less variable option when food requirements are less than the average expected reward. This foraging strategy (avoid variation) is thought to be most adaptive in stable habitats with high levels of prey. The risk prone foraging mode (high variation) is more likely to occur in seasonal habitats with low prey availability (Real and Caraco, 1986; Uetz, 1988). Risk-sensitive analysis has been applied successfully to several decision-making problems—for example, foraging (e.g., choosing a foraging habitat/food type) and sociality (e.g., stay or leave the group).

In colonial spiders, risk-prone solitary foraging occurs under conditions of low prey abundance and changes to risk-averse group foraging with a superabundance of prey (Gillespie and Caraco, 1987; Uetz, 1988; Rypstra, 1989). Such risk sensitivity reduces the variance in prey captured per individual, thus it is proposed to be an important factor in the evolution of sociality in these arachnids. What about scorpions?

The seasonal tropical environment of *Pandinus* is marked by great fluctuations in prey availability, with peaks of high insect abundance in the rainy season (chiefly pulses of swarming termites) where *Pandinus* are most active. Such factors should favor a risk-averse strategy and promote sociality. *Pandinus* may be risk-averse in their tendency to remain in family groups for an extended period of time where they can benefit from cooperative group foraging. A positive effect of this cooperation is shown by laboratory experiments of Ehrlich (1993) with young *P. imperator* and *H. fulvipes*. Individuals kept in a group grew faster than solitary siblings that had access to the same amount of food. My observations of *P. imperator* feeding on large millipedes—a "shareable" and patchy food resource (Pulliam and Millikan, 1982)—provide additional support for the advantage of communal foraging: although millipedes are heavily sclerotized, *Pandinus* chelae easily crack the cuticle of these arthropods, but only if the scorpions are large enough or attack cooperatively (see table 10.5). Subsequently, all individuals share the captured food resources. Due to heavy predation from outside their shelter, we should expect social scorpions to benefit from living in groups of relatives even after molting to the second instar. However, we finally expect a critical period, when the risks of staying are higher than the risks of leaving the natal group. This trade-off should determine the actual time of dispersal.

Beyond this, several important questions remain to be answered: how do obligate, fossorial, sit-and-wait predators exploit peaks of prey abundance in a tropical habitat? There are no reliable data on the spatio-temporal distribution of arthropod prey available to tropical "doorkeeping" scorpions or other ground-living predators (for aerial prey in tropical spiders, see references in Rypstra, 1986). Do juvenile scorpions really have the capacity to survive as solitary individuals? Is switching to risk-prone foraging a stable alternative for tropical scorpions such as *Pandinus*? It would be intriguing to compare the group size and age distribution of social scorpions along gradients of decreasing resource availability. Finally, is starvation a reliable parameter for risk sensitivity analysis in animals with low metabolic rates? Many scorpion species can fast for months (McCormick and Polis, 1990). For animals capable of surviving food deprivation for long periods, cumulative prey intake over time may be a more realistic parameter to analyze than period of starvation (Rypstra, 1989). If theory approximates reality here, the cumulative prey intake over the time should be higher in tropical than in desert scorpions. This might explain why nearly all social scorpions have a tropical distribution (notice that scorpion diversity is highest in subtropical deserts; Polis, 1990).

Foraging scorpions should also be sensitive to the risk of predation. A possible antipredator behavior of *Pandinus* and *Heterometrus* spp. may exemplify another benefit of staying together. Groups of juveniles display extraordinarily synchronized movements (Krapf, 1986): scorpions start and stop walking simultaneously, with pauses lasting only about a second. Intragroup synchrony is probably mediated by vibratory signals passing through the substrate or by near-field air currents detected by trichobothria. Synchronization of movements and close association with conspecifics may increase protection from predators hunting by sight (Endler, 1991). The Asian toad *Bufo melanostictus,* a natural predator of scorpions (K. Eisenmann, personal observation, 1978, in Sri Lanka), hesitated to snap at "synchronized" young *Heterometrus spinifer* (Mahsberg, unpublished laboratory observations). Such an increase in the predator's discrimination time can be sufficient for prey to retreat. The study of this predator–prey relationship offers insight not only to the complexity of sensory systems involved but also to the "arms race" among members of the same trophic level ("intraguild predation"; Polis and McCormick, 1986).

Scorpions' Burrows and the Evolution of Social Behavior

Scorpions possess several characteristics that facilitate the evolution of group living (Vehrencamp, 1979; Buskirk, 1981; Anderson, 1984): most species are K-selected with a relatively long life span; they have low reproductive and replacement rates (the lowest maximum intrinsic rate for daily population increase in the animal kingdom; Polis, 1990); and they exhibit viviparity, iteroparity, and extended brood care. These traits favor saturation of a suitable habitat to its carrying capacity, leaving young with few places to establish good burrow sites ("habitat saturation hypothesis"; Emlen, 1982). The costs of dispersal are great because scorpions outside their shelter suffer high predation by invertebrates and vertebrates (McCormick and Polis, 1990; Polis, 1990). This gives obvious advantage for young remaining at home in a stable environment, provided the mother allows it.

Beyond these conditions favoring sociality, many scorpions reside in a well-defined space, the central burrow (Polis, 1990). Anderson's (1984) synthesis of factors encouraging the evolution of social behavior argues that, in most social taxa, the *sine qua non* of sociality is the existence of a nest or burrow. High-quality nest or burrow sites are likely limited in number and thus subject to strong competition. Burrows and nests are often energetically expensive to construct and maintain and serve vital protective functions as refuges from adverse abiotic (e.g., climate) and biotic (e.g., predators) factors. Burrows or nests centralize conspecifics. In animals with limited sensory equipment for long-distance communication, social and cooperative behavior should most likely evolve in such localities. Examples are spiders, isopods, and insects, or some mammals (e.g., Linsenmair, 1984; Hölldobler and Wilson, 1990; Sherman et al., 1991; Avilés, 1997).

Thus, behavior related to living in burrows is likely essential to understand the evolution of sociality in *Scorpio, Heterometrus, Pandinus, Opisthacanthus, Didymocentrus,* and other scorpions (see table 10.1). Burrows for social scorpions are the analogy of spider webs for social spiders. Spider families that have evolved social behavior along the familial route (e.g., Theridiidae, Agelenidae, Eresidae) all possess sheetlike or tangled webs that can be extended or enlarged by young and other family members (Avilés, 1997).

The accumulation of prey remains in burrows indicates that probably all social scorpions consume prey within their retreats (see references in table 10.1 and McCormick and Polis, 1990). This feeding strategy minimizes high predation known to threaten surface-active scorpions (McCormick and Polis, 1990) and may facilitate food sharing between mother and offspring. Young *H. fulvipes* devoured masticated food remains their mother had dropped inside the burrow (Shivashankar et al., 1990). This is similar to passive food provisioning in the funnel-web spider *Coelotes* (see references in Foelix, 1996). These spiders communicate via vibrations transmitted in the sheet web. There is no information on whether scorpion burrows (Polis, 1990) serve as a means of information transmission comparable to spider webs, but the tubelike tunnels of scorpion burrows may be ideal structures to propagate airborne and substrate vibrations generated by prey, predators, or conspecifics. Juddering (a rapid quivering of the body back and forth), tail wagging, and stridulation are all behaviors characteristic of *Pandinus, Heterometrus,* and other scorpions (Alexander, 1958; Polis and Sissom, 1990; Shivashankar et al., 1990; Ehrlich, 1993); these could generate vibratory signals in the air or substrate and thus mediate intraspecific vibratory communication.

Maternal burrows of scorpions can be extended to accommodate offspring and other relatives. Burrows of family groups can form colonies (for examples, see Polis, 1990). Burrows may even promote interspecific associations previously unknown for scorpions: *P. imperator* and the frog *Phrynomantis microps* cohabit in *Pandinus* burrows, likely in a mutualistic relationship (Mahsberg and Linsenmair, unpublished observations). A similar partnership exists between a neotropical microhylid frog and a burrow-dwelling theraphosid spider (Cocroft and Hambler, 1989). Such interactions clearly would be difficult without a centralized location for their evolution.

CONCLUSIONS

Brood care can maximize the reproductive success of scorpions and may have evolved initially to prevent predation on young. Brood care may be one important key to the great success of these "living fossils" (scorpions date back to the Silurian and are among the earliest terrestrial invertebrates). In ancient times scorpions "ruled the world" (Jeram, 1990) and survived more than 400 million years. But why are fewer than 1% of extant scorpion species social, especially if their biology and life-history characteristic predispose them to a social existence with its many potential advantages? To answer this question, perhaps it is best to consider the costs of sociality, the disadvantages of living in groups (Wilson, 1971, 1975; Alexander, 1974; Anderson, 1984; Pulliam and Caraco, 1984; Krebs and Davies, 1993): increased intensity of competition for resources, increased probability of disease and parasite transmission, increased conspicuousness, "reproductivity effect" (larger colonies with decreased per capita reproduction), and inbreeding depression.

I focused on the general costs and benefits of two aspects of sociality in scorpions: maternal behavior and foraging. Although young individuals benefit by being in the presence of their mother and by communal foraging, they also are subject to disadvantages: aggression and occasional cannibalism during periods of food shortage and inability of juveniles to maintain a positive energy balance if the mother is killed are all factors weighing against sociality. These costs and benefits of group living are clearly not specific to scorpions. They also affect the reproductive success of other arthropods with a "scorpionlike" life history and ecology—for example, orthognath spiders (Buskirk, 1981; Foelix, 1996), some cockroaches (Nalepa and Bell, 1997), or other long-lived and burrow-inhabiting invertebrates. The study of such creatures may contribute to questions of general interest: how habitat leads to mothercare (Wyatt, 1987), how mothercare varies with environmental conditions (Clutton-Brock, 1991), and how ecological factors may foster the formation of social groups (Emlen, 1982).

Brood care and sociality in scorpions may be especially useful for experimental analysis of the costs and benefits of living singly or as a group (Krebs and Davies, 1993). One may ask why sociality is so rare in apoikogenic families (see table 10.1; Farley, this volume). How do brood care and social behavior differ in desert and tropical scorpions, and how do they differ within species occurring along habitat gradients (e.g., *P. imperator* from savannas to tropical forests)? Which factors promote tolerance or intolerance in congeneric species (e.g., *Heterometrus fulvipes* versus *H. longimanus;* Shivashankar et al., 1990). How is the availability of essential resources (food, burrows, etc.) shaped by environmental factors? Do tropical biota really promote sociality? Do mother scorpions adjust expenditures on broods in relation to offspring quality, quantity, or age? What balances sibling cannibalism and communal foraging? And how do individuals know when and where to leave the group? For many such questions scorpions are worthy and fascinating subjects for future research.

Acknowledgments I thank Karl Eduard Linsenmair and Barbara König for helpful criticism of earlier versions of the manuscript and the volume editors for improving exposition. I am grateful to Petra Eichler, Monika Ehrlich, Heike Steingen, Andrea Hilpert and Miriam Brandt for their skillful assistance and to Paul Yao Koffi and Pascal Simporé for providing additional field observations. Special thanks to Karl Eduard Linsenmair the founder of the Projet Biodiversité in the Comoé National Park/Ivory Coast. Permission to conduct field work was kindly given by the Ministère de l'Agriculture et des Ressources Animales and the Ministère de la Recherche Scientifique de Côte d'Ivoire, Abidjan. Research on *Pandinus* in Africa was generously supported by a grant of the Universitätsbund Würzburg.

References

Alcock, J. (1993) *Animal Behavior: an evolutionary approach.* Sinauer Associates, Inc., Sunderland, MA.

Alexander, A.J. (1958) On the stridulation of scorpions. *Behaviour* 12: 339–352.

Alexander, R.D. (1974) The evolution of social behavior. *Annu. Rev. Ecol. Syst.* 5: 325–383.

Anderson, M. (1984) The evolution of sociality. *Annu. Rev. Ecol. Syst.* 15: 165–189.

Avilés, L. (1997) Causes and consequences of cooperation and permanent-sociality in spiders. *In* Choe, J.C. and Crespi, B.J. (eds.) *The evolution of social behavior in insects and arachnids,* pp. 476–498. Cambridge University Press, Cambridge.

Barnard, C. (1991) Kinship and social behaviour: the trouble with relatives. *Trends Ecol. Evol.* 6: 310–312.

Barth, F.G. (1982) Spiders and vibratory signals: sensory reception and behavioral significance. *In:* Witt, P.N. and Rovner, J.S. (eds.) *Spider Communication. Mechanisms and Ecological Significance,* pp. 67–122. Princeton University Press, Princeton, NJ.

Bekoff, M. (1987) Group living, natal philopatry and Lindström's lottery: it's all in the family. *Trends Ecol. Evol.* 2: 115–116.

Benton, T.G. (1991) Reproduction and parental care in the scorpion, *Euscorpius flavicaudis. Behaviour* 117: 20–28.

Benton, T.G. (1992) Determinants of male mating success in a scorpion. *Anim. Behav.* 43: 125–135.

Birkhead, T.R. and Møller, A.P. (1992) *Sperm Competition in Birds.* Academic Press, London.

Blaustein, A.R. and O'Hara, R.K. (1982) Kin recognition in *Rana cascadae* tadpoles: maternal and paternal effects. *Anim. Behav.* 30: 1151–1157.

Brown, C.A. and Formanowicz, D.R., Jr. (1996) Reproductive investment in two species of scorpion, *Vaejovis waueri* (Vaejovidae) and *Diplocentrus linda* (Diplocentridae), from West Texas. *Ann. Entomol. Soc. Am.* 89: 41–46.

Brownell, P. and Farley, R.D. (1979) Prey-localizing behaviour of the nocturnal desert scorpion, *Paruroctonus mesaensis:* orientation to substrate vibrations. *Anim. Behav.* 27: 185–193.

Buskirk, R.E. (1981) Sociality in the Arachnida. *In* Hermann, H.R. (ed.) *Social Insects,* vol. III, pp. 281–367. Academic Press, New York.

Caraco, T. (1980) On foraging time allocation in a stochastic environment. *Ecology* 61: 119–128.

Clutton-Brock, T.H. (1991) *The Evolution of Parental Care.* Princeton University Press, Princeton, NJ.

Cocroft, R.B. and Hambler, K. (1989) Observations on a commensial relationship of the microhylid frog *Chiasmocleis ventrimaculata* and the burrowing theraphosid spider *Xenesthis immanis* in southeastern Peru. *Biotropica* 21: 2–8.

Crespi, B.J. (1992) Cannibalism and trophic eggs in subsocial and eusocial insects. *In* Elgar, M.A. and Crespi, B.J. (eds.) *Cannibalism,* pp. 176–213. Oxford University Press, Oxford.

Crump, M.L. (1992) Cannibalism in amphibians. *In* Elgar, M.A. and Crespi, B.J. (eds.) *Cannibalism,* pp. 256–276. Oxford University Press, Oxford.

Ehrlich, M. (1993) *Der Einfluß von Gruppengröße und Beuteparametern auf die postembryonale Entwicklung von Skorpionen* (Master's thesis). University of Würzburg, Würzburg, Germany.

Eichler, P. (1991) *Untersuchungen zum Gruppenverhalten und zur Brutpflege bei tropischen Skorpionen* (Scorpiones: Scorpionidae) (Master's thesis). University of Würzburg, Würzburg, Germany.

Elwood, R. (1992) Pup-cannibalism in rodents: causes and consequences. *In* Elgar, M.A. and Crespi, B.J. (eds.) *Cannibalism,* pp. 299–322. Oxford University Press, Oxford.

Emlen, S.T. (1982) The evolution of helping. I. An ecological constraints model. *Am. Nat.* 119: 29–39.

Endler, J.A. (1991) Interactions between predators and prey. *In* Krebs, J.R. and Davies, N.B. (eds.) *Behavioural Ecology: An Evolutionary Approach,* pp. 169–196. Blackwell, Oxford.

FitzGerald, G.F. (1992) Filial cannibalism in fishes: why do parents eat their offspring? *Trends Ecol. Evol.* 7: 7–10.

Foelix, R. (1996) *Biology of spiders.* Oxford University Press, Oxford.

Francke, O.F. (1982) Parturition in scorpions (Arachnida, Scorpiones): a review of the ideas. *Rev. Arachnol.* 4: 27–37.

Gillespie, R. and Caraco, T. (1987) Risk-sensitive foraging strategies of two spider populations. *Ecology* 68: 887–899.

Hölldobler, B. and Wilson, E.O. (1990) *The Ants.* Springer-Verlag, Berlin.

Hrdy, S.B. (1979) Infanticide among animals: a review, classification and examination of the implications for the reproductive strategies of females. *Ethol. Sociobiol.* 1: 13–40.

Jeram, A. (1990) When scorpions ruled the world. *New Scientist* 16: 52–55.

Krapf, D. (1986) *Verhaltensphysiologische Untersuchungen zum Beutefang von Skorpionen mit besonderer Berücksichtigung der Trichobothrien* (Doctoral dissertation). University of Würzburg, Würzburg, Germany.

Krebs, J.R. and Davies, N.B. (1993) *An Introduction to Behavioural Ecology.* Blackwell, Oxford.

Larrouy, G., Signorel, M.Ch. and Cambefort, Y. (1973) Comportement en captivité de *Pandinus imperator* C. L. Koch et naissance des jeunes. *Bull. Soc. Hist. Nat. Toulouse* 109: 346–350.

Linsenmair, K.E. (1984) Comparative studies on the social behaviour of the desert isopod *Hemilepistus reaumuri* and of a *Porcellio* species. *Symp. Zool. Soc. Lond.* 53: 423–453.

Little, C. (1990) *The Terrestrial Invasion: An Ecophysiological Approach to the Origins of Land Animals.* Cambridge University Press, Cambridge.

Lourenço, W.R. (1991) *Opisthacanthus,* genre gondwanien défini comme groupe naturel. Caractérisation des sous-genres et des groupes d'espèces. (Arachnida, Scorpiones, Ischnuridae). *Iheringia Sér. Zool.* 71: 5–42.

Machado, G. and Oliveira, P.S. (1998) Reproductive biology of the neotropical harvestman *Goniosoma longipes* (Arachnida, Opiliones: Gonyleptidae): Mating and oviposition behaviour, brood mortality, and parental care. *J. Zool. (Lond.)* 246: 359–367.

Mahsberg, D. (1990) Brood care and family cohesion in the tropical scorpion *Pandinus imperator* (Koch)(Scorpiones: Scorpionidae). *Acta Zool. Fenn.* 190: 267–272.

Mahsberg, D., Lippe, R., and Kallas, S. (1999) *Skorpione.* Natur und Tier-Verlag, Münster, Germany.

Mahsberg, D. and Warburg, M.R. (2000) Ovariuterus of *Pandinus imperator,* Koch (Scorpiones; Scorpionidae): Comparison of virgin female with mother. *Annals of Anatomy* 182: 171–174.

Martens, J. (1993) Further cases of paternal care in Opiliones (Arachnida). *Trop. Zool.* 6: 97–107.

McCormick, S.J. and Polis, G.A. (1990) Prey, predators and parasites. *In* Polis, G.A. (ed.) *The Biology of Scorpions,* pp. 294–320. Stanford University Press, Palo Alto, CA.

Mora, G. (1990) Paternal care in a neotropical harvestman, *Zygopachylus albomarginis* (Arachnida, Opiliones: Gonyleptidae). *Anim. Behav.* 39: 582–593.

Nalepa, C.A. and Bell, W.J. (1997) Postovulation parental investment and parental care in cockroaches. *In* Choe, J.C and Crespi, B.J. (eds.) *The evolution of social behavior in insects and arachnids,* pp. 26–51. Cambridge University Press, Cambridge.

Peretti, A.V. (1994) Comportamiento de relacion madre-cria de *Tityus trivittatus* Kraepelin (Scorpiones, Buthidae). *Bol. Soc. Biol. Concepción, Chile* 65: 9–21.

Pianka, E.R. and Parker, W.S. (1975) Age-specific reproductive tactics. *Am. Nat.* 109: 453–464.

Polis, G.A. (1980) The effect of cannibalism on the demography and activity of a natural population of desert scorpions. *Behav. Ecol. Sociobiol.* 7: 25–35.

Polis, G.A. (1981) The evolution and dynamics of intraspecific predation. *Annu. Rev. Ecol. Syst.* 12: 225–251.

Polis, G.A. (1990) Ecology. *In* Polis, G.A. (ed.) *The Biology of Scorpions,* pp. 247–293. Stanford University Press, Palo Alto, CA.

Polis, G.A. and Farley, R.D. (1979) Characteristics and environmental determinants of natality, growth and maturity in a natural population of the desert scorpion, *Paruroctonus mesaensis* (Scorpionida: Vaejovidae). *J. Zool.* 187: 517–542.

Polis, G.A. and Lourenço, W.R. (1986) Sociality among scorpions. *Actas X Congr. Intl. Aracnol. Jaca* 1: 111–115.

Polis, G.A. and McCormick, S.J. (1986) Scorpions, spiders and solpugids: predation and competition among distantly related taxa. *Oecologia* 71: 111–116.

Polis, G.A., Myers, C.A. and Holt, R. (1989) The ecology and evolution of intraguild predation: potential competitors that eat each other. *Annu. Rev. Ecol. Syst.* 20: 297–330.

Polis, G.A. and Sissom, W.D. (1990) Life history. *In* Polis, G.A. (ed.) *The Biology of Scorpions,* pp. 161–223. Stanford University Press, Palo Alto, CA.

Pulliam, H.R. and Caraco, T. (1984) Living in groups: is there an optimal group size? *In* Krebs, J.R. and Davies, N.B. (eds.) *Behavioural Ecology: An Evolutionary Approach,* pp. 122–147. Blackwell, Oxford.

Pulliam, H.R. and Millikan, G.C. (1982) Social organization in the nonreproductive season. *In* Farner, D.S., King, J.R. and Parkes, K.C. (eds.) *Avian Biology,* pp. 169–197. Academic Press, New York.

Real, L. and Caraco, T. (1986) Risk and foraging in stochastic environments. *Annu. Rev. Ecol. Syst.* 17: 371–390.

Rypstra, A.L. (1986) High prey abundance and a reduction in cannibalism: the first step to sociality in spiders (Arachnida). *J. Arachnol.* 14: 193–200.

Rypstra, A.L. (1989) Foraging success of solitary and aggregated spiders: insights into flock formation. *Anim. Behav.* 37: 274–281.

Rypstra, A.L. and Tirey, R.S. (1991) Prey size, prey perishability and group foraging in a social spider. *Oecologia* 86: 25–30.

Saito, Y. (1997) Sociality and kin selection in Acari. *In* Choe, J.C. and Crespi, B.J. (eds.) *The evolution of social behavior in insects and arachnids,* pp. 443–457. Cambridge University Press, Cambridge.

Schultze, W. (1927) Biology of the large Philippine forest scorpion. *Philipp. J. Sci.* 32: 375–389.

Shachak, M. and Brand, S. (1983) The relationship between sit and wait foraging strategy and dispersal in the desert scorpion, *Scorpio maurus palmatus. Oecologia* 60: 371–377.

Shaffer, L.R. and Formanowicz, D.R., Jr. (1996) A cost of viviparity and parental care in scorpions: reduced sprint speed and behavioural compensation. *Anim. Behav.* 51: 1017–1024.

Sherman, P., Jarvis, J. and Alexander, R. (1991) *The Biology of the Naked Mole-rat.* Princeton University Press, Princeton, NJ.

Shivashankar, T., Rao, N.D.D. and Veeresh, G.K. (1990) On the evolution of communal behaviour in the scorpion *Heterometrus fulvipes* (Scorpionidae). *In* Veeresh, G.K., Mallik, B. and Viraktamath, C.A. (eds.) *Social Insects and the Environment,* pp. 209–213. Oxford, New Dehli.

Sissom, W.D., Polis, G.A. and Watt, D.D. (1990) Field and laboratory methods. *In* Polis, G.A. (ed.) *The Biology of Scorpions,* pp. 445–461. Stanford University Press, Palo Alto, CA.

Stanback, M.T and Koenig, W.D. (1992) Cannibalism in birds. *In* Elgar, M.A. and Crespi, B.J. (eds.) *Cannibalism,* pp. 277–298. Oxford University Press, Oxford.

Stearns, S.C. (1992) *The Evolution of Life Histories.* Oxford University Press, Oxford.
Torres, F. and Heatwole, H. (1967) Factors influencing behavioral interaction of female parent and offspring in scorpions. *Carib. J. Sci.* 7: 19–22.
Trivers, R.L. (1972) Parental investment and sexual selection. *In* Campbell, B. (ed.) *Sexual Selection and the Descent of Man,* pp. 136–179. Aldine, Chicago.
Uetz, G.W. (1988) Risk sensitivity and foraging in colonial spiders. *In* Slobodchikoff, C.N. (ed.) *The Ecology of Social Behavior,* pp. 353–377. Academic Press, New York.
Uetz, G.W. (1989) The "ricochet effect" and prey capture in colonial spiders. *Oecologia* 81: 154–159.
Uetz, G.W. and Hieber, C.S. (1997) Colonial web-building spiders: balancing the costs and benefits of group-living. *In* Choe, J.C. and Crespi, B.J. (eds.) *The evolution of social behavior in insects and arachnids,* pp. 458–475. Cambridge University Press, Cambridge.
Vachon, M. (1973) Etude des caractères utilisés pour classer les familles et les genres de scorpions (Arachnides). *Bull. Mus. Natl. Hist. Nat. Zool.* 104: 857–958.
Vannini, M., Balzi, M., Becciolini, A., Carmignani, J. and Ugolini, A. (1985) Water exchange between mother and larvae in scorpions. *Experientia* 41: 1620–1621.
Vannini, M. and Ugolini, A. (1980) Permanence of *Euscorpius carpathicus* (L.) larvae on the mother's back (Scorpiones, Chactidae). *Behav. Ecol. Sociobiol.* 7: 45–47.
Vannini, M., Ugolini, A. and Marucelli, C. (1978) Notes on the mother-young relationship in some *Euscorpius* (Scorpiones Chactidae). *Monit. Zool. Ital.* 12: 143–154.
Vehrencamp, S.L. (1979) The roles of individual, kin and group selection in the evolution of sociality. *In* Marler, P. and J. Vandenbergh, J. (eds.) *Handbook of Behavioral Neurobiology,* vol. 3: *Social Behavior and Communication,* pp. 351–394. Plenum Press, New York.
Warburg, M.R. and Rosenberg, M. (1992) The reproductive system of a scorpion, *Nebo hierichonticus* (Simon)(Scorpiones: Diplocentridae). *Int. J. Insect Morphol. Embryol.* 21: 365–368.
Warburg, M.R. and Rosenberg, M. (1993) The female reproductive system in *Scorpio maurus fuscus* (Scorpiones; Scorpionidae). *Isr. J. Zool.* 39: 23–27.
Warburg, M.R. and Elias, R. (1998) The reproductive potential and strategy of *Scorpio maurus fuscus* (Scorpiones: Scorpionidae): Anatomical clues in the ovariuterus. *J. Zool. (Lond.)* 246: 29–37.
Wilson, E.O. (1971) *The Insect Societies.* Harvard University Press, Cambridge, MA.
Wilson, E.O. (1975) *Sociobiology, the New Synthesis.* Belknap Press, Cambridge, MA.
Wrensch, D.L. and Bruce, W.A. (1991) Sex ratio, fitness and capacity for population increase in *Pyemotes tritici* (Pyemotidae). *In* Schuster, R. and Murphy, P.W. (eds.) *The Acari,* pp. 209–221. Chapman & Hall, London.
Wyatt, T. (1987) How habitat leads to mothercare. *New Sci.* 116: 50–53.
Zeh, D.W. and Smith, R.L. (1985) Paternal investment by terrestrial arthropods. *Am. Zool.* 25: 785–805.
Zeh, J.A. and Zeh, D.W. (1990) Cooperative foraging for large prey by *Paratemnus elongatus* (Pseudoscorpionida, Atemnidae). *J. Arachnol.* 18: 307–311.

11

Reproductive Ecology

TIM BENTON

The biological answer to the question, Why does the natural world take on the form it does? was first proposed by Charles Darwin in his theory of evolution by natural selection. If a trait with an underlying genetic component arises that promotes an organism's ability to survive and/or reproduce, it will increase in frequency from one generation to the next: it will be selected. Over time this simple process can lead to the wonderful adaptations we observe in nature.

Fisher realized the fundamental evolutionary difference between the sexes. Female reproductive success is generally limited by the number of eggs a female can produce, so females may chose among males for the best sire of their young. Conversely, male reproductive success is limited by the number of females with which a male can mate, so males may compete among themselves to mate with females. These processes of sexual selection have resulted in complex adaptations: colorful plumage of some male birds, horns and fighting behavior of beetles and deer, sculptured genitalia of damselflies. Unlike many adaptations (e.g., mammalian eyes) which obviously function to enhance survival, sexually selected adaptations may seem useless, or even a handicap. Explaining the presence of these traits is a challenge for evolutionary biology.

The study of behavioral ecology (the way an animal's environment, particularly its biotic environment, shapes the evolution of its behavior) is an important field of evolutionary biology. Naturally, behavioral ecology tends to concentrate on animals that are most like us: birds and mammals. As mammals and birds have similar sensory modalities, we have some intuitive feel about the way they interact with their world. Additionally, these animals are behaviorally complex, and understanding behavioral complexity in any form might illuminate our own behavior. However, behavioral complexity is not limited to the vertebrates. In many situations arthropod models can be used to investigate areas of general evolutionary interest.

Scorpions are one such model system. On the fast–slow continuum of life histories, scorpions are, unlike most arthropods, markedly toward the slow end. They live for a long time, often have indeterminate growth, reach a large body size, and are typically iteroparous (K-selected, sensu Pianka, 1970; Polis and Farley, 1980). In this respect, they are among the most vertebrate-like of the invertebrates. Although scorpions may spend the majority of their time immobile (either foraging or in their retreats), this changes during their reproductive season. Males become vagrant and actively search for females. In some species males can be found cohabiting with females, and competition may occur among males for access to females. The courtship can be long and complex, resulting in sperm transfer by means of a stalked spermatophore deposited on the substrate.

The study of scorpion reproductive ecology (how ecology shapes reproductive behavior) is at an early stage, perhaps because few people realize the subtlety and complexity of reproduction in scorpions. The aim of this chapter is to outline this subtlety and complexity. Most published research on scorpion reproductive behavior has resulted from laboratory-based studies (but see Polis and Farley, 1979b). Although useful, these results should be regarded with caution until they can be confirmed by field observation. Scorpion behavior is often influenced by laboratory conditions, and may be abnormal. In addition, laboratory animals may differ in the timing of their reproductive season compared with wild animals (e.g., Polis and Sissom, 1990). Our knowledge of scorpion reproductive ecology comes from only a handful of studies, so this review is sometimes speculative. I hope that it stimulates interest in a group of animals that may further furnish important examples of the process of evolution.

SEASONALITY OF REPRODUCTION

Scorpions are generally sedentary, sit-and-wait predators, which, like web spiders, forage while immobile. Depending on their ecology, scorpions may forage near their retreats (surface foragers; e.g., *Paruroctonus mesaensis* Stahnke), or at the entrance of their burrow (doorkeeping scorpions; e.g., *Euscorpius flavicaudis* [deGeer], *Anuroctonus phaiodactylus* [Wood]). During the reproductive season (defined as the period in which birth and reproductive behaviors occur), typical patterns of behavior change in three important ways, as described below.

Reduction in Female Foraging Activity

In many species studied to date, birth is markedly synchronized, perhaps following a pulse of prey availability (Polis and Sissom, 1990). For example, *Vaejovis vorhiesi* Stahnke gives birth during 2 weeks in midsummer (Williams, 1969); *P. mesaensis* young first appear on the surface in early August (Polis and Farley, 1979a); *E. flavicaudis* give birth during 3–4 weeks in late summer (fig. 11.1A). Tropical species as a group seem to show no preferred seasonality, but birth may nonetheless be synchronized within species. For example, *Isometrus maculatus* (deGeer) gives birth in January and February in tropical Africa (Probst, 1972). Females often show a reduction in foraging (i.e., surface activity) as birth approaches and while the juveniles are on the their backs (fig. 11.1A), such that birth and maternal care are often difficult to observe in the field (Williams, 1969). Often the emergence of second instar young is the first indication that birth has occurred (Williams, 1969; Polis and Farley, 1979a). Where surface density is normally low, it may be difficult to identify this reduction in surface activity except in a long field study. This is well illustrated by the surface-foraging scorpion *P. mesaensis*, where the females' surface density is bimodal, with the reduction in surface density occurring at the time they give birth (Polis, 1980).

Increase in Male Surface Activity

In many species, male activity increases greatly around the time females give birth (fig. 11.1B). Instead of remaining immobile to forage, the males become

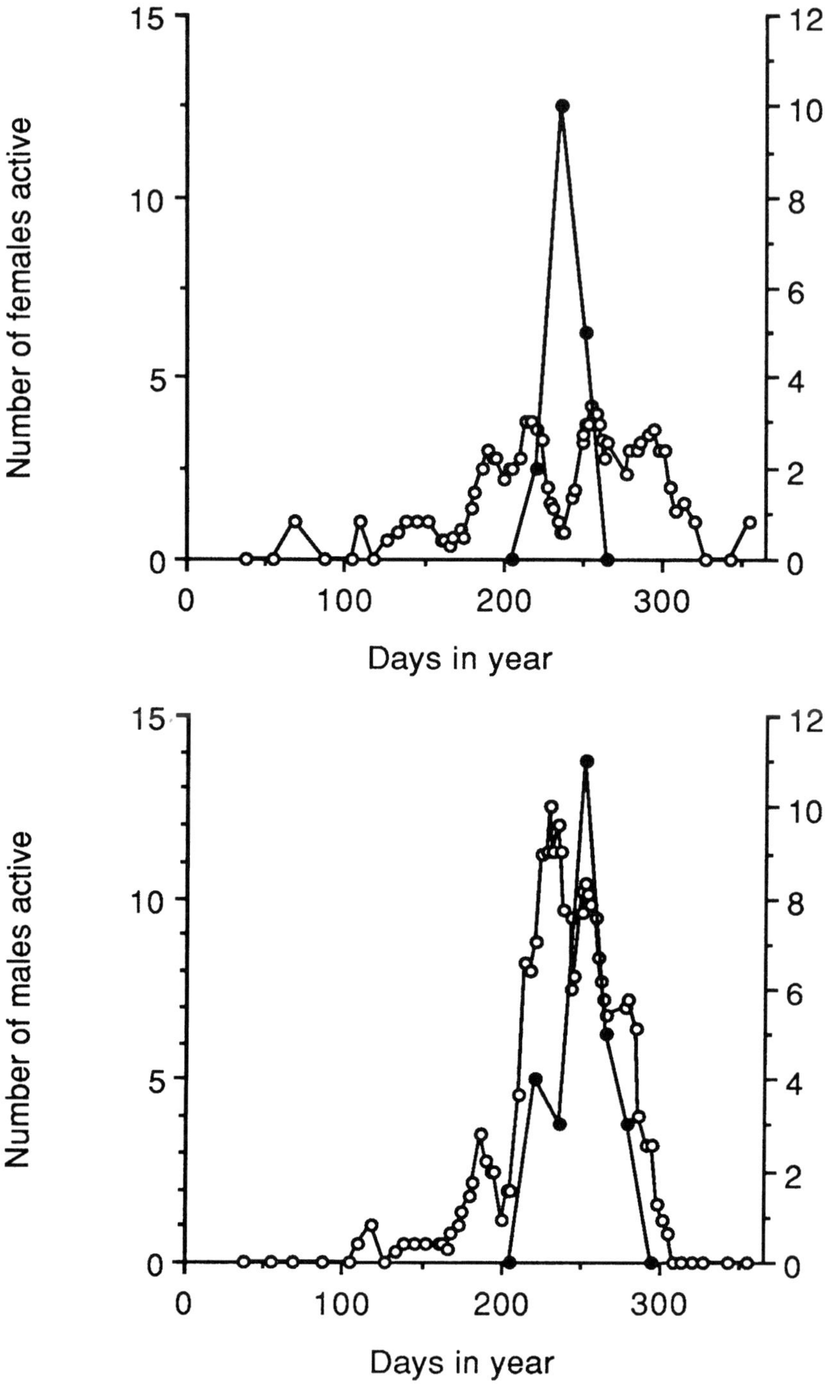

Figure 11.1 Reproductive phenology of *Euscorpius flavicaudis.* (A) Seasonality of female surface activity (open circles) and the number of births (closed circles). The number of births (for each half-month) is estimated from the number of females known to be engaged in maternal care and the time at which this ended. Note that the peak in the number of

more mobile and consequently feed less frequently. For example, in *P. mesaensis,* 48.3% of males were observed to be moving when first sighted during this period, compared with 1.8% of females and juveniles. Additionally, males abandon their retreats frequently, often nightly (Polis and Farley, 1979a; Benton, 1992a), sometimes moving great distances in a short time (Polis and Farley, 1979a). This surface activity is termed "sexual vagrancy" and is well documented in scorpions (e.g., Tourtlotte, 1974; Lamoral, 1979; Polis and Farley, 1979a). Sexual vagrancy functions to increase males' encounter rates with females with whom they can mate. The population of *E. flavicaudis* I studied occurs in an urban habitat in southeast England. These scorpions live in a large brick wall in crevices created when mortar crumbles away. During the summer, males move across the wall, entering each crack and moving on if it is empty (Benton, 1992b). At times, the scorpions appeared to "quarter" the wall: moving one way a few meters, turning, and moving back on a parallel track.

The Mating Season

Scorpions lack the sperm-storing diverticula common in many terrestrial arthropods (Walker, 1980; Farley, this volume). If sperm are stored in the lumen of the genital tract, they are likely to be flushed out when the female gives birth, necessitating reinsemination for each litter. Although this generally seems true (de Armas, 1987), Kovoor et al. (1987) described a mechanism by which some buthids can store sperm inside glandular tissue. Scorpions that are known to give birth to more than one litter following a single mating are all buthids and all are tropical (except *Centruroides exilicauda* [Wood]; Polis and Sissom, 1990). The gestation of these scorpions is much shorter than most (about 4 months rather than 12), suggesting that viable storage is not possible for those species with longer gestation.

As a result of the need to mate once per litter, two types of adult female exist during the reproductive season: those that recently gave birth and need reinsemination and virgins that recently molted to maturity. Perhaps the same patterns of prey availability that cause synchronization of births may also allow juveniles to molt to maturity. Therefore females may generally become receptive simultaneously as a direct result of periods of prey availability influencing embryonic and juvenile development. In *P. mesaensis* matings occur from late May to early October, but >80% occur in August and September following synchronized birth (Polis and Farley, 1979a). In *E. flavicaudis,* I observed a few matings early in the season, but most (86%) followed, or were consecutive with, the birthing season.

In many spiders (e.g., *Linyphia litigiosa:* Watson, 1990; *Nephila clavipes:* Vollrath, 1980; *Frontinella pyramitela:* Austadt, 1982) and some other arachnids (e.g., some mites: Thomas and Zeh, 1984) the first male to transfer sperm to a

---

births occurs simultaneously with a reduction in female surface activity. (B) The seasonality of male surface activity (open circles) and the known numbers of matings (either observed or numbers of spermatophores found) (closed circles) for each half-month. Note that male surface activity increases before the main peak in observed matings. Surface activity is smoothed over a 2-week period, the count for each night is shown as an average of the preceding and following week's counts. Data originally presented, in part, in Benton (1992b; data from England, 1988).

female will sire most of the brood (first-male sperm priority). In this situation, we might expect (1) males to compete to be the first to mate with receptive females (which sometimes occurs; see below); and (2) a selection pressure on females to discourage later males from attempting to mate with them (as they gain little from remating) (Parker, 1984). In some scorpions (e.g., *E. flavicaudis*) females become unresponsive to males after insemination. As female scorpions are generally larger than males, they can repel (or even eat: Peretti et al., 1999) the male if they are unreceptive. Additionally, females can secrete a spermatocleutrum so it becomes impossible for males to transfer sperm to them (as in *Urodacus manicatus* [Thorell]; Smith, 1966).

FINDING A MATE

Patterns of Female Receptivity

While caring for young, females of most scorpion species remain in a retreat (fig. 11.1A). Presumably this ensures the survival of newborn young (Mahsberg, this volume; Benton, 1991a) and reduces the females' predation risk, as their own motility is reduced by carrying young (Shaffer and Formanowicz, 1996). If male activity occurs purely on the surface, males are unlikely to encounter females until females resume their normal patterns of foraging after juvenile dispersal. However, in some species males actively search retreats for females. In *E. flavicaudis* where patterns of female receptivity were noted, females repelled the advances of the male until the period of maternal care was over (Benton, 1991b). Reports of males mating with females during their period of maternal care are known (Williams, 1969; de Armas, 1987; Polis and Mohnac, 1990), but are exceptional, especially in nonbuthids.

Mate Encounter

The simple scenario of a wandering male making direct physical contact with a stationary female appears naive. First, many scorpions, such as *P. mesaensis*, are surface foragers. In this species, adults emerge onto the surface to forage where, as sit-and-wait predators, they remain immobile (usually within a meter of the burrow; Polis et al., 1985). Moving prey produces vibrations in the substrate that can be detected from a considerable distance (Brownell and Farley, 1979). Females resume surface foraging after maternal care is complete and the second instars have dispersed (Polis and Farley, 1979a,b). Therefore, receptive females may sense the movement of vagrant males from a distance and so can approach them to initiate courtship (Polis and Farley, 1979a).

Increasing evidence suggests that males can determine if they are near females by chemical cues left or released by females. Initially only anecdotal evidence supported this (e.g., Alexander, 1959). However, recently Gaffin and Brownell (1992, this volume) presented quantitative evidence to show that in *P. mesaensis*, at least, males chemically recognize the past-presence of females and respond accordingly. As yet, little evidence of a volatile sex-attractant pheromone exists (Benton, 1992b; Gaffin and Brownell, 1992); instead, stimulus detection seems to depend directly on contact (probably by the male's pectines) with substrate bearing a nonvolatile signal. This signal seems to have high potency, lasting for many days. It is unclear whether the chemical is a signal produced by the female to attract males or whether males have evolved the sensitivity to identify the presence of a female by chemical traces she may incidentally leave on the surface. Whatever evolutionary route, males encountering substrates previously exposed to females ceased nondirected movements and instead localized mate search to the vicinity of the signal (Gaffin and Brownell, 1992).

As most scorpions remain near a semipermanent retreat, wandering males are likely to encounter the female's chemical cue if they approach her burrow. Upon encountering the chemical, males can begin a more localized pattern of locomotory activity to bring him close enough to encounter the female. The frequency of encounters between the sexes will thus depend on the area over which the female's chemical cues remain (and/or the chemical's volatility), the distance at which a female can detect a male's movement, the density of females, and the population's operational sex ratio.

Polis and Sissom (1990) report clustering of male *Paruroctonus luteolus* Gertsch around the burrow of a female. Individuals of this species are rarely seen, so it is unlikely that several males, searching randomly, simultaneously encounter the female's burrow. This anecdote supports the hypothesis that males can chemically detect females, perhaps in this case by a more volatile chemical than that identified by Gaffin and Brownell. In such a species, where the population density is especially low, it may obviously benefit a female to signal her presence to males for two reasons. First, she ensures she finds a mate. Second, as several males may be attracted simultaneously, they may compete among themselves to mate with her. If the winner of the competition is the strongest and most vigorous male, then, by stimulating competition, she need not choose between them. This process has been previously noted in some spiders and is called female-enhanced male–male competition (Watson, 1990). Although there is as yet no evidence of male–male competition in *P. luteolus,* it has been reported in another species of scorpion, *Euscorpius flavicaudis* (see below).

WHAT HAPPENS AFTER MALES LOCATE FEMALES?

Scorpion longevity is uncharacteristically long for invertebrates, typically being over 4 four years and ranging up to 25 years (Polis and Sissom, 1990). For example, Warburg and Elias (1998) estimate longevity of *Scorpion maurus fuscus* to be a minimum of 6 years and a maximum of 17. As a result of this longevity, most mature females in a population are likely to have mated the previous year. In *P. mesaensis* 62.5% of females were adult the previous year (Polis and Farley, 1979a). A few matings occur early in the season, presumably with virgin females, but most occur later, after the pulse of births in the population (Polis and Farley, 1979b; see fig. 11.1 for data from *E. flavicaudis*). In *E. flavicaudis,* females retreat to the back of their crevice during maternal care. No female with young was seen on the surface during my field study. During maternal care males are highly vagrant (fig. 11.1), routinely searching retreats for females. If a female is encountered by a male at this time, she repels courtship attempts by him until after the litter has molted and begun to disperse. Females will court with any male encountered 2 ± SD 1.2 days (n = 12) after the juveniles molt (Benton, 1992b).

What should a male do if he encounters a female engaged in maternal care? He has two options. He could either leave and attempt to find another female nearer to receptivity, or, as female receptivity is predictable, he could stay and wait until she becomes receptive and then mate with her. Which option the male should take depends on two main theoretical considerations (Grafen and Ridley, 1983): the sex ratio and the rate at which males encounter females. If males encounter females frequently, it may be better to search for another female nearer receptivity. This is especially true if the operational sex ratio is female biased, as it is less likely that females nearer receptivity will have already been found by other males. Conversely, if a male's encounter rate with females is low, he is unlikely to find another female nearer receptivity. Additionally, if the sex ratio is not female biased, any female nearer receptivity may already have been found by another male. So it may be in his interest to stay with the current female and wait for her to become receptive.

Scorpion species differ in the degree of female sex ratio bias and, presumably, in the encounter rate between the sexes. In surface-foraging scorpions the encounter rate may be relatively high because females leaving their burrows routinely could lay down chemical cues that could be used by males to find them. Additionally, females can detect moving animals (such as male scorpions) from a distance (Brownell and Farley, 1979). As a result, male scorpions have a high chance of encountering a female if they encounter her home range. The sex ratio of many scorpions studied is strongly female biased, often as a result of high male mortality during the mating season (e.g., Polis, 1980, 1990). Field reports generally do not mention males remaining with encountered females before mating in species that regularly leave their retreats.

In doorkeeping scorpions, on the other hand, the a priori expectation is that the encounter rate between males and females will be lower than in surface foragers. As these females rarely leave their burrows, a male may need to encounter a female's burrow before he can sense her presence. Additionally, if doorkeeping scorpions are less able to detect moving prey than surface foragers, the chances of wandering males being cannibalized will be decreased, so the sex ratio may not be so biased. In the population of *E. flavicaudis* in England, mortality due to predation was low, resulting in high survivorship for males during the period of vagrancy. I found no evidence of a volatile attraction pheromone (Benton, 1992b) and so presume that for a male to encounter a female he must encounter either her crevice or its immediate vicinity. Perhaps as a result of a sex ratio nearer unity and a relatively low encounter rate, I found that males encountering soon-to-be-receptive females often stayed with them rather than continuing their search.

Cohabitation was common and occurred before copulation. At no time other than the reproductive season were two scorpions observed in one retreat. It is difficult to assess quantitatively the occurrence of cohabitation, as a male at the entrance to a retreat can obscure the presence of a female. However, on a typical night during the mating season, at least 55% of precopulatory females were cohabiting with males (Benton, 1992b). In total, for 31 courtships observed in the field and a seminatural laboratory colony, at least 58% were preceded by cohabitation.

Upon encountering a crevice, a vagrant male would enter. If there were a female within, he would move toward her and attempt to take her pedipalpal chelae (claws) in his own. If the female were carrying young, she would respond by rapidly moving her pedipalps horizontally to and fro. Typically, the male would then retreat, turn around, and become immobile inside the entrance to the crevice, facing out. In a few cases, the female's retreat was too shallow to be occupied by both male and female. In these cases, a cohabiting male spent the day in a nearby crevice, and "commuted" each night to the female's crevice, where he remained on the surface beside the entrance.

The mean duration of cohabitation was 10.0 ± SD 9.9 nights (range 1–30, $n = 19$). Cohabitation ended in all cases when the female mated (either this was observed or a spent spermatophore was found). Of five females in crevices too shallow to allow them to retreat out of sight, four had associated males from sometime during the week before giving birth (Benton, 1992b).

The benefits of cohabitation are clear for males, but less so for females. If a male is unlikely to encounter another female, then to ensure some reproductive success he has to remain with the current female until she becomes receptive. In 15 cases of cohabitation, I saw who mated with the female, and in 14 cases this was the male currently cohabiting ($p < 0.005$). The exception to this was a male who left the soon-to-be-receptive female when a currently receptive female (one whose young had dispersed, but who had not herself been encountered by a male) passed the entrance of the crevice. He mated with the new female and did not return to the crevice. There are three possible benefits females gain from cohabitation. First, if a male waits for her to become receptive, she wastes no time or effort locating a mate. Second, if more than one male encounters the crevice, they may contest, thereby saving her the task of choosing among males (female-enhanced male competition). Third, maternal care, in this species at least, prevents predation of the young (Benton, 1991a). Perhaps the presence of an additional guarder—the male—also serves to protect her offspring.

Precopulatory cohabitation is usually associated with male–male competition and so is often termed mate guarding. It is widespread in animals (Ridley, 1983), commonly occurring in spiders (Bristowe, 1958; Vollrath, 1980; Austadt, 1984; Watson, 1990), and also some mites (Thomas and Zeh, 1984). In *E. flavicaudis,* when a male enters a crevice already occupied by a cohabiting pair, the resident male turns to face the intruder and extends his pedipalps. Upon contact, males attempt to grasp each other's claws. Usually, within 2 sec, before either scorpion obtains a grip on the other, one turns and flees. I observed over 140 such contests in the field and laboratory, and in only 2 did the contest escalate into a fight—both times between a small male intruding into a crevice occupied by a large male guarding a female. Such quick contests between *Euscorpius* males have been noted before (Angermann, 1957).

In a series of staged contests in a seminatural laboratory colony (see Benton, 1991a, for details), I found significant effects of both size and ownership status on the outcome of mate-guarding contests. The best predictor of contest outcome was the size of the pedipalpal claws. This result is to be expected in this species because their claws are their main offensive weapon; like most fossorial scorpions they tend to crush prey rather than sting it. In 80% of contests ($n = 110$), the larger clawed male won. In the 20% of cases when the smaller clawed animal won, the winner was the current resident 91% (20/22) of the time. Both relative claw size and ownership status have highly significant effects (two-way ANOVA; size: $F_{1,36} = 68.5$, $p < 0.0001$; status: $F_{1,36} = 42.7$, $p < 0.0001$). Large-clawed (and thus large-bodied) scorpions usually won contests over access to females, but if the contestants were matched in size, ownership status was the asymmetry used to decide contest outcome. Being a resident gave a scorpion an advantage equivalent to an increase in claw length of 10.6 ± SD 7.2% ($n = 11$). The rule for deciding contests (if much smaller, retreat; if closely matched, owner wins) is expected from game theoretical analyses of animal contests with asymmetric resource-holding potential (e.g., Maynard Smith and Parker, 1976) and also occurs in the grassland population of the spider *Agelenopsis aperta* (Hammerstein and Riechert, 1988).

Reports of cohabitation (and perhaps therefore mate guarding) occur in the literature. McDaniel (1968) reports one observed case of cohabitation from the field and one from the laboratory in the doorkeeping scorpion *Anuroctonus phaiodactylus.* Shachak and Brand (1983, pp. 377–8) report that in *Scorpio maurus* L. "in May and June pairs of scorpions (males and females) can be found in a single burrow, which may indicate the mating period." *Scorpio maurus* is also a doorkeeping scorpion (Williams, 1987; Benton personal observation). In the surface-foraging scorpion *Leiurus quinquestriatus* (Hemprich and Ehrenberg) males and female can sometimes be found together in the same retreat (Levy and Amitai, 1980; Benton, personal observation). Fabre (1923) reports a similar phenomenon in another buthid, *Buthus occitanus* (Amoreux).

MATING

After the preliminaries (maturing/giving birth in females; encountering a female and perhaps contesting for her in males), courtship and mating can occur. Since the mid-1950s, many papers have described courtship behavior in different species. There has not yet been a detailed behavioral study to ascertain the functions of the component behaviors in scorpion courtship. Here I present a quantitative description of the courtship behavior of one species and attempt to identify functions where possible. This particular study was conducted in the laboratory, but all behaviors observed in the laboratory were also observed in the field. Field courtships were, on average, significantly quicker than laboratory courtships (Benton, 1990), and little emphasis should be placed on the detailed structure of the courtship. However, the gross structure is largely similar in laboratory and field. Additionally, the courtship of this species is broadly similar to many of the published descriptions of courtship in other species (for a similarly detailed description of scorpion courtship, see Peretti, 1991).

Leiurus quinquestriatus occurs in North Africa and in the Middle East. I studied the animals in the field at Sede Boqer in the Negev Desert and in the labo-

Figure 11.2 Flow diagram of the courtship behavior of *Leiurus quinquestriatus hebraeus*. Data are from 30 courtships. Each circle represents a behavior, the size of which is proportional to its frequency. Arrows indicate behavioral transitions, their thickness being proportional to the transition frequencies (frequencies <10% are not shown). C = contact between male and female; J = judder; JK = judder-kiss; TOW = tug of war; mL = male-led walk; fL = female-led walk; K = kiss; mK = male kiss; fK = female kiss; P = pause; SH = shifting; AP = arm pulling; AAP = alternate arm pulling; fGO = female uptakes sperm. Phase 1 = introductory phase; phase 2 = promenade phase; phase 3 = spermatophore deposition phase. Les arbres droit and tail waving not shown as they occur concurrently with other behaviors. See text for descriptions of behaviors.

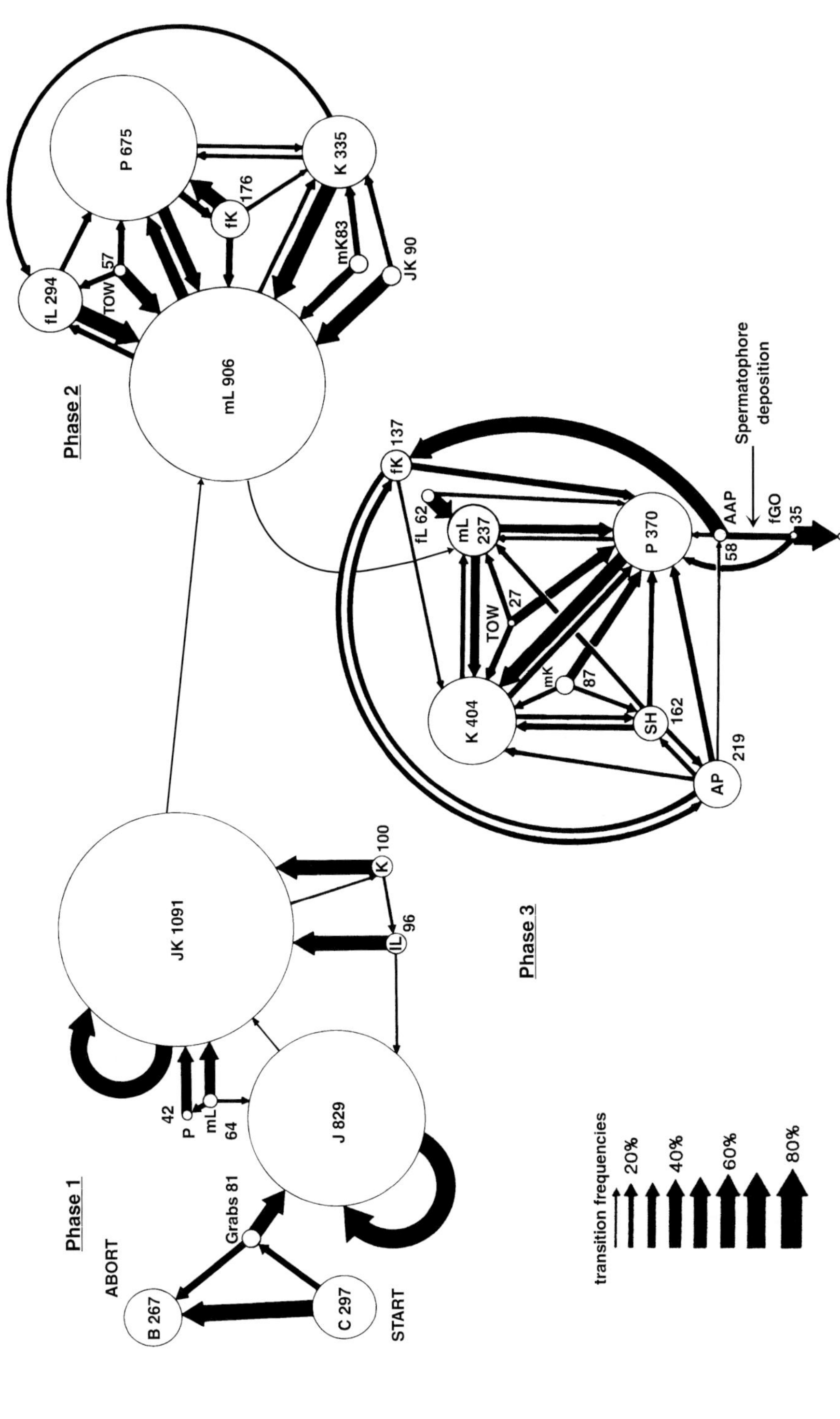

Phase 1
ABORT
START
B 267
C 297
Grabs 81
J 829
P 42
mL 64
JK 1091
K 100
lL 96
Phase 2
mL 906
fL 294
TOW 57
P 675
fK 176
K 335
mK83
JK 90
Phase 3
fK 137
fL 62
mL 237
P 370
AAP 58
Spermatophore deposition
fGO 35
PART 30
TOW 27
mK 87
K 404
SH 162
AP 219
transition frequencies
20%
40%
60%
80%

ratory. The basic quantitative data come from observations of 30 courtships recorded under red-light conditions onto videotape and later transcribed (see Benton, 1990, 1992c for details).

The courtship of *Leiurus* consists of a nonrandom sequence of 11 component behaviors. These occur in three phases, each with a distinct pattern (fig. 11.2). In the introductory phase, the animals meet and the male pacifies the female. In the promenade phase, the male leads the female around as they search for a site on which sperm transfer can occur. In the spermatophore deposition phase, sperm transfer occurs. In total, the courtships observed in the laboratory averaged 47±44 min.

Introductory Phase

A courtship attempt may result from any contact between male and female, irrespective of whether the male meets the female or vice versa. The beginning is signaled by the male gripping the female with his pedipalpal claws. This initial grip can occur around any part of her body. The female typically responds to the male by attempting to push him away with her tail, or even by trying to sting him. He responds by fending off her tail with his own, which serves to protect him from the female's sting and in some species allows him to wipe secretions from his caudal glands on the female. Such secretions perhaps function to reduce her aggression (Peretti, 1997). This tail-wrestling behavior, called *les arbres droit* by Fabre (1907), or clubbing (Polis and Farley, 1979a), especially when the scorpions are not attempting to sting, typically lasts 1.7 min in *Leiurus*. The duration of *les arbres droit* (LAD) is inversely proportional to the female's willingness to mate: hungry females fight for longer, and receptive females in the field do not always resist the male in this manner (Benton, 1990). The male responds to the female's attack by showing a gender-specific behavior, which occurs only in a courtship situation: he judders. A judder lasts 0.5–1 sec and consists of rapidly rocking backward and forward two to six times while keeping his legs immobile. As the female's attack is moderated, the judder changes in character. At the beginning, judders are rapid and of low amplitude, later they are slower, of greater amplitude, and involve the male retracting his pedipalps and pulling the female slightly toward him. This type of judder has been termed the judder-kiss (Benton, 1992c) because of its similarity to the kiss (see below). In an analysis of 35 courtships and aborted courtships, juddering was significantly associated with female aggressive behaviors ($p < 0.0001$), specifically LAD. As LAD typically occurs at the start of courtship, juddering can be thought of as an introductory/pacifying signal by the male. If the female is very receptive (as in some field-observed courtships) no LAD occurs and males do not judder. Two other descriptions of the courtship of this species (Shulov and Amitai, 1958; Abushama, 1968) report juddering by the male before female contact. This was not observed in my study (unless the male had met but lost the female) but may indicate the male's ability to identify chemical cues to the female's presence (Gaffin and Brownell, 1992). Juddering identifies gender, rather than species, as interspecies pairing may commonly occur in the laboratory, and in such situations the females may become more passive following juddering by heterospecific males. If the female becomes more passive, the male's rate of juddering slowly decreases, and he rearranges his grip to take the female's pedipalps in his own. After a mean of 6 min, the promenade phase begins.

Promenade Phase

The *promenade à deux* (Fabre, 1907) is the motile stage of courtship. This is also the most variable in length, both in laboratory and field. In the laboratory it lasted from 2.4 to 254 min (mean 29 ± SD 46, $n = 30$). The characteristic behavior during this phase is the male-led walk, during which the male walks backward with the female following. In the laboratory, I only allowed one site suitable for spermatophore deposition (a brick) in the courtship arena. The correlation between the length of the promenade phase and the time taken to encounter the brick is strong ($r_s = .891$, $n = 30$, $p < 0.001$), indicating that the assumption that the function of this phase is to find a suitable spermatophore deposition site is correct (Shulov and Amitai, 1958; Alexander, 1959; Polis and Farley, 1979a).

Two other main behaviors occur in this phase: the pause and the kiss (often in the sequence walk-pause-kiss). The pause occurs when the pair ceases all activity. I used the arbitrary criterion of a pause being a break in activity of greater than 5 sec. Pauses are not an artifact of the laboratory situation, as they have been observed in the field in *Leiurus, Euscorpius, Compsobuthus werneri* (Birula) (Benton, personal observation), and *P. mesaensis* (Polis and Farley, 1979a). However, long pauses (>1 min) occur when the courtship is not proceeding smoothly because the male cannot find a suitable site or when the female is restless. Long courtships often do not result in spermatophore deposition, and their length is generally a result of long pauses rather than more promenading.

The kiss (Fabre, 1923) (cheliceral massage of Polis and Farley, 1979a; Polis and Sissom, 1990) begins with the male retracting his pedipalps, causing the female to move toward him (though receptive females may initiate a kiss by walking toward the male first). The male slightly raises his anterior end and extends his chelicerae, which come close to and sometimes contact the female's. Simultaneously, he extends his first pair of legs under the female and rapidly moves his tarsi backward and forward over the female's genital region (termed "thrilling" by Shulov and Amitai [1958] and "feeling" by McAlister [1965]). On average kisses occurred 30 times per courtship, of which 11.2±12.6 ($n = 30$) occur during the promenade phase. Kisses last on average 6 sec in this species. Polis and Sissom (1990: p. 169) speculate that the kiss may function to "suppress the aggressive tendencies of the female." This speculation appears correct for an interesting reason.

The kiss is a displacement activity in that "it is an act that appears irrelevant or inappropriate to the context in which it occurs" (McCleery, 1983: p. 154). It appears to be derived from the behaviors associated with the actual process of sperm transfer. After the male has deposited the spermatophore, the female walks over the top of it. The spermatophore is then correctly aligned to the genital opening during a kiss (fig. 11.3 shows positioning of the spermatophore, but the postures of the animals are the same during a kiss). Kisses usually immediately follow pauses or follow a brief male-led walk after a pause. If the function of the promenade is to find a suitable site for sperm transfer and one is not im-

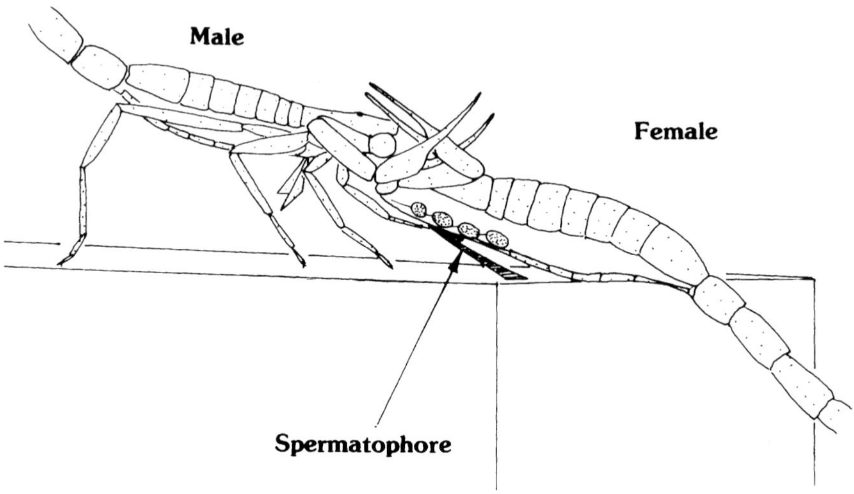

Figure 11.3 Preludes to sperm transfer in *L. quinquestriatus.* The male has deposited the spermatophore, and the female has moved over the top of it. The male uses his front legs to "thrill" the female, a process which seems to aid the correct positioning of the spermatophore. The female's legs are not drawn to aid clarity. This figure is traced from a video recording.

mediately encountered, the male may face an "approach–avoidance" motivational conflict: to stay or to flee (McFarland, 1985). This conflict may result in a pause if the "motivation to flee" balances the "motivation to continue with the courtship." A kiss could then occur through disinhibition (McCleery, 1983; Kennedy, 1985). Functionally, the male is mimicking the process of sperm transfer, "reminding" the female that he is a mate rather than potential prey. In one species I observed (*Centruroides margaritatus* [Gervais] from Costa Rica), the male begins thrilling before initiating the kiss, and thus for a few seconds his legs move forward and back over the soil surface before the female approaches. This could perhaps explain the observations of sand scraping made by several authors (e.g., Alexander, 1959): rather than functioning to clear the area for spermatophore deposition, perhaps sand scraping is a ritualization of the kiss, serving the same function. The cheliceral grip that occurs in many scorpions during courtship (Polis and Sissom, 1990) may also be derived in the same way.

Together the male-led walk, the pause, and the kiss make up 73% of the behaviors in the promenade phase. Additionally there are four others. The female-led walk occurs as a mild form of antagonistic behavior when she walks backward dragging the male with her. Tug-of-wars occur when the male and female each attempt to walk backward simultaneously. Male and female kisses also occur when either the male or female attempts to initiate a kiss without the other scorpion responding; they often occur within a pause. The female kiss (when she approaches the male until their chelicerae touch, without him having initiated a kiss) is an indication of her receptivity.

When a suitable spermatophore deposition site is encountered (something smooth and hard; e.g., hard mud, stones, wood, etc.) the character of the courtship changes and the third phase begins.

Spermatophore Deposition Phase

The spermatophore deposition phase is the least structured phase of the courtship, with behaviors occurring more frequently, and almost all behaviors being followed by any other. On average it lasted 12 ± SD 8.2 min in the 30 courtships. Several new behaviors occur. While on the suitable site the male-led walks may become very reduced, such that he moves around a centimeter or so in different directions, without causing the female to move. This is termed shifting. As sperm transfer approaches, the male reduces his movements still further and his body becomes immobile, but his pedipalps continue to move forward and backward, causing the female's to do likewise. This is "arm pulling." These behaviors seem to be associated with the male finding the exact spot on which to deposit the spermatophore. This may involve mechanoreceptors, which are found on the pectines (Carthy, 1968) and possibly the tarsi (Abushama, 1964). Another behavior (alternate arm pulling), in which the male alternatively protracts and retracts his right and left pedipalps, is a good indicator of the onset of spermatophore deposition. Also at this time, the male's tail is continually raised and lowered over his back (tail waving), a behavior that becomes more stereotyped as spermatophore deposition approaches. Both these latter behaviors also occur in *E. flavicaudis* (Benton, 1993) at the same stage of courtship and so perhaps are associated with the expulsion of the spermatophore from the paraxial organs. Perhaps rhythmic movements of the tail and pedipalps raise the local hemocoelic pressure and help to pump out the spermatophore.

The deposition of the spermatophore takes between 35 and 105 sec (cf. Shulov and Amitai, 1958). Soon after alternate arm-pulling begins, the male presses his genital opening against the substrate, which sticks the base of the spermatophore to the ground. He then gradually lifts his prosoma upward and backward (fig. 11.4). The spermatophore consists of four main regions: base, trunk, sclerotized capsule, and flagella. In the lower part of the trunk is a jelly mass, acting like a hydrostatic skeleton, on top of which is the sperm mass. Sperm exits through the capsule. The flagella remain connected to the male's genital region during sperm transfer and presumably communicate the spermatophore's position to him and perhaps keep it correctly positioned. Once the spermatophore is deposited, the female usually initiates sperm transfer by walking forward and over it (as in a female kiss). Sometimes the male has to pull her over (as in a male kiss). During the following kiss the movements of the male's first legs perhaps ensure the correct positioning of the spermatophore in the female's genital opening. Sperm transfer occurs by the male sharply pushing the female backward. This makes the spermatophore pivot at its base, which causes longitudinal compressive forces within it. As the jelly mass in the spermatophore's trunk does not compress, it acts as a plunger, pumping the sperm mass into the female (fig. 11.5).

After sperm transfer has been accomplished, the pair usually parts quickly; either the male immediately releases and flees, or the female attacks the male, causing him to release and flee. No postreproductive cannibalism was observed in this species; neither was the spermatophore ever eaten (in contrast to, e.g., *E. flavicaudis;* Benton, 1993). Following the end of the courtship, the female (or occasionally the male) may lower her ventral surface and sway from side to side, rubbing it against the ground. The function of this behavior is unclear.

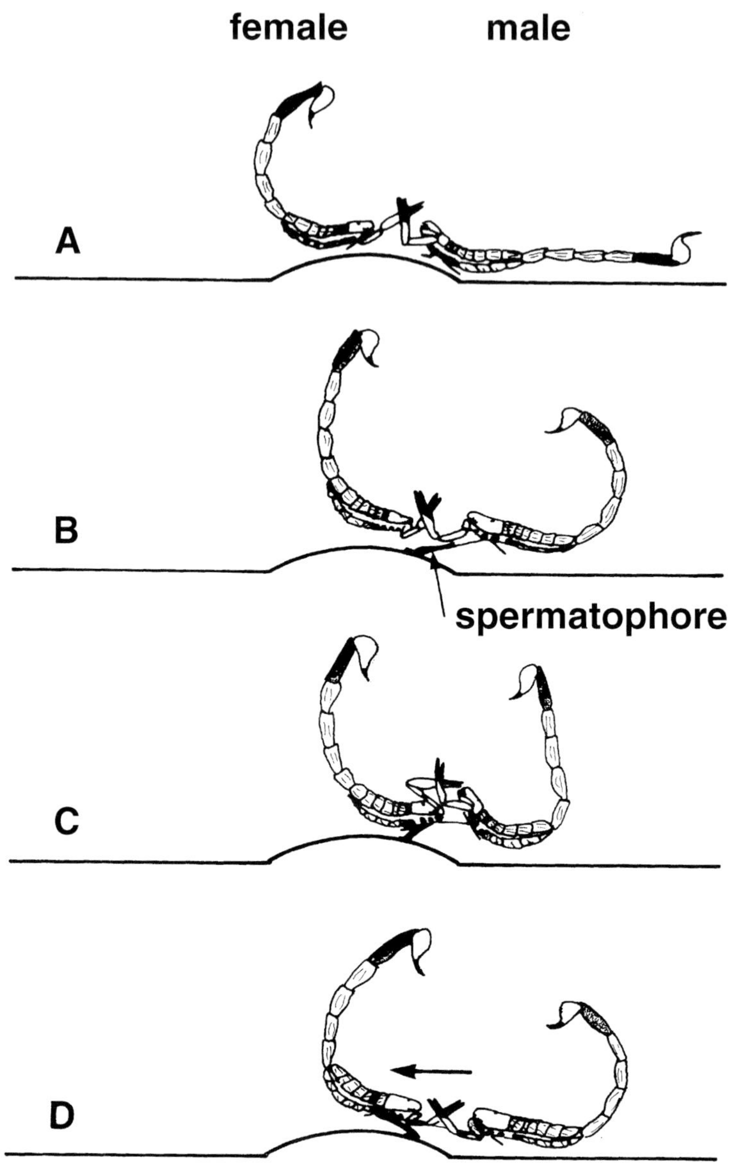

Figure 11.4 The process of spermatophore deposition and sperm transfer in *Leiurus quinquestriatus.* (A) The male begins deposition by expelling the base of the spermatophore, which he presses against the ground. (B) The male then moves backwards, drawing the spermatophore from his genital opening. (C) The female advances over the top of the spermatophore, so that the tip is located in her genital opening. (D) Once the spermatophore is correctly positioned, the male suddenly pushes the female backwards. This sets up compressive forces in the spermatophore which ejects the sperm (fig. 11.5). To aid clarity, the scorpions legs are not drawn. A–C after Shulov and Amitai (1958).

General Discussion of Courtship

Broadly speaking, scorpion courtships occur similarly across the order (the main exception seems to be in chactid scorpions; see Benton, 1993). Although experimentally little studied, the functional significance of the courtship and its components is becoming clearer.

It is possible that a behavior sequence of this complexity is a result of runaway, Fisherian sexual selection, in a similar manner to the explanation often given for the bright colors and long tails of male birds. If this were the case, females would chose males on the basis of their behavior during courtship, with the behaviors having no obvious proximal function. However, in *Leiurus* a male's mating success in the laboratory is proportional to the number of courtship attempts he makes ($r_s = .764$, $n = 11$, $p < 0.01$), not to any correlates of courtship behavior or morphology. Highly motivated males may attempt to mate with anything (males, females of other species) and are less likely to give up if the female is initially aggressive. If a male successfully pacifies a female and the promenade phase of the courtship begins, then more than 90% of the time the courtship will be completed. This means that the female is unlikely to be using the behaviors beyond the introductory phase to assess the males. Similar results were found with *Euscorpius*. We can also hypothesize functional explanations for the behaviors shown during the courtship. There seems to be little nonfunctional complexity: behaviors are either pacifying (judders, kisses), for locating a spermatophore deposition site (male-led walk, etc.), or depositing

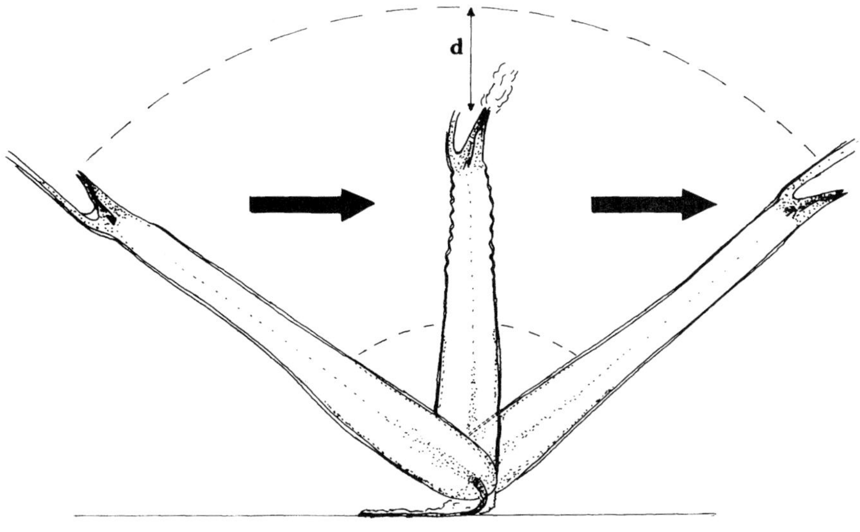

Figure 11.5 The mechanism of action of the flagelliform spermatophore of *Leiurus quinquestriatus.* The base of the spermatophore is glued down, and the top is anchored in the female's genital opening. During sperm transfer, the male suddenly pushes the female backward, causing the spermatophore to pivot at the pedal flexure. As the female does not raise her body during the push, the spermatophore is compressed by length *d.* The base of the spermatophore is filled with an incompressible jelly mass, which acts as a plunger in a syringe during the spermatophore's compression, ejecting the sperm into the female.

the spermatophore (arm-pulling, etc.). Extreme lengths of courtship occur if a suitable spermatophore-deposition site is not encountered, or the animals are unhealthy; not because the male is "showing off."

SEXUAL CANNIBALISM

Scorpions have a reputation for sexual cannibalism (Gould, 1984). In Polis and Sissom (1990, Table 4.1), 11/29 species are noted to exhibit mate cannibalism, and only 3/29 are noted not to exhibit mate cannibalism (i.e., 38–79% are sexual cannibals, depending whether the null records are considered). A recent review of this topic (Peretti et al., 1999) indicates, however, that these figures are likely to be overestimates. For three species identified as sexual cannibals in the literature, detailed investigation of the primary sources coupled with new studies showed that post-mating cannibalism was very rare (and its incidence is likely to be inflated by unnatural laboratory conditions). For another 19 species that the authors had observed courting, no cases of sexual cannibalism were observed (Peretti et al., 1999). It is thus likely that sexual cannibalism is an exception rather than the rule in scorpions.

Although rare, cannibalism of males by females in a sexual contest has been observed in the field (e.g., Polis and Farley, 1979). However, these observations do not indicate that such cannibalism is an adaptive cannibalism, where males sacrifice themselves in order to provide nutrients for their offspring (as seen, for example, in redback spiders [Forster, 1992]). The available evidence indicates that males are unwilling victims and try to escape (Peretti et al., 1999). Given scorpion longevity and the possibility of mating multiple times even within a season, male self-sacrificing behavior is unlikely to be naturally selected.

SIZE AND REPRODUCTIVE SUCCESS

There is a well-known relationship between the size of a female and her reproductive success among terrestrial arthropods (e.g., spiders: Bristowe 1958; Wise, 1976; Rubenstein, 1987; Elgar and Nash, 1988; insects: Hinton, 1981). This relationship generally applies in scorpions: larger females have more young (Benton, 1991b; Brown and Formanowicz, 1995; Myers, this volume) and/or heavier young (Brown and Formanowicz, 1995; Polis and McCormick, 1987; but see Bradley, 1984). Often male size also influences reproductive success (Vollrath, 1980; Thornhill and Alcock, 1983; Rubenstein, 1987; Watson, 1990). In *Leiurus* in the laboratory, I found no such effect. Under more natural conditions a strong effect was observed in *Euscorpius flavicaudis:* males that deposited two spermatophores in one season had on average larger claws than those that deposited one, and their claws were in turn larger than those of a random sample of scorpions that deposited no spermatophores (fig. 11.6).

The relationship between male size and reproductive success in *E. flavicaudis* exists for several reasons. Large males can win male–male contests. Large males may be less prone to cannibalism than small males (see Myers, this volume) and so may be able to be wander more in search of mates than small males (as in the spider *Nephila:* large males move from web to web, small males stay longer with unreceptive females; Vollrath, 1980). Larger males may be better at pacifying ini-

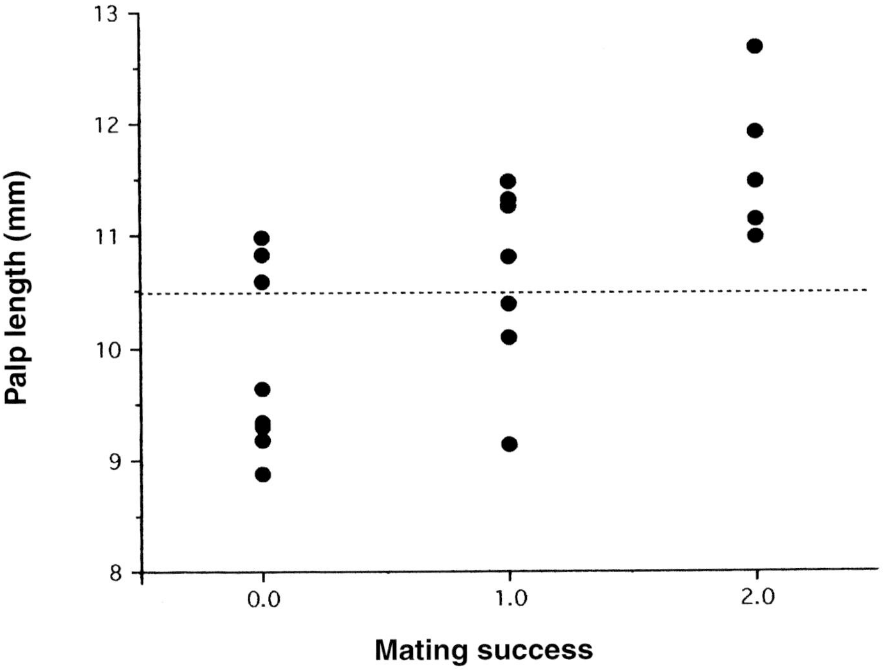

Figure 11.6 The relationship between male claw size and mating success in *Euscorpius flavicaudis.* Males that mated twice ($n = 6$) over a season in a seminatural laboratory colony had larger claws than those that mated once ($n = 8$) which, in turn, had larger claws than a random sample of males present for the whole season that did not mate ($n = 10$; $H_2 = 11.4$, $P < .005$). Claw size is represented by length (in millimeters), but a similar result occurs if any index of claw size is used. There are two instars of adult males in this population, the dotted line representing the approximate border between them (Benton, 1991b).

tially aggressive females (such as unmated females emerging after maternal care) because they are less susceptible to cannibalism and need not retreat at the first sign of female aggression. Large males may also be more able to obtain and store reserves needed to produce spermatophores—a process that takes a minimum of 4 days in *Euscorpius* and *Leiurus* and 4–8 days in a range of buthriurids (Peretti et al., 1999). For these reasons, a relationship between male size and mating success may generally exist in scorpions (see Myers, this volume).

Adult male *Euscorpius* vary markedly in size. The smallest adult male averaged 58 ± SD 9% of the size of the largest male for four measures of size. Adult males are easily distinguished from subadults and from females by the secondary sexual characteristic of a notch in the pedipalpal fingers and elongation of the pectines (Benton, 1991b). Males observed to produce spermatophores spanned the whole range of male sizes (Benton, 1991b). This size range is explained by the presence of two instars of adult male in the population: sixth and seventh instars (Benton, 1991b). On average (based on four measures of size), seventh instar males are 1.20 times larger than sixth instar males.

THE INFLUENCE OF REPRODUCTIVE ECOLOGY ON LIFE-HISTORY

The observation that there may be more than one instar of adults in an arachnid population is not new (see Myers, this volume). This form of life-history variation is common, occurring in about 75% of spider and almost 50% of scorpion species where life-history data are available (Benton, unpublished data). However, the existence of two instars of adult male in *Euscorpius* raises an interesting evolutionary conundrum. Postmaturity molts are unknown in scorpions (Polis and Sissom, 1990), so when a male matures, he also stops growing (at least exoskeletally). The paradox is, if large males enjoy a higher per-season mating success, why do any males mature at a smaller, disadvantageous size? Two observations are important in explaining why this may not be as paradoxical as it appears.

First, reproductive success is likely only to be a good measure of fitness in a restricted number of cases, as it does not take into account the timing of reproduction, the quality of the offspring, or environmental variabilty (Benton et al., 1995). For example, it may pay the scorpions to mature early and receive a lower reproductive success if:

a. the population is growing (analagous to putting your money in the bank to gain interest: the earlier the better),
b. survival is unpredictable, so an animal delaying maturity may risk death,
c. they gain more time in which to reproduce. In this case the lifetime reproductive success of early maturers may be similar to those that delay maturity. The ratio of early to late maturers would be balanced by frequency dependent selection at the evolutionarily stable state (ESS).

Second, it is notable that environments vary. In a variable environment the optimal life history may differ considerably from that which would be optimal in a constant environment (Benton and Grant, 1999), with the result that bet-hedging polymorphic phenotypes may evolve (Ellner and Sasaki, 1996). If the environment is unpredictable, the best genetic strategy may be to have a mix of early maturers (who are likely to survive, even though they breed little) and late maturers (who are more likely to die before maturity, but if they do they breed a lot). In a variable environment using environmental clues to map phenotype to the specific environmental state may be adaptive. Phenotypic plasticity is the range of phenotypes expressed by a genotype across a range of environments (Stearns, 1992) and the adaptive rule linking phenotype to environment (reaction norm) may be very complex (e.g., Stearns and Koella, 1986).

The environment is known to influence arachnid life histories (for reviews see Polis and Sissom, 1990), and it is obvious that growth rates vary with temperature and food availability (Browning, 1941; Deevey, 1949; Tanaka, 1991; Myers, this volume). However, what is not obvious is whether a fast-growing animal should mature at the same time as other animals (and be larger), or mature at the same size as other animals (and mature earlier), or whether fast- and slow-growing animals should mature at different times and sizes.

Myers (this volume) and Francke and Jones (1984) indicate that there is a strong condition-dependence to scorpion maturation. Fast-growing scorpions tend to mature early, and therefore appear to trade-off size-related reproductive gains against the fitness benefits of early reproductive success. Condition-dependent maturation is also the general result reported in the spider literature (e.g., Deevey, 1949; Vollrath, 1980; Wise, 1987). The matter is further complicated by Ramouse's (1972) study of spiders. He reared spiders under identical conditions and found that spiders ate at different rates and so matured at different times. He concluded that the feeding rate was genetically fixed.

There are therefore a number of candidate explanations for polymorphic adult sizes in scorpions. What is now needed is a rigorous experimental approach to attempt to disentangle the different causes. Do 6th instar males have similar lifetime reproductive success to 7th instar males (in which case frequency dependence could explain the polymorphism)? If reproductive success differs between the morphs, is there a genetic component to maturation strategy (in which case, it could be a bet-hedging polymorphism)? Do different morphs occur with different growth rates, and if so what is the reaction norm?

CONCLUSIONS

From these observations we may conclude that the patterns of behavior surrounding reproduction in scorpions are more complicated than previously realized. Matings generally follow female maturation or birth of young and are preceded by a period of male vagrancy. Under some situations, males may compete to mate guard females that are caring for their young. The actual courtship consists of a complicated sequence of behaviors, each with a different function. Courtship does not appear to be used by females to assess male quality. Under certain circumstances, especially if competition for mates is strong, males may benefit by delaying maturity to gain size. Therefore adult scorpion size may sometimes reflect reproductive strategy, as well as such factors as food availability.

Because of these characteristics reproductive ecology of scorpions is a fertile field for researchers with an interest in the link between ecology and evolution. Although little research has been conducted on the reproductive ecology of scorpions, it is now clear that the situation can be more complex and interesting than a simple "male meets female, then mates" scenario: reproductive success, and perhaps reproductive strategy, may depend on an individual's size; size may depend on the size of other members of the population, as well as on the environmental conditions.

The highly complex and lengthy courtship behavior of scorpions seems to be a result of various proximal factors: the need for the male to identify himself to the female and pacify her initial aggression; the need to locate a suitable site on which a spermatophore can be deposited; the need to position the spermatophore exactly and transfer the sperm successfully, and so on. There seems to be little evidence that the complex courtship allows females to assess male quality; instead, either male motivation, luck (encountering receptive females), or male–male competition determines male reproductive success.

There are many questions that are prompted by what we now know. It appears that in some species sexual selection may be quite strong, yet in others it may be weak or nonexistent. Pinpointing the ecological conditions promoting sexual selection is a realistic and worthwhile goal of future research. One of the most interesting questions that arises concerns the link between sexual strategy and life-history evolution. Natural selection occurs because of the fitness benefits accrued, but the proper definition of fitness remains elusive. Obviously, fitness is related to reproductive success, but it is also related to the time that the success is gained and also to the variance in reproductive success of the strategy over time. Scorpions that show a range of life-history tactics within populations present a useful seam that can be exploited in the investigation of linkage between life-history evolution and fitness. The fact that some scorpions mature at a time and size that apparently does not maximize their reproductive success begs a question about natural selection and the definition of fitness that is of enormous widespread interest.

Acknowledgments Thanks are due to many institutions and people for helping with this study. The Mitrani Center for Desert Ecology of the Ben Gurion University of the Negev provided me with accommodation while I collected and studied *Leiurus* in the field. The Medway Ports Authority and gardeners at Sheerness gave me permission to study *Euscorpius* on their land. The study was supported by SERC and University of Cambridge. Special thanks go to Matthew Evans and Andrew Bourke for the hours of discussion and help they gave me and especially to William Foster, my PhD supervisor, friend, and colleague. Sarah Benton, James Deutsch, Gary Polis, and Philip Brownell read the manuscript. Sarah Benton provided much emotional and material support throughout this study, in return for an absentee husband. Without her it would not have happened. Gary Polis willingly offered advice over many years. He is greatly missed.

References

Abushama, F.T. (1964) On the behavior and sensory physiology of the scorpion *Leiurus quinquestriatus* (H. and E.). *Anim. Behav.* 12: 140–153.

Abushama, F.T. (1968) Observations on the mating behavior and birth of *Leiurus quinquestriatus* (H. and E.), a common scorpion species in the Central Sudan. *Rev. Zool. Botan. Afr.* 77: 37–43.

Alexander, A.J. (1959) Courtship and mating in the buthid scorpions. *Proc. Zool. Soc.* (London) 133: 145–169.

Angermann, H. (1957) Uber Verhalten, Spermatophorenbildung und Sinnesphysiologie von *Euscorpius italicus* Herbst und verwandten Arten (Scorpiones, Chactidae). *Z. Tierpsychol.* 14: 276–302.

Austadt, S.N. (1982) First male sperm priority in the bowl and doily spider, *Frontinella pyramitela* (Walckenaer). *Evolution* 36: 777–785.

Austadt, S.N. (1984) The evolution of sperm priority patterns in spiders. *In* Smith, R.L. (ed.) *Sperm Competition and the Evolution of Animal Mating Systems,* pp. 223–247. Academic Press, New York.

Benton, T.G. (1990) The behaviour and ecology of scorpions (PhD thesis). University of Cambridge.

Benton, T.G. (1991a) Reproduction and parental care in the scorpion, *Euscorpius flavicaudis. Behaviour* 117: 1–2.

Benton, T.G. (1991b) The life history of *Euscorpius flavicaudis* (Scorpiones, Chactidae). *J. Arachnol.* 19: 105–110.

Benton, T.G. (1992a) The ecology of the scorpion *Euscorpius flavicaudis* in England. *J. Zool.* 226: 351–368.

Benton, T.G. (1992b) Determinants of male mating success in a scorpion. *Anim. Behav.* 43: 125–135.

Benton, T.G. (1992c) Courtship and mating in *Leiurus quinquestiatus* (Scorpiones; Buthidae). *In* Cooper, J.E., Pearce-Kelly, P. and Williams, D.L. (eds.) *Arachnida,* pp. 83–98. Chiron Publications, London.

Benton, T.G. (1993) The courtship behavior of the scorpion, *Euscorpius flavicaudis. Bull. Br. Arachnol. Soc.* 9: 137–141

Benton, T.G. and Grant, A. (1999) Optimal reproductive effort in stochastic, density-dependent environments. *Evol.* 53: 677–688.

Benton, T.G., Grant, A. and Clutton-Brock, T.H. (1995) Does environmental stochasticity matter? Analysis of red deer life-histories on Rum. *Evol. Ecol.* 9: 559–574.

Bradley, R.A. (1984) The influence of the quantity of food on fecundity in the desert grassland scorpion (*Paruroctonus utahensis*) (Scorpionida, Vaejovidae): an experimental test. *Oecologia* 62: 53–56.

Bristowe, W.S. (1958) *The World of Spiders.* Collins, London.

Brown, C. A. and Formanowicz, D.R. (1995) Variation in reproductive investment among and within populations of the scorpion Centruroides vittatus. *Oecolgia* 103: 140–147.

Brownell, P.H. and Farley, R.D. (1979) Prey-localizing behavior of the nocturnal desert scorpion, *Paruroctonus mesaensis: orientation to substrate vibrations. Anim. Behav.* 27: 185–193.

Browning, H.C. (1941) The relation of instar length to the external and internal environment in *Tegenaria atrica. Proc. Zool. Soc.* (London) 111: 303–317.

Carthy, J.D. (1968) The pectines of scorpions. *Symp. Zool. Soc. Lond.* 23: 251–261.

de Armas, L.F. (1987) Copula multiple en escorpiones (Arachnida: Scorpiones). *Acad. Cienc. Cuba, Misc. Zool.* 30: 1–2.

Deevey, G.B. (1949) The developmental history of *Latrodectus mactans* (Fabr.) at different rates of feeding. *Am. Midl. Nat.* 42: 189–219.

Elgar, M.A. and Nash, D.R. (1988) Sexual cannibalism in the garden spider *Araneus diadematus. Anim. Behav.* 36: 1511–1517.

Ellner, S. and Sasaki, A. (1996) Patterns of genetic polymorphism maintained by fluctuating selection with overlapping generations. *Theor. Pop. Biol.* 50: 31–65.

Fabre, J.H. (1907) *Souvenirs Entomologiques, 9me serie,* edition definitive. Delagrave, Paris.

Fabre, J.H. (1923) *The Life of the Scorpion.* A.T. de Mattos, trans. Hodder and Stoughton, London.

Forster, L.M. (1992) The stereotyped behaviour of sexual cannibalism in Latrodectus hasselti (Thorell): the Australian redback spider. *Aust. J. Zool.* 40: 1–11.

Francke O.F. and Jones, S.K. (1984) The life history of *Centruroides gracilis* (Latreille). *J. Arachnol.* 10: 223–239.

Gaffin, D.G. and Brownell, P.H. (1992) Evidence of chemical signalling in the sand scorpion, *Paruroctonus mesaensis* (Scorpionida: Vaejovida). *Ethology* 91: 59–69.

Gould, S.J. (1984) Only his wings remained. *Nat. Hist.* 93: 10–18.

Grafen, A. and Ridley, M. (1983) A model of mate guarding. *J. Theor. Biol.* 102: 549–567.

Hammerstein, P. and Riechert, S.E. (1988) Payoffs and strategies in territorial contests: ESS analysis of two ecotypes of the spider *Agelenopsis aperta. Evol. Ecol.* 2: 115–138.

Hinton, H.E. (1981) *Biology of Insect Eggs,* vol. 1. Pergamon Press, Oxford.

Kovoor, J., Lourenço, W.R. and Munoz-Cuevas, A. (1987) Conservation des spermatozoides dans les voies genitales des femelles et biologie de la reproduction des scorpions (Chelicerates). *C. Re. Acad. Sci.* (Paris), ser. 3, 10: 259–264.

Kennedy, J.S. (1985) Displacement activities and post-inhibitory rebound. *Anim. Behav.* 33: 1375–1377.
Lamoral, B.H. (1979) The scorpions of Namibia (Arachnida, Scorpionida). *Ann. Nat. Mus.* 23: 497–784.
Levy, G. and Amitai, P. (1980) *Fauna Palaestina: Arachnida I: Scorpiones.* Israel Academy of Sciences and Humanities, Jerusalem.
Maynard Smith, J. and Parker, G.A. (1976) The logic of asymmetric contests. *Anim. Behav.* 24: 159–175.
McAlister, W.H. (1965) The mating behavior of *Centruroides vittatus* Say (Arachnida: Scorpionida). *Texas J. Sci.* 17: 307–312.
McCleery, R.H. (1983) Interaction between activities. *In* Halliday, T.R. and Slater, P.J.B. (eds.) *Animal Behavior: 1, Causes and Effects,* pp. 132–165. Blackwell Scientific, Oxford.
McDaniel, M.M. (1968) Notes on the biology of California scorpions. *Entomol. News* 79: 278–284.
McFarland, D.J. (1985) *Animal Behaviour.* Longman, Essex.
Parker, G.A. (1984) Sperm competition and the evolution of animal mating strategies. *In* Smith, R.L. (ed.) *Sperm Competition and the Evolution of Animal Mating Systems,* pp. 2–55. Academic Press, New York.
Peretti, A.V. (1991) Comportamiento de apareamiento de *Zabius fuscus* (Thorell) (Buthidae, Scorpiones). *Bol. Soc. Biol. Concepcion, Chile,* 62: 123–146.
Peretti, A.V. (1997) Relacion de las glandulas caudales de machos escorpiones Bothriuridae con el comportamiento sexual (Scorpiones). *Rev. Arachnol.* 12: 31–41.
Peretti, A.V., Acosta, L.E. and Benton T.G. (1999) Sexual cannibalism in scorpions: fact or fiction? *Biol. J. Linn. Soc.,* 68: 485–496.
Pianka, E.R. (1970) On r- and K- selection. *Am. Nat.* 104: 592–597.
Polis, G.A. (1980) The significance of cannibalism on the demography and activity in a natural population of desert scorpions. *Behav. Ecol. Sociobiol.* 7: 25–35.
Polis, G.A. (1990) Ecology. *In* Polis, G.A. (ed.) *The Biology of Scorpions.* Stanford University Press, Palo Alto, CA.
Polis, G.A. and Farley, R.D. (1979a) Characteristics and environmental determinants of natality, growth and maturity in a natural population of the desert scorpion, *Paruroctonus mesaensis* (Scorpionida: Vaejovidae). *J. Zool.* 187: 517–542.
Polis, G.A. and Farley, R.D. (1979b) Behavior and ecology of mating in the cannibalistic scorpion, *Paruroctonus mesaensis* Stahnke (Scorpionida: Vaejovidae). *J. Arachnol.* 7: 33–46.
Polis, G.A. and Farley, R.D. (1980) Population bioloy of a desert scorpion: survivorship, microhabitat and the evolution of life history strategy. *Ecology* 61: 620–629.
Polis, G.A. and McCormick, S.J. (1987) Intraguild predation and competition among desert scorpions. *Ecology* 68: 332–343.
Polis, G.A., McReynolds, C.N. and Ford, R.G. (1985) Home range geometry of the desert scorpion *Paruroctonus mesaensis. Oecologia* 67: 273–277.
Polis, G.A. and Mohnac, M. (1990) Mating by female scorpions while still carrying young. *J. Arachnol.* 18: 363–364.
Polis, G.A. and Sissom, W.D. (1990) Life history. *In* Polis, G.A. (ed.) *The Biology of Scorpions,* pp. 161–223. Stanford University Press, Palo Alto, CA.
Probst, P.J. (1972) On the reproduction of the scorpion *Isometrus maculatus* and on the development of its venom glands. *Acta Tropica* 29: 1–87.
Ramousse, R. (1973) Body, web-building and feeding characteristics of males of the spider *Araneus diadematus* (Araneae: Araneidae). *Psyche* 80: 22–47.

Ridley, M. (1983) *The Explanation of Organic Diversity.* Oxford University Press, Oxford.

Rubenstein, D.I. (1987) Alternative reproductive tactics in the spider *Meta segmentata. Behav. Ecol. Sociobiol.* 20: 229–237.

Shachak, M. and Brand, S. (1983) The relationship between sit and wait foraging strategy and dispersal in the desert scorpion, *Scorpio maurus palmatus. Oecologia* 60: 371–377.

Shaffer, L.R. and Formanowicz, D.R. (1996) A cost of viviparity and parental care in scorpions: reduced sprint speed and behavioural compensation. *Anim. Behav.* 51: 1017–1024.

Shulov, A. and Amitai, P. (1958) On mating habits of three scorpions: *Leiurus quinquestriatus* H. et E., *Buthotus judaicus* E. Sim. and *Nebo hierichonticus* E. Sim. *Arch. Inst. Pasteur d'Algerie* 36: 351–369.

Smith, G.T. (1966) Observations on the life history of the scorpion *Urodacus abruptus* Pocock (Scorpionidae) and an analysis of its home sites. *Austral. J. Zool.* 14: 383–398.

Stearns, S.C. (1992) *The Evolution of Life Histories.* Oxford University Press, Oxford.

Stearns, S.C. and Koella, J. (1986) The evolution of phenotypic plasticity in life-history traits: predictions for norms of reaction for age- and size-at-maturity. *Evolution* 40: 893–913.

Tanaka, K. (1991) Diapause and seasonal life-cycle strategy in the house spider, *Achaeranea tepidariorum* (Aranea, Theridiidae). *Physiol. Entomol.* 16: 249–262

Thomas, R.H. and Zeh, D.W. (1984) Sperm transfer and utilisation strategies in Arachnids; ecological and morphological constraints. *In* Smith, R.L. (ed.) *Sperm Competition and the Evolution of Animal Mating Systems,* pp. 180–216. Academic Press, New York.

Thornhill, R. and Alcock, J. (1983) *The evolution of insect mating systems.* Harvard University Press, Cambridge.

Tourtlotte, G.I. (1974) Studies on the biology and ecology of the Northern Scorpion, *Paruroctonus boreus* (Girard). *Great Basin Natural.* 34: 167–179.

Vollrath, F. (1980) Male body size and fitness in the web-building spider *Nephila clavipes. Z. Tierpsychol.* 53: 61–78.

Walker, W.F. (1980) Sperm utilisation strategies in nonsocial insects. *Am. Nat.* 115: 780–799.

Warburg, M.R. and Elias, R. (1998) The reproductive potential and strategy of *Scorpio maurus fuscus* (Scorpiones: Scorpionidae): anatomical clues in the ovariuterus. *J. Zool, Lond.* 246: 29–37.

Watson, P.J. (1990) Female-enhanced male competition determines the first mate and principal sire in the spider *Linyphia litigiosa* (Linyphiidae). *Behav. Ecol. Sociobiol.* 26: 77–90.

Williams, S.C. (1969) Birth activity of some North American scorpions. *Proc. Calif. Acad. Sci.* 37: 1–24.

Williams, S.C. (1987) Scorpion bionomics. *Annu. Rev. Entomol.* 32: 275–295.

Wise, D.H. (1976) Variable rates of maturation of the spider, *Neriene radiata* (*Linyphia marginata*). *Am. Midl. Nat.* 96: 66–75.

Wise, D.H. (1987) Rearing studies with a spider exhibiting a variable phenology: no evidence of substantial genetic variation. *Bull. Br. Arachnol. Soc.* 7: 107–110.

12

Population and Community Ecology of Desert Scorpions

GARY A. POLIS

Ecology is a rapidly developing field with a healthy dose of controversy. As in other fields, theory and speculation has advanced far more rapidly than empirical work, leaving a vacuum that can only be filled by experimental work. Scorpions possess a number of characteristics that make them good models to test and advance ecological theory (see Introduction to this volume). Thus, it is relatively easy to collect data and to experimentally manipulate individuals and entire populations. Research on scorpions has contributed to many areas of population, behavioral, and community ecology, including life history theory (Polis and Farley, 1980; Bradley, 1986), the evolution and ecology of age-structured populations (Polis 1984a, 1988; McCormick and Polis, 1990), the dynamics of cannibalism (Polis, 1980a, 1981, 1984b), and intraguild predation (Polis and McCormick 1986b, 1987; Polis et al., 1989), foraging theory (Polis, 1980b; Bradley, 1988), patterns and processes in biogeography (Due and Polis, 1986), and the patterns and processes affecting community and food web structure (Bradley, 1983; Polis, 1991a).

This chapter evaluates the ecological importance of scorpions and how research on them has advanced our general understanding of ecology. "Ecological importance" can be expressed in terms of either energetics or nutrient cycling (the quantity of matter and energy flowing through a population or functional group) or regulation of community structure and dynamics (the impact on diversity, distribution, and abundance of other populations). Although scorpions occur in nearly all nonboreal habitats, desert scorpions are a particularly important group in arid ecosystems in terms of both energetics and dynamics. They are diverse, some taxa are abundant, and, as a group, they form a large proportion of the biomass of all desert arthropods and often exceed the biomass of all desert vertebrates. Their success arises partially from the suite of physiological and ecological traits that preadapt them to the low and unpredictable food availability of deserts.

ENERGETICS AND NUTRIENT CYCLING

The importance of desert scorpions for energy and nutrient cycling is a function of the quantity of prey biomass captured. The amount of captured prey biomass is a function of predator density, population biomass, metabolism, and efficiency of energy transfer. To present a rough approximation of the relative importance of scorpions to the flow of energy and nutrients in desert ecosystems,

Table 12.1 Estimated average density and diversity of major taxa of desert macrofauna reported from the literature

Group	Density/ha[a]	Diversity[b]	n^c
Mammals	43 ± 130	28.5 ± 21.3	14
Birds	0.6 ± 1.1	46.7 ± 27.3	14
Lizards	36.8 ± 43	12 ± 4.6	12
Insects	35,100 ± 54600	592 ± 796	10
Termites	4,025,000 ± 2,793,000	9.1 ± 15	7
Isopods	241,810 ± 336,850	1.3 ± 0.8	7
Millipedes	1,150 ± 214	2.4 ± 2.1	7
Spiders	3,220 ± 8,800	54.4 ± 26.7	31
Solpugids		9.9 ± 6.7	37
Scorpions	3,210 ± 3,500	7.1 ± 2.5	58

[a]Density statistics were standardized to a per hectare basis. Note that data on ants were insufficient to include here, and no data are available on solpugid density. Means are reported ± SD. Note that these statistics are rough approximations (see Polis and Yamashita, 1991, for details).

[b]Diversity is the mean number of species per taxon from local sites in different deserts.

[c]Sample size.

I compiled from the literature data on diversity, density, and biomass of various groups of desert taxa (see Polis, 1990, 1991b; Polis and Yamashita, 1991, for additional details). These data are not without large uncertainties: for example, studies are not usually conducted in areas where the focal taxon is absent or rare, nor are rare species studied as frequently or as well as common ones. Consequently, the statistics presented here overestimate density and biomass and should be taken as an approximation of actual parameters. However, these data are instructive and suggest that scorpions could be a major link in the flow of energy through desert communities.

How do desert scorpions measure against other groups of consumers in terms of diversity, density, population biomass, metabolism, and energy transfer? First, scorpions are relatively speciose in deserts. On the average, 7.1 species co-occur in deserts throughout the world (range: 2–16 sympatric species, typically, 5–9 species; Polis, 1990; see table 12.1 for comparison with other desert taxa). Furthermore, populations are often dense, averaging >3200 individuals/ha with several species maintaining populations of 5000 to >10,000/ha (e.g., Shorthouse, 1971; Lamoral, 1978; Polis and Farley, 1980; Bradley, 1986; Polis and McCormick, 1986a; Polis, 1990). On average, scorpions are reportedly more dense than all other macroscopic animal taxa in these deserts except "insects," termites, and isopods (ants are undoubtedly also more dense, and there are insufficient data to evaluate spiders; table 12.1). Because scorpions are among the largest of all terrestrial arthropods (adults of most desert species weigh approximately 0.5–10 g; Polis and Farley, 1980), these high densities produce large estimates of standing biomass (= density of individual species × mass of individual animal). Each species population of desert scorpion averaged 7.15 kg/ha. Only termites (and probably ants) support a greater population biomass per species or per taxon (from the ordinal level down) than scorpions (table 12.2).

Table 12.2 Estimated population biomass for various taxa

Taxon	Population biomass (kg/ha)[a] per species	per taxon	*n*
Mammals	1.40	39.9	29
Birds	0.02	0.9	25
Lizards	0.57	6.8	46
Insects	0.88	521.2	31
Termites	12.45	113.4	7
Isopods	9.91	12.8	2
Millipedes	1.15	2.8	2
Spiders	0.13	7.1	4
Scorpions	7.15	50.8	17
All vertebrates		47.7	
All arthropods		708.1	
Total macrofauna		755.8	

[a]The population biomass per species is the average wet weight of one species per hectare. These statistics are reported or calculated as the product of average mass of an individual times its population density. Population biomass per taxa is the product of the population biomass per species and average diversity of that taxon. Data on ants were insufficient to include here. Additional information is reported in Polis and Yamashita (1991). *N* refers to the number of species in a particular taxon for which biomass data exist. Note that these statistics are approximations.

When the biomass of all scorpions living sympatrically in desert areas is calculated (=mass of individual species × average number of sympatric species), scorpions as a group have a greater biomass (= 50.8 kg/ha) than all other taxa except termites (=113.4 kg/ha) and the sum of all other insects (=521.2 kg/ha; table 12.2). Note also that the population biomass of scorpions is higher than all vertebrates combined (47.7 kg/ha). Overall, scorpions make up an estimated 6.7% of the biomass of all macrofauna species combined, 7.1% of the biomass of all macroarthropods, and 106% of the biomass of all vertebrates.

Thus, by abundance, scorpions appear to be important conduits for energy transfer in deserts. However, two characteristics of their biology mitigate this potentially important ecological role. First, they exhibit the highest ecological (production) efficiencies (percent assimilated energy incorporated into new biomass) of all taxa that live in the desert (table 12.3). Second, scorpions display very low metabolic rates relative to other poikilotherms and endotherms (table 12.4). These features powerfully adapt scorpions to succeed in deserts because they ensure optimal use of food resources. However, the net effect is to decrease the amount of energy uptake and transfer. Thus, a gram of scorpion does not process as much food as a gram of arthropodivorous vertebrate. Overall, then, desert scorpions are less important in energy and nutrient cycling than their diversity, abundance, and biomass would suggest.

How much energy do desert scorpions process? We have two estimates: Polis (1988) calculated that average populations of *Paruroctonus mesaensis* used more than 9000 g of prey/ha/year. The Australian *Urodacus yaschenkoi* requires about 8000 g/ha/year, the equivalent of about 100,000 ants or 30,000

Table 12.3 Ecological production efficiency of various animal taxa[a]

Group	P/A %
Insectivorous mammals	0.86
Birds	1.29
Small mammals	1.51
Other mammals	3.14
Homeotherms	3.1
Fish	9.8
Social insects	10.3
Terrestrial invertebrates[b]	25.0
Solitary herbivorous insects	38.8
Solitary detritivorous insects	47.0
Solitary carnivorous insects	55.6
Spiders	
Scorpions	68.2

[a]Production efficiency (P/A%) is equal to the proportion of assimilated energy incorporated into new biomass (=production/assimilation). Highest efficiencies are found in carnivorous insects, spiders and scorpions.

[b]Terrestrial invertebrates do not include insects or arachnids (primarily from Humphreys, 1979; see Polis and Yamashita, 1991, for further details).

Table 12.4 Metabolic rates of various animal taxa

Taxon	Metabolic Rate (ml O/g/r)
Homeotherms (basal rates)	
Mammals	0.07–7.4
Shrew	7.4
Rodents	1.80
Elephant	0.07
Birds	2.3–4.7
Poikilothermsa[a] (at 25°C)	
Insects	1.665 ± 1.25
Spiders	0.92 ± 0.92
Scorpions	0.057 ± 0.048

[a]The sample size for insects is 82 species; for spiders, 8 species; for scorpions, 7 species (see Polis and Yamashita, 1991, for further details).

medium sized spiders eaten per hectare per year (Shorthouse, 1971; Marples and Shorthouse, 1982). It is uncertain exactly how such figures compare to those for vertebrates. Such data would be useful because they would permit evaluation of the relative importance of specific taxa to such processes as energy transfer, nutrient cycling, and consumer–resource dynamics. Such information is required for a complete picture of ecosystem and community structure.

INFLUENCE ON COMMUNITY STRUCTURE AND DYNAMICS

A second measure of ecological importance of a taxon is the degree to which it influences the dynamics, distribution, and abundance of other taxa. In theory, scorpions can influence desert communities as predators by affecting characteristics of their prey (i.e., abundance, diversity, body size), as prey of other predators, and as competitors of other arthropodivores.

A central goal of ecological studies is to determine the factors that regulate distribution and abundance of species. To approach this question, researchers often focus on groups of similar species that use similar resources, especially resources that may be limiting and in demand (e.g., food, space). Such species groups are called guilds (a group of species that use resources in a similar, potentially competitive way). For example, guilds of granivores (birds, ants, rodents) all eat seeds, regardless of specific mode of resource acquisition. Guilds of desert arthropodivores consist of such taxonomically diverse taxa as lizards, some birds, centipedes, predaceous insects, spiders, solpugids, and scorpions (Polis and Yamashita, 1991).

The study of guild structure typically involves the descriptions of the diversity and abundance of species members and the spatial and temporal patterns and resource use (i.e., niche characteristics). Such descriptions do not reveal the processes that produce the observed patterns. An alternative approach is to determine which guild members interact and how such interactions significantly shape guild structure. Interactions among guild members can range from cooperation and mutualism to competition and even predation. Intraguild predation is a ubiquitous interaction among many assemblages of potential competitors but has received little formal attention from either theorists or empiricists. One theme of my research has been to delineate guild pattern and process and to analyze the characteristics and dynamics of intraguild predation (Polis and McCormick, 1987; Polis et al., 1989; Polis and Holt, 1992). Scorpions have been remarkably good tools for this purpose, in large part because of the ease that diet can be determined. Many scorpions eat prey (other scorpions, spiders, and solpugids) that also potentially compete with scorpions (Polis, 1990, 1991a,b; Polis and Yamashita, 1991).

In the following sections I present information on four studies from three desert ecosystems. In each example scorpions frequently eat species in the same guild of arthropodivorous predators, and such intraguild predation significantly affects the distribution, abundance, and population dynamics of these potential competitors. Large amounts of data can be collected, interactions (e.g., feedings) quantified relatively easily, and individuals or whole populations manipulated experimentally. For example, in the first study, field data were collected on >130,000 individuals, >2000 feedings, and >6000 individual scorpions were manipulated in controlled field experiments. These studies illustrate why scorpions are so amenable for ecological research.

Interactions among Scorpions in the Coachella Valley

Four species of desert scorpion co-occur in sandy habitats on the floor of the Coachella Valley (Riverside County, California). Three of these species are in the family Vaejovidae (*Paruroctonus mesaensis* Stahnke, *P. luteolus* Gertsch and *Vaejovis confusus* Stahnke); one (*Hadrurus arizonensis* Ewing) is a member of the family Iuridae (Polis and McCormick, 1986a, 1987). *Paruroctonus mesaensis* constitute >95% of all individuals and occur at densities in the range of 0.2–0.5 individuals/ha; the other three species are relatively rare. Each species requires 2–5 years to mature, and each is composed of several distinctly sized year classes. For example, over the 2 years it requires for sexual maturation, *P. mesaensis* increases 60–80 times in weight from 0.03 g (instar 2) to 2.0–2.5 g (nongravid adults). The year classes and species overlap to various degrees in use of insect and arachnid prey; average overlap among all species is moderate to high (0.67 [prey size] and 0.43 [prey taxa]). Thus, if food is limiting and scorpion foraging depresses the availability of prey for other foragers (Polis and McCormick, 1987), scorpions potentially compete for food.

Extensive intraguild predation occurs among the four species in this system; table 12.5 presents a matrix of who eats whom. Several factors characterize this intraguild predation:

- Each species was both an intraguild predator and prey. Such mutual predation occurs simply because the predator scorpion was always larger regardless of the species combination involved ($n \geq 170$ scorpion–scorpion predations). Thus, scorpions of all species are vulnerable as they grow from small juveniles to full-size adults, and predatory reversals (mutual predations) are common. For example, young *P. mesaensis* scorpions are eaten by relatively larger adult *P. luteolus* and *Vaejovis confusus;* adult *P. mesaensis* prey on the (now) relatively smaller adults of these species. Thus, age/size is a key determinant of intraguild preda-

Table 12.5 Scorpion-scorpion predation in the Coachella Valley[a]

	Prey				
Predator	*Hadrurus arizonensis*	*Paruroctonus luteolus*	*Paruroctonus mesaensis*	*Vaejovis confusus*	Total Scorpions
Hadrurus	**0.0**	6.3	12.5	3.1	21.9
P. luteolus	0.0	**33.3**	6.7	6.7	46.7
P. mesaensis	0.4	0.5	**5.0**	8.3	14.2
Vaejovis	0.0	0.0	8.0	**4.0**	12.0

[a]The entries represent percentage of the diet that each species forms as prey for each of the four other scorpion species. The bolded type values (bold type) represents intraspecific predation (cannibalism) (from Polis and McCormick, 1987).

tion—adults are the predators, and immature individuals are the prey significantly more frequently than expected by chance.

- The most common species (*P. mesaensis*) was the predator in 91% of all intraguild predations observed. Its average overlap in prey use (0.44) with other scorpion species was second highest.
- Intraguild predation, at least by *P. mesaensis,* is significantly more frequent when prey availability is low: when <1% of the population was feeding, heterospecifics formed 35% of all diet items. In contrast, when the percent feeding was >5%, only 3.2% of all prey were other species of scorpion.
- Mortality caused by intraguild predation was generally an inverse function of the density of both *P. luteolus* and *V. confusus;* this resulted because much of the surface activity of these two species occurred during (less productive) periods when *P. mesaensis* was absent from the surface.
- Intraguild predation is important in scorpion population dynamics. When analyzed as percent mortality of small species (here defined as the total number of individuals eaten by *P. mesaensis* divided by the total number ever observed), intraguild predation by *P. mesaensis* killed 8% and 6% of all *P. luteolus* and *V. confusus* ever observed.

Can such high rates of mortality cause the rarity of these prey species? It is only possible to approach this question using field experiments. Removal of >6000 *P. mesaensis* (=3.2 kg) from 300 (100=m^2) quadrats over a 29-month period demonstrated that the rarity of these species is caused substantially by intraguild predation from *P. mesaensis*. Both *P. luteolus* and *V. confusus* (but not *H. arizonensis*) increased significantly (600% and 135%) in removal of *P. mesaensis* as compared with 60% in control quadrats. The rarity of the largest species (*H. arizonensis*) may be a result of a bottleneck in adult recruitment; predation by *P. mesaensis* killed >10% of all newborn *H. arizonensis* observed during the study. In general, the age structure of each of the three rarer species was significantly different in removal areas: first-year juveniles were 1.75–2.85 more abundant in removals versus controls. This suggests that the numerical response by these species would be even more dramatic if this experiment continued beyond its 29-month period. Note that a plausible alternative hypothesis exists: removal of *P. mesaensis* relaxed exploitation competition and thus allowed the observed increases in density. A robust test failed to detect competition in this system (Polis and McCormick, 1986b, 1987).

Thus, intraguild predation by *P. mesaensis* significantly depressed the abundance of the rarer species. Does intraguild predation likewise affect their distribution? Because predation is apparently a key factor in the population dynamics of these species, natural selection is expected to favor adaptations that reduce the probability that an individual will encounter its predator. Indeed, prey scorpions often avoid places and times that their predators frequent or where the probability of predation is high (see Polis and McCormick, 1987). Typically, the large predatory entity (e.g., scorpion species and/or age class) occurs in productive periods and microhabitats, whereas smaller entities coexist by spatial segregation in a heterogeneous habitat and by temporal displacement.

The temporal and spatial distribution of smaller age classes and species of Coachella Valley scorpions reflect avoidance (in ecological and/or evolutionary time) of larger age classes and species. The overall distribution of *P. luteolus* and *V. confusus* tend to place these species on the surface during times (in winter, late fall) and in places (off sand) characterized by relatively low surface populations of adult *P. mesaensis*. These times and microhabitats support significantly less prey than those used by adult *P. mesaensis;* consequently, *P. mesaensis* has a feeding rate (2.95%) significantly greater than all other species combined (1.70%). Further, the minority of *P. luteolus* and *V. confusus* that forage when and where *P. mesaensis* is active suffer a disproportionately greater chance of being eaten by *P. mesaensis*. Intraspecific predation (cannibalism) has produced similar patterns of temporal distribution, feeding, and mortality patterns among age classes of *P. mesaensis* (Polis, 1980, 1984a).

Thus, intraguild predation is an important factor for limiting abundance and shaping the distribution of these scorpions. Many other assemblages of desert scorpions exhibit similar patterns of distribution, abundance, and intraguild predation, suggesting that scorpion–scorpion predation is a major process shaping these populations (Polis and McCormick, 1987; Polis, 1990; but see Bradley, 1988).

Interactions among Scorpions, Spiders, and Solpugids in the Coachella Valley

Scorpions also frequently eat competitors other than other scorpions. For example, the diet of the scorpion *P. mesaensis* consisted of 8% spiders and 14% solpugids (Polis and McCormick, 1986b). Does such intraguild predation significantly affect the distribution and abundance of these unrelated taxa? Spiders responded in the above experimental removal of >6000 *P. mesaensis* by doubling their density in removal quadrats as compared to controls. Surprisingly, the abundance of solpugids and all insects combined did not increase significantly (all $p > .05$) in the removal plots. There are three possible explanations for this unresponsiveness: (1) although there was significantly less intraguild predation in removal quadrats, individual solpugids and/or insects dispersed away from these areas. The explanation is likely true for widely foraging solpugids and is possible but unlikely for the more sessile insects. (2) The increase of spiders and smaller scorpions compensated for decreased numbers of removed *P. mesaensis* by eating surplus arthropods. This explanation is unlikely because the biomass increase of spiders and scorpions represented <10% of the removed *P. mesaensis* biomass. (3) Scorpions simply exerted little impact on insect populations. This explanation, difficult though it is to accept, is nonetheless a possibility; *P. mesaensis* may take such a small proportion of all insects that its removal does not affect insect density. The overall conclusion from this study is that intraguild predation by scorpions significantly limits the abundance of desert spiders.

Interactions between Scorpions and Spiders in the Namib Desert

Predation by scorpions on spiders also appears to be a key determinant of spider densities in the Namib Desert (Polis and Seely, unpublished data, 1988, 1989). Both the scorpion *Uroplectes otjimbinguensis* (Karsch) (Buthidae) and the spider *Gandanimeno echinatus* (Purcell) (Erisidae) live under loose bark on larger *Acacia* trees and are the major arboreal predators of insects there. Populations of *G. echinatus* are severely reduced when they occur together on *Acacia erioloba* trees with the scorpion *U. otjimbinguensis.* This system exemplifies how predator–prey interactions are complicated greatly by patterns of local distribution. Not all suitable patches (*Acacia* trees) contain the full array of local species capable of existing within the patch. Trees grow along the banks of dry rivers and become less dense and more sporadic with increasing distance from the riverbed. The local abundance of scorpions and spiders on each tree is a function of differential dispersal, extinction, and a predator–prey relationship, with *U. otjimbinguensis* scorpions eating *G. echinatus* spiders. Although both species are found on trees in the dry Kuiseb riverbed, *U. otjimbinguensis* is, with no exceptions, the numerically dominant species (10 to >50 scorpions/trees) and *G. echinatus* relatively uncommon (5–20/tree). This occurs because scorpions are effective predators on these spiders.

However, spider and scorpion densities are increasingly variable on isolated trees farther from the river and along smaller washes entering the river. On some trees close to the river, abundance patterns are similar to those found in the riverbed. However, some trees have no scorpions and great numbers of spiders (50–400/tree), some trees have neither scorpions nor spiders, and some, no scorpions and few (<20) spiders. Overall, trees without scorpions support significantly more spiders (112.3 ± 60.6, $n = 21$) than trees with scorpions (24.5 ± 14.3, $n = 20$; $p < .001$; only experimental trees scored). This variation in abundance and distribution exist because neither *U. otjimbinguensis* nor *G. echinatus* disperse far from the river, yet spiders disperse farther than scorpions. Dense spider populations occur only in more isolated trees where scorpions are absent. In trees quite distant from the river, neither species occur.

Intraguild predation by *U. otjimbinguensis* scorpions on *G. echinatus* spiders appears to play a major role in establishing such patterns of distribution and abundance. This was demonstrated experimentally by adding *Uroplectes* to scorpion-free trees ($n = 11$) over a 1-year period. *Gandanimeno echinatus* populations were reduced to 42% of that on control trees ($n = 12$; $p < .001$). Removal of scorpions from trees ($n = 8$) also produced a highly significant, threefold increase in *G. echinatus* as compared to control trees ($n = 12$) with their full complement of scorpions.

These experiments showed that intraguild predation was concentrated on young spiders and significantly altered age distributions, skewing the population toward more adults when scorpions are present ($p < .001$) for each of the following comparisons. The smallest size class of spiders on experimental trees represented 49% of the population ($n = 728$ spiders) 1 year after scorpions were removed, compared to only 31% ($n = 221$) on control trees where scorpions remained. Similarly, the smallest size class formed 48% of all spiders ($n = 903$) on nonexperimental trees where scorpions were not present, compared to 34% ($n = 820$) on those trees to which scorpions were added.

Thus, semideterministic biotic interactions (scorpion predation decreases spider abundance) combined with differential dispersal against a background of varying isolation of patches (trees in and away from the river) are major determinants of the distribution, abundance, and age structure of these species. In general, historical and stochastic dispersal events in patchy environments are a paramount factor explaining the distribution and abundance of predators and their prey and species of competitors (Polis, 1991b; Polis and Yamashita, 1991).

Such conditions can produce local extinctions or great variance in abundances via deterministic biotic interactions, but they promote overall coexistence. Such situations are probably normal among many species living in heterogeneous desert. Many such assemblages occur in patches; differential dispersal, local extinctions, and "hide-and-seek" dynamics are undoubtedly extremely important in determining the exact structure of local assemblages. Unfortunately, little research has focused on these processes. The system with scorpions and spiders on *Acacia* trees is well suited to analyze such processes and represents another example of the use of scorpions to advance our comprehension of ecological processes.

Interactions between Scorpions and Spiders on Islands in the Gulf of California

Island ecosystems permit studies of predator–prey interactions in spatially structured populations. Spider and scorpion populations on small (<1 km^2) islands are one to three orders of magnitude more dense than on larger islands (approximately 1 to >1000 km^2) or the mainland, and significant negative relationships occur between island size and density for each of these taxa. For example, the scorpion *Centruroides exilicauda* (Buthidae) is 2 to >25 times more abundant on smaller islands, and orb spiders in the genera *Argiope* and *Metepeira* (Araneidae) and *Latrodectus* (Theriididae) can be 2 to >100 times more abundant there. Three major variables probably explain this variance in spider abundance: (1) the presence of scorpion predators (often absent from small islands), (2) the greater dispersal and colonizing ability of spiders relative to scorpions, and (3) differential energy flow from the ocean to land (flow is much greater to small islands).

This research has been conducted for several years (1989–1994) in the Midriff area of the Gulf on 41 island and 6 mainland sites between Bahia de Los Angeles (Baja California del Norte, Mexico) and Guaymas (Sonora, Mexico) (Polis and Hurd, 1996). Scorpion and spider abundance were quantified at each site; spiders were counted on >4000 cacti (one sample unit) and >8000 m^2 of supralittoral shoreline (another sample unit). Insect abundance was estimated at these sites for >1000 trap days.

Small islands exhibit significantly greater secondary production of insect prey and various predators (spiders, scorpions and lizards) than larger islands and mainland areas. However, high productivity is not from in situ primary production by terrestrial plants. The evidence strongly suggests that imported production from the ocean is the prime source of productivity (Polis and Hurd, 1997). Many islands are desert islands with limited primary productivity from terrestrial plants. Several of the most productive small islands support 2–28 plants (individuals, not species) yet 50 to >1000 spiders can occur on a single cactus.

Small islands are more productive and support higher numbers of arthropods than large islands and mainland areas for two reasons. First, shoreline constitutes a relatively larger proportion of land area on small islands because the perimeter:area ratio of an island decreases with size (perimeter is a linear function; area, a square function). Consequently, smaller islands receive relatively more nutrient and energetic input from marine drift (shore wrack—algal detritus and animal carrion). Second, small islands lack predators of nesting marine birds, making them favorable locations for large colonies of pelicans, gulls, petrels, and terns. These colonies import much marine material in the form of dead tissue (eggs, chicks, adults) and fish scraps.

Marine drift and material from nesting birds are eaten by many terrestrial and semiterrestrial arthropods, which become food resources for arthropodivorous predators. Moreover, dipteran parasites are dense consumers of nesting birds and form an important conduit of energy from the sea to spiders. Trapping shows that significantly more insects occur in supralittoral zones and around nests than other areas on these islands and on islands with colonies of marine birds compared to those without colonies ($p < .001$, all comparisons; Polis and Hurd, 1997; also see Due and Polis, 1985). Dietary analysis confirms that such detritivorous and parasitic arthropods act as conduits of energy from the sea when consumed by large populations of terrestrial spiders, scorpions, and/or lizards. For example, spiders are two to five times more dense on islands with colonies of marine birds versus those without colonies.

Predation is also a critically important variable affecting the abundance of these taxa. On Gulf Midriff islands, spider abundance is significantly lower in the presence of scorpions and lizards (particularly *Centruroides exilicauda* and *Uta stansburiana;* 1990: 0.19 spiders/m^3 cactus; 1991: 2.5/m^3) versus their absence (63.4/m^3 $p < .005$). The important impact of scorpions, but not lizards, on spider abundance is suggested by the analysis of 10 islands near Guaymas in Sonora. Here spider density is 38 times lower (0.41/m^3 cactus) on all islands with *C. exilicauda* ($n = 8$) compared to similarly sized islands with lizards but without scorpions (15.3/m^3 cactus; $n = 2$). This suggests that scorpions, but not lizards, depress the abundance of spiders.

Another important variable in this system is the relative colonization ability of spiders and scorpions. Incidence functions (percent occurrence on islands classified by area) describe the ability of taxa to colonize and remain successfully on islands of different sizes. Incidence functions suggest that relative ability to colonize small islands can be roughly ranked: spiders > *Uta* lizards > scorpions > predators of seabird nests > songbirds. Colonization would be deterministic if a particular taxon were always present or absent for all islands of a particular size class. In fact, incidence values for smaller island size classes are neither 0 nor 100%, suggesting that presence or absence of a particular taxon is somewhat stochastic. Thus some small islands support high densities of spiders when scorpions are absent, whereas other similarly sized islands exhibit low densities of spiders when scorpions are present.

An integration of predation, productivity, and colonizing abilities is required to understand the distribution and abundance of spiders. Each of these factors varies more or less regularly with island size: as size increases, secondary productivity and the importance of differential colonizing ability decrease as predation increases. Statistical dissection by multivariate analysis reveals the relative contribution of each of these factors to the observed variance in an independent variable, here spider density. This analysis (table 12.6) shows that spider density is a significant positive function of prey availability, and it is significantly depressed in the presence of scorpions. Lizards are a seemingly unimportant factor, accounting for little of the variance in spider abundance on Midriff Islands.

In summary, the productivity of these habitats sets potential maximal population size of spiders. Small islands are much more productive per unit surface area than larger islands because of the greater input of marine material from

Table 12.6 Multivariate regression to reveal the factors that influence spider abundance on 19 islands in the Midriff area of Gulf of California in 1990 and 1991[a]

	df	SS	MS	F	$p > F$
1990 Season					
Regression	4	3.25	0.81	8.67	.0008
Error	15	1.42	0.09		
Total	19	4.65			
Perimeter:area ratio				10.3	.0058
Scorpion presence				4.29	.0559
Lizard presence				0.32	.5827
Cactus volume				1.20	.2902
Total R^2 of this model = .70					
1991 Season					
Regression	3	3.13	1.04	11.01	.0004
Error	16	1.51	0.09		
Total	19	4.65			
Perimeter:area ratio				9.06	.0083
Scorpion presence				5.80	.0285
Lizard presence				3.76	.9687
Total R^2 of this model = .674					

[a]The independent variable is spider density on cactus. The dependent variables include lizard presence, scorpion presence, perimeter : area ratio of island, mean cactus volume/island and prey availability/island. The most significant models (both with $p < .001$) explain 67–70% (total R^2) of the variability of spider abundance among different islands; they show that perimeter:area ratio and scorpion presence are the two significant determinates of spider abundance.

shore drift and marine birds. Colonizing ability establishes species combinations on islands such that larger islands support a greater diversity of predators. The realized abundance of spiders is limited by intraguild predation. If scorpion predators are absent, spiders are dense on small, high-productivity islands. When scorpions are present, spider density is lower (but still higher than on large islands and the mainland), and the density of scorpions is relatively high. As island size increases, productivity decreases as nest predators eliminate bird colonies, and their input in the form of marine detritus decreases with the island perimeter/area ratio. As island size increases (with commensurate decreases in productivity and increased predation), the abundance of spiders decreases until abundance on large islands approaches that of the mainland. Strong predation from many sources occurs on the mainland where productivity is relatively low, and consequently spider populations are diminished. The same processes combine to produce the observed variability in populations of other taxa, particularly scorpions and lizards.

CONCLUSIONS

This chapter evaluated the ecological importance of scorpions and other competing arachnid predators in deserts. Ecological importance was first considered in terms of population energetics (the quantity of energy and mass flowing through scorpion populations) and then in terms of the regulation of community structure and dynamics (how intraguild predation by scorpions influences the distribution and abundance of their competitors/prey). Studies of energy flow in these systems lead to three conclusions: (1) scorpions are diverse and abundant in deserts, and they occupy a relatively large proportion of animal biomass in desert communities relative to vertebrates and other arthropods. (2) The relative success of scorpions is due to a suite of autecological traits that are particularly suited to the harsh and variable climatic conditions of deserts. Very low metabolism and very high assimilation efficiency (leading to low energy requirements) and their tolerances to water, thermal and starvation stresses preadapt scorpions to prosper where unpredictable and marginal productivity and food resources prevail. (3) These traits of low metabolism and high assimilation efficiencies also diminish the impact of scorpions on energy flow and potential interactions with prey and competitors. Thus the importance of scorpions relative to homeothermic vertebrates is less than an analysis of density and biomass suggest: scorpions require and process prey in quantities relatively low to their biomass. Nevertheless, because of their considerable biomass, scorpions are significant conduits of energy flow in deserts.

Research on the interactions among scorpions and on interactions between scorpions and spiders strongly suggests that scorpions can play key roles in the structure and dynamics of the communities in which they live. These studies showed that intraguild predation by scorpions was a major force determining the temporal and spatial distribution, abundance, and age structure of populations of their competitors and prey. However, as an important caveat, these interactions must be viewed in the context of the environment in which they occur. Dispersal ability, spatial structure, and productivity are just some of the possible important factors that moderate the predator–prey interaction between scorpions and their intraguild prey.

The goal of ecology is to integrate these factors to produce a synthetic understanding of the processes and dynamics that structure natural communities. Scorpions are particularly suited for this task and will continue to be a productive research model to advance theoretical and empirical studies in ecology.

Acknowledgments I have been extraordinarily fortunate to have been accompanied and supported by an insightful, dedicated, and enthusiastic corps of coworkers over the years. An incomplete list of some of the many who have contributed their time, energy, and enthusiasm to this work includes Denise Due, Victor Fet, Rick Fleck, Steve Hurd, Todd Jackson, Steve Ladwig, Sharon Lee-Polis, Sharon McCormick, Chris Myers, Dan Polis, Steve Polis, Mike Quinlan, Mike Rosen, Mary Seely, Tracey Wadsworth, and Tsunemi Yamashita. Careful reading of this manuscript by Philip Brownell and Bill Humphreys greatly clarified my writing and thoughts. To all these people, I express my warm and sincere gratitude. Financial support was provided by the National Science Foundation, Earthwatch Foundation, the International Scholar Exchange (Fulbright) Program, the Natural Science Committee, and University Research Council of Vanderbilt University.

References

Bradley, R. (1983) Complex food webs and manipulative experiments in ecology. *Oikos* 41: 150–152.

Bradley, R.A. (1986) The relationship between population density of *Paruroctonus utahensis* (Scorpionida: Vaejovidae) and characteristics of its habitat. *J. Arid Environ.* 11: 165–171.

Bradley, R.A. (1988) The influence of weather and biotic factors on the behaviour of the scorpion (*Paruroctonus utahensis*). *J. Anim. Ecol.* 57: 533–551.

Due, A.D. and Polis, G.A. (1985) Biology of the intertidal scorpion, *Vaejovis littoralis. J. Zool.* 207: 563–580.

Due, A.D. and Polis, G.A. (1986) Trends in scorpion diversity along the Baja California peninsula. *Am. Nat.* 128: 460–468.

Humphreys, W.F. (1979) Production and respiration in animal populations. *J. Anim. Ecol.* 48: 427–454.

Lamoral, B. (1978) Soil hardness, an important and limiting factor in burrowing scorpions of the genus *Opisthophthalmus* C.L. Koch, 1837 (Scorpionidae, Scorpionida). *Symp. Zool. Soc. Lond.* 42: 171–181.

Marples, T.G. and Shorthouse, D.J. (1982) An energy and water budget for a population of arid zone scorpion *Urodacus yaschenkoi* (Birula 1903) *Austral. J. Ecol.* 7: 119–127.

McCormick, S.J. and Polis, G.A. (1990) Prey, predators and parasites. *In* Polis, G.A. (ed.) *The Biology of Scorpions,* pp. 294–320. Stanford University Press, Palo Alto, CA.

Polis, G.A. (1980a) The significance of cannibalism on the population dynamics and surface activity of a natural population of desert scorpions. *Behav. Ecol. Sociobiol.* 7: 25–35.

Polis, G.A. (1980b) Seasonal and age specific variation in the surface activity of a population of desert scorpions in relation to environmental factors. *J. Anim. Ecol.* 49: 1–18

Polis, G.A. (1981) The evolution and dynamics of intraspecific predation. *Annu. Rev. Ecol. System.* 12: 225–251.

Polis, G.A. (1984a) Age structure component of niche width and intraspecific resource partitioning: can age groups function as ecological species? *Am. Nat.* 123: 541–564.

Polis, G.A. (1984b) Intraspecific predation and "infant killing" among invertebrates. *In* Hausfater, G. and Hrdy, S. (eds.) *Infanticide: Comparative and Evolutionary Perspectives,* pp. 87–104. Aldine, New York.

Polis, G.A. (1988) Exploitation competition and the evolution of interference, cannibalism and intraguild predation in age/size structured populations. *In* Persson, L. and Ebenmann, B. (eds.) *Size Structured Populations: Ecology and Evolution,* pp. 185–202. Springer-Verlag.

Polis, G.A. (1990) Ecology. *In* Polis, G.A. (ed.) *The Biology of Scorpions,* pp. 247–293. Stanford University Press, Palo Alto, CA.

Polis, G.A. (1991a) Complex trophic interactions in deserts: an empirical critique of food web theory. *Am. Nat.* 138: 123–155.

Polis, G.A. (1991b) Desert communities: an overview of patterns and processes. *In* Polis, G.A. (ed.) *The Ecology of Desert Communities,* pp. 1–26. University of Arizona Press, Tucson.

Polis, G.A. and Farley, R.D. (1980) Population biology of a desert scorpion: survivorship, microhabitat and the evolution of life history strategy. *Ecology* 61: 620–629.

Polis, G.A. and Holt, R.D. (1992) Intraguild predation: the dynamics of complex trophic interactions. *Trends Ecol. Evol.* 7: 151–154.

Polis, G.A. and Hurd, S. (1996) Subsidized consumers and apparent trophic cascades: food web effects of allochthonous flow between habitats. *In* Polis, G.A. and Winemiller, K. (eds.) *Food Webs: Integration of Patterns and Dynamics,* pp. 275–285. Chapman & Hall, London.

Polis, G.A. and McCormick, S.J. (1986a) Patterns of resource use and age structure among species of desert scorpions. *J. Anim. Ecol.* 55: 59–73.
Polis, G.A. and McCormick, S.J. (1986b) Scorpions, spiders and solpugids: predation and competition among distantly related taxa. *Oecologia* 71: 111–116.
Polis, G.A. and McCormick, S.J. (1987) Intraguild predation and competition among desert scorpions. *Ecology* 68: 332–343.
Polis, G.A., Myers, C.A. and Holt, R. (1989) The ecology and evolution of intraguild predation: Potential competitors that eat each other. *Annu. Rev. Ecol. Systematics* 20: 297–330.
Polis, G.A. and Yamashita, T. (1991) The ecology and importance of predaceous arthropods in desert communities. *In* Polis, G.A. (ed.) *The Ecology of Desert Communities,* pp. 180–222. University of Arizona Press, Tucson.
Shorthouse, D. (1971) *Studies on the Biology and Energetics of the Scorpion, Urodacus yaschenkoi* (Birula, 1904) (Ph.D. dissertation). Australian National University, Canberra.

13

Evolution of Geographic Variation in Body Size and Life-History Traits

CHRIS MYERS

Body size is a fundamental trait affecting virtually every aspect of an organism's biology. Size influences the metabolic and structural characteristics of individuals and their ability to accommodate cold, heat, water, and other environmental stresses (Bergmann, 1847; Oliver et al., 1979; Calder, 1984), size is often linked to life-history traits of organisms such as fecundity, age at maturity, and life span (Cole, 1954; Gadgil and Bossert, 1970; Oliver et al., 1979; Roff, 1981; Rose and Charlesworth, 1981a,b; Speiss and Speiss, 1986), and size influences ecological interactions by affecting an individual's status as potential competitor, predator, or prey (Morse, 1974; Wilbur, 1975; Peters, 1983; Polis, 1984; Werner and Gilliam, 1984; Ebenman and Persson, 1988). It is no accident that a rich tradition of theoretical and empirical science continues to grow around questions of how body size influences an individual's life history, population structure, and community organization.

Most species do not contain adults of uniform size. Researchers study size differences to gain ecological and evolutionary insights, but discovering the cause of variance, even within a single species, is confounded by the many environmental and selection forces acting now and in the past to shape this character. For example, natural selection may cause size variation by acting indirectly on genetically correlated characters such as fecundity or age of maturity; this makes interpretation especially difficult (Via and Lande, 1985; Stearns and Koella, 1986). Despite the difficulties, previous research has yielded several explanations for variation in body size. The most prevalent of these are listed briefly here in the form of six nonmutually exclusive hypotheses:

1. Smaller species or genotypes should be favored at low levels of food availability because they require less energy to survive (Case, 1978; Lynch, 1980; Lawlor, 1982; Persson, 1985; but see Gerritsen, 1984; Lindstedt and Boyce, 1985).
2. Variation in body size is due to differences in the size of available prey; large prey are more efficiently handled by large predators (Schoener, 1969; Case and Cody, 1983).
3. Size is determined by size-selective predation on the species exhibiting size differences (Case and Cody, 1983).
4. Warmer temperatures may favor larger body size in ectotherms if growth rates are increased (Stearns and Koella, 1986; but see Voorhies, 1996).

5. Small morphotypes gain a reproductive advantage by reaching maturity in a shorter amount of time (e.g., Cole, 1954; Stearns and Koella, 1986).
6. Size may be related to intraspecific social factors such as sexual selection or aggression (Case and Cody, 1983; see Benton, this volume).

These hypotheses are testable using appropriate models for descriptive and experimental investigations.

Scorpions are useful models for studying size variation for several reasons. As discussed in the Introduction to this volume, scorpions are readily observed in the field because of their fluorescent cuticle and habits as sessile predators. The ease of locating them at night and of monitoring marked individuals over time enable ecological measurements of exceptional detail. Scorpions are often abundant in desert ecosystems, where they are important intermediate-level predators, an ecological role in which size-related predatory and competitive advantage is likely significant. Most important, many scorpion species display geographic variation in body size on several spatial scales. On a regional scale, for example, McCormick and Polis (1986) found a 20% difference in body size among adult *Paruroctonus mesaensis* separated by only 15 km, and Scorza (1954) found a 70% increase in size among adult *Rhopalurus laticauda* over a 525 m increase in elevation. On a larger scale, several scorpion species show differences in adult body size across their continental range (Williams, 1980; Koch, 1981; see Polis and Sissom, 1990).

Perhaps the most striking example of geographic variation in size among scorpions occurs in the arboreal *Centruroides exilicauda*. This species is variable throughout its range, but nowhere more so than near La Paz, Baja California del Sur, Mexico, the setting of the present study. Adults from populations north of La Paz are more than twice as large by weight than adults from populations south of La Paz (about 15 km distant). These differences occur along a gradient or cline where adult size declines toward the south. On near offshore islands, gigantism of *C. exilicauda* is evident in individuals three times larger than mainland conspecifics. This natural and somewhat orderly progression provides a rare opportunity to examine the causes and consequences of variation in adult size within a single species.

From and evolutionary perspective, a particularly important question concerns the origins and maintenance of size variation of *Centruroides* populations within their geographical range. Since Darwin (1859) first argued that geographically variable traits hold clues to the process of speciation, evolutionary biologists have examined clines to determine how differences originate and how they are sustained. From theoretical considerations, Haldane (1948) and Fisher (1950) demonstrated that stable, geographical clines could arise where a balance between gene flow and selection exists. Empirical investigations now seek to identify ecological and historic factors that operate to maintain clinal variation through evolutionary processes (Clarke, 1966; Endler, 1973, 1977; Slatkin, 1973; White, 1978; Lande, 1982; Barton and Hewitt, 1981; Mallet, 1986). In recognition of the need for better experimental models in which to test these concepts, I present in this chapter a summary of my studies of the *Centruroides* system and its implications for the possible ecological and evolutionary causes of size variation in the context of the six hypotheses listed above.

SPECIES AND SITE CHARACTERISTICS

Centruroides exilicauda Wood (Family Buthidae) is an abundant and important intermediate-level predator in desert environments of Baja California and Sonora, Mexico, the associated islands in the Sea of Cortez, and the U. S. states of Arizona, New Mexico, and Utah. *Centruroides exilicauda* most commonly occurs on a variety of trees, woody shrubs, and cacti; less frequently it resides on or under rocks or ground debris. It is a generalist predator, eating arthropod prey of widely variable size and even small lizards on rare occasion. This is the first extended ecological investigation of this species (Myers, 1990).

Considerable differences in size, color, and pectinal tooth counts exist among geographic races of this scorpion (Williams, 1980), with the most extreme differences occurring near La Paz, Baja California del Sur, Mexico (fig. 13.1). Adults from populations north of La Paz are more than twice as large, by weight, than typical adults from populations south of La Paz (about 15 km distant). Northern populations are pale; southern populations are striped. Adults from the middle of the cline are of intermediate size and may be either striped or pale. The "giant" *C. exilicauda* from islands near La Paz (e.g., Espiritu Santo and Partida) constitute the extreme of the size cline at three times the weight of mainland adults. Patterns of geographic variation among populations of *C. exilicauda* are most apparent at a relatively localized scale; no consistent latitudinal or longitudinal trends in morphological differentiation are known to exist at the scale of the entire Baja California peninsula (Walker, 1973).

The general climate of the study area is hot and dry. Mean yearly rainfall from 1970 to 1985 was 23.4 cm in La Paz (in the middle of the size cline) and 19.9 cm at Pichilingue (at the northern, coastal end of the cline). Rainfall was concentrated in July through September. Mean annual temperature over the same period was 24.7°C in La Paz (absolute minimum 2°C, maximum 42.5°C) and 26.9°C in Pichilingue (absolute minimum 3.5°C, maximum 41°C).

PATTERNS OF GEOGRAPHIC VARIATION AND THEIR ENVIRONMENTAL CORRELATIONS

Morphological variation among *C. exilicauda* populations near La Paz and adjacent islands are likely to be influenced by environmental variables such as food availability and the density of potential competitors. I identified the factors most consistently associated with size differences along the cline through simple and multiple correlation analysis.

Methods

Twenty sites along the mainland cline and four sites on Isla Espiritu Santo were surveyed in 1989 (fig. 13.1). At each site, four 216-m line transects were established using fluorescent string, and the abundance of *C. exilicauda* was estimated by walking these transects at night with ultraviolet lights to detect scorpions within 3 m. In this way, 1296 m^2 were surveyed per transect, for a total of 5184 m^2 at each site; 724 *C. exilicauda* were recorded in total in 1989, 463 on transects. A multiyear mark–recapture study in coastal and inland quadrants

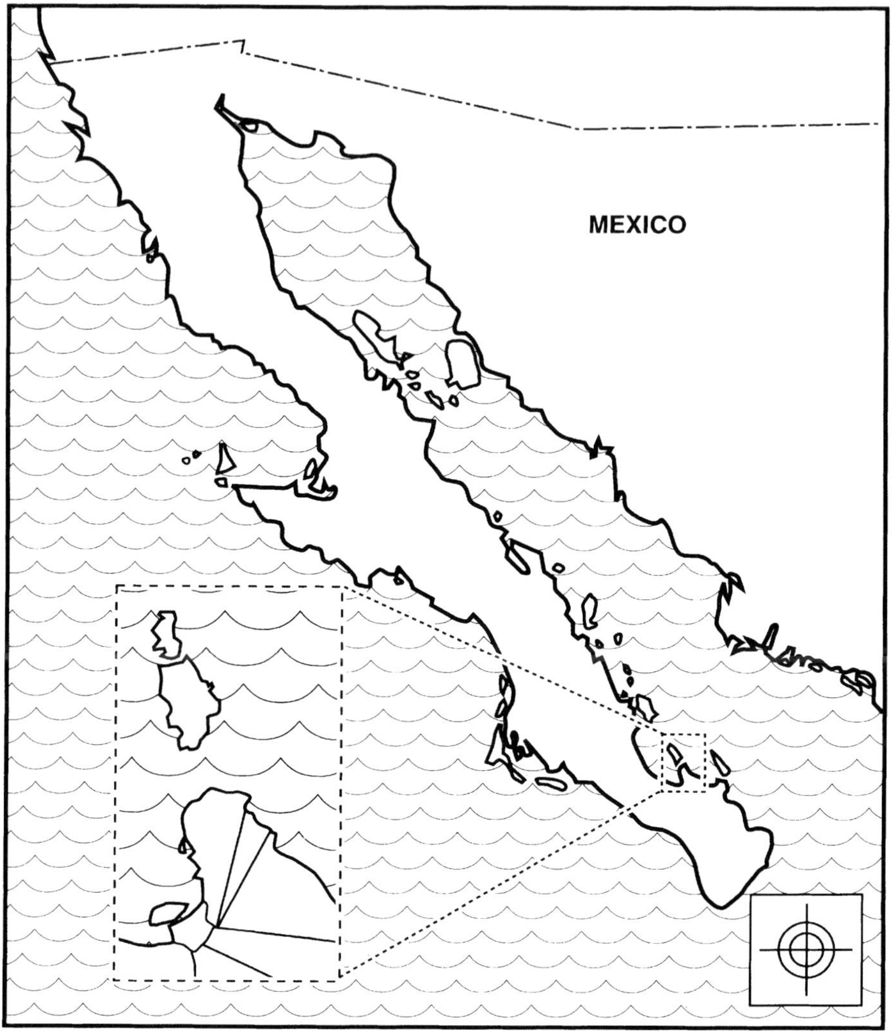

Figure 13.1 Map of study area showing intraspecific morphological variation among populations of *C. exilicauda* along the mainland cline and nearby islands. Map of study areas near La Paz, in Baja California, Mexico show roads as dark lines originating from La Paz. Single capital letters on the map label survey sites in order of north–south longitudinal distance. Pie charts show the percentage of striped individuals (black shading) versus pale, concolorous individuals (white areas) collected at each site.

revealed no significant differences in nightly surface activity or recapture rates of scorpions from these areas (Myers, 1990). Thus, single-night surveys were judged sufficient to estimate relative abundance. All site locations along the cline were randomly distributed with respect to the date of survey. Because strong moonlight can hinder the detection of scorpions by ultraviolet light, no surveys were attempted when more than half a moon was visible.

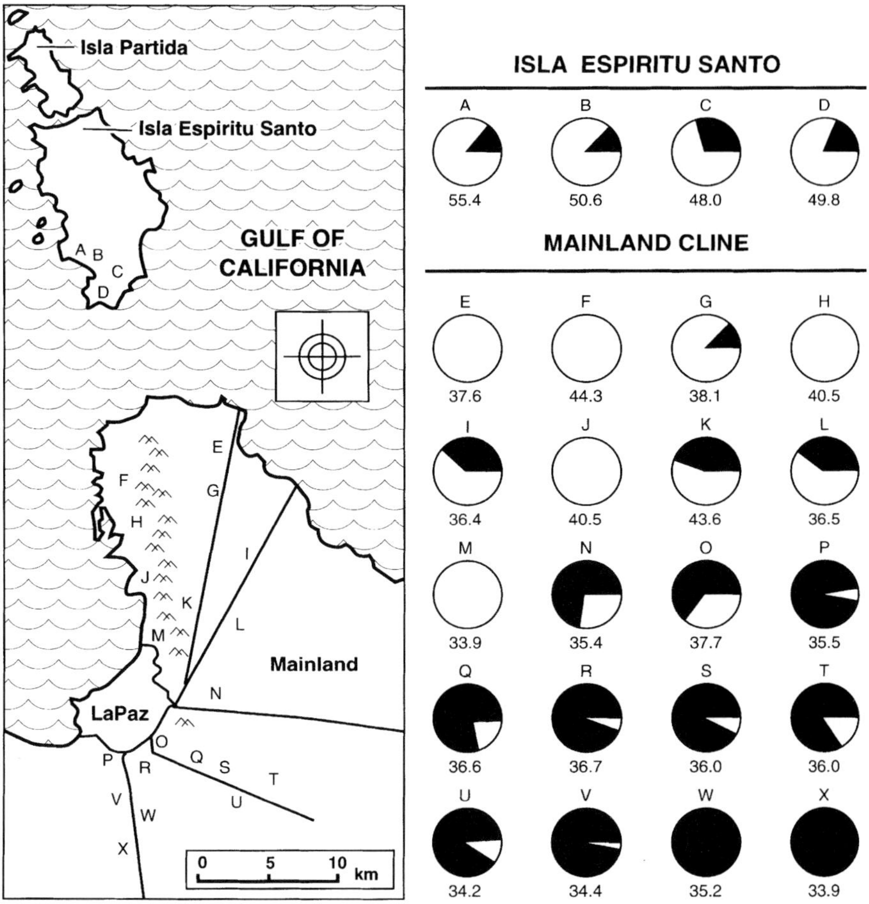

Letters above each pie chart refer to the site labels. Numbers below each pie chart give the mean overall length, in millimeters, of adult females. Body size measurements (prosoma + mesosoma + metasoma) listed were estimated, for illustrative reasons, from carapace length using linear regression of total body length on carapace length based on more than 200 individuals from the study area (total length = 9.12 × carapace length − 3.84: r^2 = .92, $p < .0001$).

Only females with a carapace length of at least 3.4 mm were considered "potential adults" and included in the estimates of mean adult size at each site. This was the minimum carapace length of the 77 females captured during the study that were known to be mature (i.e., females carrying young on their back or with developing embryos visible). Feeding rates were estimated directly from observed feeding events. Potential prey availability was estimated using 7.6 cm × 15.2 cm sticky traps. Eight traps per transect were attached to vegetation along each transect for a total 736 traps across the cline and on the island (limits in time and traps prohibited setting insect traps at site E; fig. 13.1). The percentage of vegetation cover was estimated by recording line intercept lengths for all woody vegetation within four evenly spaced 10-m segments along each line transect, for a total of 160 m of line surveyed at each site.

I used spatial autocorrelation analysis to examine geographic patterns of phenotypic differentiation (Cliff and Ord, 1973; Sokal and Oden, 1978a,b). Spatial autocorrelation determines whether values of a variable of interest at one site are independent of values at neighboring sites. If neighboring sites are similar for a given variable, the autocorrelation coefficient is positive; dissimilar neighbors produce negative coefficients. Trends in similarity between sites as a function of distance are illustrated in correlograms, which depict separate autocorrelation coefficients calculated for a range of distance classes (Sokal and Oden, 1978a,b). Moran's I, the autocorrelation coefficient used below, ranges from −1 to 1.

Patterns of Size Variation

Accurate maps illustrating how a species varies across a landscape can yield significant insights into the species ecology and evolutionary history. A map of the geographic variation in mean adult size and percentage of striped individuals in *C. exilicauda* populations is given in fig. 13.1. Although coincident clines are generally evident for body size and the percentage of striped *C. exilicauda* in mainland populations, evidence suggests little or no genetic basis to the correlation between size and the percentage of striped forms along the mainland cline (Myers, 1990). For example, small *C. exilicauda* are not always striped. Giant striped *C. exilicauda* occur on Espiritu Santo and Partida, and relatively large striped individuals occur about 30 km southeast of the study area.

Figure 13.2 displays geographic patterns of mean adult size for mainland populations of *C. exilicauda* as a function of distance between sites. All sites within each distance class (the abscissa) were considered neighbors for calculating the autocorrelation coefficient (Moran's I) for that class. The correlogram shows that similarity in mean adult size among sites decreases as the distance between sites increases (little significance should be attached to the last distance class because few site pairs occur in this class). The correlogram pattern is characteristic of a cline (Sokal and Oden, 1978a,b) where morphological differentiation is a function of intersite distance.

Environmental Correlations

The univariate correlations shown in table 13.1 indicate which environmental factors may best explain variation in size and striping along the cline. Despite the strong size–stripe correlation, ecological factors appear to affect these traits differently. Size is particularly associated with measures of food abundance and feeding rates, whereas striping is not; there are also differences in the correlations with mean cover of live and dead vegetation. No correlations exist between mean adult size in *C. exilicauda* and measures of either conspecific or heterospecific densities. These data suggest that exploitation competition is not a major force governing adult size in this system.

Because univariate correlations do not separate direct from indirect effects, multivariate analysis was also used to examine the relationship between size and the environmental variables listed in table 13.1. Stepwise multiple regression reveals that feeding rate is the variable that best explains clinal variation

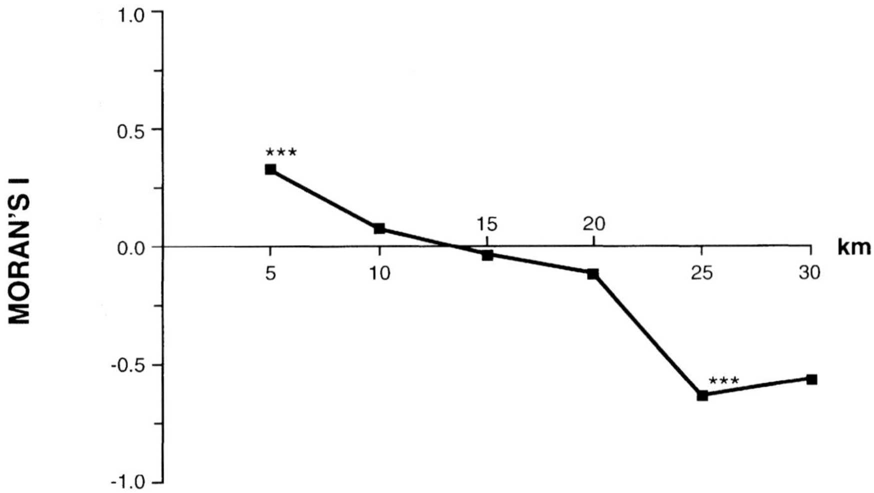

Figure 13.2 Correlogram showing geographic patterns of mean adult size for populations of *C. exilicauda* on the mainland as a function of intersite distance. All sites within each distance class (abscissa) were considered neighbors for calculating the autocorrelation coefficient (Moran's I) for that class. Distance class marks are upper bounds and represent longitudinal distance (e.g., the autocorrelation coefficient at 10 km on the abscissa was calculated using mean adult size values for all site combinations with an intersite longitudinal distance between >5 and 10 km). The correlogram pattern is characteristic of a cline (Sokal and Oden, 1978a,b). ***Indicates autocorrelation coefficients significantly different from 0 ($p < .001$); i.e., scorpions from sites near each other (longitudinal distance ≤ 5 km) are significantly similar in size; scorpions from sites that are far from each other (longitudinal distance between >20 and 25 km) are significantly dissimilar.

in adult size of *C. exilicauda* ($p < .001$). Prey abundance was positively correlated with adult size in the univariate analysis (table 13.1) but was not significant in the multivariate analysis due to a high positive correlation between prey abundance and *C. exilicauda* feeding rates ($r^2 = .63$, $p = .002$). Thus, *C. exilicauda* were found feeding more frequently at sites with high prey availability (estimated from sticky traps), and large adult size is associated with high feeding rates. On the island Espiritu Santo (where giant scorpions occur) the average abundance of potential prey was significantly higher than on the mainland (insects caught/sticky trap/h = 0.103 on the mainland; 0.209 on the island; $t = -4.04$, df = 21, $p < .001$). This gives further evidence for a causal relationship between food abundance and body size exists in this system.

The positive relationship between size and feeding rates supports the first hypothesis. Size variation does not appear to be due to differences in the size of available prey (hypothesis 2). Univariate (table 13.1) and multivariate correlations failed to reveal a significant relationship between scorpion size and the size of potential prey captured on sticky traps. Due to the difficulty of observing predators feeding on *C. exilicauda* in the field, hypothesis 3 (size selective predation) cannot be eliminated by this data set. The possible importance of temperature (hypothesis 4), reproductive advantage (hypothesis 5), and intraspecific interactions (hypothesis 6) are discussed below.

Table 13.1 Relevant variable descriptions recorded at each site and their univariate correlation with the mean adult size and percentage of striped *C. exilicauda*

Variable Label	Description	Size	% Striped
% Feeding	Proportion of *C. exilicauda* observed feeding	.551*	−.027
Prey abundance	No. of arthropods/sticky trap/h	.510*	−.259
Prey length	Mean length of arthropods captured on sticky traps	.451	−.440
Cactus	Mean cactus (all species) cover/transect	.323	−.379
Tree	Mean tree (all species) cover/transect	−.204	.323
Shrub	Mean shrub (all species) cover/transect	.257	−.140
Live vegetation	Mean cover of all live vegetation/transect	−.473*	.404
Dead vegetation	Mean cover of all dead vegetation/transect	.310	−.546*
Conspecifics	Total number of *C. exilicauda* seen along transects	−.106	.166
Heterospecifics	Total number of other scorpions seen along transects	.166	.291
S-N (km)	Distance along a south–north line, beginning at 0 km for southernmost site of the cline	.550*	−.924***
Size	Mean carapace length of adult (>3.4 mm) female *C. exilicauda*	—	−.627**
% Stripe	% of individual *C. exilicauda* striped	−.627**	—

$^*p < .05$, $^{**}p < .01$, $^{***}p < .001$.

EXPERIMENTAL ANALYSIS OF THE CAUSES AND CONSEQUENCES OF SIZE VARIATION

Controlled experiments are often needed to reach reliable conclusions about factors that influence size. A key objective in this study was to test whether size variation among *C. exilicauda* has a heritable (genetic) component or is determined merely by environmental conditions. This was investigated experimentally by rearing *C. exilicauda* from different populations in the same environment and measuring their size at maturity. Experiments were also used to investigate how size differences affect mating and aggression; that is, does size act as a potential barrier to reproduction and migration, thus maintaining size variation across the cline?

Genetic Basis of Size Variation

To describe development and to investigate the genetic basis for size variation, offspring from gravid *C. exilicauda* were collected from inland, coastal, and island sites, then reared under the same laboratory conditions. For these experiments, small, gravid *C. exilicauda* from the southern, inland portion of the mainland cline (near site X in fig. 13.1), larger gravid females from the northern, coastal portion of the cline (near site F), and giant gravid females from Partida Island were collected in the field in late August 1988 and brought to the laboratory at Vanderbilt University, Nashville, Tennessee. To minimize possible differences in offspring due to initial maternal condition, females were fed cockroaches ad libitum within 1 day after capture.

Fifty-two second instar young (30 males, 22 females) were removed from the backs of 14 mothers at 14 sites (6 inland, 5 coastal, 3 island). Approximately four offspring were raised from each per brood (mean $\pm$ SD $= 3.7 \pm 1.2$), placed in plastic jars, and randomly assigned to one of four food–temperature regimes: (1) high food, 25°C (most young were raised under these conditions, which served as the "common garden"), (2) low food, 25°C, (3) low food, 31°C, and (4) high food, 31°C. All young were kept in environmental chambers that remained dark except when scorpions were fed (following the methods of Francke and Jones, 1982). *Centruroides exilicauda* in high food treatments were fed ad libitum; all scorpions in the high food treatment appeared physically gorged, and roaches placed with scorpions one week were frequently found alive the next. Scorpions in low food treatments were fed one roach approximately every 2 weeks. For analyses of final adult size, reproductive maturity was ascertained by dissection or by production in males of a spermatophore. Scorpions are not known to molt after reaching reproductive maturity (Francke and Jones, 1982; Polis and Sissom, 1990).

Centruroides exilicauda *reared in a common environment*

The evidence strongly suggests that size variation among populations of *C. exilicauda* is genetically based. Even when raised from birth under common environmental conditions, *C. exilicauda* from inland, coastal, and island populations reach maturity at different instars and different sizes (fig. 13.3). Of the seven inland offspring to reach maturity, five matured in the fourth instar, two in the fifth. No inland offspring attained the sixth instar. Of four coastal offspring, two matured in the fifth instar, two in the sixth. Only offspring from the island population reached seventh instar (three matured in the sixth instar, two in the seventh). Adult size differences among populations were statistically significant (inland versus coastal, $t = -5.05$, df $= 9$, $p < .001$; coastal versus island, $t = -3.2$, df $= 5.6$, $p = .01$).

Size variation was correlated with differences among populations in the number of molts required to reach maturity rather than differences in size at birth or growth rate (Dyar's constant; also see Benton, this volume). With few exceptions, young of the same instar were generally of equivalent size, even if they were from different populations (fig. 13.3); birth weight and maternal size were not correlated. A few significant differences in size within instars did occur, but these were not maintained past the third instar.

Correlations in size between parent and offspring further support a genetic basis for size variation (fig. 13.4). Under similar rearing conditions the young of large females matured at a larger size than the young of small females. Parent–offspring correlations are significant for carapace length (a direct measure of size) and for pectinal tooth counts (an indirect measure of size). Pectinal tooth count is positively correlated with size in field-caught scorpions in this study;

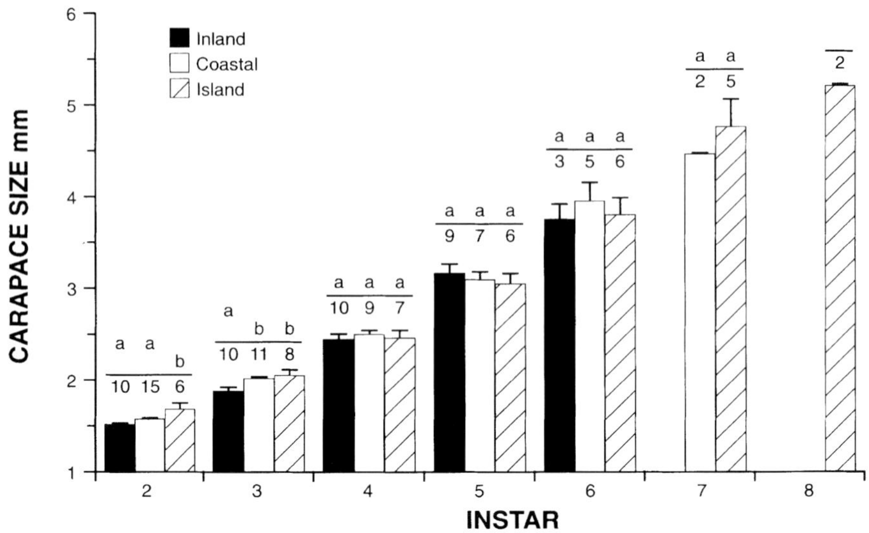

Figure 13.3 Comparison of carapace size (±SE) during development for inland (black bars), coastal (white bars), and island (cross-hatched bars) populations. All animals were raised from birth in a common laboratory environment (25°C, food ad libitum). Letters above horizontal lines are results from a Duncan's multiple range test. Intra-instar comparisons sharing the same letter are not significantly different ($\alpha > 0.05$). Sample size is shown below the horizontal line (sample size for each population may vary across instars due to damaged, unusable molts as well as mortality).

it is also a trait, fixed at birth, that is not influenced by postmaturation environmental conditions and may thus be useful to estimate maximum potential growth of scorpions in this system. All offspring at maturity were slightly smaller than their mothers (a result also exhibited in *C. gracilis;* Franke and Jones, 1982), an indication that growth conditions are better in the field.

Food, temperature, and some life-history considerations

Faster growth should produce larger size at maturity (hypotheses 4 and 5). Factors shown to affect growth rates among *C. exilicauda* in this study were food and temperature, with food probably being the major determinant of growth rates in the field. On average, *C. exilicauda* raised at 25°C under ad libitum food conditions reached fourth instar (the earliest instar of reproductive maturity observed) in 199 days ($n = 29$). Scorpions reared at the same temperature under low food conditions required an average of 423 days to reach fourth instar ($n = 9$). Those raised under ad libitum food conditions and 31°C reached fourth instar in 112 days ($n = 5$). The line transect study indicated that differences in food affect growth rates among *Centruroides* in the field more than temperature differences. Temperature over the 30-km cline does not vary greatly, with mean annual temperature only 2.2°C lower in La Paz than in the northern end of the

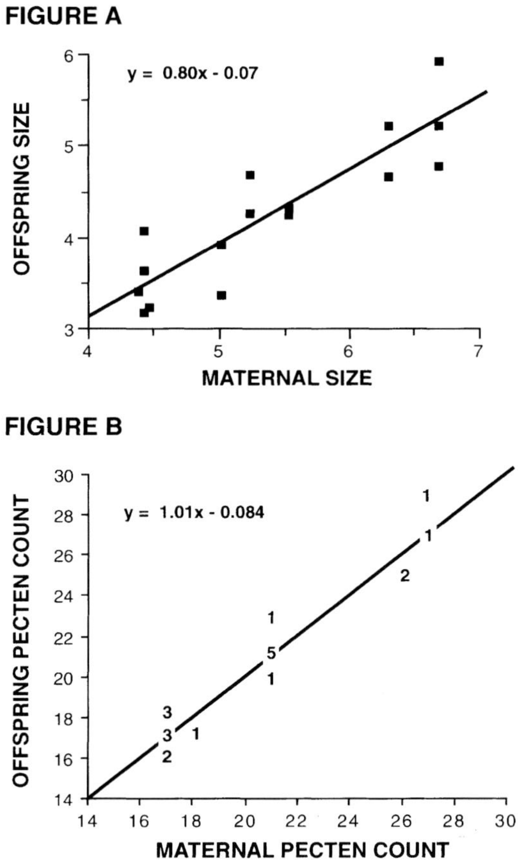

Figure 13.4 Regression analysis of offspring size versus parent size for key morphological features of scorpions. (A) Carapace size of adults (mm; r^2 = .78, df = 15, $p < .0001$) and (B) pectinal tooth count (right pecten; r^2 = .94, df = 19, $p < .0001$) for inland, coastal, and island populations and mature offspring raised from birth in a common laboratory environment (25°C, food ad libitum). Pectinal tooth count is positively correlated with body size. Numbers on regression line show number of data points at a given location.

cline (1970–1985). In contrast, observed feeding rates in the field ranged from 0% (inland) to 21% (coastal) in the line transect study.

Data from observed births revealed a strong correlation between maternal size and number of offspring (r^2 = .69, df = 41, $p < .0001$; mean brood size = 15.95 ± 7.00). Gravid females from coastal populations may bear almost twice the number of young as inland females (coastal = 18.19 ± 4.74; inland = 10.12 ± 3.00). If small (inland) adults mature at the same age as adults from the coastal site, their small size confers a reproductive disadvantage. In the rearing experiments, however, small adults reached maturity at fourth instar in 199 days on average, whereas large adults raised under the same conditions averaged of 471 days to maturity at sixth instar. Thus, it is likely that small adults can potentially mate and bear young in the first reproductive season following their birth, while large *C. exilicauda* cannot. If so, early maturation at the fourth instar confers a reproductive advantage (i.e., a higher intrinsic rate of population increase) to smaller scorpions despite their small brood size. Additional field measurements of growth rates and the number of broods per year would facilitate analyses of the relationship between size, population growth rates, and total fitness.

Intraspecific Interactions

Large size may be selected for despite delayed maturity if it confers an advantage in sexual or agonistic encounters with conspecifics (hypothesis 6; Case and Cody, 1983). The most common agonistic encounters in this system probably involve competition for home sites. *Centruroides exilicauda* do not burrow, so they must find suitable homesites (e.g., in dead wood) to escape diurnal heat and predators. I observed marked adults using the same home site for more than a year. Results from artificial home-site experiments and reciprocal transplants in the field show that *C. exilicauda* can detect small differences in home-site quality and that large adults can outcompete smaller adults for the best refuges (Myers, 1990; Benton, this volume).

Size differences may also result in size-mediated assortative mating and cannibalism. General field observations indicate that large adults sometimes treat smaller adults as potential prey rather than as mates. I conducted experiments to assess the nature of these cannibalistic interactions because of their implications for maintenance of size differences across the cline. The results are summarized in table 13.2. A total of 103 crosses within and between *C. exilicauda* from inland, coastal, and island sites were attempted between 1988 and 1990 by placing mature pairs in plastic tubs for 2 days, with homotypic pairs (=same sized, n = 57) and heterotypic pairs (=different sized, n = 46). The number of males and females in each size class were also approximately equal (e.g., the number of coastal male by inland female crosses was similar to the number of coastal females by inland males). No sex-related differences were detected for either the propensity to mate or to cannibalize.

The percentages of matings and cannibalisms during the mating experiments are shown in fig. 13.5 for small (inland), large (coastal), and giant (island) adult *C. exilicauda*. Within these staged matings, homotypic pairs are more likely to treat each other as potential mates ($p < .05$), whereas the larger adults of het-

Table 13.2 Chi-square significance tests for assortative mating and differences in rates of cannibalism for crosses within and between populations of small (southern) and large (northern) adult *C. exilicauda* from the mainland cline

Cross[a]	n	% Mating	χ^2	% Cannibalism	χ^2
Homotypic	45	40.0	4.30*	4.4	6.07*
Heterotypic	25	16.0		24.0	

[a]Homotypic combinations: either small × small or large × large cross; heterotypic combinations: small × large cross.

*$p < .05$.

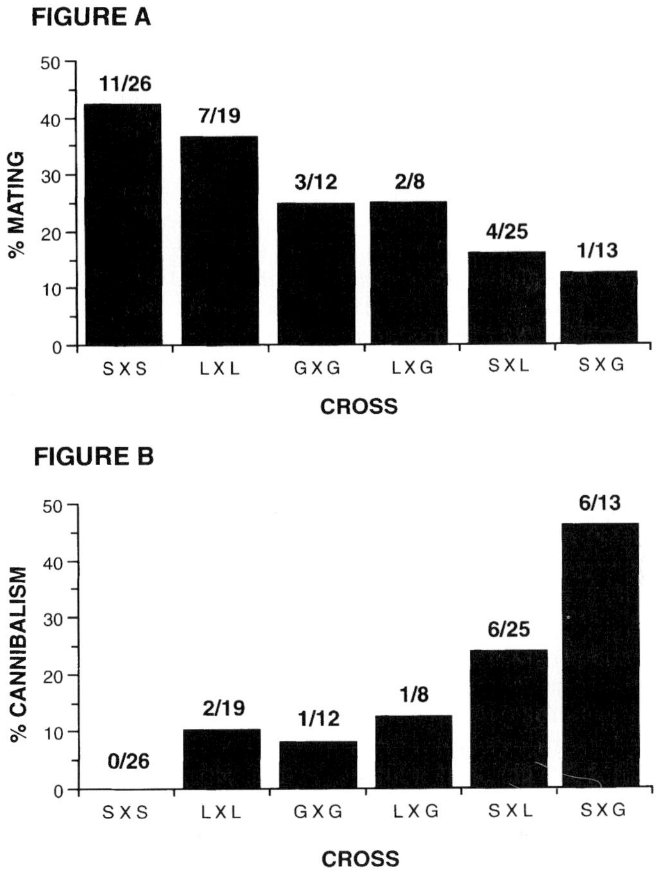

Figure 13.5 Percentage of successful matings and cannibalisms during mating experiments between *C. exilicauda* from mainland and island populations. (A) Successful matings (indicated by spermatophore deposition). (B) Percentage of cannibalism. S = small adult (from southern portion of the mainland cline); L = large adult (from the north end of cline); G = giant adult (from island). Numbers above each bar show the number of successful crosses (left number) and the total number of crosses attempted (right number). There were no observed cases of mating behavior before cannibalism.

erotypic pairs often treat smaller adults as potential prey (i.e., 24% of attempts resulted in cannibalism). Most cannibalism occurred between giant and small or large and small crosses. Large and giant scorpions were more compatible. No cannibalisms occurred between scorpions that had mated.

The tendency for *C. exilicauda* to chose similar-sized mates may reduce gene flow across the cline: This is significant from an evolutionary perspective because understanding limits to gene flow is crucial for understanding patterns of genetic variation and mechanisms of speciation (Hewitt, 1988). Evolutionary theory requires at least a partial barrier to gene flow for the long-term maintenance of genetic differences; without barriers, gene flow eventually homogenizes populations. One mechanism proposed for resisting gene flow across clines is natural selection against offspring of genetically dissimilar parents at the center of the cline (offspring of dissimilar parents are less fit). In this model, selection favors the evolution of assortative mating. However, because only a small percentage of organisms in the center of a cline are under selective pressure, this model has been considered insufficient to explain the existence of clines (Endler, 1977). Thus, even if offspring from genetically differentiated parents are less fit (there is no evidence for this among *C. exilicauda*), the distinctions are not sufficient to account for the observed pattern. Typically, researchers who cite hybrid inferiority also indicate the influence of environmental gradients and other factors (Bert and Arnold, 1996; Smith et al., 1996; but see Deharveng et al., 1998). In another model, Lande (1982) argues that selection on a secondary sexual characteristic, combined with simultaneous selection for mate choice, can lead to rapid character divergence along a cline. This model invokes simultaneous selection of two genetically independent traits—it relies on coincidence.

This study indicates that selection on size also affects mate choice in *C. exilicauda,* which may illustrate a mechanism more favorable to the origin and/or maintenance of morphological differences across a cline (as well as the probability of future speciation) than those proposed in the literature. It allows assortative mating to evolve without prior postmating reproductive disadvantage and with selection on only one trait which consequently affects mating. Relatively few studies of clinal variation have focused on predacious, cannibalistic species. This study suggests that the role of size in maintaining genetically differentiated populations and the probability of speciation may be underestimated, particularly for species with mating behaviors strongly affected by body size.

CONCLUSIONS

Investigations of morphological variation among geographic races have provided insights into mechanisms of character divergence and speciation. Clines have been studied intensively to understand how divergence is maintained in contiguous populations in spite of the homogenizing effects of gene flow. In this chapter, I focused on possible factors responsible for, and the ecological implications of, the dramatic clinal variation in adult body size and island gigantism in the arboreal scorpion *Centruroides exilicauda* Wood. Differences among populations of *C. exilicauda* raised from birth indicate a genetic basis to size variation. Correlations indicate high feeding rates and prey availability are the most consistent environmental factors associated with large size. Mating, home site, and rearing experiments coupled with field observations reveal that adult size strongly affects age at maturity, brood size, the ability to win intraspecific contests for home sites, likelihood of cannibalism, and mate choice.

To understand the evolution of size differences among adult *C. exilicauda,* it is useful to consider historical factors operating in this system. The Baja peninsula is believed to have arisen by gradual cleavage from mainland Mexico. General physical features and climate were probably recognizable after the mid-Cretaceous orogeny (100 MYA; Durham and Allison, 1960), but changes in sea level wrought changes in the landscape throughout the Cenozoic. Systemitists of reptiles, amphibians (Savage, 1960), and scorpions (Williams, 1980) have cited the complex geologic history of Baja California as a major factor contributing to high degrees of genetic differentiation and speciation on the peninsula (see Case and Cody, 1983). Mainland populations of *C. exilicauda* undoubtedly experienced some isolation during the complex paleogeological history of Baja, and it is not difficult to imagine how this might have affected the differentiation of its regional populations.

Although geographic isolation may have set the stage for morphological differentiation among *C. exilicauda,* its maintenance through time is likely the product of extant, size-mediated ecological and behavioral factors. These factors include (1) selection against large adults in the southern portion of the cline due to lowered prey abundance and feeding rates in the south (food and scarcity and slower growth rates are expected to favor small size); (2) selection against small adults in the northern portion of the cline due to cannibalism by larger conspecifics (smaller *C. exilicauda* are more likely to lose intraspecific agonistic encounters); and (3) size-related assortative mating. Rearing experiments indicate that size variation among *C. exilicauda* is due, in part, to genotypic differences. The proximate cause of variation in adult size appears to be differences in the number of molts to maturity rather than to differences in size at birth or growth rate.

Geographic variation among *C. exilicauda,* the unique qualities of scorpions as a group, and the ease with which they can be accurately measured in the field and laboratory is proving to be useful in the study of body size evolution and its ecological implications. There are, of course, many unanswered questions, but the accessibility of this genus make it a most attractive system among scorpions for continued study.

Acknowledgments This chapter is dedicated in fond memory to my friend and advisor, Dr. Gary A. Polis, who introduced my to the deserts of Baja California, Mexico. Thanks to Drs. Susan Brawley, Bill Eickmeier, David McCauley, Richard Porter, and the many good people who assisted me in the field, particularly Denise Due, Amy Rosemond, Dave Sissom, Michael Whitlock, Tsunemi Yamashita, Todd Jackson, and Meg Nicholson. I am especially grateful to Lynne Born Myers. This work was partially supported by NSF Doctoral Dissertation Improvement Grant BSR-8514439.

References

Barton, N.H. and Hewitt, G.M. (1981) Hybrid zones and speciation. *In* Atchley, W.R. and Woodruff, D.S. (eds.) *Evolution and Speciation.* Cambridge University Press, Cambridge.

Bergmann, C. (1847) Ueber die Verhaltnisse der Warmeokonomie der Thiere zu ihrer Grosse. *Gottinger Studien,* Part 1: 595–708.

Bert, T.M. and Arnold, W.S. (1995) An empirical test of predictions of two competing models for the maintenance and fate of hybrid zones: Both models are supported in a hard-clam hybrid zone. *Evolution* 49: 276–289.

Calder, W.A. (1984) *Size, Function and Life History.* Harvard University Press, Cambridge, MA.

Case, T.J. (1978) A general explanation for insular body size trends in terrestrial vertebrates. *Ecology* 59: 1–18.

Case, T.J. and Cody, M. (eds.) (1983) *Island Biogeography of the Sea of Cortez.* University of California Press, Berkeley.

Clarke, B.C. (1966) The evolution of morph-ratio clines. *Am. Nat.* 100: 389–402.

Cliff, A.K. and Ord, J.K. (1973) *Spatial Autocorrelation.* Pion, London.

Cole, L.C. (1954) The population consequences of life history phenomena. *Q. Rev. Biol.* 29: 103–137.

Darwin, C. (1859) *The Origin of Species.* D. Appleton, London.

Deharveng, L., Bedos, A. and Gisclard, C. (1998) Environmental factors, microgeographic patterns of endemism and hybrid zones in *Monabella grassei* (Insecta: Collembola: Neanuridae). *Biol. J. Linn. Soc.* 64: 527–554.

Durham, J.W. and Allison, E.C. (1960) The geologic history of Baja California and its maritime faunas. *Syst. Zool.* 9: 47–91.

Ebenman, B. and Persson, L. (1988) *Size-structured Populations.* Springer-Verlag, New York.

Endler, J.A. (1973) Gene flow and population differentiation. *Science* 179: 243–250.

Endler, J.A. (1977) *Geographic Variation, Speciation and Clines.* Princeton University Press, Princeton, NJ.

Fisher, R.A. (1950) Gene frequencies in a cline determined by selection and diffusion. *Biometrics* 6: 353–361.

Francke, O.F. and Jones, S.K (1982) The life history of *Centruroides gracilis* (Scorpiones, Buthidae). *J. Arachnol.* 10: 223–239.

Gadgil, M. and Bossert, W. (1970) Life history consequences of natural selection. *Am. Nat.* 104: 1–24.

Gerristen, J. (1984) Size efficiency reconsidered: a general foraging model for free-swimming aquatic animals. *Am. Nat.* 123: 450–467.

Haldane, J.B.S. (1948) The theory of a cline. *J. Genet.* 48: 277–284.

Hewitt, G.M. (1988) Hybrid zones—natural laboratories for evolutionary studies. *Trends Ecol. Evol.* 3: 158–167.

Koch, L.E. (1981) The scorpions of Australia: aspects of their ecology and zoogeography. *In* Keast, A. (ed.) *Ecological Biogeography of Australia,* pp. 875–884. W. Junk, The Hague.

Lande, R. (1982) Rapid origin of sexual isolation and character divergence in a cline. *Evolution* 36: 213–223.

Lawlor, T.E. (1982) The evolution of body size in mammals: Evidence from insular populations in Mexico. *Am. Nat.* 119: 54–72.

Lindstedt, S.L. and Boyce, M.S. (1985) Seasonality, fasting, endurance and body size in mammals. *Am. Nat.* 125: 873–878.

Lynch, M. (1980) Predation, enrichment and the evolution of cladoceran life histories: at theoretical approach. *In* Kerfoot, W.C. (ed.) *Evolution and Ecology of Zooplankton Communities,* pp. 367–376. University Press of New England, NH.

Mallet, J. (1986) Hybrid zones of Heliconius butterflies in Panama and the stability and movement of warning color clines. *Heredity* 56: 191–202.

McCormick, S.J. and Polis, G.A. (1986) Comparison of the diet of *Paruroctonus mesaensis*at two sites. *Proc. 9th Intl. Congr. Arachnol.* 167–171.

Morse, D.H. (1974) Niche breadth as a function of social dominance. *Am. Nat.* 108: 809–826.

Myers, C.A. (1990) Ecological and evolutionary implications of size variation in the scorpion *Centruroides exilicauda* (Wood) (Ph.D. dissertation). Vanderbilt University, Nashville, TN.

Oliver, J.D., Holeton, G.F. and Chua, K.E. (1979) Overwinter mortality of fingerling smallmouth bass in relation to size, relative energy stores and environmental temperature. *Trans. Am. Fish. Soc.* 108: 130–136.

Persson, L. (1985) Asymmetrical competition: are larger animals competitively superior? *Am. Nat.* 126: 261–266.

Peters, R.H. (1983) *The Ecological Implications of Body Size.* Cambridge University Press, Cambridge.
Polis, G.A. (1984) Age structure component of niche width and intraspecific resource partitioning by predators: can age groups function as ecological species? *Am. Nat.* 123: 541–564.
Polis, G.A. and Sissom, W.D. (1990) Life history. *In* Polis, G. (ed.) *The Biology of Scorpions,* pp. 161–223. Stanford University Press, Palo Alto, CA.
Roff, D.A. (1981) On being the right size. *Am. Nat.* 118: 404–422.
Rose, M.R. and Charlesworth, B. (1981a) Genetics of life history in *Drosophilia melanogaster.* I. Sib analysis of adult females. *Genetics* 97: 173–186.
Rose, M.R. and Charlesworth, B. (1981b) Genetics of life history in Drosophilia melanogaster. II. Exploratory selection experiments. *Genetics* 97: 187–196.
Savage, J.M. (1960) Evolution of a peninsular herpetofauna. *Syst. Zool.* 9: 184–212.
Schoener, T.W. (1969) Models of optimal size for solitary predators. *Am. Nat.* 122: 240–285.
Scorza, J.V. (1954) Sistematica, distribucion geografica y observaciones ecologicas de algunos alacranes encontrados en Venezuela. *Mem. Soc. Ciencias Nat. La Salle* (Caracas) 14(38): 179–214.
Slatkin, M. (1973) Gene flow and selection in a cline. *Genetics* 75: 733–756.
Smith, D.A., Owen, D.F., Gordon, F.J. and Lowis, N.K. (1997) The butterfly *Danaus chrysippus* (L.) in East Africa: polymorphism and morph-ratio clines within a complex, extensive and dynamic hybrid zone. *Zool. J. Linn. Soc.* 120: 51–78.
Sokal, R.R. and Oden, N.L. (1978a) Spatial autocorrelation in biology. I. Methodology. *Biol. J. Linn. Soc.* 10: 199–228.
Sokal, R.R. and Oden, N.L. (1978b) Spatial autocorrelation in biology. 2. Some biological applications of evolutionary and ecological interest. *Biol. J. Linn. Soc.* 10: 229–241.
Speiss, E.B. and Speiss, L.D. (1986) Selection for rate of development and gene arrangement frequencies in Drosophilia persimilis. II. Fitness properties at equilibrium. *Genetics* 53: 695–708.
Stearns, S.C. and Koella, J.C. (1986) The evolution of phenotypic plasticity in life-history traits: predictions of recation norms for age and size at maturity. *Evolution* 40: 893–913.
Via, S. and Lande, R. (1985) Genotype-environment interaction and the evolution of phenotypic plasticity. *Evolution* 39: 505–523.
Voorhies, W.V. (1996) Bergmann size clines: A simple explanation for their occurrence in ectotherms. *Evolution* 50: 1259–1264.
Walker, J.M. (1973) A study of variability in the scorpion *Centruroides exilicauda* (Wood) (Scorpionida: Buthidae) (M.A. thesis). San Francisco State University, CA.
Werner, E.E. and Gilliam, J.F. (1984) The ontogenetic niche and species interactions in size-structured populations. *Annu. Rev. Ecol. Syst.* 15: 393–425.
White, M.J.D. (1978) *Modes of Speciation.* W. Freeman, San Francisco, CA.
Wilbur, H.M (1975) The evolutionary and mathematical demography of the turtle *Chrysemys picta. Ecology* 56: 64–77.
Williams, S.C. (1980) Scorpions of Baja California, Mexico and adjacent islands. *Occas. Pap. Calif. Acad. Sci.* 135: 1–127.

14

Historical Biogeography

A. DENISE DUE

Biogeography is the study of the distribution of plants and animals. For nearly two centuries, scientists, naturalists, and conservationists have sought to understand biogeographic patterns exhibited by various taxa throughout the world. These studies range from examinations of the distribution and numbers of species in entire biotas on a global scale to investigations of a single species at a local level. A primary goal of contemporary biogeography is to determine the processes that produce the observed patterns. Obviously, entire continental or marine areas cannot be studied experimentally to discern what processes lead to large-scale distributional patterns. A reasonable alternative is to select a representative area for detailed description and experimentation. This was the reason for the success of *The Theory of Island Biogeography* by MacArthur and Wilson (1967), a popular and influential work in analytical biogeography. Island biogeographical studies flourished following this advance in theory, but the range of taxa considered were limited, first favoring birds (Diamond, 1969, 1972, 1975; Simberloff, 1970), then eventually, plants, reptiles, and mammals (e.g., Schoener, 1968; Wilcox, 1978; Lomolino, 1982, 1983; Cody, 1983; Lawlor, 1983; Murphy, 1983). Although Wilson's early research focused on the island biogeography of ants, invertebrates generally continue to be ignored in biogeographic studies. The research on Bahaman spiders by Spiller and Schoener (e.g., 1988) is a notable exception.

MacArthur and Wilson's equilibrium theory of island biogeography predicts that species distributions should exist in an equilibrium state resulting from the ongoing stochastic processes of immigration and extinction. As biogeographical studies advance, the observed distributional patterns of some taxonomic groups depart significantly from patterns predicted by the equilibrium theory (Brown, 1971, 1978; Dueser and Brown, 1980; Case, 1983; Lawlor, 1983; Kadmon and Pulliam, 1995; Torres and Snelling, 1997; Thiollay, 1998). For example, taxa with limited dispersal capabilities or specialized habitat requirements tend not to exhibit equilibrium distributional patterns. Moreover, regions with a varied geological history produce unpredicted species distributions. A paleogeographic explanation, the historical legacy hypothesis, has been proposed to explain observed disequilibria. Historical biogeographers attempt to explain these discrepancies by reevaluating the importance of evolutionary and geological history on present-day patterns. This requires biogeographers to study well-described taxonomic groups in discrete areas that have a relatively well understood geological history.

One of the island systems that meets the requirements for historical biogeographical analyses is the Gulf of California. Studies there benefit from a reasonable knowledge of the complex geological history of the region and a comprehensive collection of biogeographical data for numerous taxa other than invertebrates (Case and Cody, 1983). The records for the scorpion fauna on mainland Baja California (Mexico) and Gulf Islands are the most extensive in the world. For this region they form one of the most complete databases assembled for any taxonomic group. The major effort in faunistic study was by Williams (1980), whose studies were continued by Due and Polis (1985; Due, 1992). In total, there are 63 species recorded over the Baja California Peninsula and 67 islands in the Gulf. In this chapter, I show that scorpion fauna on islands in the Gulf of California present an exceptional system for studies of historical biogeography.

SCORPIONS AND BIOGEOGRAPHICAL STUDIES

Several characteristics of scorpions make them useful tools in biogeographical studies: near-worldwide distribution, moderate diversity, moderately well-known taxonomic status, intermediate dispersal capabilities, and adaptive use of habitat. Scorpions are moderately diverse animals of cosmopolitan distribution within southern temperate and equatorial regions (Polis, 1990). They form a significant component of continental biotas (Due and Polis, 1985; Polis, 1990), particularly deserts (Polis, this volume), with appreciable diversity on islands. Ecologically, they occur in varied habitats, from tropical forests to sandy deserts (Polis, 1990). Approximately 1200 extant species are now described. The number of species in a given geographical area is generally small enough (3–16) to render biogeographic patterns interpretable (Due and Polis, 1985; Polis, 1990).

Scorpions have limited locomotive and dispersal abilities (Polis, 1990), roughly comparable to weakly flying or nonflying insects (Yamashita and Polis, 1995; Yamashita and Fet, this volume). They lack intrinsic means of long-distance dispersal (e.g., they do not fly or "balloon" like many spiderlings and are not phoretic). However, scorpions are resilient to sea water and long-term submergence, making it feasible for those living on wood or other floating debris to raft over wide stretches of ocean (Due and Polis, 1985). In recent times, human commerce has also greatly accelerated long-distance dispersal. The latter mechanism has contributed to the spread of *Isometrus maculatus*, *Euscorpius* spp., *Centruroides* spp., and a few *Vaejovis* spp. that inhabit the intertidal zone in coastal areas (Due and Polis, 1985; Due, 1992). Long-distance dispersal by rafting is much less likely, and its proposal to explain distributional disjunctions remains speculative.

Perhaps the most important characteristic of scorpions for historical biogeography is their fidelity to particular environments and apparent inability of many to disperse through unfavorable habitats. For example, many species of *Paruroctonus* are obligate psammophiles never found beyond the sandy soils of dunes where they can establish burrows (Yamashita and Polis, 1995; Yamashita and Fet, this volume). In North America, species of *Uroctonus*, the *Vaejovis nitidulus* group, *Serradigitus*, and other groups only inhabit crevices in vertical rock surfaces or take refuge among stones on rocky slopes. Similarly, Caribbean *Tityus* tend to live in montane forests and are believed to be poor dispersers (Santiago-Blay, 1988). Adaptations to such specialized habitats reduce or eliminate dispersal into areas outside the normal range. Not surprisingly, generalists (e.g., *Centruroides* spp.) are the most successful dispersers. Thus, although the taxon as a whole can utilize a wide variety of habitats, scorpions tend to be habitat specialists at the species level. This fidelity to habitat over time predisposes scorpions to be useful for biogeographical analysis.

EQUILIBRIUM THEORY OF ISLAND BIOGEOGRAPHY AND SCORPIONS ON ISLANDS IN THE GULF OF CALIFORNIA

MacArthur and Wilson's equilibrium theory of island biogeography was one attempt to understand the relationship between species number and (island) area. Both earlier and later biogeographical studies demonstrate a general trend of increasing number of species with increasing area (e.g., Darlington, 1957; Preston, 1962; MacArthur and Wilson, 1967; Diamond and Mayr, 1976; Simberloff, 1976; Rusterholz and Howe, 1979; Lomolino, 1984). A mathematical model presented by Preston (1960, 1962) describes the species–area relationship by plotting species number and area on a log-log scale (referred to as the species-area curve). This relationship is described by the equation $S = kA^z$, where S is the number of species, A is the area, k is the y-intercept, and z is the slope of the curve. Preston (1962) calculated a theoretical value of 0.26 to describe the lognormal distribution (the basis of the mathematical model). This number corresponds to the slope of the species-area curve (z in the equation) and is referred to as a z value. This is the most widely accepted model to explain the species–area relationship and is incorporated into the equilibrium theory.

In most cases, empirical z values fall within a range around Preston's predicted value of 0.26 (0.20–0.35; MacArthur and Wilson, 1967). Considering that sampling error and other factors cause variation around the curve, investigators have come to accept that z values within the range of 0.20–0.35 represent species–area relationships consistent with the equilibrium theory.

Equilibrium theory certainly encompasses more than just mathematical formulations. MacArthur and Wilson attribute the observed biogeographical patterns to the processes of immigration and extinction. Equilibrium theory contends that species richness on an island represents a dynamic equilibrium between ongoing, stochastic immigration and extinction. Four basic assumptions are made: (1) immigration decreases as species number increases; (2) extinction increases with increased species number; (3) immigration is dependent on the degree of isolation; and (4) extinction is dependent on island area. Based on these assumptions, equilibrium theory has three basic predictions: (1) species richness is positively correlated with island area (size); (2) species richness is negatively correlated with isolation of the island (distance from the colonizing source for new species); and (3) the equilibrium number of species on an island remains constant as species composition changes over time (due to stochastic immigration and extinction; this is called "species turnover").

The biogeographical data for scorpions on islands in the Gulf of California were analyzed with respect to these predictions of equilibrium theory. The species-area curve for these scorpions (fig. 14.1) shows a significant positive correlation between area and species richness, but its z value of 0.16 falls outside the accepted range of 0.20–0.35 consistent with equilibrium theory. Lower z values are the result of either fewer species being added with increasing island

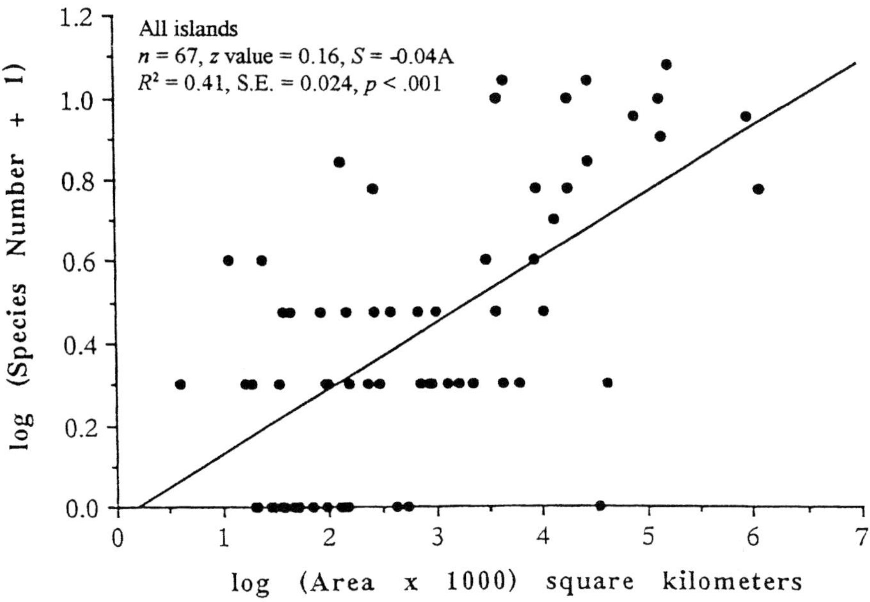

Figure 14.1 Scorpion species-area curve for all islands in the Gulf of California. The *z* value (0.16) is significantly different from the hypothetical value (0.26) from equilibrium theory ($t = 6.68$, $p < .01$, df = 65). The correlation is significant ($r = .637$, $p < .01$, df = 65).

area or more species than expected on smaller islands. An archipelago not sufficiently isolated from the mainland (colonizing source) would yield such results. This idea is supported by the data that do not show a correlation between number of scorpion species and distance to islands (fig. 14.2). Thus, scorpions on Gulf of California islands do not fully conform to these two predictions of equilibrium theory.

Consideration of alternative explanations to the equilibrium theory is warranted. Characteristics of this particular archipelago (e.g., wide range in island age, various island origins, an abundance of small, nearshore islands) may render equilibrium unattainable in ecological time, thus the insular fauna may be in disequilibrium. Ongoing ecological, geological, or evolutionary processes may influence immigration and extinction rates to a point that species distributions are due to deterministic rather than stochastic events as predicted by equilibrium theory (Torres and Snelling, 1997; Perry et al., 1998). The Baja scorpion distribution is not an isolated case of systems in disequilibrium. Earlier studies of mammalian communities demonstrated a correlation of species diversity with area but not isolation (Brown, 1971, 1978; Dueser and Brown, 1980). Brown's (1971) study of mammals on mountaintop "islands" found an unusually high species number for the area (supersaturation). This was attributed to extremely high rates of immigration when islands were connected by favorable environment (land bridges) during the Pleistocene period and unexpectedly low extinction rates since isolation. This scenario suggests that departure from equilibrium is due to historical effects that are not yet overcome by present-day pro-

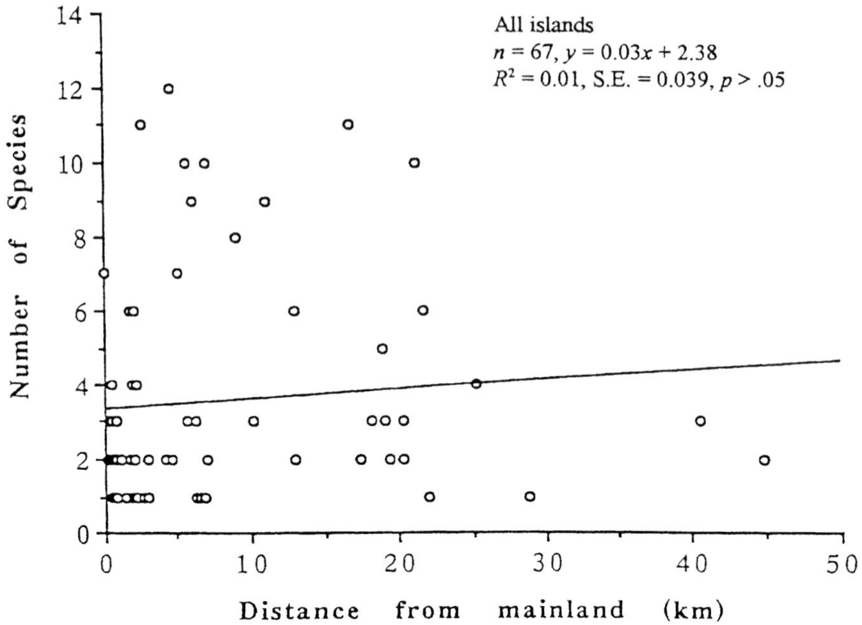

Figure 14.2 Relationship between scorpion species richness and isolation. Species number is a slightly positive function of isolation (distance from the mainland). Correlation: $r = .102$, $p > .05$, df = 65. Equilibrium theory predicts a negative correlation.

cesses. Dueser and Brown (1980) explain disequilibrium of rodents on Virginia's Barrier Islands as the result of low, present-day rates of immigration due to limited dispersal ability. Situations such as these prompted the development of the historical legacy hypothesis (Case and Cody, 1983, 1987). This alternative may be particularly useful when applied to archipelagos with a varied geological history or to taxa with limited dispersal ability. The scorpion fauna on islands in the Gulf of California lends itself well to analysis of this alternative hypothesis.

HISTORICAL ANALYSIS OF THE BIOGEOGRAPHICAL PATTERNS OF SCORPIONS ON ISLANDS IN THE GULF OF CALIFORNIA

Equilibrium theory assumes that historical (paleogeographic) events are erased as insular faunas reach an equilibrium due to current, stochastic immigration and extinction. The historical legacy hypothesis proposes that immigrations occurred in waves (e.g., prior to separation of land-bridge islands from the mainland) and that colonization ceased once an island became isolated from the mainland. Extinctions are considered to be deterministic due to inadequate or insufficient habitat that results from isolation. These processes minimize the importance of species turnover (contra the continual turnover in equilibrium theory).

In contrast to the simplicity of the equilibrium theory, the historical legacy hypothesis also considers specific characteristics of a taxon's biology. The likelihood that a species will occur on a particular island most likely is strongly influenced by taxon-specific traits (e.g., minimum habitat requirements, dispersal abilities). The development of a historical hypothesis to explain observed species richness and distribution is thus more difficult and detailed than for the equilibrium hypothesis. Furthermore, its interpretation varies among archipelagos and taxonomic groups, making conclusions much less general.

Specific characterization of islands based on their geological history (island origin/age) is likewise implicit to understand the predictions of the historical legacy hypothesis. Land-bridge (continental) islands are fragments of the mainland area isolated by rising sea levels or subsidence of surrounding areas. The fauna and flora of land-bridge islands are detached fragments of the mainland biota. Oceanic islands (sensu stricto) have risen out of the ocean and were never connected to the mainland. The biota of oceanic islands is derived from overwater dispersal and subsequent in situ speciation. Islands of different origin that exhibit disequilibrium would be expected to approach equilibrium from different directions. Land-bridge islands would initially contain more species than expected and equilibrium would be attained by a decline in species richness, a process known as relaxation (extinctions exceed immigrations). Oceanic islands would initially support no species and equilibrium would be approached via immigration.

Thus island history and ecological/evolutionary processes are the basis for predictions of the historical legacy hypothesis. These predictions include (1) land-bridge islands are species rich (supersaturated) compared to oceanic islands; (2) species numbers on land-bridge islands (if supersaturated) decline with time; (3) Holocene volcanic islands are species poor compared to old, oceanic islands; (4) physical characteristics (island area, habitat heterogeneity, and isolation) significantly influencing species numbers are different for land-bridge and oceanic islands; and (5) the degree of endemism varies with island origin (i.e., land-bridge or oceanic) and island age (time since isolation). Scorpion distribution on islands in the Gulf of California can be used to test these predictions.

The 67 islands included in this study were divided into two categories based on island origin following the division of Murphy (1983): land-bridge islands (younger islands isolated by Pleistocene sea level changes) and deep-water oceanic islands (including Holocene volcanic islands arising de novo and old, deep-water land-bridge islands that have been isolated from the mainland long enough to be effectively oceanic for the taxa under consideration). Islands were also grouped by approximate age and placed into categories representing youngest to oldest: Holocene, Pleistocene, Pliocene, and Miocene.

Species Richness of Land-bridge versus Oceanic Islands

Figure 14.3 shows the species-area curves for scorpions on land-bridge and oceanic islands. Although the curves are not statistically different from each other, simple visual inspection shows that overall, land-bridge islands now support a larger number of species than oceanic islands. Species numbers on each type of island of comparable size are consistently higher on land-bridge islands (5–10 species versus 1–7 species on oceanic islands). Because these islands appear to be in disequilibrium in terms of the equilibrium theory, at least two possible explanations exist. Land-bridge islands may still have an excess of species,

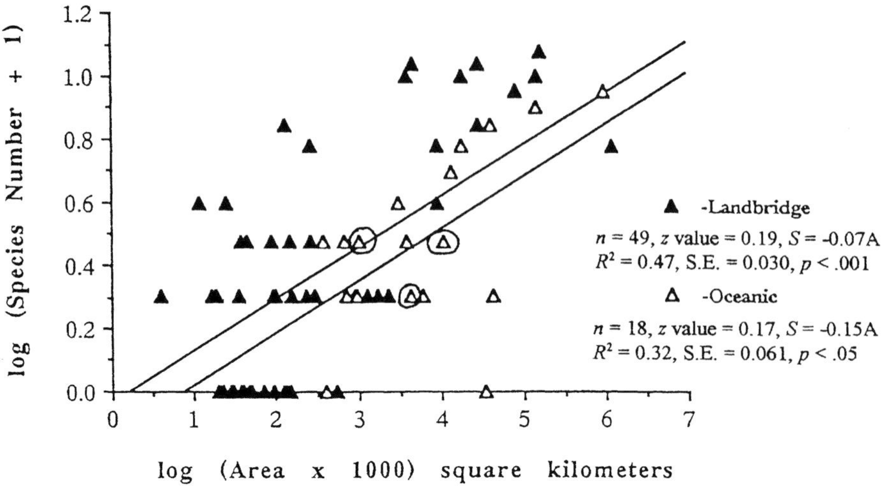

Figure 14.3 Species-area curves of scorpions on land-bridge and oceanic islands. Points representing Holocene volcanic islands are circled. The legends match the right-hand ends of their corresponding curves. Both correlations are significant. The z values are not significantly different from each other ($t = 1.20$).

relicts from the mainland before isolation. Alternatively, oceanic islands may not yet have acquired an equilibrium number of species. The observation that land-bridge islands are species rich cannot be solely attributed to the supersaturation phenomenon without proof that species numbers are declining with time (i.e., relaxation).

Relaxation of Supersaturated Land-bridge Fauna

Wilcox (1978) proposed that, after accounting for area and latitude effects, the residual species diversity should be negatively correlated with age if the fauna is declining toward an equilibrium number of species from a historic period of species superabundance. Wilcox demonstrated such a decline for insular reptiles on 17 land-bridge islands in the Gulf of California. However, a similar analysis for scorpions on 12 of the same islands revealed only an insignificant negative correlation. This result suggests that the time required for relaxation of species number is relatively short for scorpions, making supersaturation an unsatisfactory explanation for the greater number of species presently observed on land-bridge islands.

Species Richness of Holocene Volcanic Islands versus Old Oceanic Islands

If land-bridge islands are not supersaturated, perhaps oceanic islands are species poor due to low immigration rates. It is difficult to measure immigration rate over a short period of time. However, if true volcanic islands are species poor compared to old oceanic islands, low immigration rates might be inferred.

Only three islands in the Gulf (Raza, Tortuga, and Encantada Grande [San Luis]) are reported to be of Holocene volcanic origin, and each supports only one or two scorpion species. The data points representing these islands are indicated on the oceanic species-area curve (fig. 14.3). These points are not extreme outliers; they fall within the same range on the curve as the points indicating species number on comparably sized old oceanic islands. This suggests that the volcanic islands are not species-poor compared to old oceanic islands. It appears that scorpion dispersal ability is adequate for present-day immigration rates to overcome the historical aspect of island origin.

Variables Significantly Influencing Species Numbers on Islands

Physical characteristics, such as habitat heterogeneity and isolation, might dictate the colonization success of species on islands. This may be especially true for scorpions and similar taxa that exhibit species-specific affinity to particular habitats or dispersal ability limited to short distances. The effect of physical variables on scorpion species richness was analyzed by stepwise multiple regressions for all islands together and for land-bridge and oceanic islands separately (table 14.1). The independent variables included three estimates of habitat heterogeneity—namely, island area, number of plant species, elevation, and one estimate of isolation (distance from island to nearest mainland). Only factors related to habitat heterogeneity explain a significant portion of the species–area relationship. Over all islands, the number of plant species appears most important. Land-bridge and oceanic island species richness also is influenced by area and elevation, respectively. Distance (isolation) exhibits little, if any, effect on the number of scorpion species. The absence of an isolation effect on species richness is consistent with the observation that land-bridge and oceanic islands included in this analysis do not conform well to the predictions of equilibrium theory.

Table 14.1 Regression analysis for three groups of islands using four independent variables: log(area), number of plant species, distance from mainland, and elevation

Variable	Contribution to R^2	F value	p
All islands[a]			
No. of plant species	0.514	43.29	.0001
log(area)	0.033	2.86	.0984
Elevation	0.046	4.41	.0421
Distance	0.022	2.41	.1463
Total	0.615		
Land bridge			
log (area)	0.668	46.37	.0001
Oceanic			
Elevation	0.560	20.39	.0004

[a]Land bridge and oceanic.

Table 14.2 Endemism as a function of island origin and island age[a]

	Land bridge	Oceanic	Young	Old
No. of endemic species	3	3	2	5
Avg. % endemism	2.0	5.0	1.0	9.0
Range (%)	0–40	0–50	0–11	0–50
No. of islands	36	16	38	14
t test (df = 50)	$t = 1.07$ $p > .05$		$t = 2.95$ $p < .01$	

[a]Similarity or difference between percent endemism for each association is determined by *t* tests. There are no differences in percent endemism between land bridge and oceanic islands. Endemics are significantly more frequent on older compared to younger islands.

Endemism

The final prediction of the historical legacy hypothesis is that the degree of endemism differs for islands of different origin and age. Six of 34 island scorpion species (18% overall) are endemic. For land-bridge and oceanic islands, 3 of 28 species (11%) and 3 of 23 species (13%) are endemic, respectively. However, when percent endemism for individual islands is averaged (as done for reptiles by Case, 1983), the degree of endemism is much lower: 2% on land-bridge and 5% on oceanic islands. Regardless of how it is calculated, endemism is relatively low for scorpions compared to other taxa (e.g., mammals, 16% on land-bridge and 69% on oceanic islands; reptiles, 5% and 35%, respectively [Case and Cody, 1983]). Importantly, endemism does not differ between land-bridge and oceanic islands (table 14.2). It was also suggested that endemism may be correlated with island size, isolation, and age (Mayr, 1965; Case, 1983; Gilpin et al., unpublished manuscript). A positive correlation between endemism and island size was demonstrated by Mayr (1965) and supported by Gilpin et al. (unpublished manuscript), who suggested that smaller populations on smaller islands would likely go extinct before endemic status was reached. When correlation coefficients were determined for scorpion endemism and each of the above factors, the relationship between endemism and island size was slightly negative ($p > .05$), whereas significant correlations exist between endemism and island isolation ($p < .05$) or island age ($p < .01$). The significant difference between percent endemism on young (1%) and old (9%) islands (table 14.2) is also evidence of the relationship between endemism and island age.

Taxonomic Comparisons

The historical legacy hypothesis was evaluated previously for other taxonomic groups—specifically reptiles, amphibians, mammals, and plants (Case, 1983; Lawlor, 1983). Table 14.3 shows taxonomic comparisons of these groupings and how well they accommodate the historical legacy hypothesis. Notably, the various taxa show a gradation in the importance of historical factors on biogeographic patterns. Generally, this gradation is also reflected in the dispersal ability of the taxon. The biogeographic patterns of groups with more limited dispersal ability are better explained by historical factors, whereas species distributions of highly mobile taxa conform closely to the predictions of equilib-

Table 14.3 Questions and answers from the historical legacy hypothesis applied to various taxonomic groups in the Gulf of California

Taxa	Are land bridge islands species rich relative to oceanic islands?	Are holocene volcanic islands species poor compared to old oceanic islands?	Does number of species decrease with distance?		Endemism (%)	
			Land bridge	Oceanic	Land bridge (young)	Oceanic (old)
Plants	No[a]	No	No	No	0	2
Land birds[b]	No	No	No	No	0	0
Scorpions	No	No	No	No	2 (1)[d]	5 (9)
Land reptiles[c]	Yes	No	No	No	5	35
Lizards[c]	Yes	No	No	Yes	0	47
Mammals	Extremely	Slightly	Slightly	Yes	16	69

[a]"Yes" and "no" imply statistical significance or its absence at $p < .05$. "Slightly" implies a small effect that may or may not be statistically significant, and "extremely" implies a large effect.

[b]Land birds were not included in the comparative studies with scorpions but are included here for reference. It is interesting to note the similarities between land birds and scorpions.

[c]These two taxa were considered as one in the comparative study with scorpions.

[d]Values in parentheses indicate the percent endemism of scorpions if islands are considered as young and old, rather than land bridge and oceanic. Modified from Case and Cody (1987).

rium theory. Biogeographic patterns of insular plant distribution are largely unaffected by the historical factors listed here. Mammals, as generally poor dispersers, conform more closely to the predictions of historical legacy hypothesis and conform little to equilibrium theory.

Reptiles, amphibians, and scorpions fall between the extremes set by plants and mammals, exhibiting patterns consistent with both the equilibrium theory and historical legacy hypothesis. One important departure from equilibrium theory is indicated by the progressive increase in endemism of scorpions as islands age, suggesting that an equilibrium number is not maintained by dynamic immigrations and extinctions (species turnover). Instead, the scorpion fauna appears to be more stable over time, allowing speciation to occur. This suggests that immigration and extinction may not be stochastic processes as predicted by equilibrium theory, but deterministic processes.

DIRECTIONS FOR FUTURE RESEARCH

Any one model of biogeographic diversity rarely provides a satisfactory explanation of the patterns for a wide range of taxa. It is not surprising, therefore, that scorpion distributions on islands in the Gulf of California are not fully explained by either the equilibrium theory or the historical legacy hypothesis. It is their failure to conform completely to theory that provides some insight into the actual processes that may shape distributional patterns. Their limited dispersal ability and the progressive increase of endemism with island age suggest that immigration and/or extinction may not be stochastic but deterministic for scorpion fauna. Characteristics of the scorpion taxa and the history of the Baja archipelago can be further evaluated to better understand the processes that underlie observed distributions. For example, data on species-specific presence and absence on the islands should be examined for recognizable patterns indicating a deterministic element to colonization success. If some species always or never occur together, or if some only occur on islands of a minimum size, then habitat use is likely to be correlated with distribution and biogeographic patterns. In this way, detailed niche analysis may eventually indicate the likelihood of colonization success and coexistence of certain species.

References

Brown, J.H. (1971) Mammals on mountaintops: nonequilibrium insular biogeography. *Am. Nat.* 105: 467–478.

Brown, J.H. (1978) The theory of insular biogeography and the distribution of boreal birds and mammals. *Great Basin Nat. Mem.* 2: 209–227.

Case, T.J. (1983) The reptiles: Ecology. Pp. 159–209. *In* Case, T.J. and Cody, M.L. (eds.) *Island Biogeography in the Sea of Cortez,* University of California Press, Berkeley.

Case, T.J. and Cody, M.L. (eds.) (1983) *Island Biogeography in the Sea of Cortez,* University of California Press, Berkeley.

Case, T.J. and Cody, M.L. (1987) Testing theories of island biogeography. *Am. Sci.* 75: 402–411.

Cody, M.L. (1983) The land birds. *In* Case, T.J. and Cody, M.L. (eds.) *Island Biogeography in the Sea of Cortez,* pp. 210–245. University of California Press, Berkeley.

Darlington, P.J. (1957) *Zoogeography: The Geographical Distribution of Animals.* John Wiley and Sons, New York.
Diamond, J.M. (1969) Avifaunal equilibria and species turnover on the Channel Islands of California. *Proc. Natl. Acad. Sci. USA* 64: 57–63.
Diamond, J.M. (1972) Biogeographic kinetics: estimation of relaxation times for avifaunas of southwest Pacific islands. *Proc. Natl. Acad. Sci. USA* 69: 3199–3203.
Diamond, J.M. (1975) Assembly of species communities. *In* Cody, M.L. and Diamond, J.M. (eds.) *Ecology and Evolution of Communities,* pp. 342–444. Belknap Press, Cambridge, MA.
Diamond, J.M. and Mayr, E. (1976) Species-area relation for birds of the Solomon archipelago. *Proc. Natl. Acad. Sci. USA* 73: 262–266.
Due, A.D. (1992) Biogeography of scorpions in Baja California, Mexico and an analysis of the insular scorpion fauna in the Gulf of California (Ph.D. dissertation). Vanderbilt University, Nashville, TN.
Due, A.D. and Polis, G.A. (1985) Biology of the intertidal scorpion, *Vaejovis littoralis. J. Zool.* 207: 563–580.
Dueser, R.D. and Brown, W.C. (1980) Ecological correlates of insular rodent diversity. *Ecology* 61: 50–56.
Kadmon, R. and Pulliam, H.R. (1995) Effects of isolation, logging and dispersal on woody-species richness of islands. *Vegetatio* 116: 63–68.
Lawlor, T.E. (1983) The mammals. *In* Case, T.J. and Cody, M.L. (eds.) *Island Biogeography in the Sea of Cortez,* pp. 265–289. University of California Press, Berkeley.
Lomolino, M.V. (1982) Species-area and species-distance relationships of terrestrial mammals in the Thousand Island Region. *Oecologia* 54: 72–75.
Lomolino, M.V. (1983) Island biogeography immigrant selection and mammalian body size on islands (Ph.D. dissertation). State University of New York at Binghamton.
Lomolino, M.V. (1984) Mammalian island biogeography: effects of area, isolation and vagility. *Oecologia* 61: 376–382.
MacArthur, R.H. and Wilson, E.O. (1967) *The Theory of Island Biogeography.* Princeton University Press, Princeton, NJ.
Mayr, E. (1965) Avifauna: turnover on islands. *Science* 150: 1587–1588.
Murphy, R.W. (1983) The reptiles: origin and evolution. *In* Case, T.J. and Cody, M.L. (eds.) *Island Biogeography in the Sea of Cortez,* pp. 130–158. University of California Press, Berkeley.
Perry, G., Rodda, G.H., Fritts, T.H. and Sharp, T.R. (1998) The lizard fauna of Guam's fringing islets: Island biogeography, phylogenetic history, and conservation implications. *Glob. Ecol. Biogeogr. Lett.* 7: 353–365.
Polis, G.A. (1990) Ecology. *In* Polis, G.A. (ed.) *The Biology of Scorpions,* pp. 247–293. Stanford University Press, Palo Alto, CA.
Preston, F.W. (1960) Time and space and the variation of species. *Ecology* 41: 611–627.
Preston, F.W. (1962) The canonical distribution of commonness and rarity: Part I. *Ecology* 43: 185–215.
Rusterholz, K.A. and Howe, Y.W. (1979) Species-area relations of birds on small islands in a Minnesota lake. *Evolution* 33: 468–477.
Santiago-Blay, J.A. (1988) A fossil scorpion *Tityus geratus,* new species (Scorpiones: Buthidae) from Dominican amber. *Hist. Biol.* 1: 345–354.
Schoener, T.W. (1968) Sizes of feeding territories among birds. *Ecology* 49: 123–141.
Simberloff, D. (1970) Taxonomic diversity of island biotas. *Evolution* 24: 23–47.
Simberloff, D. (1976) Experimental zoogeography of islands: effects of island size. *Ecology* 57: 629–648.

Spiller, D.A. and Schoener, T.W. (1988) An experimental study of the effect of lizards on web-spider communities. *Ecol. Monogr.* 58: 57–77.

Thiollay, J.M. (1998) Distribution patterns and insular biogeography of South Asian raptor communities. *J. Biogeogr.* 25: 57–72.

Torres, J.A. and Snelling, R.R. (1997) Biogeography of Puerto Rican ants: A non-equilibrium case? *Biodivers. Conserv.* 6: 1103–1121.

Wilcox, B.A. (1978) Supersaturated island faunas: a species-age relationship for lizards on post-Pleistocene land-bridge islands. *Science* 199: 996–998.

Williams, S.C. (1980) Scorpions of Baja California, Mexico and adjacent islands. *Occas. Pap. Calif. Acad. Sci.* 135: 1–127.

Yamashita, T. and Polis, G.A. (1995) A test of the central-marginal model using sand scorpion populations (*Paruroctonus mesaensis, Vaejovidae*). *J. Arachnol.* 2: 60–64.

15

Molecular Approaches to Biogeography

TSUNEMI YAMASHITA
VICTOR FET

Over the past several decades, the most powerful tools for taxonomic and biogeographic studies of organismal diversity have included morphological and molecular means of analysis. This evolution in methodology has significantly improved our understanding of the genetic variability and taxonomic relationships of modern scorpions and promises to reveal aspects of their biogeographic history and factors controlling speciation within the group. In this chapter we summarize how morphological (phenotypic), allozyme (protein), and nucleotide sequence information obtained from native populations of scorpions can be used to construct a more accurate picture of their systematics, biogeography, and evolutionary history. Although morphological characteristics continue to be the most easily assessed indication of natural groupings within the order, taxonomic delineation is also accurately assessed through molecular analysis of genetic characters. Allozyme analysis was the first molecular technique used to quantify genetic variation and geographic differentiation (Lewontin and Hubby, 1966; Ferguson, 1980; Murphy et al., 1990), and it clearly established the importance of molecular methods for testing hypotheses formulated from morphological data. Recently, techniques that measure nucleotide (DNA) differences among populations have become favored tools for quantification of genetic structure and taxonomic affinities. Using a combination of these molecular approaches, we may now define in detail the populational and ecological processes that create genetic variation among scorpion populations and begin to reconstruct the events leading to their current geographic distribution.

INTRASPECIFIC VARIATION IN SCORPIONS

Until recently, scorpion taxonomists described variation among populations using only morphological criteria. The most useful traits for comparison are listed here, along with examples of how they have been applied in classical taxonomic studies.

Morphosculpture describes the shape and size of keels, crests, denticles, and other details of cuticle features. Hirsuteness defines the number, size, and color of bristles, other than trichobothria, found on the pedipalps, walking legs, and metasoma of scorpions. Hirsuteness has been used to separate subspecies of *Vaejovis punctipalpi cerralvensis* Williams (Vaejovidae) from Isla Cerralvo (Gulf of California), which differs from mainland *V. p. punctipalpi* (Wood) only by the telson hirsutiness (Williams, 1980). In some species coloration varies intensity in pattern, especially for melanization of metasomal segments. Color may express phenotypic plasticity but may as well be a selected trait, as in the black variant of *Centruroides exilicauda* living on black volcanic islands in the Gulf of California (Williams, 1980). Number of pectinal teeth gives a meristic (countable) character that is variable in all scorpions. Teeth number usually ranges from 10 to 30, with mean number varying between sexes and geographic locations. Using this character, Kinzelbach (1975) confirmed the separation of *Mesobuthus gibbosus gibbosus* (Brullé) (Buthidae; southern Balkan Peninsula and northern Aegean) from *M.g. anatolicus* (Schenkel) (southern Aegean, Cyprus, and Turkey). Finally, the fine sensory hairs lining segments of the pedipalp, the trichobothria, give a species-constant and easily described character for taxonomic differentiation (Vachon, 1974; Fet and Rechkin, 1990). In Mediterranean *Euscorpius,* trichobothrial number and patterns differed within the species' range, and even within one population (Bonacina, 1980, 1983; Fet, 1986), indicating that it may be a sensitive indicator of genetic diversity.

The delineation of populations and subspecies by these morphological criteria can generate a confusing taxonomy. For example, in *Euscorpius* (Euscorpiidae) the primary criterion for establishing subspecies is number and location of trichobothria on the pedipalp. By this criterion, *Euscorpius carpathicus* (L.) includes 26 distinguishable subspecies from the shores and islands of the Mediterranean (fig. 15.1; Hadzi 1931; di Caporiacco, 1950; Curcic, 1972; Kinzelbach, 1975; Valle, 1975; Vachon and Jaques, 1977; Bonacina, 1980, 1983; Fet, 1985, 1997; Fet and Rechkin, 1990). Similarly, *Centruroides exilicauda* (Wood) (Buthidae) occurs from Sonora and Baja California, Mexico through Arizona to Utah, where its populations vary greatly in morphology, color, size, life history, and venom toxicity. This species was originally described as four different species (Gertsch, 1958), but an extensive morphological analysis of *C. exilicauda* from Baja California (Walker, 1973) failed to delineate subspecies due to extensive phenotypic variation without apparent geographic patterns.

Figure 15.1 Distribution of *Euscorpius carpathicus* (Euscorpiidae).

Currently, taxonomists remain undecided as to whether *C. exilicauda* is one highly variable species or a complex of similar species (Williams, 1980; W.D. Sissom, personal communication). These demarcations are of more than academic interest because "*C. sculpturatus* (Ewing)" (Sonora and Arizona) is potentially lethal to humans, but *C. exilicauda* (Baja California) is not. Without more dependable and uniform criteria for estimating intraspecific variation in scorpions, their taxonomic associations will remain uncertain. By this reasoning, genetic techniques become the only accurate means of analyzing variation within and between some populations and species groupings.

ALLOZYME ANALYSIS: A CASE STUDY OF *PARUROCTONUS MESAENSIS*

One means of quantifying genetic variability in populations of scorpions is through allozyme electrophoresis. This method differentiates among populations or individuals by distinguishing genetically variable enzymes (allozymes) on the basis of their relative electrophoretic mobility on a gel substrate. We applied this method to resolve genetic differences in populational structure of the sand scorpion, *Paruroctonus mesaensis* Stahnke (Vaejovidae), an abundant, psammophilic species inhabiting sand dunes and sandy substrates of the southwestern United States and Mexico (fig. 15.2). Geographically, these sandy habitats are interrupted by mountain ranges that limit or prevent migration of *P. mesaensis,* thus creating terrestrial "islands" within the larger geographic range of their distribution (approximately 2400 km^2, from Southern California to Arizona, Baja and Sonoran Mexico) (Williams, 1980; Haradon, 1983). By morphological criteria, these potentially isolated populations show variability in number of pectinal teeth, cuticular denticles on the pedipalps, adult size, and color (Haradon, 1983), but there is no clear distinction of subspecies.

As with other species, the genetic structure of *Paruroctonus* populations is likely affected by their abilities to disperse throughout the landscape. Good dispersers spread gametes over a broader area than poorly mobile species, which are more likely to form differentiated and isolated subpopulations. Dispersal abilities of most scorpion species are probably limited also by their ecology. Strongly psammophilic scorpions, such as *P. mesaensis,* possess morphological features (e.g., "sand shoes") that both adapt and restrict them to sandy soils (Polis, 1990b), thus limiting their migration over mountains and rocky substrates. Such ecological circumstances favor the formation of many endemic species or even genera on sand dunes (e.g., *Vejovoides longiunguis* in Baja California) (Pavlik, 1989; Polis, 1990b). Dependence of some scorpions on semipermanent burrows is another ecological factor that influences their dispersal because foraging and reproductive activities are usually confined to within a few meters of this refuge (Polis, 1980, 1990b). Females probably remain in the same burrow for many years, while mature males move tens to hundreds of meters during the mating season. In one study (Polis et al., 1986), marked, mature male *P. mesaensis* moved at least 35 m from their burrows and were occasionally observed hundreds of meters from sandy soils. Thus, male scorpions are much more vagile than females and probably provide the greatest conduit for gene flow and introgression among adjacent demes (Polis and Farley, 1979, 1980; Benton, this volume).

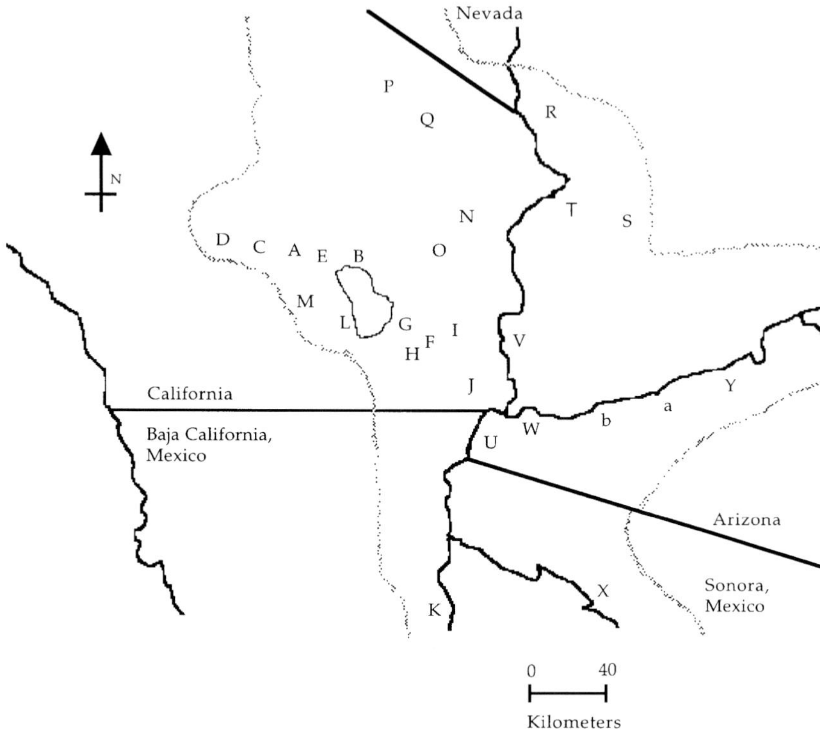

Figure 15.2 Desert areas of California, Nevada, Arizona, Sonora (Mex.), and Baja California Norte (Mex.) where populations of *Paruroctonus mesaensis* were collected. The scorpion's approximate geographical range boundary is shown by the stippled line. A, Thousand Palms; B, North Shore; C, GAT; D, Windy Point; E, Avenue 44; F, Glamis; G, Calipatria; H, Glamis West; I, Glamis East; J, Ogilby; K, San Felipe; L, Salton Sea; M, Borrego Springs; N, Rice; O, Palen; P, Kelso North; Q, Kelso; R, Needles; S, Bouse; T, Parker; U, Yuma Desert; V, Yuma Army; W, N. Fontana; X, Rocky Point; Y, Agua Caliente; Z, Phoenix; a, Dateland; and b, Mohawk.

If the psammophilic and burrowing habits of *P. mesaensis* prevent extensive dispersal and encourage formation of isolated populations, several predictions can be made and experimentally tested concerning the genetic structure of *P. mesaensis* populations:

Prediction 1: Medium-level average heterozygosity. The proportion of heterozygous individuals for an enzyme locus is a measure of a population's genetic variation, defined as heterozygosity (Hartl and Clark, 1989). Allozyme electrophoresis allows one to distinguish heterozygotes from homozygotes and thus to calculate average heterozygosity as the mean value for all enzyme loci distinguished by electrophoresis. Heterozygosity ranges from 0.03 to 0.31 (Ferguson, 1980), with small, isolated populations showing the lowest values (Nei, 1987; Crouau-Roy, 1989). Gene flow is constrained by isolation, and genetic drift acts to reduce the numbers of alleles in small populations (Nei, 1987;

Spiess, 1989). *P. mesaensis* is a poor disperser, so its populations are expected to experience little gene flow and, thus, low values of heterozygosity. However, the population density of this species is high, often >2000 individuals/ha (Polis and Yamashita, 1991). This large population size should balance allele reduction through genetic drift and thus support an median level of heterozygosity for *P. mesaensis.*

Prediction 2: Relatively high allele number. When a new population is established by a small number of individuals, it will possess only a fraction of the genetic variability of the source population (Mayr, 1954; Cook, 1991). This "founder effect," or catastrophic reductions in population size caused by natural events, can force populations through genetic bottlenecks where allelic diversity per locus is diminished (Spiess, 1989). In *P. mesaensis,* population sizes are sufficiently large to preclude founder effects and genetic bottlenecks. Thus, we predict allele number to be stable and relatively high.

Prediction 3: Medium to high level of population subdivision (F_{ST}). F_{ST} values quantify genetic substructure of populations in values that range from 0 (no genetic differentiation, or panmixia) to 1 (complete isolation of subpopulations and no gene flow) (Wright, 1978; Nei, 1987; Barrowclough, 1983; Larson et al., 1983). Most organisms capable of extended dispersal (i.e., birds, flying insects) have F_{ST} values near zero: those with reduced dispersal capabilities (i.e., amphibians) show values near one (Barrowclough, 1983). Sand scorpions are poor dispersers, so *P. mesaensis* is predicted to have medium to high values of F_{ST}.

Prediction 4: Small neighborhood size. The neighborhood of a population is the geographic area within which conspecifics experience a high probability of gamete exchange. Mobile species have large neighborhood sizes, and poor dispersers, such as *P. mesaensis,* should have small neighborhood sizes.

Methods of Allozyme Analysis

We studied five polymorphic and three monomorphic loci from 28 populations across the range of *P. mesaensis* habitation. These eight genetic markers were the only interpretable loci of 25 enzyme loci that were assayed. For each locus 30 (± 6.5) individuals from each population were assayed using a standard protocol (Richardson et al., 1986; Yamashita, 1993). To measure genetic variation within a population, we used estimates of average heterozygosity, percent polymorphism, and the mean number of alleles in each population (Nei, 1987; Pasteur et al., 1988). To measure variation among populations, we calculated Wright's *F* statistic values and Modified Rogers' genetic distance with the Biosys-1 computer program (Swofford and Selander, 1989). Wright's *F*-statistic is commonly used to estimate population subdivision (Wright, 1978; Barrowclough, 1983; McCauley and Eanes, 1987; Hartl and Clark, 1989; Spiess, 1989). The significance of the *F* values was tested by a chi-square contingency analysis (Workman and Niswander, 1970; Caccone and Sbordoni, 1987). Genetic distance data were analyzed via the unweighted pair-group method using arithmetic averages (UPGMA) clustering analysis (Sneath and Sokal, 1975; Nei, 1987; Swofford and Olsen, 1990). Because data from closely related populations were often difficult to discern by clustering, we also performed multidimen-

Table 15.1 Estimates of genetic variability for 28 populations of *Paruroctonus mesaensis*

Population	Mean heterozygosity/ locus (SD)	Mean allele number/ locus (SD)
TP4	0.168 (0.065)	1.63 (0.18)
North Shore	0.181 (0.081)	1.75 (0.25)
GAT	0.146 (0.053)	1.75 (0.25)
Windy Point	0.103 (0.064)	1.50 (0.27)
Ave 44	0.146 (0.069)	1.50 (0.19)
Glamis	0.153 (0.073)	1.75 (0.25)
Calipatria	0.205 (0.088)	1.63 (0.26)
Glamis West	0.133 (0.064)	1.63 (0.26)
Glamis East	0.187 (0.091)	1.75 (0.25)
Ogilby	0.167 (0.074)	1.63 (0.26)
San Felipe	0.146 (0.058)	1.88 (0.30)
Salton Sea	0.181 (0.085)	1.75 (0.25)
Borrego Springs	0.160 (0.073)	1.75 (0.25)
Rice	0.190 (0.078)	1.75 (0.25)
Palen	0.119 (0.061)	1.63 (0.26)
Kelso North	0.142 (0.090)	1.50 (0.27)
Kelso	0.084 (0.055)	1.63 (0.26)
Needles	0.097 (0.056)	1.50 (0.19)
Bouse	0.131 (0.072)	1.63 (0.26)
Parker	0.114 (0.042)	1.75 (0.25)
Yuma Desert	0.176 (0.072)	2.00 (0.38)
Yuma Army	0.136 (0.057)	1.63 (0.26)
N. Fontana	0.166 (0.080)	1.63 (0.26)
Rocky Point	0.121 (0.056)	1.63 (0.26)
Agua Caliente	0.140 (0.080)	1.50 (0.27)
Phoenix	0.078 (0.060)	1.38 (0.18)
Dateland	0.115 (0.068)	1.50 (0.19)
Mohawk	0.110 (0.053)	1.63 (0.26)
Mean	0.143 (0.033)	1.65 (0.13)

sional scaling (MDS) analysis (Wilkinson, 1987; Wilkinson et al., 1992), which maps distances among data points onto several dimensions (Schiffman et al., 1981; Lessa, 1990). To investigate geographical patterns in allele frequencies (Sokal and Oden, 1978b; Bocquet-Appel and Sokal, 1989), we performed a spatial autocorrelation analysis (Wartenberg, 1989). This test determines whether the observed value of a variable at one location is independent of values of the same variable at neighboring locations (Sokal and Oden, 1978a; Sokal, 1979).

Genetic Structure and Differentiation of Sand Scorpion Populations

Analysis of allozyme variability revealed that most sand scorpion populations are not genetically isolated from one another. Average heterozygosity for all the populations was 0.14 ± 0.03 (n = 28; range 0.08–0.20) (table 15.1). This high value was obtained in part because relatively few monomorphic loci were re-

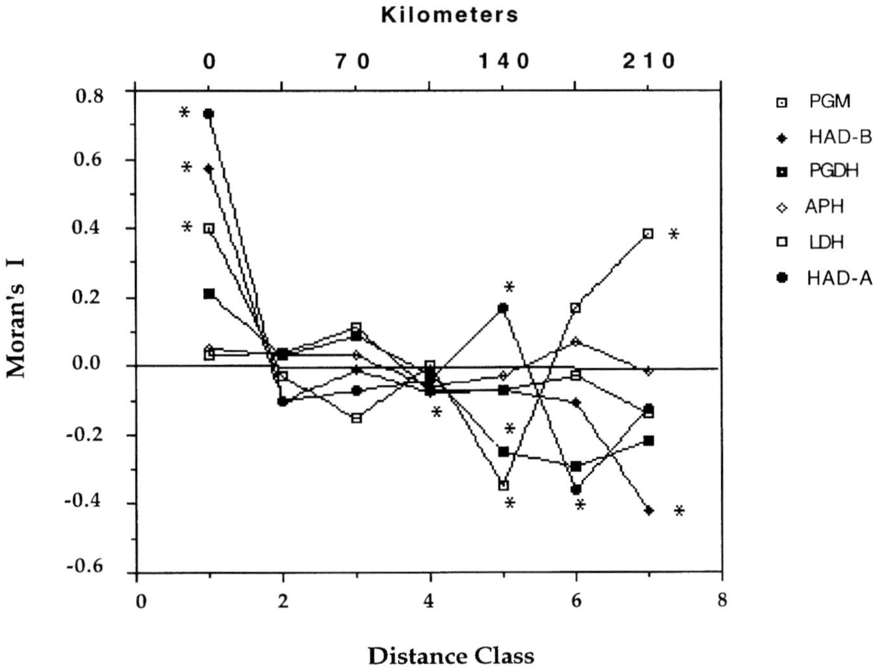

Figure 15.3 Spatial autocorrelation showing association among allozyme frequencies. The initial decline in the autocorrelation coefficients indicates a small neighborhood size. Significant autocorrelation coefficients ($p < .05$) are marked with an asterisk.

solved. Nevertheless, these values are consistent when compared among *P. mesaensis* populations (fig. 15.3). The percent polymorphism for all populations ranged from 37.5% to 62.5% (mean 53.1 ± 8.7), and no population was completely monomorphic for the five polymorphic loci. The mean allele number for each locus in each population (table 15.1) ranged from 1.38 to 2.00 (mean 1.65 ± 0.13). Seventy-five percent (89/119) of all loci examined did not deviate significantly from Hardy-Weinberg equilibrium. The only populations with loci deviating significantly from Hardy-Weinberg predictions were those with small population sizes or on margins of their range (e.g., Phoenix and Bouse).

None of the genetic markers showed the effects predicted for reproductively isolated populations. Three of eight loci were monomorphic for all populations, and no locus was fixed for unique alleles in any population (completely monomorphic). However, unique alleles existed in two populations. For example, in the San Felipe and Yuma Desert populations, the *PGM* (phosphoglucomutase) gene was represented by three and four alleles, respectively; other populations contained two alleles. Two of the five polymorphic loci (*ACPH* in 25 populations and *LDH* [lactate dehydrogenase] in 26 populations) exhibited frequencies >0.90 for the most common allele (i.e., most individuals were homozygous for these loci). Thus only three loci (*HAD* [3-hydroxybutyrate dehydrogenase], *PGM,* and *PGDH* [phosphogluconate dehydrogenase]) exhibited substantial polymorphisms.

Table 15.2 Wright's *F* statistics and heterogeneity χ^2 of six loci for all populations of *P. mesaensis* ($n = 28$)[a]

Locus	F_{IS}	F_{IT}	F_{ST}	χ^2 ($p < .001$)
PGM	−0.022	0.164	0.182	504.60
HAD	0.178	0.341	0.199	525.09
PGDH	0.453	0.555	0.187	294.35
APH	−0.026	0.012	0.036	134.01
LDH	−0.110	−0.030	0.072	111.88
Mean	0.170	0.315	0.175	

[a]The heterogeneity χ^2 tests the significance of heterozygosity in gene frequencies among populations. F_{IS} is the reduction in heterozygosity of individuals due to inbreeding within a subpopulation; F_{IT} is the reduction in heterozygosity of an individual compared to the total population; and F_{ST} is the reduction in heterozygosity of subpopulations. *PGM,* phosphoglucomutase; *HAD,* 3-hydroxybutyrate dehydrogenase; *PGDH,* phosphogluconate dehydrogenase; *APH,* acid phosphotase; *LDH,* lactate dehydrogenase.

As summarized in table 15.2, the genetic differentiation among populations of *P. mesaensis* was low, with a mean F_{ST} value for all polymorphic loci of 0.175. Mean values for F_{IS} and F_{IT} were 0.170 and 0.315, respectively. The chi-square test showed the F_{ST} value are significant, indicating that gene frequencies differ among populations (Workman and Niswander, 1970; Caccone and Sbordoni, 1987).

The clustering analysis summarized in fig. 15.4 shows that many populations are close in genetic distance, especially those in the center of the range for *P. mesaensis.* Genetic distance values ranged from 0.047 to 0.398, with most values below 0.300. These low distance values made it difficult to associate populations, as many branched near the terminal nodes. Geographically marginal populations appeared at the periphery of the MDS graph shown in fig. 15.5, with three populations (Bouse, Phoenix, and Needles) being isolated from the others. The Coachella Valley populations on the western margin of the *P. mesaensis* range are generally distinct from the other clusters. Thus, the high level of UPGMA clustering near the terminal nodes and the large number of populations at the center of the MDS graph support the conclusions of the F_{ST} analysis that populations are not well differentiated. If populations were more isolated, genetic distances and F_{ST} values would be higher.

Although *P. mesaensis* populations have small neighborhood size (fig. 15.3), they exhibit medium levels of average heterozygosity within populations and limited genetic differentiation among populations. These results suggest that these populations are not strongly isolated from each other or genetically depauperate.

Variation within populations

The values determined here for average heterozygosity in *P. mesaensis* (mean = 0.143) are higher than those reported by others (Ferguson, 1980), probably because the total number of loci resolved for our analysis was lower than in most

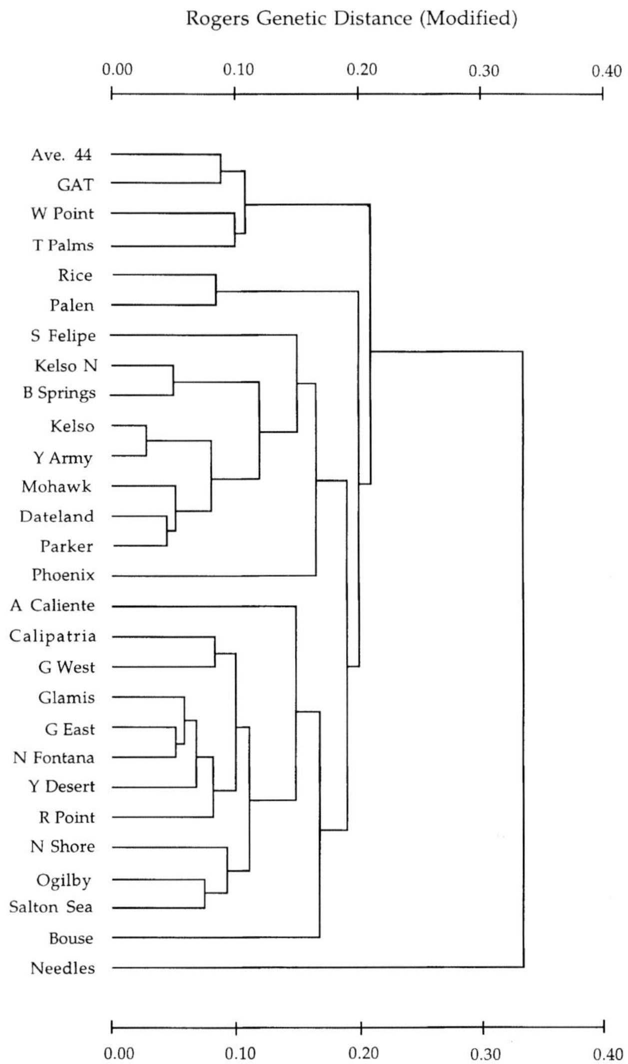

Figure 15.4 Unweighted pair-group method. Arithmetic averages clustering of 28 populations with modified Rogers' genetic distance.

studies. However, comparison of average heterozygosity among populations is a reliable indicator of genetic diversity because the same loci are scored for each population. The percent polymorphism for all scorpion populations (mean = 53.1%) is higher than most published data (mean = 35.0% ± 0.17, n = 32; Ferguson, 1980). The number of loci in the analysis also affects this variable because, as the number of monomorphic loci increases, the percent polymorphism will decrease. The mean number of alleles resolved for each population (1.65 ± 0.13) is similar to that found by Sullivan (1985) for montane chipmunks (1.30)

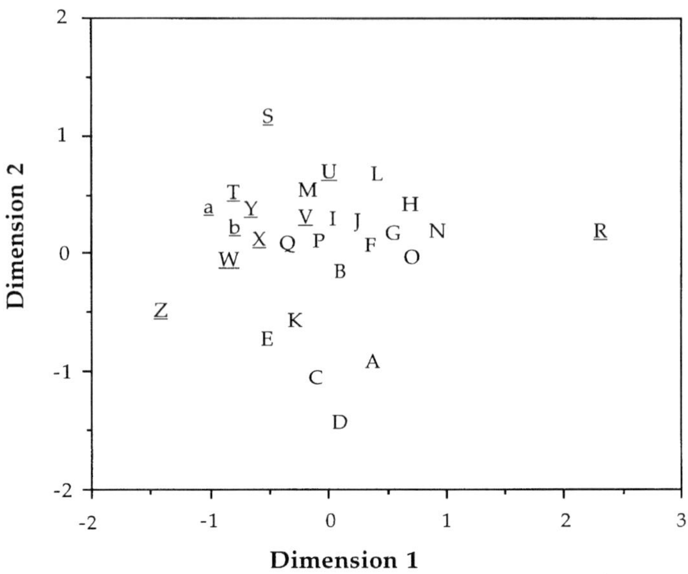

Figure 15.5 Two-dimensional scaling plot of genetic distances among scorpion populations. The underlined letters represent Arizona populations. Population designations as in fig. 15.2.

and Crouau-Roy (1989) for troglobitic beetles (1.20). It should be noted that allele number depends on sample size (Nei, 1987), because rarer alleles are found as sample size increases.

Variation among populations

The F_{ST} value among populations (0.175) lies between values known for highly dispersed and sessile organisms. This result does not support our prediction that scorpion populations are significantly isolated from one another. Instead, many populations contain a considerable level of genetic variability, which generally decreases for the marginal populations.

The moderately high levels of genetic variability within populations suggest that most populations either experienced gene flow, were colonized by a large number of individuals, or were large in size. Each condition prevents fixation of alleles in a population and slows differentiation among populations. The absence of fixed allelic differences implies these populations were neither isolated nor constrained in size during establishment. Long isolation periods and small population sizes tend to cause a random loss of alleles (Nei, 1987).

A low F_{ST} value (0.175) also implies that populations of *P. mesaensis* are not genetically isolated. Species that disperse great distances have low F_{ST} values, whereas those with fragmented populations express higher values (Barrowclough, 1983); for example, F_{ST} ranges from 0.006 in fish to 0.906 in plethodon salamanders. However, the genetic separation of populations may be greater than we have calculated because Wright's *F* statistics assume equilibrium between drift and migration. If equilibrium is not reached, these statistics underestimate population differentiation at equilibrium (Wright, 1978; Slatkin, 1985; Preziosi and Fairbairn, 1992). It is likely that our *F* values underestimate differentiation because the Sonoran Desert is less than 10,000 years old (Betancourt et al., 1990).

The genetic distance analysis shows that most populations cluster centrally on the MDS plot and therefore are genetically similar. Most populations that lie on the periphery of the MDS plot are marginal for *P. mesaensis'* geographic range. Nevertheless, some geographically peripheral populations, such as Kelso, are central on the MDS graph. This population, although slightly lower in average heterozygosity, exhibits no large reduction in allele number or unique allelic frequencies to separate it from other populations.

The separation of the Coachella Valley from other populations on the MDS plot possibly reflects an isolation event between the Coachella Valley and the Algodones. Historically, a large (5700 km^2) lake (Lake Cahuilla) occupied the southern Coachella Valley, the Salton Sea, and the Imperial Valley (Waters, 1983), where it filled four times during the period 1300–500 BP, each time lasting about 100 years. Thus, until approximately 400 years ago (Norris and Norris, 1961; Waters, 1983), the lake may have formed an effective barrier to gene flow, consequently separating populations in the northern Coachella Valley from those in the Algodones Dunes. This scenario is consistent with populational differentiations observed for the desert tortoise and the pocket gopher (Patton and Yang, 1977; Lamb et al., 1989).

Eighteen of the central populations and two of the isolated populations (Kelso and Needles) showed no reduction in allele number, an indication of a large number of founding colonists (Nei, 1987). Alternately, this pattern could arise if gene flow introduces alleles into a population, thereby maintaining genetic diversity (Slatkin, 1987). Movements of sandy substrates in alluvial systems can expand ranges for *P. mesaensis* as it moves along temporary sand corridors to establish new populations. Such corridors could be created by the Colorado River sediment depositing in the lower Colorado River region and near the Gulf of California (Coachella Valley, Algodones, and Yuma). Dune formation and movement in this region may have enhanced scorpion dispersal. Migration to isolated dunes along the Colorado and Salt Rivers probably took place when floods and arid climate cycles created temporary corridors. These ephemeral corridors isolated populations as shores and river sediment changed. Likewise, during the late Wisconsin glacial period (12,000–14,000 years ago), the range of *P. mesaensis* most likely contracted to northern Sonora (Mexico), the lower Colorado River delta, and the lower Gila River valley (western Arizona) (Betancourt et al., 1990). The populations in this region probably exchanged genetic material frequently because soils there were primarily sandy alluvium (Crosswhite and Crosswhite, 1982). As the climate became arid and sand dunes habitats increased, *P. mesaensis* likely accompanied mobile sand dunes to new areas. Small dunes can move up to 20 m per year, and larger dunes systems at much as 5 m annually (Sharp, 1978).

As these studies clearly demonstrate, allozyme analysis is a useful technique for assessing genetic structure and biogeographic history of sand scorpion populations. However, more direct measures of genetic variability, using DNA sequence information, have several advantages over allozyme analysis, including greater accuracy, greater sensitivity, and immunity from misinterpretation. In the next section we illustrate how applications of mitochondrial DNA analysis can be used to refine our measurements of genetic variability and taxonomic relationships of scorpions.

TECHNIQUES AND PERSPECTIVES OF MITOCHONDRIAL DNA ANALYSIS

The 1980s brought rapid development of two molecular techniques useful to evolutionary biologists: the use of animal mitochondrial DNA (mtDNA) as a genetic marker and polymerase chain reaction (PCR). The mtDNA molecule is relatively small, does not have introns, and is believed to be a more reliable indicator of a species' history ("phylogeography") than variation in many other characters (Avise et al., 1987). Maternally inherited haploid genotypes (haplotypes) of mitochondrial DNA often produce a high historical biogeographic signal (Riddle and Honeycutt, 1990). In some animals, mtDNA patterns are highly concordant with those of allozymes (e.g., in rodents) (Tegelström et al., 1988; Ashley and Wills, 1989); other taxa show striking differences (e.g., the American oyster; Reeb and Avise, 1989). Usually mtDNA is analyzed by the restriction fragment length polymorphism (RFLP) analysis of the entire molecule (Dowling et al., 1990).

Isolation and RFLP screening of mtDNA often is technically difficult in small invertebrates because mtDNA yield is usually low: it requires living animals (DNA not degraded), ultracentrifugation, and radioactive end labeling. A newer, highly efficient technique, PCR (Arnheim et al., 1990; Innis et al., 1990), is based on the selective enzymatic amplification of a DNA region flanked with two known conserved primers. Polymerase chain reaction yields microgram amounts of sequence-specific DNA fragments in a few hours. Moreover, PCR can use DNA from living and dead animals (e.g., alcohol-preserved museum specimens; Arnheim et al., 1990) and fossils (DeSalle et al., 1992).

Phylogenies of species and populations are commonly inferred from absence or presence of restriction sites or from the alignment of mtDNA sequences (Avise et al., 1987; Nei, 1987, 1991; Hillis et al., 1996). Two common approaches are used: (1) the distance approach compares pairs of haplotypes in terms of genetic distances (usually followed by a clustering analysis such as UPGMA); (2) the cladistic analysis, which treats absence/presence of restriction sites or nucleotides as independent characters. Computerized cladistic algorithms are available, including several parsimony programs (Swofford, 1993; Felsenstein, 1995).

Relatively few analyses of mtDNA have been conducted on arthropods (Simon et al., 1991, 1994). Several arachnid mtDNA studies have been published (Croom et al., 1991; Smith and Brown, 1991; Gillespie et al., 1994). The first restriction map of scorpion mtDNA was constructed by Smith and Brown (1991) for *Hadrurus arizonensis* Ewing (Iuridae). The 12S RNA gene of mtDNA from an Australian scorpion, *Liocheles waigiensis* (Gervais) (Ischnuridae), was sequenced by Ballard et al. (1992). We have used PCR routinely to amplify several mtDNA genes from several genera and families of scorpions (Poindexter and Fet, 1992; Fet and Vezzetti, 1994; Gantenbein et al., 1999).

Methods

We used PCR to amplify mtDNA from the following scorpion species: *Hadrurus concolor* Stahnke, *H. pinteri* Stahnke (Iuridae), *Centruroides exilicauda* Wood, and *C. vittatus* (Say) (Buthidae). Total (nuclear and mitochondrial) DNA was extracted following Simon's (1991) protocol. An aliquot of total extracted DNA was amplified using synthetic oligonucleotide PCR primers derived from the mtDNA sequence for *Drosophila yakuba* (Simon, 1991; Simon et al., 1994). Several regions of the mtDNA molecule were amplified successfully after we optimized primer and enzyme concentrations and temperature/time parameters of the PCR reaction.

Polymerase chain reaction was performed using standard techniques, slightly modified for scorpions (Gantenbein et al., 1999). Figure 15.6 represents a schematic order of mtDNA genes for *D. yakuba* (Clary and Wolstenholme, 1985; C. Simon, personal communication) as well as the position of PCR primers we used. We used eight universal primers derived from *D. yakuba* (Clary and Wolstenholme, 1985, 1987; Simon et al., 1991), one primer specific for *Drosophila* (Simon et al., 1991), and one specific for the spider genus *Tetragnatha* (Tetragnathidae) (Croom et al., 1991). The total of about 2500 base pairs (bp) of scorpion mtDNA (about 18% of the entire molecule) was obtained by PCR, which includes the entire control (A + T-rich) region, both mitochondrial ribosomal RNA genes (12S and 16S; fig. 15.6), as well as part of the cytochrome oxidase I gene (CO I).

For RFLP analysis, 28 restriction enzymes were used to search for restriction sites in mtDNA fragments obtained by PCR: *Alu*I, *Apa*I, *Bam*HI, *Bgl*I, *Bgl*II, *Cfo*I, *Dra*I, *Dde*I, *Eco*RI, *Eco*RV, *Eco*47, *Hind*III, *Hinf*I, *Hpa*II, *Kpn*I, *Msp*I, *Nsi*I, *Pst*I, *Pvu*II, *Rsa*I, *Sal*I, *Sau*3AI, *Sma*I, *Spe*I, *Ssp*I, *Taq*I, VspI, and *Xho*I. The strategy of this RFLP screening was (1) to find a restriction site, (2) to check the presence of this site in mtDNA among the most distant populations (*C. exilicauda* from Baja California versus Arizona), and (3) if a variable site is found, to search for its presence in all available populations.

For sequencing, we used amplified fragments of cytochrome oxidase I (CO I) and 12S rRNA genes of mtDNA of *Centruroides exilicauda* and *C. vittatus*. The

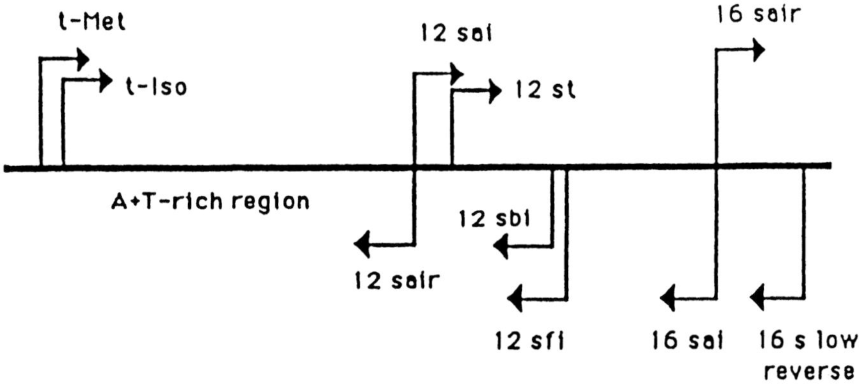

Figure 15.6 A scheme of mtDNA polymerase chain reaction primers. Primer positions are based on *Drosophila yakuba* mtDNA genome.

PCR product was purified using Millipore microfuge filters and tested by electrophoresis on 2% agarose gel. Several fragments of CO I and 12S rRNA genes were directly sequenced by dideoxynucleotide sequencing method according to standard protocols (Innis et al., 1990; Simon, 1991). A total of 1648 bp of DNA sequences of *Centruroides* was analyzed.

Results

Length polymorphism

The length polymorphism (ca. 300 bp) was discovered between the control region mtDNA of *H. concolor* and *H. pinteri* (800 bp) and *C. exilicauda* (1,100 bp). This fact is not surprising because considerable variation in mtDNA control region length is documented in different animals (Boyce et al., 1989). Furthermore, the Iuridae and Buthidae scorpion families are not closely related (Fet and Rechkin, 1990; Sissom, 1990). The length of the amplified portion of 12S rRNA gene is the same in *Hadrurus* and *Centruroides* (ca. 400 bp).

RFLP data

In a comparative study of *H. concolor* (Bahia de Las Animas, Baja California Norte) and *H. pinteri* (Danzante Island, Baja California Sur), we screened the control region (ca. 800 bp) for restriction sites. In *H. concolor,* eight restriction sites were found for five enzymes. Of these sites, only the *Hinf*I site was found in *H. pinteri;* however, *H. pinteri* possessed two different restriction sites for *Hae*III. Thus, these two species can be distinguished by a single diagnostic RFLP digest (with *Hae*III) of the 800-bp control region. This digest yields three fragments of 450, 200, and 150 bp in *H. concolor* and three fragments of 350, 270, and 180 bp in *H. pinteri.* This diagnostic approach is more reliable than a simple detection of presence or absence of the restriction site because it requires an enzyme to give a positive result (i.e., to digest DNA).

In *C. exilicauda,* our preliminary search for the control region and 12S rRNA gene (total of ca. 1500 bp) yielded 24 restriction sites for the 14 restriction enzymes. Fourteen restriction sites were variable among populations from Baja California and Arizona. Nine variable sites were detected within the control region; five sites were within the 12S rRNA gene. We concentrated on the analysis of the 400-bp PCR fragment of the 12S rRNA gene (primers *12 sai–12 sbi*), selecting five variable restriction sites within the 12S rRNA gene as markers (*Bam*HI, *Hinf*I, *Sau*3AI, *Taq*I, and *Xho*I). In search of RFLPs, we amplified and screened 400-bp PCR fragments from 31 *C. exilicauda* individuals from seven localities.

Our data fall into two groups characterized by the presence of specific restriction sites. In the preliminary screening of the mtDNA control region, five restriction sites for *Alu*I, *Bgl*II, *Dra*I, *Sau*3A1, and *Taq*I were found in a Baja population (Isla Pescador), whereas the site for *Vsp*I was detected in an Arizona population (Tucson). In a more detailed screening of the 12S region, the *Hinf*I site was found in all Baja populations but was absent from Arizona populations.

Among the Baja populations, considerable RFLP differences were also detected within *C. exilicauda.* Presence or absence of at least seven restriction sites varies within *C. exilicauda*. The *Xho*I site in the 12S gene is unique for only one of the seven Baja populations (Dolores del Sur); therefore, even a small sample of genetic data yielded a high level of variation detectable by RFLP analysis. By expanding this analysis to more genetic data, phylogenetic trees of populations can be constructed and models of population structure tested.

Sequence data

Figure 15.7 shows four scorpion sequences of a 344-bp fragment of mitochondrial *CO I* gene of *Centruroides exilicauda* and *C. vittatus*. We aligned sequences using *Drosophila yakuba* mtDNA sequence (Clary and Wolstenholme, 1985); the studied fragment corresponds to nucleotide positions 2225 to 2568; no insertions or deletions were found. Of 1376 bp sequenced, A + T content was 64.5% (adenine: 18.9%; thymidine: 45.6%), and C + G content was 35.5% (cytosine: 11.5%; guanine: 24.0%). Of 344 sites, 58 (16.9%) were polymorphic; however, only 13 (3.7% of the total number) of these sites were in the first two codon positions.

A preliminary phylogenetic tree for four *CO I Centruroides* sequences (PAUP program, version 3.1.1; Swofford, 1993) did not present an informative phylogeny: this cladogram had a triple branching point that cannot be resolved within this set of data. It was evident, however, that Baja populations (Pescador and Danzante) are the most related, that the Sonora population (Puerto Penasco) differs strongly from those of Baja, and that both the Sonora population and the Baja cluster differ from the outgroup species, *C. vittatus*.

Figure 15.8 represents the distance matrix for four *CO I* scorpion sequences obtained via PAUP. Although sequence divergence alone is not a precise indication of phylogenetic relationships, it is instructive to note that scorpions of the genus *Centruroides* exhibit overall pairwise nucleotide distance in the *CO I* gene from 2.9% to 12.5%. Such a level is common for protein-coding mtDNA genes in congeneric animals.

We obtained additional preliminary data by comparing two 12S rRNA gene sequences of *C. exilicauda* from Baja and Arizona populations, aligned with a corresponding sequence of *Drosophila yakuba* (fig. 15.9). Pairwise nucleotide divergence between two scorpion sequences was 10.3%. Given that Croom et al. (1992) detected 3–13% sequence divergence in 12S rRNA gene among congeneric *Tetragnatha* spiders, the discovered level of divergence in scorpions may be high enough to indicate the existence of two different species.

We can conclude that even a small sample of sequences yielded considerable variation. Evidently, a considerable genetic differentiation exists over the range of *C. exilicauda*. The obvious direction of the search is to extend our analysis to many more populations and DNA sequences of different variability.

Discussion

The RFLP variation we observed comes from the 12S rRNA gene of scorpion mtDNA. Ribosomal RNA genes were believed to be conserved, although they contain highly variable regions called short-range stems (Simon, 1991, 1994). Recent studies revealed considerable inter- and intraspecific variation in the 12S rRNA gene (Carr et al., 1986; Phillips et al., 1991). Croom et al. (1991) de-

```
D.yak  TAATTTCTCATATTATTAGACAAGAATCTGGTAAAAAGGAAACTTTCGGTTCTTTAGGAATAATC
C-258  TGATTTCACATATTGTAAGTCATCATATGGGTAAGAAGGAGCCTTTTGGTTCTTTGGGCATGGTT
C-263  TAGTTTCTCACATTGTTAGTCATCATATAGGAAAGAAGGAACCTTTTGGTTCTTTGGGTATGGTT
C-296  TGATTTCTCATATTGTTAGCCATCATATAGGAAAGAAGGAACCTTTTGGTTXTTTGGGTATGGTT
C-265  TAGTTTCTCATATTGTTAGACATCATGTAGGTAAGAGGGAGCCTTTTGGTTCTTTAGGAATGGTT

D.yak  TATGCTATACTTGCTATTGGATTATTAGGATTTATTGTTTGAGCTCATCATCATTTTACAGTTGG
C-258  TATGCTATAGTTGCTATTGGTGGATTAGGTTTTATTGTTTGGGCTCATCATATATTTACTGTGGG
C-296  TATGCTATAGTTGCTATTGGGGGGTTAGGTTTTATTGTTTGGGCTCATCATATGTTTACTGTAGG
C-263  TATGCTATAGTTGCTATTGGGGGGTTAGGTTTTATTGTTTGGGCTCATCATATGTTTACTGTAGG
C-265  TATGCTATGGTTGCTATTGGGGGTTTGGGGCTTATTGTATGAGCACATCATATGTTTACTGTGGG

D.yak  AATAGACGTTGATACACGAGCTTATTTTACTTCTGCTACTATAATTATTGCGGTTCCTACAGGAA
C-258  TATGGATGTAGATACTCGGGCTTATTTTACTGCTGCTACAATGATTATTGCTGTTCCTACGGGTA
C-296  AATGGATGTAGATACTCGGGCTTATTTTACTGCTGCTACAATAATTATTGCTGTTCCTACGGGTA
C-263  AATGGATGTGGATACTCGGGCTTATTTTACTGCTGCTACAATAATTATTGCTGTTCCTACGGGTA
C-265  TATGGATGTGGATACTCGAGCTTATTTTACTGCTGCTACTATGATTATTGCTGTCCCTACTGGAA

D.yak  TTAAAATTTTTAGATGATTAGCTACTTTACATGGAACTCAACTTTCTTATTCTCCAGCTATTTTA
C-258  TTAAGATTTTTAGTTGATTAGCCACACTTCATGGGTCTTATTTTGTTTTTAGTTCTGCTATGTTG
C-296  TTAGGATTTTTAGTTGATTGGCTACTTTACATGGGTCTTATTTTGTTTTTAGTTCTGCTATGCTA
C-263  TTAGGATTTTTAGTTGGTTGGTTACTTTACATGGGTCTTATTTTGTTTTTAGTTCTGCTATGCTA
C-265  TTAAGATTTTTAGTTGATTGGCGACTTTGCATGGGTCTTATTTTGTTTTTAATTCTGCTATGTTG

D.yak  TGAGCTTTAGGATTTGTTTTTTTATTCACAGTAGGAGGATTAACAGGAGTTGTATTAGCTAATTC
C-258  TGGGCCTTGGGTTTTATTTTTCTTTTTACTGTTGGGGGATTGACGGGTGTAATTTTGGCTAACTC
C-296  TGGGCTTTGGGTTTTATTTTTCTTTTCACTGTTGGGGGATTAACTGGTGTAATTTTAGCTAATTC
C-263  TGGGCTTTGGGTTTTATTTTTCTTTTCACTGTTGGAGGATTGACTGGTGTAATTTTAGCTAATTC
C-265  TTGGCTTTAGGTTTTATTTTTCTTTTTACTGTAGGAGGATTGACTGGTGTGGTTTTGGCTAATTC

D.yak  ATCAGTTGATATTATTTTA
C-258  TTCTTTGGATATTGTTTTA
C-296  TTCTTTGGATATTGTTTGA
C-263  TTCTTTGGATATTGTTTTA
C-265  TTCTTTATATATTGTTTTA
```

Figure 15.7 Alignment of cytochrome oxidase I sequences. Aligned are 344 bp of *Drosophila yakuba* (D. yak; Clary and Wolstenholme, 1985), and scorpions (our data) *Centruroides exilicauda* from northern Sonora (Puerto Penasco; C-258), Baja California Sur (Isla Danzante, C-263), and Baja California (Isla Pescador; C-296), and *C. vittatus* (Jefferson County, Missouri, USA; C-265).

tected 3–13% sequence divergence in this gene among *Tetragnatha* congeneric Hawaiian spiders. Our observations show as much as 30% divergence between the 12S sequence in two scorpion genera, *Liocheles* (Ischnuridae) (Ballard et al., 1992) and *Centruroides* (Buthidae); therefore, the 12S rRNA gene has sufficient variability to allow comparisons at the population and species levels. Most recently, Gantenbein et al. (1999) published the first scorpion mitochondrial DNA phylogeny based on the 16S rRNA gene (conserved 5′-end) for the genus

	1	2	3	4
1 (C-258)	—	0.087	0.087	0.125
2 (C-263)	30	—	0.029	0.125
3 (C-296)	30	10	—	0.122
4 (C-265)	43	43	42	—

Figure 15.8 A genetic distance matrix for four CO I *Centruroides* sequences. Below diagonal: absolute distances; above diagonal: mean distances.

Euscorpius (Euscorpiidae). For four congeneric morphospecies, the sequence divergence was from 2% to 13%.

The RFLP analysis of PCR products of mtDNA provides taxonomists with new characters to infer phylogenies as well as basic diagnosis and description of variation. A preliminary analysis of a highly variable but morphologically indivisible species, *C. exilicauda,* demonstrates the value of mtDNA markers. Considerable genetic differentiation, due to gene flow limitations, is observed in this species; however, no morphologically consistent differences exist between Baja and Arizona populations (Williams, 1980). Here we report the first evidence of a simple diagnostic character (*Hinf*I restriction site of 12S r RNA region found in Baja but absent from Arizona). Potential diagnostic sites are restriction sites for *Alu*I, *Bgl*II, *Dra*I, *Sau*3A1, and *Taq*I (only found in Baja) and for *Vsp*I (only detected in Arizona). Our preliminary sequence comparisons from both *CO I* (fig. 15.7) and 12S rRNA (fig. 15.9) mitochondrial genes seem to confirm the existence of genetic separation between Baja and Arizona/Sonora populations of *Centruroides exilicauda.* These lineages may prove to be candi-

```
D.yak  AACCTGTTTTGTAATCGATAATCCACGATGGACCTTACTTAAATTTGTAATCAGTTTA
L.wai  AACCTGTTTCTTAATTGATAATCCACATCATATCTTACCTTATTTTT----GTTTGTA
C-115  AATCTGTTTTATAATCGACAATACACGATAAATCTTACTTGGATTAA---TGCTTGTA
C-103  AACCTGTTTTATAATCGATAATACACGACAAATCTTACTTAAGTTGA---GGCTTGTA

D.yak  TATACCGTCGTTATC-AGAATATTTTATAAGAATAATAATATTCAATAATTTTAATAA
L.wai  TAT--CGTTGTTTTA-AG-ATA---ATGTATATTAACTATAATCAAAATTAAAA-GTT
C-115  TAT--CGTTGTGGTA----ATA---ATGTTTTATAATAATTATTAGGGGTAATTTTAT
C-103  TAT--CGTTGTGATA----ATA---ATGTTTTATAATAATTATTAGGGGTAATTTTAT

D.yak  AAATTTATATCAGATCAAGGTGTAGCTTATATT
L.wai  TTTAAAAAATCAAATCAAGGTGCAGCTTAATAT
C-115  TTTT-GATATCAGATCAAGGTTCAGTTTATGTT
C-103  TTTT-GATATCAGATCAAGGTTCAGTTTATACT
```

Figure 15.9 Alignment of 12s rRNA mtDNA sequences. Aligned are sequences from *Drosophila yakuba* (D. yak; 148 bp; Clary and Wolstenholme, 1985), and scorpions *Liocheles waigiensis* (L.wai; 137 bp; Scorpionidae; Ballard et al., 1993), and 135-bp fragments for *Centruroides exilicauda* from Baja California (Bahia de los Angeles; C-115) and from Arizona (Tuscon; C-103)(our data).

dates for valid sibling species. A detailed analysis using outgroup species will allow us to make decisions about the taxonomic status of all populations which now compose *C. exilicauda.*

In another case, an RFLP analysis of two morphologically distinct species that belong to the same genus (*Hadrurus concolorous* and *H. pinteri*), we found two molecular characters of diagnostic value. This technical approach allows one to use short (in this case 800 bp) PCR-amplified fragments of mtDNA in the same fashion as total mtDNA molecule (ca. 15,000 bp) for a rapid assessment of genetic heterogeneity between species and to detect hybrid forms. More information can be derived from direct sequencing of PCR fragments and by comparing nucleotide sequences from different populations.

CONCLUSIONS

Modern molecular techniques permit us to assess genetic variation for scorpions, to construct phylogenies, and to test hypotheses in historical biogeography. Scorpions are readily available, abundant in nature, well studied taxonomically, moderately diverse at the genus/species level, and often geographically differentiated into discrete populations. Existing patterns of variation within and among scorpion populations and estimates of gene flow can be elucidated with allozyme analysis. Assessment of mitochondrial DNA variation via RFLP analysis, PCR amplification, and DNA sequencing provides another system of molecular markers to estimate genetic variation among populations. Thus, molecular approaches to the study of scorpion systematics and evolution is now poised to address several important questions of general interest to biologists. What are the relative contributions of historical events and gene flow to variation within populations, and to what extent does gene flow constrain differentiation and affect speciation? How is genetic variation related to morphological variation, and how far do clades sharing common ancestry evolve after separation? The opportunity to answer these and many other long-standing question poses an exciting challenge for this ancient and adaptable organism.

Acknowledgments We thank people who introduced us to molecular research and advised thereafter: Grigory Dymshits (Siberian Academy of Sciences, Russia), Brigid L. M. Hogan, Karen Lyons, David McCauley, and Nathan Schiff (Vanderbilt University), and Shintaro Nomura (Institute of Microbiology, Tokyo, Japan). We are grateful to Gary A. Polis, W. David Sissom, Chris Myers, Marshall Hedin, and Philip Sudman for their advice and discussions. We thank Brian Poindexter, Jeffrey Metts, Robert Vezzetti, and Aravinda Rao for their assistance with mtDNA projects. We also thank Marshall Hedin, Gary and Steve Polis, and Mike Rosen for their assistance with collecting scorpions in the field. The allozyme research was supported by a dissertation improvement award from the Vanderbilt University Graduate School to T.Y. The mtDNA studies were supported by grants from Loyola University (New Orleans) and Theodore Roosevelt Memorial Foundation to V.F.

References

Arnheim, N., White, T. and Rainey, W. (1990) Application of PCR: organismal and population biology. *Bioscience* 40: 174–182.

Ashley, M. and Wills, C. (1989) Mitochondrial DNA and allozyme divergence patterns are correlated among island deer mice. *Evolution* 43: 646–650.

Avise, J.C., Arnold, J., Ball, R.M., Bermingham, E., Lamb, T., Neigel, J.E., Reeb, C.A. and Saunders, N.C. (1987) Intraspecific phylogeography: the mitochondrial DNA bridge between population genetics and systematics. *Annu. Rev. Ecol. System.* 18: 489–522.

Ballard, J.W.O., Olsen, G.J., Faith, D.P., Odgers, W.A., Rowell, D.M. and Atkinson, P.W. (1992) Evidence from 12S ribosomal RNA sequences that onychophorans are modified arthropods. *Science* 258: 1345–1348.

Barrowclough, G.F. (1983) Biochemical studies of microevolutionary processes. *In* Brush, A.H. and Clark, G.A. (eds.) *Perspectives in Ornithology,* pp. 223–270. Cambridge University Press, Cambridge.

Betancourt, J.L., Van Devender, T.R. and Martin, P.S. (1990) Synthesis and prospectus. *In* Betancourt, J.L. Van Devender, T.R. and Martin, P.S. (eds.) *Packrat Middens: The Last 40,000 Years of Biotic Change,* pp. 435–447. University of Arizona Press, Tucson.

Bocquet-Appel, J.P. and Sokal, R.R. (1989) Spatial autocorrelation analysis of trend residuals in biological data. *Syst. Zool.* 38: 333–341.

Bonacina, A. (1980) Sistematica specifica e sottospecifica del complesso "*Euscorpius germanus*" (Scorpiones, Chactidae). *Riv. Mus. Sci. Nat.* (Bergamo) 2: 47–100.

Bonacina, A. (1983) Note di alcuni scorpioni di Romania. *Riv. Mus. Sci. Nat.* (Bergamo) 5: 3–10.

Boyce, T., Zwick, M. and Aquadro, C. (1989) Mitochondrial DNA in the pine weevils: size, structure and heterogeneity. *Genetics* 123: 825–836.

Caccone, A. and Sbordoni, V. (1987) Molecular evolutionary divergence among North American cave crickets. I. Allozyme variation. *Evolution* 41: 1198–1214.

Carr, S.M., Ballinger, S.W., Derr, J.N., Blankenship, L.H. and Bickham, J.W. (1986) Mitochondrial DNA analysis of hybridization between sympatric white-tailed deer and mule deer in West Texas. *Proc. Natl. Acad. Sci. USA* 83: 9576–9580.

Clary, D.O. and Wolstenholme, D.R. (1985) The mitochondrial DNA molecule of *Drosophila yakuba:* nucleotide sequence, gene organization and genetic code. *J. Mol. Evol.* 22: 252–271.

Clary, D.O. and Wolstenholme, D.R. (1987) *Drosophila* mitochondrial DNA: conserved sequences in the A+T-rich region and supporting evidence for a secondary structure model of the small ribosomal RNA. *J. Mol. Evol.* 25: 116–125.

Cook, L.M. (1991) *Genetic and Ecological Diversity: The Sport of Nature.* Chapman and Hall, London.

Crosswhite, F.S. and Crosswhite, C.D. (1982) The Sonoran Desert. *In* Bender, G.L. (ed.) *Reference Handbook on the Deserts of North America,* pp. 163–295. Greenwood Press, Westport, CT.

Croom, H.B., Gillespie, R.G. and Palumbi, S.R. (1991) Mitochondrial DNA sequences corresponding to a portion of the RNA of the small ribosomal subunits of *Tetragnatha mandibulata* and *T. hawaiensis* (Araneae, Tetragnathidae). *J. Arachnol.* 19: 210 -214.

Crouau-Roy, B. (1989) Population studies on an endemic troglobitic beetle: geographic patterns of genetic variation, gene flow and genetic structure compared with morphometric data. *Genetics* 121: 571–582.

Curcic, B.P.M. (1972) Considerations upon the geographic distribution and origin of some populations in the genus *Euscorpius* Thorell (Chactidae, Scorpiones). *Rap. Com. Int. Medit.* (Monaco) 21: 83–88.

DeSalle, R., Gatesy, J., Wheeler, W. and Grimaldi, D. (1992) DNA sequences from a fossil termite in oligo-Miocene amber and their phylogenetic implications. *Science* 257: 1933–1936.

di Caporiacco, L. (1950) Le specie e sottospecie del genere *Euscorpius* viventi in Italia e in alcune zone confinanti. *Atti. Acc. Naz. Lincei* 8: 159–230.

Dowling, T.E., Moritz, C. and Palmer, J.D. (1990) Nucleic acids II: Restriction site analysis. *In* Hillis, D.M. and Moritz, C. (eds.) *Molecular Systematics,* pp. 250–317. Sinauer, Sunderland, MA.

Felsenstein, J. (1988) Phylogenies from molecular sequences: inference and reliability. *Annu. Rev. Genet.* 22: 521–565.

Felsenstein, J. (1991) *PHYLIP (Phylogeny Inference Package),* version 3.4. University of Washington, Seattle.

Felsenstein, J. (1995) PHYLIP (Phylogeny Inference Package). Version 3.57c. University of Washington, Seattle, WA.

Ferguson, A. (1980) *Biochemical Systematics and Evolution.* J. Wiley and Sons. New York.

Fet, V. (1986) Notes on some *Euscorpius* (Scorpiones, Chactidae) from Greece and Turkey. *Riv. Mus. Sci. Nat.* 9: 3–11.

Fet, V. (1989) A catalogue of scorpions (Chelicerata: Scorpiones) of the USSR. *Riv. Mus. Sci. Nat.* 1988 13: 73–171.

Fet, V. (1997) A noe on *Euscorpius carpathicus* (Scorpiones: Chactidae) from the Crimea. *J. Arachnol.* 25: 106–108.

Fet, V. and Poindexter, B.J. (1992) Genetic variation of mitochondrial DNA in *Centruroides exilicauda* (Scorpiones: Buthidae) as revealed by the polymerase chain reaction. *Am. Zool.* 32(5): 136A.

Fet, V. and Rechkin, D.V. (1990) Scorpion trichobothriotaxy: a principal component analysis. *Riv. Mus. Sci. Nat.* 1989 14: 191–206.

Fet, V. and Vezzetti, R.M. (1994) PCR amplification of mitochondrial DNA from preserved scorpions. *Ancient DNA Newsl.* 2(1): 14.

Gantenbein, B., Büchi, L., Braunwalder, M.E. and Scholl, A. (1998) The genetic population structure of *Euscorpius germanus* (C.L. Koch) (Scorpiones: Chactidae) in Switzerland. In: Selden, P.A. (ed.), *Proceedings of the 17th European Colloquium of Arachnology,* pp. 33–40. Edinburgh 1997, 14–18 July, 1997.

Gantenbein, B., Fet, V., Largiader, C. and Scholl, A. (1999) First DNA phylogeny of *Euscorpius* Thorell, 1876 (Scorpiones: Euscorpiidae) and its bearing on taxonomy and biogeography of this genus. *Biogeographica* 75: 49–65.

Gertsch, W.J. (1958) Results of the Puritan-American Museum expedition to Western Mexico. 4. The scorpions. *American Museum Novitates* 1903: 1–20.

Gillespie, R.G., Croom, H.B. and Palumbi, S.R. (1994) Multiple origins of a spider radiation in Hawaii. *Proc. Natl. Acad. Sci. USA* 91: 2290–2294.

Hadzi, J. (1931) Der Artbildungprozess in der Gattung *Euscorpius* Thorell. *Arch. Inst. Zool. Ital.* 16(1–2): 356–362.

Haradon, R.M. (1983) *Smeringurus,* a new species of *Paruroctonus* Werner (Scorpiones, Vaejovidae). *J. Arachnol.* 11: 251–270.

Hartl, D.L. and Clark, A.G. (1989) *Principles of Population Genetics,* 2nd ed. Sinauer, Sunderland, MA.

Hillis, D.M., Moritz, C. and Mable, B.K. (eds.) (1996) *Molecular Systematics* (2nd edition). Sinauer, Sunderland, MA.

Innis, M., Gelfand, D., Sninsky, J. and White, T. (eds.) (1990) *PCR Protocols: A Guide to Methods and Applications.* Academic Press, San Diego, CA.

Kinzelbach, R. (1975) Die Skorpione der Agäis. Beiträge zur Systematik, Phylogenie und Biogeographie. *Zool. Jahrb. Syst.* 102: 12–50.

Lamb, T., Avise, J.C. and Gibbons, J.W. (1989) Phylogeographic patterns in mitochondrial DNA of the desert tortoise (*Xerobates agassizi*) and evolutionary relationships among the North American gopher tortoises. *Evolution* 43: 76–87.

Larson, A., Wake, D.B. and Yanev, K.P. (1983) Measuring gene flow among populations having high levels of genetic fragmentation. *Genetics* 106: 293–308

Lessa, E.P. (1990) Multidimensional analysis of geographic genetic structuire. *Syst. Zool.* 39: 242–252.

Lewontin, R.C. and Hubby, J.L. (1966) A molecular approach to the study of genic heterozygosity in natural populations, II. Amount of variation and degree of heterozygosity in natural populations of *Drosophila pseudoobscura. Genetics* 54: 595–609.

Mayr, E. (1954) Change of genetic environment and evolution. *In* Huxley, J., Hardy, A.C. and Ford, E.B. (eds.) *Evolution as a Process,* pp. 157–180. George Allen & Unwin, London.

McCauley, D.E. and Eanes, W.F. (1987) Hierarchical population structure analysis of the milkweed beetle, *Tetraopes tetraopthalmus* (Forster). *Heredity* 58: 193–201.

Murphy, R.W., Sites, J.W., Jr., Buth, D.G. and Haufler, C.H. (1990) Proteins I: Isozyme electrophoresis. *In* Hillis, D.M. and Moritz, C. (eds.) *Molecular Systematics,* pp. 45–126. Sinauer Associates, Sunderland, MA.

Nei, M. (1987) *Molecular Evolutionary Genetics.* Columbia University Press, New York.

Norris, R.M. and Norris, K.S. (1961) Algodones Dunes of southeastern California. *Bull. Geol. Soc. Am.* 72: 605–620.

Pasteur, N., Pasteur, G., Bonhomme, F., Catalan, J. and Britton-David, J. (1988) *Practical Isozyme Genetics.* Ellis Horwood, London.

Patton, J.L. and Yang, S.Y. (1977) Genetic variation in *Thomomys bottae* pocket gophers: Macrogeographic patterns. *Evolution* 31: 697–720.

Pavlik, B.M. (1989) Phytogeography of sand dunes in the Great Basin and Mojave deserts. *J. Biogeogr.* 16: 227–238.

Phillips, C.J., Pumo, D.E., Genoways, H.H., Ray, P.E. and Briskey, C.A. (1991) Mitochondrial DNA evolution and phylogeography in two Neotropical fruit bats, *Arbiteus jamaicensis* and *Arbiteus lituratus. In* Mares, M.A. and Schmidly, D.J. (eds.) *Latin American Mammalogy,* pp. 97–123. University of Oklahoma Press, Tulsa.

Poindexter, B.J. and Fet, V. (1992) PCR amplification of mitochondrial DNA provides genetic markers for biogeographic study of desert scorpions. *ASB Bull.* 39: 100–101.

Polis, G.A. (1980) Seasonal patterns and age-specific variation in the surface activity of a population of desert scorpions in relation to environmental factors. *J. Anim. Ecol.* 49: 1–18.

Polis, G.A. (ed.) (1990a) *Biology of Scorpions.* Stanford University Press, Stanford.

Polis, G.A. (1990b) Ecology. *In* Polis, G.A. (ed.) *The Biology of Scorpions,* pp. 247–293. Stanford University Press, Stanford.

Polis, G.A. and Farley, R.D. (1979) Characteristics and environmental determinants of natality, growth and maturity in a natural population of the desert scorpion, *Paruroctonus mesaensis* (Scorpionidae: Vaejovidae). *J. Zool.* 187: 517–542.

Polis, G.A. and Farley, R.D. (1980) Population biology of a desert sand scorpion: survivorship, microhabitat and the evolution of life history strategy. *Ecology* 61: 620–629.

Polis, G.A. and Yamashita, T. (1991) The ecology and importance of predaceous arthropods in desert communities. *In* Polis, G.A. (ed.) *The Ecology of Desert Communities,* pp. 180–222. University of Arizona Press, Tucson.

Preziosi, R.F. and Fairbairn, D.J. (1992) Genetic population structure and levels of gene flow in the stream dwelling waterstrider, *Aquarius* (=Gerris) remigis (Hemiptera: Gerridae). *Evolution* 46: 430–444.

Reeb, C. and Avise, J.C. (1989) A genetic discontinuity in a continuously distributed species: mitochondrial DNA in the American oyster, *Crassostrea virginica. Genetics* 124: 397–406.

Richardson, B.J., Baverstock, P.R. and Adams, M. (1986) *Allozyme Electrophoresis: A Handbook for Animal Systematics and Population Studies.* Academic Press, San Diego, CA.

Riddle, B.R. and Honeycutt, R.L. (1990) Historical biogeography in North American arid regions: an approach using mitochondrial DNA phylogeny in grasshopper mice (genus *Onychomys*). *Evolution* 44: 1–15.

Schiffman, S.S., Reynolds, M.L., and Young, F.W. (1981) *Introduction to Multidimensional Scaling: Theory, Methods, and Applications.* Academic Press, New York.

Sharp, R.P. (1978) The Kelso Dune complex. *In* Greeley, R., Womer, M.B., Papson, R.P. and Spudis, P.D. (eds.) *Aeolian Features of Southern California: A Comparative Planetary Geology Guidebook,* pp. 54–63. NASA Office of Planetary Geology, Washington, DC.

Simon, C. (1991) Molecular systematics at the species boundary: exploiting conserved and variable regions of the mitochondrial genome of animals via direct sequencing from amplified DNA. *In* Hewitt, G.M., Johnson, A.W. and. Young, J.P.W (eds.) *Molecular Techniques in Taxonomy,* pp. 33–72. Springer, NATO Advanced Studies Institute, Berlin.

Simon, C., Francke, A. and Martin, A. (1991) The polymerase chain reaction: DNA extraction and amplification. *In* Hewitt, G.M., Johnson, A.W. and Young, J.P.W. (eds.) *Molecular Techniques in Taxonomy,* pp. 329–356. Springer, NATO Advanced Studies Institute, Berlin.

Simon, C., Frati, F., Beckenbach, A., Crespi, B., Liu, H. and Flook, P. (1994) Evolution, weighting, and phylogenetic utility of mitochondrial gene sequences and a compilation of conserved polymerase chain reaction primers. *Ann. Entomol. Soc. Amer.* 87: 651–701.

Sissom, W.D. (1990) Systematics, biogeography and paleontology. *In* Polis, G.A. (ed.) *The Biology of Scorpions,* pp. 64–160. Stanford University Press, Stanford.

Slatkin, M. (1985) Gene flow in natural populations. *Annu. Rev. Ecol. System.* 16: 393–430.

Smith, D.R. and Brown, W.M. (1991) A mitochondrial DNA restriction enzyme cleavage map for the scorpion, *Hadrurus arizonensis* (Iuridae). *J. Arachnol.* 19: 85–87.

Sneath, P.H. and Sokal, R.R. (1973) *Numerical Taxonomy.* W.H. Freeman, San Francisco, CA.

Sokal, R.R. (1979) Ecological parameters inferred from spatial correlograms. *In* Patil, G.P. and Rosenzweig, M. (eds.) *Contemporary Quantitative Ecology and Related Ecometrics,* pp. 167–196. International COOP Publishing House, Fairland, MD.

Sokal, R.R. and Oden, N.L. (1978a) Spatial autocorrelation in biology 1. Methodology. *Biol. J. Linn. Soc.* 10: 199–228.

Sokal, R.R. and Oden, N.L. (1978b) Spatial autocorrelation in biology 2. Some biological implications and four applications of evolutionary and ecological interest. *Biol. J. Linn. Soc.* 10: 229–249.

Spiess, E.B. (1989) *Genes in Populations.* John Wiley and Sons, New York. 774 pp.

Sullivan, R.M. (1985) Phyletic, biogeographic and ecologic relationships among montane populations of least chipmunks (*Eutamias minimus*) in the southwest. *Syst. Zool.* 34: 419–448.

Swofford, D.L. (1993) *PAUP: phylogenetic analysis using parsimony.* Version 3.1.1. Smithsonian Institution, Washington, DC.

Swofford, D.L. and Selander, R.B. (1989) *Biosys-1: A Computer Program for the Analysis of Allelic Variation in Population Genetics and Biochemical Systematics. User's Manual.* Illinois Natural History Survey, Springfield.

Tegelström, H., Wyoni, P., Gelter, H. and Jaarola, M. (1988) Concordant divergence in proteins and mitochondrial DNA between two vole species in the genus *Clethrionomys. Biochem. Genetics* 26: 223–237.

Vachon, M. (1974) Etudes des caracteres utilises pour classer les familles et les genres du Scorpions (Arachnides). I. La trichobothriotaxie en arachnologie. Sigles trichobothriaux et types de trichobothriotaxie chez les Scorpions. *Bull. Mus. Nat. Hist.* sér. 3 140: 857–958.

Vachon, M. and Jaques, M. (1977) Recherches sur les Scorpions appartenant ou deposes au Museum d'Histoire naturelle de Geneve. II. Contribution a la connaisance de l'ancienne espece *Scorpius banaticus* C. L. Koch 1841, actuellement consideree comme synonyme de *Euscorpius carpathicus* (Linne 1767) (Fam. des Chactidae). *Rev. Suisse Zool.* 84: 409–436.

Valle, A. (1975) Considerazioni intorno alle sottospecie di *Euscorpius carpathicus* (L.) (Scorpiones, Chactidae). *Ateneo Parmense Act. Nat.* 11: 209–234.

Walker, J.M. (1973) A study of variability in the scorpion *Centruroides exilicauda* (Wood) (Scorpionida: Buthidae) (M.A. thesis). San Francisco State University, San Francisco, CA.

Wartenberg, D. (1989) *SAAP-A Spatial Autocorrelation Analysis Program.* Exeter Software, Inc., Setauket, NY.

Waters, M.R. (1983) Late Holocene lacustrine chronology and archaeology of ancient Lake Cahuilla, California. *Quatern. Res.* 19: 373–387.

Wilkinson, L. (1987) *SYSTAT: The System for Statistics.* Systat Inc., Evanston, IL.

Wilkinson, L., Hill, M. and Vang, E. (1992) *SYSTAT: Statistics,* version 5. SYSTAT, Inc., Evanston, IL.

Williams, S.C. (1980) Scorpions of Baja California, Mexico and adjacent islands. *Occas. Papers Calif. Acad. Sci.* 135: 1–127.

Workman, P.L. and Niswander, J.D. (1970) Population studies on southwestern Indian tribes. II. Local genetic differentiation in the Papago. *Am. J. Human Gen.* 22: 24–49.

Wright, S. (1978) *Evolution and Genetic of Populations,* vol. 4. *Variability within and among Natural Populations.* University of Chicago Press, Chicago.

Yamashita, T. (1993) Genetic and morphological variation in scorpion populations on habitat islands (Ph.D. dissertation). Vanderbilt University, Nashville, TN.

16

Paleontology

ANDREW J. JERAM

Today's scorpion fauna reaches its peak of diversity in desert habitats (Polis, 1990), but fossil evidence indicates that the earliest scorpions were fully aquatic. Around 410 MYA the scorpions began a dramatic shift in biology, successfully making the transition from water to land. Once on land they proceeded to colonize a wide range of terrestrial environments, although this took place over tens of millions of years, culminating in the desert scorpions of the present day.

Scorpions were a significant component of the early land biota and are also the only terrestrial arthropod group that has immediate and recognizable aquatic ancestors. This means that scorpions are the best fossil arthropods in which to study the process of terrestrialization. As might be expected, this involved not only morphological adaptations designed to cope with different physical conditions encountered on land, but also significant changes in their behavior and ecology.

Paradoxically, although arthropods abound on land today, the fossil remains of early terrestrial arthropods are rare (Rolfe, 1980, 1985). The first part of this chapter examines why this is so and also how scorpions have contributed to our understanding of this sparse record. The remainder of the chapter discusses some of the fossil evidence for the changes that took place in scorpions as they invaded the land and became progressively better adapted to it.

EARLY TERRESTRIAL ARTHROPODS: NATURE OF THE RECORD

Scorpions

The scorpion fossil record is better than those of other arachnids (Selden, 1988). Nevertheless, complete scorpion fossils (body fossils) are rare and usually far too poorly preserved to reveal much more than gross morphology. The oldest fossil scorpions look superficially like modern ones, and this led early workers to the conclusion that their biology and habits were also similar. Subsequently it has been established that they were, in fact, aquatic (e.g., Rolfe and Beckett, 1984; Kjellesvig-Waering, 1986). Until recently, paleontologists have been unable to tell which of the later Paleozoic (Devonian and Carboniferous) scorpions were aquatic or terrestrial. Now the long-standing question of whether scorpions with ventral abdominal plates (see below) had gills or book lungs has been partly resolved (Jeram, 1990a). As a result the scorpion fossil record can be reassessed. This task has been aided by the systematic recovery of scorpion cuticle fragments from Upper Paleozoic rocks and by the recent discovery of some extraordinarily well-preserved Carboniferous scorpions.

Dispersed Arthropod Cuticles

Paleobotanists routinely disaggregate or dissolve ancient sediments to recover fragmentary plant remains (sometimes referred to as "fossil tea leaves"), and they have occasionally found arthropod cuticle fragments among their material (reports reviewed by Bartram et al., 1987). One such chance discovery has led to the description of a diverse terrestrial arthropod fauna of Middle Devonian age from Gilboa, New York State (Shear et al., 1984). The Gilboa assemblage has revolutionized our perception of the components and organization of terrestrial ecosystems that pre-date the invasion of land by vertebrates, and the possibility that cuticle assemblages like this might be found at other localities has opened up a new aspect of paleobiology (Shear, 1990).

That the fragmentary remains of scorpions could be obtained by disaggregating some Carboniferous shales has been known since the last century (Dunlop, 1898), but it was not until the painstaking work of Wills (1910, 1947) on a Triassic scorpion cuticle assemblage that any attempt was made to use such fragments. Since the discovery of the Gilboa fauna revived interest in Paleozoic cuticle assemblages, it has become increasingly apparent that scorpion cuticles are far more common than those of other arthropods. More often than not they are the only group represented in an assemblage (Jeram, 1989). When examined by transmitted light microscopy, the quality of cuticle preservation is excellent, and the fossil material looks just like flattened modern arthropod cuticle (see fig. 16.5A). Every detail of the external morphology can be studied, and because the arthropod integument functions not only as an exoskeleton but also as the organism's interface with its environment, their remains yield information not usually available from fossils. For example, the radical reorganization of sensory, respiratory, and locomotory systems required as scorpions adapted to life on land can be followed through time, and changes in the ecological role and behavior of scorpions are also indicated by morphological developments. Efforts to use the dispersed cuticle record of scorpions, and indeed of other arthropods, are still in their infancy. If the lessons learned from scorpion remains are an indication of what can be achieved in other groups, then some fascinating insights into the origins and evolution of these early land animals will be forthcoming as dispersed cuticle studies come of age.

The Record of Early Terrestrial Arthropods

Because nonmineralized arthropod cuticle is easily biodegradable, terrestrial arthropods are only preserved under unusual sedimentary circumstances (Allison and Briggs, 1991). Most of what we know about early land arthropods is therefore derived from just a handful of fossil sites (table 16.1) at which the diversity and number of specimens found is reasonably good. In the Upper Carboniferous (Pennsylvanian) there are numerous arthropod-bearing sites of lesser importance which augment the overall picture. Scorpions deviate from this distribution pattern. Although their remains are usually present at these key

Table 16.1 Stratigraphic distribution of some major Palaeozoic terrestrial arthropod faunas.

Million years BP	System	Series/stage	Site	Preservation style	Terrestrial arthropods	Reference
290	Permian					
		Stephanian	Montceau-les-Mines, France	Nodule	Insects, millipedes, arthropleurids, scorpions, arachnids	Rolfe et al. (1982)
			Writhlington, England	Compression	Insects, phalangiotarbids, trigonotarbids	Todd (1991)
			Mazon Creek, Illinois	Nodule	Insects, millipedes, centipedes, arthropleurids, scorpions, uropygids, spider, phalangiotarbids, anthracomartids, trigonotarbids, ricinulids	Nitecki (1979)
310	Carboniferous	Westphalian	Coseley, England	Nodule	Insects, millipedes, scorpions, uropygids, ?spiders, phalangiotarbids, arthracomartids, trigonotarbids, ricinulids	Jeram (personal observation)
		Namurian	Hagen-Vorhalle, Germany	Compression	Insects, millipedes, trigonotarbids, ricinulid, uropygid	Brauckmann (1988)
			(Earliest pterygote insects)	Compression		
		Visean	East Kirkton, Scotland	Lacustrine limestone	Scorpions, opilionids, millipedes, (also earliest reptile and land-going amphibians	Rolfe et al. (1990)
360		Tournaisian				
		Famennian				

		Frasnian	(Earliest tetrapods)			Jarvik (1980)
		Givetian	Gilboa, New York	Cuticle assemblage	See table 16.2	Shear, (1990)
	Devonian	Eifelian				
		Emsian	Alken-an-der-Mosel, Germany	Compression	Arthropleurids, trigonotarbids	Størmer, (1976)
		Pragian	Rhynie Chert, Scotland	Hot-spring chert deposit	Trigonotarbids, mites, collembolans, ?centipedes	Jeram et al. (1990)
410		Lochkovian				
		Pridoli	Ludford Lane, England	Cuticle assemblage	Centipedes, trigonotarbids	Jeram et al. (1990)
		Ludlow				
	Silurian	Wenlock	(Oldest probable vascular plants)	Compression		Edwards and Feehan (1980)
440		Llandovery				
	Ordovician					
		(Mid)	(Oldest evidence for possible land plants)	Spores		Gray et al. (1982)

sites, they also occur as isolated finds, without any associated terrestrial arthropod fauna. Because it is difficult to envisage an ecosystem in which scorpions are the only arthropods, it is reasonable to assume that the fossil record does not accurately reflect the original arthropod community, only a highly selected aspect of it.

Preservational bias

To interpret the fossil record intelligently, an understanding of its inherent bias is a prerequisite. If scorpions are selectively preserved, what implications does this have for our interpretation of fossil arthropod faunas? For example, do the three Devonian faunas (table 16.1) really represent original communities? The style of preservation in each case is quite different, so the contents of the faunas may depend as much on which arthropods are most likely to be preserved under each specific set of preservational conditions as on the composition of the original community. This point is emphasized here because there has been a tacit assumption on the part of paleontologists that arthropod cuticles are fairly uniform in structure and composition from group to group (Neville, 1975; Dalingwater, 1986), giving all arthropods with nonmineralized cuticle approximately the same probability of preservation.

The Mazon Creek fauna from the Upper Carboniferous of Illinois contains a rich assemblage of terrestrial arthropods which are preserved within sideritic (ironstone) nodules (see essays in Nitecki, 1979). It has been extensively studied and is generally regarded as the most complete Paleozoic terrestrial arthropod fauna yet found. It is therefore possible to use this fauna as a yardstick against which to measure the completeness of other faunas of similar age and environmental setting (e.g., Briggs and Gall, 1990). Immediately below the nodule-bearing shale of the Mazon Creek area lies the Colchester (no. 2) coal seam. Samples of this coal reacted with fuming nitric acid yield abundant scorpion cuticles from the acid-resistant residues. At least five species, all of which also occur in the overlying Mazon Creek fauna, have been recognized. No other arthropod cuticle remains are present, in stark contrast to the great diversity of the nodule fauna. This suggests that there is strong preferential preservation of scorpion cuticle in the coal. It could be argued that the cuticle-recovery process selectively removes nonscorpion cuticle, but when shales from immediately above coal seams are treated by less destructive means, such as disaggregating them in water, the results are the same (Bartram et al., 1987).

When the available data are tabulated (table 16.2), it becomes clear that the number of taxa represented in cuticle assemblages varies considerably from site to site. However, the cuticle of some groups is preserved more frequently than others, and it is possible to rank them in order of the relative preservation potential of their cuticle. Several of the assemblages are from sites that have yielded terrestrial arthropod body fossils, and there are invariably differences between the cuticle and body fossil assemblages of these sites. For example, the Lower Carboniferous East Kirkton site yields only scorpion and eurypterid cuticles from rock samples, yet body fossils of millipedes and an opilionid have also been found (Wood et al., 1985; Rolfe, 1988). The Hamilton Quarries (Kansas) locality has yielded abundant insects and millipedes (Hannibal, 1988), but again, only scorpion and eurypterid cuticles occur in the cuticle assemblage. These findings illustrate the point that cuticle assemblages will only contain

Table 16.2 Composition of Palaeozoic arthropod cuticle assemblages[a]

Assemblage	A	B	C	D	E	F	G	H	I	J	K	L	M	N
Silurian														
1. Ludlow, England	+	+	?	+					+		+	+		
Devonian														
2. Hudwick Dingle, England	+	+							+			+		
3. Rhynie chert, Scotland			◆	+					+				◆	
4. Cottonwood Canyon, Wyoming	+	+												
5. Gilboa Mudstone, New York	+	+	+	+	+		?	+	+		+		+	
6. Escuminac Bay, Canada	+	+												
Carboniferous														
7. East Kirkton, Scotland	+	+	+			◆				◆				
8. Dunbar, Scotland		+												
9. Melville Pit, Scotland	+	+												
10. Goyt's Coal, England		+												
11. Bustey Coal, England		+												
12. Wigan 5'coal, England		+												
13. Swillington, England		+	+	+										
14. Barnsley coal, England		+	+											
15. Whitehaven pit, England		+												
16. Colchester Coal, Illinois		+												
17. Lone Star Lake, Kansas		+	+	+	+		+							
18. Hamilton Quarries, Kansas	+	+					◆				◆			◆

[a]Data from Jeram (1989) with additions. Minor Carboniferous assemblages containing only scorpion ± eurypterid cuticle have been ommitted. Assemblages are in stratigraphic order and taxa are as follows: A, eurypterids; B, scorpions; C, mites; D, trigonotarbids; E, spiders; F, opilionids; G, uropygids; H, pseudoscorpions; I, centipedes; J, millipedes; K, arthropleurids; L, "kampecarids"; M, apterygote insects; N, pterygote insects. (+) Indicates cuticle present in assemblage; (◆) indicates that body fossils occur at site, but no cuticle preserved.

arthropods with cuticles well enough preserved to be isolated from the matrix. Although pterygote insect fossils are relatively common in the Upper Carboniferous, the cuticle does not retain sufficient structural integrity to be isolated and if present at all is usually represented only by an amorphous residue which disintegrates when removed from the rock matrix. That most scorpion body fossils have well-preserved cuticle (Kjellesvig-Waering, 1986) emphasizes the high preservation potential of that material relative to the cuticle of other terrestrial arthropods.

Cuticle preservation

The reason arthropod cuticle is preserved differentially probably lies in the nature of the cuticle itself. It may well be that structural organization or composition of cuticle varies from group to group in such a way as to make some cuticles more resistant to breakdown by saprotrophs than others. It is noteworthy that Schofield (this volume) has found unusually high levels of divalent metal complexed in specific regions of scorpion cuticle, where it may act to preserve its integrity and fine structure. Variable resistance to the coalification process caused by raised temperatures and pressures on deep burial must also be a factor. There have been no detailed structural studies of dispersed arthropod cuticles, but preliminary observations, particularly from scorpion cuticles, suggest avenues future investigations might explore.

All of the fossil cuticles discussed here are organic in composition. In many of their physical characteristics they seem to have altered little from their original state, although it has been suggested that they may consist largely of kerogens derived from the polymerization of cuticle breakdown products (Selden et al., 1991; Stankiewicz et al., 1997). Generally, the higher the temperatures and pressures cuticles have been subjected to, the darker in color they are. This is due to the progressive coalification of cuticle, but even within a single assemblage the cuticles of some groups (e.g., scorpions) can remain relatively unaltered, whereas those of others show a much greater degree of alteration (Jeram et al., 1990). Fossil scorpion cuticle is always extremely thin. For example, the cuticle of a large (40 cm long) Lower Carboniferous scorpion is no more than 10–12 μm thick on any part of the body. By contrast, adult cuticle of the extant scorpion *Hadrurus arizonensis* is approximately 100 μm thick (Filshie and Hadley, 1979), and that of *Pandinus imperator,* 80 μm (Kennaugh, 1959). Such insubstantial cuticle would not be strong enough to support the mass of a 40 cm scorpion (see Dalingwater, 1985), so it is likely that only a limited thickness of the original cuticle has been preserved. Scanning electron microscopy (SEM) of fossil scorpion cuticle surfaces and internal structure (figs. 16.1A–C) indicates that a layer analogous to the hyaline exocuticle of Recent scorpions is preserved (Bartram et al., 1987). No trace of the broad laminae characteristic of typical arthropod procuticle (fig. 16.1d) has been observed in any organically preserved fossil material. It is likely that the exocuticular layer found in fossil scorpions is homologous to hyaline exocuticle, although this cannot be demonstrated conclusively because there are differences in the internal structure between fossil material and hyaline exocuticle of the distantly related Recent scorpions (e.g., figs.16.1C,D).

The lack of normal procuticle in any fossil scorpion material may explain the absence of certain groups (e.g., pterygote insects) from Upper Carboniferous assemblages. Furthermore, it suggests that we should not necessarily expect to find these groups in older cuticle assemblages, even if they were present in the original communities. Those groups represented by dispersed cuticle remains presumably had resistant exocuticle, although the causes of this resistance, be they structural or compositional, may not show as extreme or obvious a manifestation as the hyaline exocuticular layer characteristic of scorpions. One puzzling aspect of the earliest cuticle assemblages is the presence of centipedes (figs. 16.1D,E; Shear and Bonamo, 1988; Jeram et al., 1990). Their cuticle is not found in even the most diverse Carboniferous assemblages, although the nodule faunas indicate their presence (Mundel, 1979). Why centipede cuticles fail to appear in the post-Devonian record is unclear, but it cautions against assuming that cuticle characteristics in arthropod clades have remained constant over geologic time and also reminds us that the study of modern taxa may not hold all the answers when it comes to Paleozoic arthropod cuticles. However, the fact that fossil cuticles do have differing preservation potentials suggests that it may be worthwhile to look afresh at modern arthropod cuticles with this in mind.

Differential preservation has acted as a filter between the original terrestrial arthropod community and the observable fossil assemblage. Variation among taxa in cuticle survivability, coupled with a range of preservation mechanisms that each introduce their own biases, means that reconstructing early terrestrial

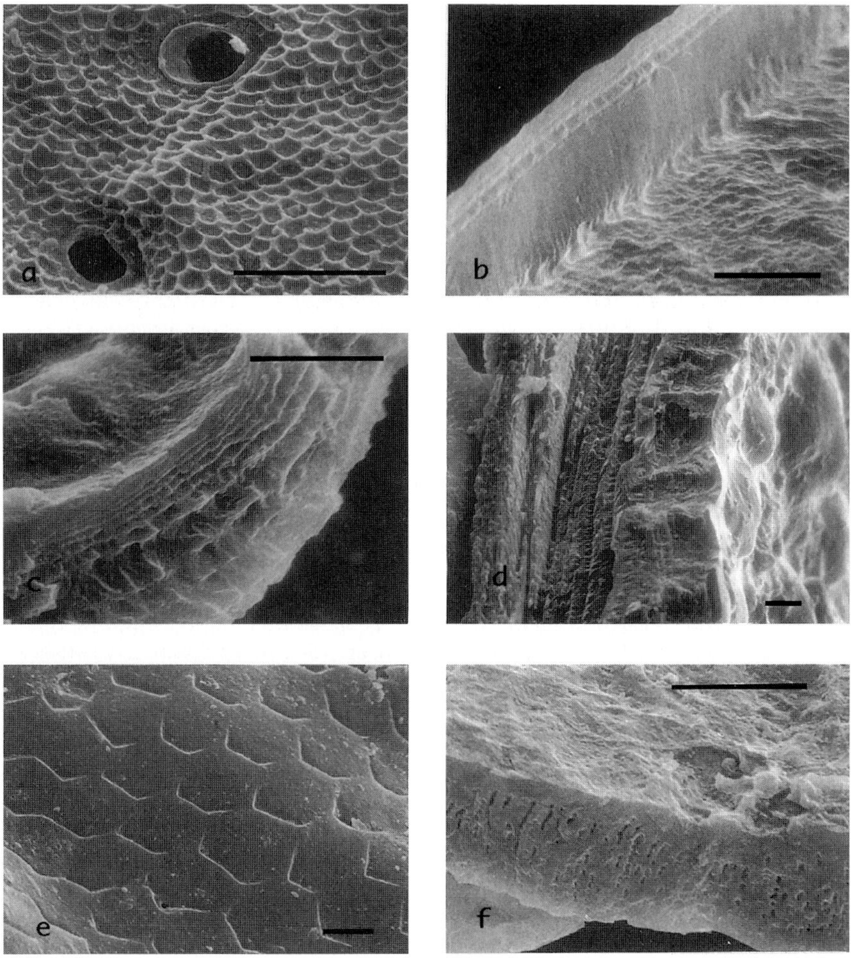

Figure 16.1 Details of fossil scorpion cuticle for comparison with recent scorpion and fossil myriapod cuticle. (A) Scanning electron micrograph of tergite cuticle from an Upper Carboniferous eoscorpiid shows well-preserved surface features such as the reticulate ornament, setal follicles, and cuticular duct (top left). Scale bar = 100 μm. (B) Torn section through cuticle of a large Upper Carboniferous eoscorpiid. There is a finely laminated outer region and thicker nonlaminate inner region. No trace of epicuticle or nonhyaline procuticle has been found in any organically preserved fossil material. (C) Torn section through podomere cuticle of a large Lower Carboniferous mesoscorpion. Laminae increase in thickness down through the cuticle, and there is no descrete nonlaminate inner region. (D) Torn section through tergite cuticle of a recent *Euscorpius* (top to right). Thin hyaline exocuticle, 4 μm thick, forms the outermost layer resolvable in this micrograph. Broad lamellae of the inner exocuticle and endocuticle, absent in fossil cuticle, form the great bulk of the cuticle thickness. (E) Tergite cuticle surface of an Upper Silurian scutigeramorph centipede. (F) Torn section through centipede or arthropleurid cuticle from the same assemblage as panel E. There is an outer structureless region and an inner region traversed by tightly spiraling ducts. As with fossil scorpion cuticles, no lamelate procuticle is preserved. Scale bar = 10 μm for B–F.

ecosystems will not be straightforward. The dispersed cuticle record is an untapped source of high-quality data, but as yet we are only just learning how to interpret it.

SCORPION TERRESTRIALIZATION

The colonization of land is a process that began at least as early as the Middle Ordovician (e.g., Gray et al., 1982; evidence reviewed by Shear, 1990), and which still continues today (Little, 1983, 1990). However, fossil evidence suggests that there was a phase of arthropod terrestrialization during the Silurian and Devonian that resulted in the establishment of terrestrial ecosystems containing vascular plants, arthropods, and tetrapods by the Lower Carboniferous (e.g., Rolfe et al., 1990). Some mollusks attained a terrestrial grade of existence by the Upper Carboniferous (Solem, 1985), but since then there have been few successful attempts to colonize the land by primarily aquatic animals.

There is obviously a continuum between fully aquatic organisms and those that would generally be regarded as fully terrestrial. Much of the soil biota, for example, depends on either surface films of water over soil particles or on interstitial air being saturated with respect to water; so although they live on land, much of their biology is still essentially aquatic. Other organisms may be unable to effectively regulate water loss and so need very damp habitats, or they may be tied to water by reproductive requirements. For the purposes of this discussion, terrestrial animals are regarded as those that live surrounded by a medium of air rather than water. In addition to the problem of evaporative water loss, animals moving onto land face a variety of other difficulties due to the contrasting physical properties of water and air. Some of these require biomechanical solutions, such as the modification of locomotory and some sensory systems to cope with the lower density of air. Others are physiological (e.g., regulating water loss), or behavioral (e.g., prey capture, reproduction). The fossil record gives a better insight into how scorpions made this transition than any other group of invertebrates; nevertheless the picture remains incomplete because only changes that had an impact on the external morphology of scorpions can be detected.

Fortunately, morphology to a great extent reflects function (Manton, 1977), so careful analysis of integumental morphology allows the paleontologist to suggest the most probable function (or functions) of any particular structure. Functional morphology can be applied to fragmentary remains as well as complete fossil scorpions. The structure of an isolated podomere, for example, can be as useful for suggesting when scorpions invaded land as a complete specimen, even if we do not know what the rest of the scorpion looked like or to what taxon the isolated podomere belongs.

In the discussion below, scorpion evolution is approached from the perspective of changing function through time, rather than trying to unravel the phylogeny of Paleozoic scorpions. The reasons for doing this include the need to integrate data from the dispersed cuticle record with that from the body fossil record. Also, in spite of Kjellesvig-Waering's (1986) efforts to sort out the classification of fossil scorpions, we are still a long way from a stable classification which properly reflects phylogeny (Stockwell, 1989; Jeram, 1994; Jeram, 1998). Finally, there is the question of transitional forms. Although it is argued here that Silurian scorpions were aquatic and that the Devonian and Carboniferous faunas contain both aquatic and terrestrial taxa, no taxa are specifically identified as being transitional (amphibious in this context). This is because taxa, or morphological features in isolation, that are adapted to perform at their maximum efficiency either in water or in air are more easily identified than those in which efficiency is compromised by the need to function in both media. It may not be possible to recognize genuinely amphibious fossil scorpions without a detailed knowledge of their respiratory structures. At present this is beyond the resolution of the fossil record, so we can only suggest that individual fossil scorpion taxa show a greater or lesser degree of adaptation to life on land. Nevertheless, this is sufficient to derive a good idea of how scorpions invaded the land.

Respiration

Modern scorpions possess four pairs of book lungs which lie above the ventral sternites and communicate with the air via stigmata on the face of the sternites. However, nearly all Paleozoic scorpions lack stigmata. Instead, the ventral abdominal wall is covered by four or five platelike ventral appendages. Unfortunately, until recently efforts to detect respiratory structures above these plates proved fruitless (e.g., Wills, 1925, 1959, 1960). The ventral appendages of eurypterids (Blatfüsse) are attached to the body wall only at their anterior ends (Waterston, 1975), and this is hypothesized to be the primitive state for scorpions. Although apparent respiratory structures have been observed in some eurypterids, their true nature is unclear (Selden, 1985). In the absence of preserved respiratory structures which would provide an unequivocal indication of whether Paleozoic scorpions achieved gaseous exchange in water or air (i.e., gill books or book lungs), some authors have tended to assume that gills were present above the abdominal plates (e.g., Kjellesvig-Waering, 1986), while others have argued for terrestriality on the basis of scorpion functional morphology (Selden and Jeram, 1989; Stockwell, 1989).

Available fossil evidence suggests that all Silurian scorpions were aquatic (Rolfe and Beckett, 1984), and it is likely that their respiratory structures were similar to those of eurypterids or the living aquatic merostome *Limulus*. How scorpion book lungs were derived from the primitive aquatic structures is therefore the outstanding question. The recent discovery of several Lower Carboniferous scorpions with respiratory structures preserved has provided some answers to this question (Selden and Jeram, 1989; Jeram, 1990a). These scorpions are typical Lower Carboniferous forms, with protolobosternous abdominal plates. Kjellesvig-Waering (1986) considered such scorpions to be aquatic, but the new material has unequivocal book lungs positioned above the abdominal plates, and all other autecological indicators suggest a terrestrial habit for them. The pendulum of opinion has therefore swung again, with nearly all Carboniferous, and some Devonian, scorpions now being considered fully terrestrial (Selden and Jeram, 1989; Stockwell, 1989).

The book lungs lie in approximately the same position relative to the underlying sclerite as those of modern scorpions (Jeram, 1990a); however, no stigmata are present, and the posterior end of each book lung opens directly into a large atrium which overlies the posterior third of each abdominal plate. Details of the book-lung lamellae are not known, as these decayed before fossilization, but at

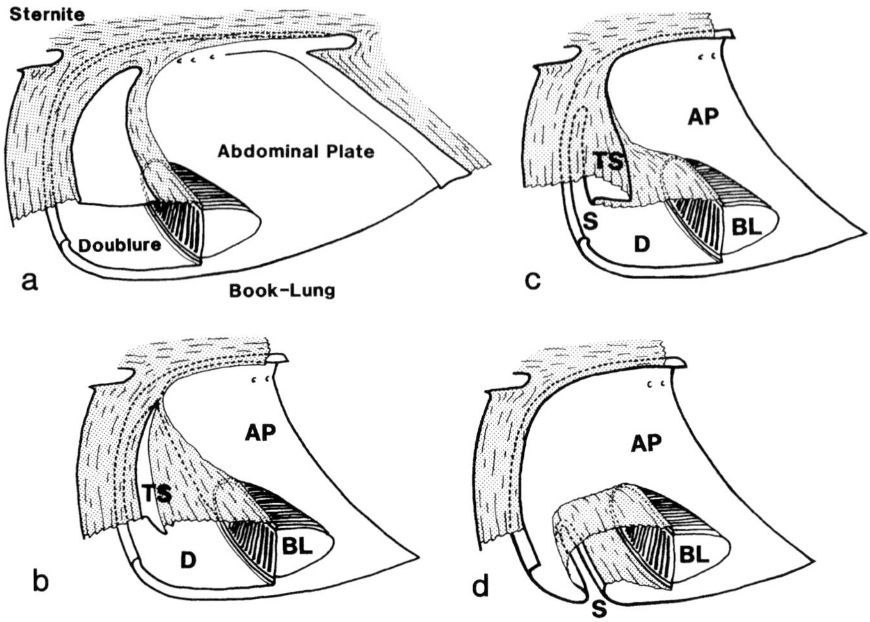

Figure 16.2 The development of intrasternite stigmata. At least two scorpion lineages developed intrasternite stigmata by the progressive suturing-on of abdominal plates. This sequence of cut-away diagrams incorporates evidence from fossil mesoscorpions and suggests how suturing-on could have occurred. (A) Left side of a reconstructed abdominal plate and book lung from a Lower Carboniferous mesoscorpion (dorsal view; after Jeram, 1990a). (B) Theoretical intermediate stage with a reduction of atrial size and lateral suturing of the true sternite to the abdominal plate doublure. (C) Reconstruction of *Mesophonus*-type abdominal plate with marginal stigma (Triassic). (D) Migration of stigmata to a more central position on the ventral plate and further reduction in atrial chamber volume leads to the situation found in recent scorpions. Intersegmental membrane is stippled. AP, ventral wall of abdominal plate; TS, true sternite; S, stigma; BL, book lung; D, doublure of abdominal plate.

the atrial end of each book lung an unusual structure is preserved. The atrial end of each lamella is strengthened by a bar of thicker cuticle, and each bar is fused at top and bottom to an annulus of cuticle, so that the whole structure forms a rigid grille supporting the lamellae. The dorsal wall of the atrium is formed by a sternite and the ventral wall by the abdominal plate. In their anterior two-thirds, the abdominal plates are attached at their edges and actually form the ventral abdominal wall. Air entered the atrium along the posterior margin of the abdominal plates because their posterior thirds were not sutured onto the body wall (fig. 16.2A). Well preserved abdominal plates from Carboniferous cuticle assemblages are constructed in the same manner (unpublished observation).

In eurypterids, the Blatfüsse were moveable, and water could circulate over the respiratory structures above them. If the primitive state in scorpions was like that seen in eurypterids, then the immobilization of abdominal plates by suturing them onto the body wall can be viewed as a critical step toward the internalization of respiratory structures. Progressive suturing-on of the abdominal plates led to a reduction in the size of the atrium, and, eventually, the loss of the true sternites as they became redundant (fig. 16.2). Such a situation is found in present-day uropygids. Discrete stigmata possibly formed in the final stages of suturing-on and then migrated onto the posterior margins of the abdominal plates as suturing was completed to give an arrangement like that described in the Triassic scorpion *Mesophonus* (Wills, 1947). The only subsequent changes necessary to arrive at the state seen in modern scorpions was the migration of stigmata to a more central position on the ventral plates and a reduction in the atrial size related to this.

The gill lamellae of *Limulus* are kept in shape by a strengthening bar of cuticle along their distal margins. Land-going crabs also have similar structures to prevent gill lamellae from collapsing together and so reducing the surface area available for gaseous exchange (Little, 1983). It is probable, therefore, that the precursors of book lungs were gills internalized above abdominal plates to prevent excessive water loss from the respiratory surface, and which were also structurally modified to maintain an optimal surface area for gaseous exchange in air. The earliest direct evidence of scorpion respiratory structures comes from a fragmentary Lower Devonian specimen (Shear et al., 1996). The specimen has book lung lamellae with reticulate surface sculpture similar to modern buthids. The reticulation helps to make thin lamellae more rigid and maintains an air gap between adjacent lamellae. More derived modern scorpion families have papillae on lamellar surfaces that perform the same function (fig. 16.3).

Fossil scorpions demonstrate conclusively that the sternites of modern scorpions are not true sternites at all, but abdominal appendages sutured onto the ventral body wall (contra Kjellesvig-Waering, 1986). The sternites and book lungs of other terrestrial chelicerates (i.e., arachnids) may have been derived by the same process. Although we cannot be sure how many chelicerate lineages originally colonized land, scorpions clearly did it independently of others. Furthermore, scorpions with marginal stigmata have recently been recognized

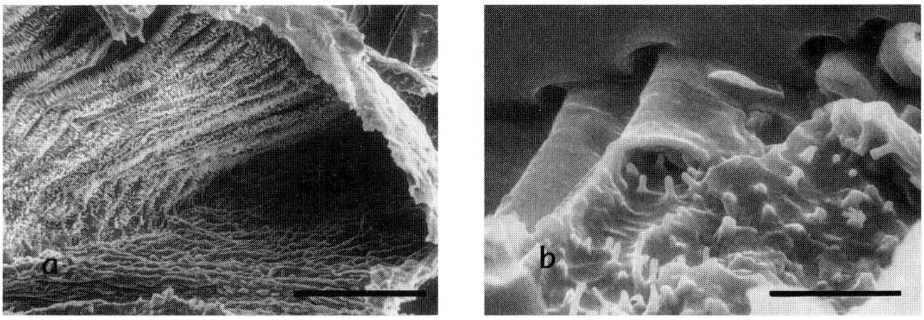

Figure 16.3 The book lungs of modern scorpions have undergone over 350 million years of refinement on land, and few vestiges of their origins as gills remain. (A) Book-lung atrial chamber of *Euscorpius*. Note the distal ends of lamellae are not rigid and grade directly into atrial wall cuticle. Scale bar = 100 μm. (B) Dried hemolymph fills the lumen of lamellae, and where lamellae are torn the papillae, which maintain an air gap between adjacent lamellae, can be seen. Same specimen as panel A. Scale bar = 10 μm.

in the Devonian Gilboa cuticle assemblage, and these are stem-group members of the clade containing all modern scorpions (Jeram, in preparation). Triassic mesophonids which also have marginal stigmata (Wills, 1947) belong to a separate large clade containing the bulk of Carboniferous scorpions. Intrasternite stigmata must therefore have been derived at least twice in scorpions, and current evidence suggests that immobilization of abdominal plates occurred by the Lower Devonian.

Locomotion

Manton (1977) drew attention to the locomotory mechanisms of arthropods and in particular to the mutually exclusive biomechanical requirements of legs constructed either for speed or for strength. Legs modified to perform at extremes of either of these locomotory modes can be recognized fairly easily in fossils and have been used by paleontologists as evidence for aquatic or terrestrial modes of life (e.g., Størmer, 1976; Selden and Jeram, 1989; Jeram et al., 1990). Unforunately, the requirements of an arthropod for speed or strength do not necessarily correlate with locomotion in water or on land. Nevertheless, the physical properties of these two media are significantly different, with each having particular influences on locomotory modes. A large number of factors are involved in the mechanics of arthropod locomotion, and these are discussed at length in the works of Manton and others (e.g., Manton, 1952, 1977; Herreid and Fourtner, 1981) Only those factors directly relevant to terrestrializing scorpions are examined here.

Scorpions lack appendages adapted for swimming like those of eurypterids and many crustaceans (see Selden, 1984). Functional morphology of fossil scorpions indicates that locomotion was achieved by walking. The density of water poses particular problems for benthic arthropods, and two of these, drag and traction, may have imposed limits on the speed of locomotion of aquatic scorpions. Anyone who has tried running across the bottom of a swimming pool while neck-deep in water will be familiar with these factors. The amount of drag incurred by an arthropod moving through water is proportional to its velocity, making rapid walking in water particularly costly in terms of energy consumption. It also means that more force must be put into the propulsive backstroke of each leg in order for a scorpion to maintain the same velocity in water as in air. The backstroke will only propel the scorpion forward if the end of the leg in contact with the substrate remains stationary. Traction may be achieved by maximizing the surface area of leg tip (often by use of tarsal modifications such as claws and cuticular projections) in contact with the substrate, but a second significant contributor to traction is the downward force exerted by the leg onto the substrate. This force has two components. Part is exerted by leg musculature during the backstroke, and the remainder is a gravitational component proportional to the arthropod's mass. If the vertical force produced by leg muscles is too large, the scorpion will be pushed clear of the substrate (i.e. jump) and traction is lost. The maximum speed of locomotion is therefore in part dependent on the arthropod's mass because this may limit the amount of traction that can be achieved. Aquatic arthropods with heavily mineralized exoskeletons (e.g., decapod crustaceans, trilobites) have significantly greater mass in water than scorpions of similar size, giving them a greater potential for rapid walking in water than scorpions. For aquatic scorpions, therefore, the odds are stacked heavily in favor of relatively slow movement.

The leg structure of Silurian scorpions is consistent with this analysis. Each leg has six cylindrical podomeres (plus the coxa and post-tarsus), which are all relatively short. The joints are not well known, but by analogy with the walking legs of eurypterids (Selden, 1981; Briggs et al., 1991), it is likely that the angle of flexure on each joint was small and stride lengths consequently short. The terminations of Silurian scorpion legs are quite variable. In protoscorpions the post-tarsus is eurypteridlike, and this can be regarded as a digitigrade condition. The paleoscorpions have a reduced post-tarsus combined with a greater development of the telotarsal spurs (homologous to the claws of modern scorpions) which in some taxa approach the condition seen in Recent scorpions (both Silurian orders, sensu Stockwell, 1989). Some of the paleoscorpions may have had true plantigrade stances, but although the telotarsus is often elongated there is scant evidence for its inferior surface contacting the substrate. The legs of Silurian scorpions, and some later aquatic (on the basis if independent evidence) forms, are therefore constructed for relatively slow movement.

Terrestrial scorpions show a number of adaptations for locomotion on land. Because their bouyancy in air is so much less than in water, traction on substrates increases greatly at the same time drag is significantly reduced. Faster locomotion is therefore feasible, and this is reflected by leg morphology. Podomeres are laterally compressed, which adds to their mechanical strength when loaded (Prange, 1977). The condyles at joints are more robustly constructed, with joints achieving much larger angles of flexure. Podomeres exhibit differentiation in length, particularly in posterior legs. These changes allow greater stride lengths because strides can be overlapped. Traction is improved by use of the inferior telotarsal surface as a large foot, in combination with inferior telotarsal spines and more highly developed claws. Dispersed cuticle assemblages show that some taxa had this combination of characters by the Lower Devonian (Lochkovian stage), and this, along with other evidence, suggests that fully terrestrial scorpions existed by that time (Selden and Jeram, 1989). The possible presence of plantigrade stances in some Silurian scorpions, which were almost certainly aquatic, illustrates the difficulty in using this character as a sole indicator of terrestriality. Whether they had a true plantigrade stance or not, changes that occurred in scorpion locomotory systems to improve walking in water effectively preadapted scorpions for walking on land.

Prey Detection

There is no doubt that the aquatic scorpions were predators like their modern descendants, employing large chelate pedipalps as their prime means of capturing prey (Rolfe and Beckett, 1984). It is less certain that the stinger was also used for this purpose, but well-preserved material from a late Silurian cuticle assemblage has paired venom channels near the tip of the aculeus (unpublished observations), so it is known that these scorpions were venomous. Feeding behavior in recent scorpions has been reviewed by McCormick and Polis (1990).

Mechanoreceptors

Prey location by modern scorpions involves specialized sensory systems such as tarsal sensilla and pedipalpal trichobothria, but apparently not visual sensilla. Tarsal hairs have been implicated in the detection of compressional waves

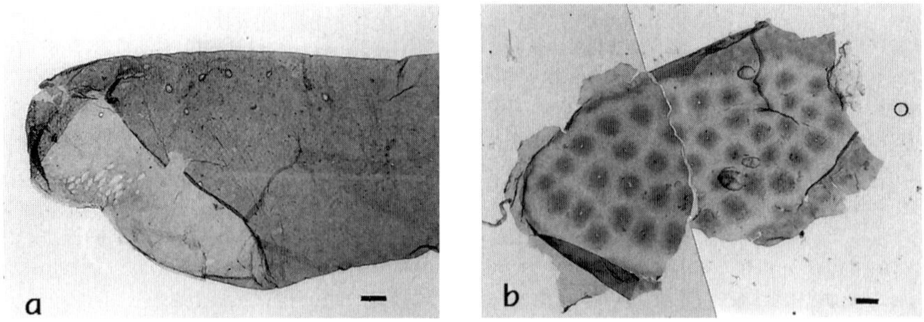

Figure 16.4 Dispersed cuticles of scorpions and changes in their sensory structures through time. (A) Juvenile eoscorpiid femur from the Upper Carboniferous, with a cluster of slit sensilla (light areas) near the prolateral margin of the distal end. (B) Scorpions show a consistent trend toward the reduction in numbers of lateral ocelli. This eoscorpiid lateral eye group from the Upper Carboniferous has 40 ocelli preserved. Scale bars = 100 μm.

generated by prey activity and transmitted through the substrate (Brownell and Farley, 1979), and similar hairs are present in fossil scorpions from at least the Middle Devonian. However, it is not possible to demonstrate that the tarsal setae of fossils had the same sensory function as their modern counterparts. Tarsal slit sensilla have not yet been recognized in fossils, although they can occasionally be found on other podomeres (fig. 16.4A), but trichobothria identical to those of modern scorpions occur in one family of Carboniferous scorpions, and it is possible that other Paleozoic scorpions also had functional trichobothria. Clearly there are limits to how reliably functions can be ascribed to cuticular sensilla in fossil scorpions, particularly where they occur in clades only distantly related to living forms. Nevertheless, dispersed cuticle remains provide a unique opportunity for investigating these systems in fossils which has not yet been exploited.

The cuticular sensilla of Silurian scorpions are practically unknown, but aquatic arthropods are generally well endowed with a variety of sensilla, and it is probable that some of these required only minor modifications to function in air. Trichobothria, for example, may have been independently derived in a number of terrestrializing lineages from the sensory hairs of aquatic arthropods which detect water movements (Selden and Jeram, 1989).

Visual systems

Scorpion visual systems have changed significantly since the Silurian. The lateral eye groups of most Paleozoic scorpions contain more than 20 ocelli (fig. 16.4B), and Kjellesvig-Waering (1986) estimated that in Silurian scorpions up to 1000 ocelli are present, constituting true compound eyes like those of eurypterids. The Paleozoic record reveals a consistent trend toward reduction in number of lateral ocelli. The anatomy of modern scorpion lateral eyes suggests that this reduction may have been achieved by fusion of ocelli into bigger units (Weygolt and Paulus, 1979), each presumably containing more rhabdomes. It is reasonable to speculate that the fusion process resulted in lateral ocelli with greater individual light sensitivity, but also a loss of overall image quality. Because only exocuticle is preserved in fossil scorpions, the imaging ability of their compound eyes cannot be measured, but the bulbous shape of those eyes and their position on the anterior corners of the carapace indicate that the rhabdomal axes intersected, so stereoscopic vision was theoretically possible.

The median eyes of Silurian scorpions are very small and located on a raised ocellar node close to the carapace anterior margin. It is unlikely that they were any better at image formation than the median eyes of modern scorpions (see Locket, this volume). The visual system of Silurian scorpions is similar to those of some predatory arthropods, rather than being adapted for detection of predators as has been suggested for some trilobites (e.g., Clarkson, 1979). Vision was therefore probably an important means of prey location in aquatic scorpions, and because they could neither swim nor pursue prey at speed, these early scorpions were probably ambush predators, operating rather like modern dragonfly larvae. If this is the case, then the behavioral pattern has persisted in spite of the move onto land.

Feeding

With the exception of some mites, all terrestrial chelicerates are liquid feeders. In modern scorpions food is mashed up by the chelicera and taken into a preoral cavity, where some extraintestinal digestion occurs. The liquid mixture of digestive juices and partly digested prey is then ingested by the sucking action of the pharynx. Indigestible solids are expelled from the preoral cavity. This method of feeding is efficient for terrestrial chelicerates because little indigestible material is ingested, reducing the amount of water that is lost from the body via excretion. In contrast, aquatic chelicerates (e.g., *Limulus*) lack a preoral cavity, and no extraintestinal digestion occurs. Prey is dismembered by the chelicera, and further comminution is achieved by coxal gnathobases. This is possible because the coxae are movable and the medial ends of the coxae are modified with teeth or spines into an efficient shredding apparatus which processes food as it passes back between the coxae. Mashed up, but nevertheless solid, food is ingested.

The preoral cavity of Recent scorpions is formed by the cheliceral bases (above), pedipalp coxae (laterally), and platelike extensions of the anterior two pairs of walking-leg coxae, termed "coxal apophyses" (see below). Coxal apophyses are absent in Silurian scorpions, so they had no preoral cavity. The walking-leg coxae may have been moveable (Kjellesvig-Waering, 1986), but well-developed coxal gnathobases like those of contemporaneous eurypterids have not been observed. Instead, comminution of prey into ingestible particles was carried out by the chelicera, which are larger than those of any modern scorpion. Furthermore, they are significantly bigger, more powerful, and with a more elaborate dentition, than those of eurypterids or *Limulus*. Kjellesvig-Waering (1986) noted the presence of rudimentary coxal apophyses in one Silurian scorpion and in several Lower Devonian taxa. At least one Lower Devonian scorpion (*Branchioscorpius*) had coxal apophyses and a preoral chamber as well developed as Recent scorpions, and a preoral cavity is present in nearly all Carboniferous scorpions (fig. 16.5B). Available evidence therefore suggests that scorpions adopted extraintestinal digestion simultaneously with their move onto land.

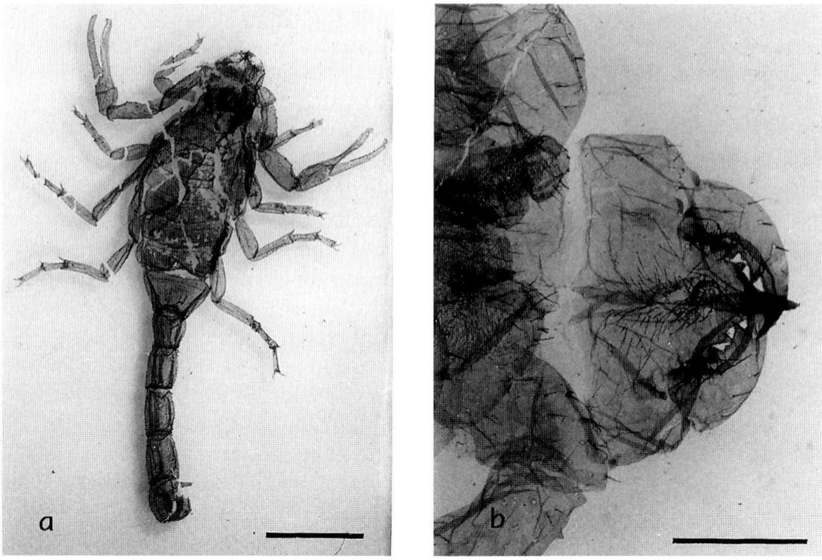

Figure 16.5 Lower Carboniferous scorpions from the East Kirkton Quarry near Edinburgh, Scotland. Exceptionally well-preserved scorpions can be etched free from the rock matrix, revealing fine structural detail not normally obtainable from typical scorpion body fossils. (A) Fourth stadium juvenile mesoscorpion. Scale bar = 5 mm. (B) Detail of chelicerae, proximal end of pedipalps, and anterior end of oral tube. Third stadium juvenile mesoscorpion. Scale bar = 1 mm.

Body Size

Paleontologists have tended to assume that the majority of fossil scorpions were adult because their usual size range corresponds with that of adult Recent scorpions. This assumption does not stand up to close scrutiny because the fossils may be derived from molted exoskeletons or dead individuals. If the survivorship curves of Paleozoic scorpions even remotely resemble those of modern arthropods, the fossil record should be heavily biased in favor of juveniles. Because scorpion body fossils are rare, sample sizes are too small to yield meaningful size-distribution data. Cuticle assemblages can provide an almost unlimited source of material, so large samples can be built up that more accurately reflect size-distribution patterns in the original communities. However, because the thin cuticle of early instars is more susceptible to decomposition than that of older ones, they may be underrepresented in cuticle assemblages.

Figure 16.6 shows the variation in scorpion maximum size (body length) through the Upper Paleozoic. Only the larger specimens from each fossil site are plotted, and most of the figures given for body fossils are estimates that take into account distortions caused by fossilization. Few Silurian scorpions have been found, so it would be unwise to comment on the maximum size for that interval. Dispersed cuticle assemblages from the Late Silurian indicate that large scorpions (>30 cm length) were present by then, but no substantial assemblages have yet been obtained from older rocks. Data are sparse for the Devonian too, but do indicate that several taxa approached a meter in length (Kjellesvig-Waering, 1986). The Carboniferous record is far better and shows large scorpions persisting in the Lower Carboniferous, but by the Upper Carboniferous (the best sampled interval) maximum body size had decreased to approximately 30 cm.

Just why several lineages of Paleozoic arthropods achieved giant proportions (relative to their modern counterparts) is unclear (Briggs, 1985). Some authors have expressed doubts that very large scorpions could be fully terrestrial, particularly during molting (Kjellesvig-Waering, 1986; Briggs et al., 1991), but unequivocally terrestrial scorpions reached lengths of around 70–80 cm in the Lower Carboniferous (Jeram, 1990b). Evidence for terrestriality in the larger Devonian scorpions is less clear cut because the material is more fragmentary, but whatever the case, there was certainly a reduction in the maximum body sizes attained by scorpions through the Carboniferous.

The pros and cons of large body size in scorpions are reviewed by Myers (this volume), but there are aspects to this subject on which fossil scorpions can contribute. Recent scorpions form a monophyletic group that originated in the lat-

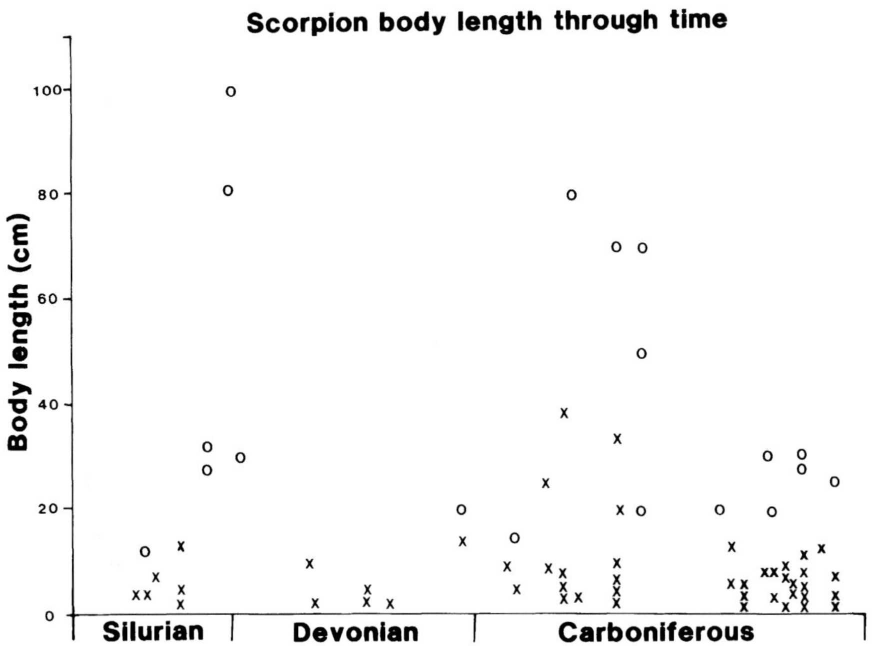

Figure 16.6 The decrease in maximum body size of scorpions through the Palaeozoic. Crosses are estimates of body length based on published records of reasonably complete scorpion fossils. Circles are estimates based on fragmentary remains, including some unpublished dispersed cuticle data. Small scorpions have a higher chance of being preserved complete than large ones, so fragmentary remains provide most of the records of big scorpions. All complete specimens >10 cm in length are included, but in assemblages yielding many small individuals, only a few representative specimens are plotted.

est Carboniferous or Permian. The stem group from which modern scorpions were derived has several Carboniferous members and one in the Middle Devonian (Jeram, unpublished observation). This means that the split between the clade containing recent scorpions (the Orthosterni sensu lato) and their sister group (Paleosterni, sensu Stockwell, 1989) had already occurred by the Middle Devonian. The significance of this is that all Paleozoic orthosterns are <5 cm in length, but subsequently size diversification has taken place to generate the range of body sizes seen in Recent scorpions. Most other Devonian and Carboniferous scorpions attained much greater body sizes, but selective pressure in the late Paleozoic shifted toward small scorpions. There is a loose correlation between the diversification of terrestrial tetrapods and the reduction of scorpion body size in the Carboniferous, although no causal relationship can be demonstrated. Given that there are size limits imposed on terrestrial arthropods by the biomechanics of their cuticle (Dalingwater, 1985) and that the skeletal organization of vertebrates allows them to greatly exceed these limits (Alexander, 1968), it is possible that the reduction in scorpion body size was due to selection pressure applied by large predatory tetrapods (see discussion in Shear and Kukalová-Peck, 1990; see also Vermeij, 1987). The maximum body size of Recent scorpions is probably also due to ecological factors rather than to any absolute biomechanical or biophysical limits to size.

In arthropods growth can only be achieved by ecdysis, and as arthropods get larger the energy cost involved in generating a new cuticle increases significantly (e.g., Célérier, 1986). Arthropods are also particularly vulnerable to predators during each ecdysial phase. The larger modern scorpions molt 6–8 times (Polis and Sissom, 1990), with an average progression ratio of 1.28 ± 0.04 (Polis and Farley, 1979). However, a morphometric analysis of scorpion pedipalps from Upper Carboniferous cuticle assemblages suggests that the progression ratio in larger species may have been as high as 1.4, with only 8 molts required to reach a length of 30 cm (Jeram, unpublished data). Such large progression ratios would clearly be advantageous for very big scorpions, but in smaller species the higher proportion of their energy budget invested in cuticle production at each molt might outweigh any benefits.

Cryptozoic Habits

If, as fossil evidence suggests, aquatic scorpions used vision as a major means of prey detection, then it is reasonable to assume that they were active by day and may have concealed themselves at night to avoid nocturnal predators. Recent scorpions are predominantly nocturnal in their periods of activity. During the day they seek shelter beneath rocks or vegetation, and many species construct substantial burrows. Advantages of this pattern of behavior are avoidance of diurnal predators, minimization of cuticular water loss, and avoidance of high surface temperatures (Hadley, 1990). This behavior is most highly developed in desert scorpions which are subject to harsh conditions and high diurnal predation pressure, but much less marked in tropical species, some of which are active diurnally (Toye, 1970).

The Paleozoic scorpions that have been preserved were inhabitants of low-altitude, equatorial, flood-basin forest. The late Paleozoic scorpions are almost all of the "errant" ecomorphotype, and only one taxon, *Garnettius,* exhibits any adaptation for burrowing. Fossil floras associated with scorpion remains indicate that environmental conditions were warm, wet, and humid, so cuticular water loss was probably not a great problem. Predation pressure is more likely to have driven scorpions toward nocturnal activity. During the Devonian, scorpions were by far the largest predators on land, as the late Devonian tetrapods were aquatic. Intraguild predation would therefore have been the most significant source of predation pressure, although juveniles, and adults of smaller species, may also have been subject to predation by arachnids and centipedes. Small species (e.g., orthosterns) are most able to adopt cryptozoic habits because the larger a scorpion is, the less chance it has of finding a suitable refuge, and the more difficult it is to construct a burrow. The Carboniferous orthosterns were probably nocturnal because they show the greatest lateral eye reduction, posterior displacement of median eyes, and development of pedipalp sensilla. Although there is a trend toward lateral eye reduction in other Paleozoic terrestrial clades, it is unclear whether any of them fully adopted nocturnal habits, particularly as the lateral eye reduction is compensated for in many taxa by enlargement of the median eyes which were forward-facing and positioned near the anterior margin of the carapace.

The Upper Carboniferous saw two ecological challenges that had an impact on the direction of scorpion evolution. The diversification of terrestrial tetrapods undoubtedly introduced a new tier of predation pressure, particularly on the larger scorpions, and climatic changes in the late Carboniferous and Permian reduced the flood basin habitats of arborescent lycophytes and pteridosperms, at least in the Euramerican area, which has provided all late Paleozoic scorpion fossils. Although relics of the Paleozoic fauna may have lingered as late as the Jurassic in some regions (Kjellesvig-Waering, 1986), biogeographic evidence suggests that the orthostern scorpions underwent a significant radiation in the Permian. The subsequent success of the orthosterns can be attributed to their adoption of nocturnal habits, smaller adult body sizes, and burrowing behavior.

Acknowledgments I thank Paul Selden, John Dalingwater, Dianne Edwards, Bill Shear, and Ian Rolfe for their help and encouragement freely given during the course of work on dispersed cuticle assemblages. This chapter was prepared during tenure of an NERC Research Grant to P. Selden, D. Edwards, and A.J.J. Thanks also to Drs. Jane Gray and Arthur Boucot for many helpful suggestions for improvement of the manuscript.

References

Alexander, R.M. (1968) *Animal Mechanics.* Sidgewick and Jackson, London.

Allison, P.A. and Briggs, D.E.G. (1991) The taphonomy of soft-bodied animals. *In* Donovan, S.K. (ed.) *The Processes of Fossilization,* pp. 121–140. Belhaven Press, London.

Bartram, K.M., Jeram, A.J. and Selden, P.A. (1987) Arthropod cuticles in coal. *J. Geol. Soc.* (London) 144: 513–517.

Brauckmann, C. (1988) Hagen-Vorhalle, a new important Namurian Insecta-bearing locality (Upper Carboniferous; FR Germany). *Entomol. Gen.* 14: 73–79.

Briggs, D.E.G. (1985) Gigantism in Palaeozoic arthropods. *Spec. Pap. Palaeontol.* 33: 157.

Briggs, D.E.G., Dalingwater, J.E. and Selden, P.A. (1991) Biomechanics of locomotion in fossil arthropods. *In* Rayner, J.M.V. and Wootton, R.J. (eds.) *Biomechanics and Evolution,* pp. 37–57. Cambridge University Press, Cambridge.

Briggs, D.E.G. and Gall, J.C. (1990) The continuum in soft-bodied biotas from transitional environments: a quantitative comparison of Triassic and Carboniferous Konservat-Lagerstätten. *Paleobiology* 16: 204–218.

Brownell, P.H. and Farley, R.D. (1979) Prey localizing behaviour of the nocturnal desert scorpion, *Paruroctonus mesaensis:* orientation to substrate vibrations. *Anim. Behav.* 27: 185–193.

Célérier, M.-L. (1986) Energy cost of exuviae production of spiders. *Bull. Br. Arachnol. Soc.* 7: 77–82.

Clarkson, E.N.K. (1979) The visual system of trilobites. *Palaeontology* 22: 1–22.

Dalingwater, J.E. (1985) Biomechanical approaches to eurypterid cuticles and chelicerate exoskeletons. *Trans. R. Soc. Edinb. Earth Sci.* 76: 359–364.

Dalingwater, J.E. (1986) Chelicerate cuticle structure. *In* Nentwig, R. (ed.) *Ecophysiology of Spiders,* pp. 3–15. Springer-Verlag, Berlin.

Dunlop, R. (1898) A fossil scorpion from Airdrie, with historical note. *Ann. Kilmarnock Glenfield Ramblers* 2: 60–61.

Edwards, D. and Feehan, J. (1980) Records of *Cooksonia*-type sporangia from the late Wenlock strata in Ireland. *Nature* 287: 41–42.

Filshie, B.K. and Hadley, N.F. (1979) Fine structure of the cuticle of the desert scorpion *Hadrurus arizonensis. Tissue Cell* 11: 249–262.

Gray, J., Massa, D. and Boucot, A.J. (1982) Caradocian land plant microfossils from Libya. *Geology* 10: 197–201.

Hadley, N.F. (1990) Environmental physiology. *In* Polis, G.A. (ed.) *The Biology of Scorpions,* pp. 321–340. Stanford University Press, Palo Alto, CA.

Hannibal, J.T. (1988) Quarries yield rare Palaeozoic fossils. *Geotimes* 33: 10–13.

Herreid, C.F. and Fourtner, C.R. (eds.) (1981) *Locomotion and Energetics in Arthropods.* Plenum Press, New York.

Jarvik, E. (1980) *Basic Structure and Evolution of Vertebrates,* vols. 1 and 2. Academic Press, London.

Jeram, A.J. (1989) Terrestrial arthropod Lagerstätten: the significance of cuticle preservation. *Abstr. 28 Intl. Geol. Congr., Washington* 2: 121.

Jeram, A.J. (1990a) Book-lungs in a Lower Carboniferous scorpion. *Nature* 343: 360–361.

Jeram, A.J. (1990b) When scorpions ruled the world. *New Sci. Lond.* 126 (1721): 52–55.

Jeram, A.J. (1994) Scorpions from the Visean of East Kirkton, West Lothian, Scotland, with revision of the infraorder Mesoscorpionina. *Trans. R. Soc. Edinb. Earth Sci.* 84: 283–299.

Jeram, A.J. (1998) Phylogeny, classification, and evolution of Silurian and Devonian Scorpions. 17–31. *In* P.A. Selden (ed.) *Proceedings of the 17th European Colloquium of Arachnology, Edinburgh, 1997.* British Arachnological Society, Burnham, Beeches, Bucks.

Jeram, A.J., Selden, P.A. and Edwards, D. (1990) Land animals in the Silurian: arachnids and myriapods from Shropshire, England. *Science* 250: 658–661.

Kennaugh, J. (1959) An examination of the cuticles of two scorpions, *Pandinus imperator* and *Scorpiops hardwickii. Q. J. Microsc. Sci.* 100: 41–50.

Kjellesvig-Waering, E.N. (1986) A restudy of the fossil Scorpionida of the world. *Palaeontogr. Am.* 55: 1–287.

Little, C. (1983) *The Colonisation of Land.* Cambridge University Press, Cambridge.

Little, C. (1990) *The Terrestrial Invasion.* Cambridge University Press, Cambridge.

Manton, S. (1952) The evolution of arthropodan locomotory mechanisms. Part 2. General introduction to the locomotory mechanisms of the Arthropoda. *J. Linn. Soc. (Zool.)* 42: 118–166.

Manton, S. (1977) *The Arthropoda: Habits, Functional Morphology and Evolution.* Clarendon Press, Oxford.

McCormick, S.J. and Polis, G.A. (1990) Prey, predators and parasites. *In* Polis, G.A. (ed.) *The Biology of Scorpions,* pp. 294–320. Stanford University Press, Palo Alto, CA.

Mundel, P. (1979) The centipedes (Chilopoda) of the Mazon Creek. *In* Nitecki, M.H. (ed.) *Mazon Creek Fossils,* pp. 361–378. Academic Press, New York.

Neville, A.C. (1975) *Biology of the Arthropod Cuticle.* Springer, Berlin.

Nitecki, M.H. (ed.) (1979) *Mazon Creek Fossils.* Academic Press, New York.

Polis, G.A. (1990) Ecology. *In* Polis, G.A. (ed.) *The Biology of Scorpions,* pp. 247–293. Stanford University Press, Palo Alto, CA.

Polis, G.A. and Farley, R.D. (1979) Characteristics and environmental determinants of natality, growth and maturity in a natural population of the desert scorpion, *Paruroctonus mesaensis* (Scorpionida: Vaejovidae). *J. Zool.* 187: 517–542.

Polis, G.A. and Sissom, W.D. (1990) Life history. *In* Polis, G.A. (ed.) *The Biology of Scorpions,* pp. 161–228. Stanford University Press, Palo Alto, CA.

Prange, H.D. (1977) The scaling and mechanics of arthropod exoskeletons. *In* Pedley, T.J. (ed.) *Scale Effects in Animal Locomotion,* pp. 169–181. Academic Press, London.

Rolfe, W.D.I. (1980) Early invertebrate terrestrial faunas. *In* Panchen, A.L. (ed.) *The Terrestrial Environment and the Origin of Land Vertebrates,* pp. 117–157. Academic Press, London.

Rolfe, W.D.I. (1985) Early terrestrial arthropods: a fragmentary record. *Phil. Trans. R. Soc. Lond.* B 309: 207–218.

Rolfe, W.D.I. (1988) Early life on land—the East Kirkton discoveries. *Earth Sci. Conserv.* 25: 22–28.

Rolfe, W.D.I. and Beckett, E.C.M. (1984) Autecology of Silurian Xiphosurida, Scorpionida, Crustacea, Cirripedia and Phyllocarida. *Spec. Pap. Palaeontol.* 32: 27–37.

Rolfe, W.D.I., Durant, G.P., Fallick, A.E., Hall, A.J., Large, D.J., Scott, A.C., Smithson, T.R. and Walkden, G.M. (1990) An early terrestrial biota preserved by Visean vulcanicity in Scotland. *Geol. Soc. Am. Spec. Pap.* 244: 13–24.

Rolfe, W.D.I., Shram, F.R., Pacaud, G., Sotty, D. and Secretan, S. (1982) A remarkable Stephanian biota from Montceau-Les-Mines, France. *J. Paleontol.* 56: 426–428.

Selden, P.A. (1981) Functional morphology of the prosoma of *Baltoeurypterus tetragonopthalmus* (Fisher) (Chelicerata: Eurypterida). *Trans. R. Soc. Edinb. Earth Sci.* 72: 9–48.

Selden, P.A. (1984) Autecology of Silurian Eurypterids. *Spec. Pap. Palaeontol.* 32: 39- 54.

Selden, P.A. (1985) Eurypterid respiration. *Phil. Trans. R. Soc. Lond.* B 309: 219–226.

Selden, P.A. (1988) The arachnid fossil record. *Br. J. Entomol. Nat. Hist.* 1: 15–18.

Selden, P.A. and Jeram, A.J. (1989) Palaeophysiology of terrestrialisation in the Chelicerata. *Trans. R. Soc. Edinb. Earth Sci.* 80: 303–310.

Selden, P.A., Shear, W.A. and Bonamo, P.M. (1991) A spider and other arachnids from the Devonian of New York and reinterpretations of Devonian Araneae. *Palaeontology* 34: 241–281.

Shear, W.A. (1990) The early development of terrestrial ecosystems. *Nature* 351: 283–289.

Shear, W.A. and Bonamo, P.M. (1988) Devonobiomorpha, a new order of Chilopoda from the Middle Devonian of New York State and the phylogeny of centiped orders. *Am. Mus. Nov.* 2972: 1–30.

Shear, W.A., Bonamo, P.M., Grierson, J.D., Rolfe, W.D.I., Laidlow Smith, E. and Norton, R.A. (1984) Early land animals in North America: evidence from Devonian age arthropods from Gilboa, New York. *Science* 224: 492–494.

Shear, W.A., Gensel, P.G. and Jeram, A.J. (1996) Fossils of large terrestrial arthropods from the Lower Devonian of Canada. *Nature* 384: 555–557.

Shear, W.A. and Kukalová-Peck, J. (1990) The ecology of Palaeozoic terrestrial arthropods: the fossil evidence. *Can. J. Zool.* 68: 1807–1834.

Solem, A. (1985) Origin and diversification of pulmonate land snails. *In* Tiveman, E.R. and Clarke, M.R. (eds.) *The Mollusca,* vol. 10, *Evolution,* pp. 269–293. Academic Press, London.

Stankiewicz, B.A., Briggs, D.E.G. and Evershed, R.P. (1997) Chemical composition of Paleozoic and Mesozoic fossil invertebrate cuticles as revealed by pyrolysis-gas chromatography/mass spectrometry. *Energy and Fuels* 11: 515–521.

Stockwell, S.A. (1989) Revision of the phylogeny and higher classification of scorpions (Chelicerata) (Ph.D dissertation). University of California, Berkeley.

Størmer, L. (1976) Arthropods from the Lower Devonian (Lower Emsian) of Alken an der Mosel, Germany. Part 5: Myriapoda and additional forms, with general remarks on fauna and problems regarding invasion of land by arthropods. *Senkenberg. Leth.* 57: 87–183.

Todd, J. (1991) A forest litter animal community from the Upper Carboniferous? Notes on the association of animal body fossils with plants and lithology in the Westphalian D Coal Measures at Writhlington, Avon. *Proc. Geol. Assoc.* (London) 102: 179–184.

Toye, S.A. (1970) Some aspects of the biology of two common species of Nigerian scorpions. *J. Zool.* 162: 1–9.

Vermeij, G. (1987) *Evolution and Escalation.* Princeton University Press, Princeton, NJ.

Waterston, C.D. (1975) Gill structures in the Lower Devonian eurypterid *Tarsoptella scotia. Foss. Strata* 4: 241–254.

Weygolt, P. and Paulus, H. (1979) Untersuchungen zur Morphologie, Taxonomie and Phylogenie der Chericerata. I, Morphologische Untersuchungen. *Z. Zool. Syst. Evol. Forsch.* 17: 85–116.

Wills, L.J. (1910) On the fossiliferous Lower Keuper rocks of Worcestershire, with descriptions of some of the plants and animals discovered therein. *Proc. Geol. Assoc. (London)* 21: 249–331.

Wills, L.J. (1925) The morphology of the Carboniferous scorpion Eobuthus, Fritsh. *J. Linn. Soc.* 36: 87–89.

Wills, L.J. (1947) A monograph of British Triassic scorpions. *Palaeontogr. Soc. Monogr.* 100, 101: 1–137.

Wills, L.J. (1959) The external anatomy of some Carboniferous 'scorpions'. Part 1. *Palaeontology* 1: 261–282.

Wills, L.J. (1960) The external anatomy of some Carboniferous 'scorpions'. Part 2. *Palaeontology* 3: 276–332.

Wood, S.P., Panchen, A.L. and Smithson, T.R. (1985) A terrestrial fauna from the Scottish Lower Carboniferous. *Nature* 314: 355–358.

17

Radioresistance of Scorpions

MAX GOYFFON
VINCENT ROMAN

The first indication that scorpions had unusually high resistance to ionizing radiation came in 1960 from nuclear weapons testing by France in the Sahara; near ground-zero the only animals to survive the detonations were scorpions (*Androctonus* sp.) and some Coleopteran beetles (*Pimelia* sp.). The resistance of these animals to irradiation was soon confirmed in laboratory tests (Vachon et al., 1963, 1964), and repeated in the field by Allred (1973) at the Nevada test site in the United States. Since that time several studies have sought to identify the characteristics of animals that determine susceptibility to ionizing radiations (Goyffon, 1975, 1983). This chapter is a synthesis of experimental findings from studies of scorpions and a discussion of their potential usefulness for future radiobiological research.

BIOLOGICALLY DAMAGING RADIATIONS

Ionizing irradiation damages biological structure at different levels of organization, depending on type and energy of the radiation. Radiation consisting of charged particles, such as electrons or positrons (β^- or β^+ particles) or heavier protons and helium nuclei (alpha particles), can produce direct ionization of molecules through coulombic interaction and dislodgment of orbital electrons. Indirect ionization is generally caused by electromagnetic radiations of short wavelength (e.g., X-rays or γ-rays) that excite atoms to eject electrons with sufficient energy to cause many more secondary ionizations in other molecules (indirectly ionizing radiations). Energetic neutrons also cause ionization by transferring kinetic energy to the nuclei of low mass number, causing them to recoil as positive ions. In particular, lower-energy neutrons may be absorbed by the nucleus, forming an unstable compound nucleus, which then relaxes to stability by emission of an ionizing photon or charged particle.

Given the minute energies contained in ionizing radiations, it is not immediately evident how they induce significant biological injury through highly localized action on single molecular or atomic targets. Two potential mechanisms are commonly cited: the direct effect or target theory postulates that cells are killed by lesion of critical, radiation-sensitive targets, such as DNA or chromosomal structures involved in cell division. Such targets are likely to be sparsely distributed within cells, so theory (originally developed to predict the outcome of artillery bombardment) predicts that irradiation should damage tissues at random and with all-or-nothing consequences. However, for acute exposures to radiation, the size of target calculated by theory is larger than the size of typical cells, producing only sublethal dosage at the cell level. Nevertheless, it is well known that dividing cells are more sensitive to radiation exposure, especially during the S-phase of the mitotic cycle when DNA replication occurs (Kiefer, 1990). If DNA is a relatively sensitive target of ionizing radiation, cells (or organisms) with a higher DNA content are predicted to have greater radiation sensitivity than those with lower content.

The indirect effect theory of irradiation damage postulates that radiolysis of water forms highly reactive free radicals such as $H^{\cdot}$ and $OH^{\cdot}$, which then react chemically with cells to affect critical structures or metabolic pathways. Oxygen amplifies water radiolysis by creating new radical species such as superoxide anion ($O_2^{\cdot}$), peroxyl radical ($HO_2^{\cdot}$) and highly toxic hydrogen peroxide (H_2O_2). In this way the primary solvent of life can act as a radiosensitizing agent. Radioprotective agents, such as mercaptoethylamine (an aminothiol), can act as free-radical scavengers by donating labile hydrogen atoms to neutralize these hyperreactive molecular species. Animals with good free-radical scavenging mechanisms should show greater radioresistance.

Other factors affecting the damage produced by irradiation are the route of administration (internal versus external), the dose rate and distribution (whole-body or partial irradiation). External irradiations (X-rays, γ-rays, neutrons, and β-emitting radioisotopes) damage tissue primarily by contact or short-distance exposure to the organism's epidermis. Internal irradiations arise from ingested (or injected) radionuclides that can damage tissue through isotopic decay or chemical transformation of radionucleides into substances that interfere with metabolism. Toxicity of internal irradiance depends on the energy and type of radioactive decay and the rate of biological clearance or elimination from the body. Thus, α-emitters, such as plutonium and its daughter product americium, are more noxious than β-emitters such as radiostrontium or tritium which, in turn, are more harmful than γ-emitters such as radiocesium or radiocobalt. Only the latter radiation emitters will be taken in account concerning scorpions.

With these potential sources of variable effect in mind, it is important for experimentalists to use accurate radiation sources that deliver ionizing radiation at a constant, known rate. For low dosages, electrically powered X-ray sources are commonly used; higher energy exposures ($>$1 million electron volts or MeV) require a linear accelerator (generally impractical) or irradiance by radioisotopes. For the latter approach, rapid changes in dose rate (due to isotopic decay) must be avoided by using isotopes with long half-lives. For studies of scorpion radioresistance, ^{60}Co (half-life of 5.3 years) has been most practical. Cobalt-60 delivers two γ-rays at 1.33 MeV and 1.17 MeV at a rate that decreases only about 1% per month. For a detailed description of ionizing radiations and for broader discussion of radiation effects on biological tissues and methods of administration, see Kiefer (1990).

RADIORESISTANCE OF SCORPIONS IN COMPARISON TO OTHER ORGANISMS

Resistance (or sensitivity) of organisms to ionizing radiations is widely variable and difficult to compare across taxa. The most common index of radioresistance is to estimate the effect of short-term exposure on longevity. The effective lethal dose (LD) of radiation is usually expressed as the dose required to kill 50% of an irradiated population in a tractable period—for example, 8, 15, or 30 days. In mammals, these times correlate with physiological and biochemical process that lead to death, for example, by destruction of the lympho-hematopoietic stem cells of bone marrow (fatal in about 30 days), or lesions of the central nervous system caused by high doses (fatal within 8 days). By convention, the lethal dose that kills 50% of the irradiated population by 30 days, noted LD_{50}/30 days, is commonly used for comparison between animals. Some studies report mean time for survival of 50% of the irradiated population (LT_{50}) regardless of the death mechanism. Willard and Cherry (1975) have normalized radiosensitivity data for species with different life spans by reporting the ratios of LT_{50} to mean lifetime. Even this index is unsatisfactory for accessing radioresistance of species with brief imago (adult) stages (e.g., insects) or for species with developmental stages with different radiosensitivity.

Table 17.1 shows a comparison of radioresistance for terrestrial animals exposed to acute, whole-body radiation using the common measure of LD_{50}. Homeothermic vertebrates are generally the most sensitive animals, and poikilothermic vertebrates are more sensitive than representative invertebrates (mollusks, arthropods). The arthropods, as represented by imago adult stages of

Table 17.1 Comparison of sensitivity to ionizing radiation for selected species[a]

	LD_{50} (Gy)
Arthropoda (*Drosophila*)	
egg (3 h)	2
egg (4 h)	5
egg (7 h)	8
pupa	28
imago adult	640
Mollusca	
Snail	80–200
Triton	30
Vertebrates	
Frog	7
Tortoise	15
Chicken	6–8
Sheep	1.6
Pig	2.5
Dog	2.7
Rabbit	8.4
Mouse	9.0
Rat	9.0
Guinea pig	2.6
Monkey	3.0–5
Human	4.5–5

[a]Data compiled from Bacq and Alexander (1961); Goyffon (1973).

A

B

C

Figure 17.1 Scorpions from Saharan North Africa used in studies of radioresistance. A, *Androctonus mauretanicus* (Pocock, 1902), a black scorpion endemic to Morocco. B, *Androctonus australis* (L.) hector (C.L. Koch, 1839), a yellow scorpion from Algeria that is among the most dangerous species to humans. C, *Androctonus amoreuxi* (Audoin et Savigny, 1826), another yellow scorpion closely resembling *A. australis.*

Drosophila (table 17.1), show the greatest radioresistance, exceeding the resistance of humans by two orders of magnitude.

The unusually high radioresistance of arthropods as a group may relate to peculiar characteristics of their biology, most notably the absence of rapidly dividing cells in the adult stage. Arthropods have different hemopoietic organs from vertebrates, and other tissues in the adult stage are weakly proliferative, except for the gonads. Ducoff (1972) described two categories of lethal effects from X-ray and γ-ray exposures in insects: one that was dose dependent and strongly age dependent, and one that was dose independent over a wide range, then decreasing resistance rapidly above a specific threshold. Scorpions appear to belong to this latter category of radioresistant arthropods.

For studies of radioresistance in scorpions, we used three species of *Androctonus* (fig. 17.1) captured from different regions of Saharan North Africa (collected by the Pasteur Institutes of Algeria, Tunis, Tunisia, and Casablanca, Morocco. *A. amoreuxi* lives in the most arid regions of the Sahara; *Androctonus australis* (L.) lives in arid and semiarid areas, usually near oases; *Androctonus mauretanicus* (Poc.) is endemic to Morocco. *Androctonus* is preferred for our experiments because of their large size (commonly 5–10 g, 8–10 cm long), ease of maintenance, and availability. For metabolic and physiological experiments, blood samples are easily taken from dorsal heart by innocuous punctures, and electrophysiological potentials can be recorded through implanted electrodes using standard instrumentation (Goyffon et al., 1980). These attributes permit investigations of the causes of injury and death induced by irradiations.

For the initial experiments with scorpions (Vachon et al., 1963, 1964) the LD_{50} /30 days was used for comparison to other animals (table 17.1). The number of scorpions tested and the time of assay was minimized by applying the cumulative method of Behrens and Karber (1935). Groups of 10 animals were whole-body irradiated by an arithmetic progression of doses from 100 Gy to 1200 Gy (1 Gy = 100 rads). In a second series of experiments, for exposure to γ-rays, a ^{60}Co irradiator was used at a dose rate of 11 Gy/min (Schantz and Goyffon, 1970). Typically, six specimens were placed in separate plastic boxes where they received identical, acute, whole-body exposures. All experiments were conducted at laboratory temperature of 22°C.

The LD_{50} for species of the same genus can differ significantly. Species from progressively drier habitats (*A. mauretanicus* < *A. australis* < *A. amoreuxi*) appear to have progressively greater resistance to irradiation, but there were no differences between sexes or instars of different developmental stage. Temperature strongly influenced these results: if *A. australis* were acclimated to 34°C, their optimal body temperature for molting and parturition (Goyffon, 1975), and maintained at that temperature after exposure, the LD_{50}/30 days value decreased to 250 Gy. The temperature coefficient (Q_{10}) for this increase in lethality of irradiation is approximately 2, suggesting that a metabolically dependent process is involved. Other investigators have noted that elevated metabolic activity generally increases radiosensitivity of other animals (Bacq and Alexander, 1961).

It is also noteworthy that scorpions irradiated at LD_{50}/30 days dosage and maintained at 22°C do not survive beyond 6 months, whereas those receiving lower doses (150–200 Gy, depending on the species) survive this period. With such distinct differences in survivorship near LD_{50}/30 days dosage, survival time statistics (LT_{50}) may be a more useful index of radioresistance.

PHYSIOLOGICAL AND METABOLIC EFFECTS OF RADIATION ON SCORPIONS

Mortality studies necessarily fail to illuminate the mechanisms of radiation damage because death of an organism can result from many causes acting independently or in concert. A more revealing approach to mechanistic studies is to examine progressively simpler levels of biological organization, from behavioral to molecular levels of function. Several characteristics of scorpions favor multilevel analysis. At the highest level of complexity scorpions have several highly stereotypical behaviors related to development (molting), feeding, and courtship; normal expressions of these behaviors depend on highly integrated functions of neural, endocrine, and other organ systems so they are likely to show discrete evidence of dysfunction after irradiation exposure. At a lower level of complexity are physiological processes controlling internal homeostasis—for example, the levels of essential metabolites, ions, and water in hemolymph. The mechanisms of homeostasis generally involve fewer regulatory elements than fully integrated behaviors, so it is of interest to compare their sensitivity to irradiation. Finally, if ionizing radiation affects cellular structure at random, then cells and the physiological processes controlled by single cells (e.g., muscle contraction), should show evidence of deterioration. In the sections below we show that each of these levels of biological function are affected by irradiation.

Effects on Molting Behavior

Scorpions molt regularly under physiological conditions of temperature and nutrition. Molting of the cuticle involves an increase of the hemolymph volume and a progressive distension of the pleural membranes (Goyffon, 1975). The duration of intermolt stages increases with age, from days or weeks for the young instars to several months for the adult stages. Each intermolt is divided into four phases: (1) the postmolt phase, lasting a few days, is characterized by a flat abdomen. (2) In the second phase, the pleural membranes appear as a thin line. The duration of this phase and the following ones vary considerably, depending on available food and temperature variations. (3) During the third phase, the abdomen is dilated. The pleural membranes are as wide as the tergites. (4) The fourth stage is the premolt, which lasts several days only, during which time the scorpion is motionless and ceases to feed. The abdomen is distended, and the compressed digestive gland is visible through the transilluminated cuticle.

The effects of irradiation vary for animals according to the different stages of molting. At the $LD_{50}/30$ days dose level, irradiation at the premolt stage stimulates one final molt, indicating that the cellular processes mediating ecdysis remain intact, at least initially. Subsequent molts do not occur unless the intensity of irradiation at the premolt stage is decreased to less than about 150 Gy (a sublethal dose). The copper content of the exuvia is constant with or without irradiation (88 μg/g of dry matter; Goyffon, 1977). In the other stages, irradiation blocks definitively the molting.

Effects on Nutrient Homeostasis

Glucose, soluble lipids, and amino acids are the principal circulating nutrients in scorpion hemolymph, and the digestive gland is the primary organ of nutrient storage (Schantz et al., 1969). By monitoring changes in distribution of these nutrients between the organs of circulation and storage, the effect of irradiation on a homeostatic process (regulation of nutrient mobilization) can be monitored. Therefore, it is possible to detect an increase in the energetic consumption linked to the radiorestoration processes, something similar to a fever in irradiated mammals.

When scorpions are deprived of food, the concentration of glucose in the hemolymph (glycemia) remains more or less stable for at least 3 weeks, confirming that it is under homeostatic control (Serfaty, 1942; Goyffon and Martoja, 1983). In the first days after irradiation at a sublethal exposure of 160 Gy, glucose content of the hemolymph declines significantly from 0.45 ± 0.04 g/l to 0.33 ± 0.02 g/l, then returns to normal values after about 3 weeks (Goyffon et al., 1969). Glycogen content of the digestive gland (4–5 g per 100 g of fresh tissue) is not affected during this time. It appears, therefore, that irradiation temporarily disrupts regulatory processes that control the breakdown or utilization of energetic carbohydrates. With supralethal doses (500 Gy), glycemia values are highly variable, and the digestive gland glycogen content decreases to 2.5 g per 100 g of fresh tissue 10 days after irradiation (Goyffon, 1975). The digestive gland is also rich in phospholipids (89 mg/g of proteins) in comparison to hemolymph, where lipid content is less than 1 g/l (Goyffon and Dupont, 1973). The total amount of lipids and their distribution remained unchanged up to 600 Gy irradiation.

Among hemolymph amino acids in scorpions, only two showed significant changes after irradiation. Changes in the level of taurine are of special interest because it is relatively abundant in the organic fluids of many invertebrates. In mammals, irradiation releases taurine into the blood (hypertaurinemia) and urine (hypertaurinuria), probably by radiolysis of sensitive cells (Bocquet and Fromageot, 1967). Taurine may have a radiocurative effect (Abe et al., 1968), and scorpion hemolymph contains large amounts of free taurine (>40% of total free amino acids). When scorpions are irradiated at sublethal doses, the aminoacidemia falls from 2.13 μmole/ml to 1.77 μmole/ml. Taurinemia decreases from 0.90 (42% of the total amount) to 0.56 μmole/ml (32% of the total amount). Thus taurinemia fails to increase as expected compared to mammals, and may even decrease faster than total aminoacidemia within the 4 days after irradiation. This result can be interpreted as an absence of cell lysis, or at least negligible cytolysis (Schantz and Goyffon, 1970).

The concentration of alanine increases significantly with irradiation, but the significance of this change is unknown. The stability of citrulline, ornithine, and the presence of ammonia (150–250 μmole/ml) and absence of urea imply that urea cycle function is not disturbed by irradiation.

Effects on Cellular Functions

Cell division

Hemocytes and gonad cells show relatively continuous mitoses during intermolt periods and thus allow the best comparisons of sensitivity between scorpion and mammalian cells. Scorpion hemolymph contains variable numbers of cells (16,000 ± 7,900 cells/μl), three to four types of which are described (Deloince et al., 1969; Gysin et al., 1969; Ravindranath, 1974). Mitoses are rarely

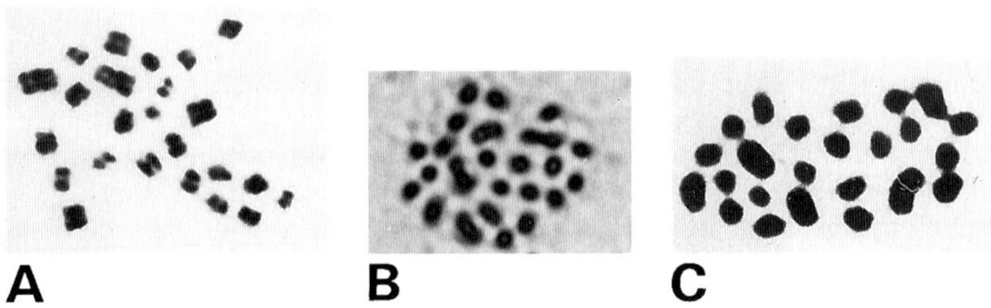

Figure 17.2 Karyotypes of buthid scorpions. A, *Androctonus mauretanicus.* B, *Androctonus australis* (L.) hector. C, *Androctonus amoreuxi.*

observed histologically, but the large variation in hemolymph cell content suggests high mitotic activity (Morel, 1974), even though the location of hematopoietic organs is not clearly identified. These experiments were carried out to evaluate the cellular radiosensitivity that is classically increased higher in cells that divide actively.

After irradiation, the hemocyte count falls rapidly in 5 days to 2100 ± 1500 cells/μl. When the irradiation is sublethal (<200 Gy), hemocytemia returns slowly to normal values within 3–4 weeks. For lethal irradiations (>300 Gy), hemocytemia remains very low at 1000–3000 cells/μl. This early drop in the hemocyte count is accompanied by an increase in cytosolic activity of malic dehydrogenase (MDH) activity, which is maximal after 24 hr and recovers by the fifth day (Goyffon, 1975). This MDH peak could signal hemocytolysis that does not significantly alter levels of free taurine in the hemolymph because of the relative abundance of taurine.

Scorpion testes are also continually cytogenic at all developmental stages, a condition similar to mammalian testes. Chromosomal karyotypes are relatively simple to obtain for buthid scorpions, so this simple histological procedure can be used to assay the effects of irradiation on cell division (fig. 17.2). Two days after exposure at ≥150 Gy, mitoses or meioses cannot be detected in the form of metaphasic plates in chromosome smears. For sublethal irradiations (<160 Gy), the metaphasic plates gradually reappear, and normal frequency and appearance is established after about 6 weeks. No chromosomal breakage was observed, but chromosomes of scorpions are small and punctiform, making breakage difficult to detect. After lethal irradiation at 400 Gy, cell divisions cease altogether. Gonadal cells, like hemocytes, begin to show irreversible and lethal damage at a threshold dose of 150–160 Gy. This threshold is comparatively high on the scale of animal sensitivities (table 17.1).

Effects on electrical excitability of cells

The "scorpion illness" described in irradiated scorpions by Niaussat et al. (1963) is characterized by a loss of motor coordination and instability. To determine whether irradiation caused lesions in electrically excitable tissues such as muscle or the central nervous system, spontaneous electrical activity was recorded using a relatively simple and sensitive method for detecting perturbations in cell functions. The nervous system controlling cardiac rhythm of scorpions is the most convenient patterned motor output for assessment of radiation effects. Electrocardiograms (EKG) can be recorded with two electrodes implanted dorsally in the heart between the fourth and the fifth tergites (100–150 μV amplitude). This EKG reflects the nerve activity that generates the heart beat frequency at 1–2 Hz. Irradiation as high as 1000 Gy caused no apparent change in EKG rhythmicity, indicating (Fig. 17.3) that the cellular processes involved in cell electrical excitability are highly resistant to radiation-induced damage. This phenomenon has been observed in other invertebrates (Goyffon, 1975).

Electrodes implanted near the principal median eyes of scorpions record spontaneous electrical activity (SEA) of the brain, in correspondence to an electrical encephalogram (EEG) in vertebrates (Goyffon et al., 1975). The origin of the SEA is unknown, but it possesses a cholinergic determinism and is controlled by antagonistic catecholaminergic and serotonergic systems (Goyffon, 1978; Goyffon et al., 1980). The SEA is characterized by regular potentials of 20–60 μV, with a frequency of 3–6 Hz. It is easily differentiated from the EKG (60–150 μV, <3 Hz) or from the SEA of the ventral nervous chain (4–8 Hz, with irregular amplitude between 3 and 75 μV; Venkatachari, 1971). After sublethal irradiations (<150 Gy), no modifications in the SEA were recorded, and the motor behavior of the animals remains normal. For lethal doses (400–600 Gy), locomotor rhythm disturbances are evident, and these are accompanied by desynchronization of the SEA rhythm. When these aberrations appear, scorpions

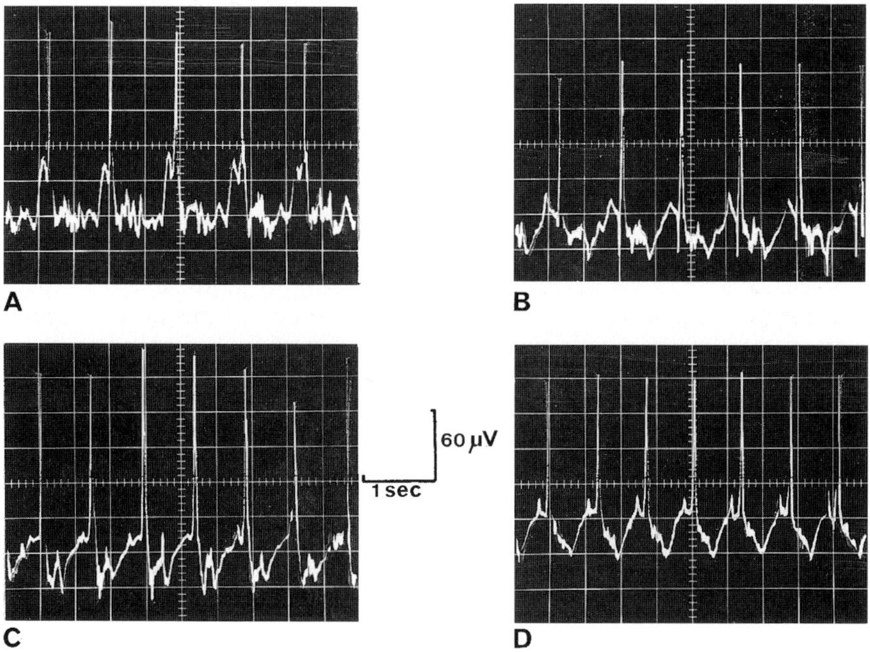

Figure 17.3 Electrocardiograms of *Androctonus mauretanicus.* A, Before irradiation. B, 24 hours after irradiation at 400 Gy. C, 24 hours after irradiation at 600 Gy. D, 24 hours after irradiation at 1000 Gy.

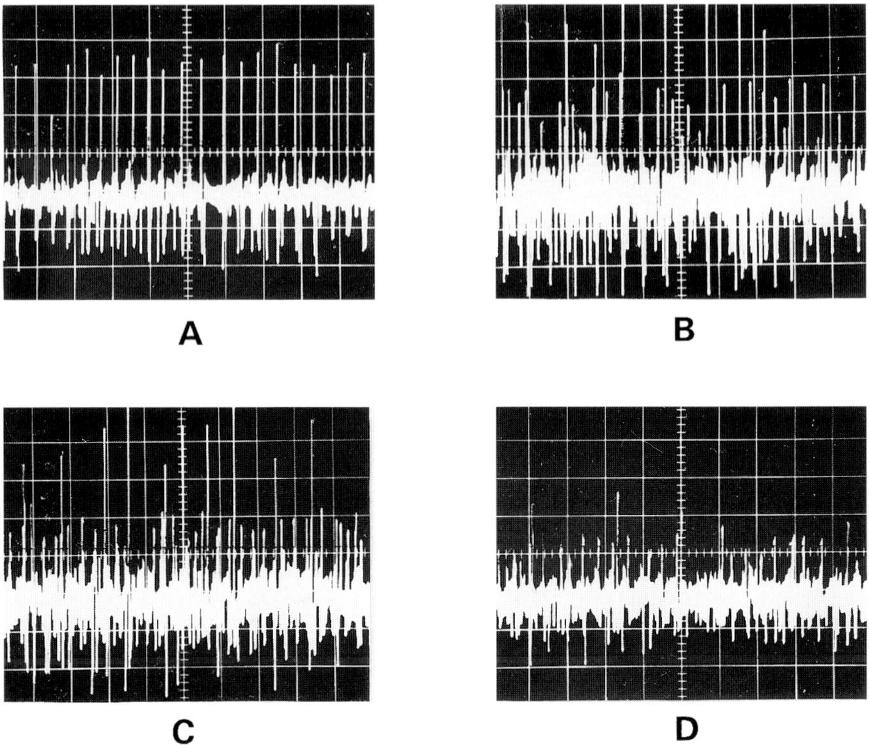

Figure 17.4 Spontaneous electrical activity recorded from the supraesophageal ganglion near the eyes after total body irradiation of *Androctonus mauretanicus* at 600 Gy. A, Before irradiation. B, 4 hours after irradiation. C, 12 hours after irradiation. D, 24 hours after irradiation.

usually survive only a few weeks or months. For nonmammalian vertebrates, EEG alterations also appear at lethal doses, but these doses are 30–40 times lower than for scorpions.

DISCUSSION

Scorpions appear to be resistant to a variety of ionizing radiations. Neutron irradiance of *Androctonus australis* (5:1 ratio mixed neutron/gamma exposure) as high as 120 Gy does not alter their life expectancy. Buthids of the genus *Centruroides* are slightly less radioresistant (killed by 115 Gy radiation exposure; Mazzotti et al., 1961), and for *Pandinus* (Scorpionidae), the LD_{50}/30 days is about 100 Gy (Goyffon, unpublished data). Variability in radioresistance between species seems to be correlated to their resistance to other environmental stress factors such as temperature, dehydration, fasting (Goyffon, 1983), and bacterial infections (Morel, 1974; Cociancich et al., 1993). In particular, the desert species *A. amoreuxi* from the most arid areas of Sahara, shows the highest radioresistance observed in scorpions. Low metabolic rate and low oxygen consumption may play a role in scorpion radioresistance. Oxygen consumption can be as low as 25 μl $hr^{-1}g^{-1}$ (Dresco-Derouet, 1973; Goyffon, 1975; Van Aardt, 1991), among the lowest consumption rates measured for any animal. It is possible that such low rates of metabolism favor radioresistance, as does the low rate of cell division observed in these animals.

The presence of molecules capable of neutralizing oxides or peroxides resulting from the indirect effects of ionizing radiations (Goyffon, 1973) is another factor favoring radioresistance. Catalase neutralizes hydrogen peroxide, and highly active catalase is found in scorpion digestive gland (Lamy and Goyffon, 1969). Scorpion digestive gland catalase neutralized by aminotriazole was no less radioresistant. Circulating hemocyanin may protect against indirect effects of irradiation by complexing with oxygen, thus reducing the quantity of free O_2 and its radiosensitizing effect. Hemocyanin also possesses a triple antioxidant effect (Queinnec et al., 1990, 1991), having enzymatic activities for catalase, peroxidase, and superoxide dismutase. Hemocyanin can efficiently neutralize the indirect effects of ionizing radiation caused by peroxide and superoxide radicals created in the aqueous medium. However, these factors do not account for other aspects of resistance, especially the restoration of normal cell divisions in gonads after an irradiation as high as 200 Gy. Given their unusually high resistance to ionizing radiation in the form of X-rays, γ-rays, or neutrons, scorpions will continue to be convenient and useful models for comparative studies of ionizing irradiation-induced damage and biological repair mechanisms.

References

Abe M., Takahashi, M., Takeushi, K. and Fukuda, M. (1968) Studies on the significance of taurine in radiation injury. *Radiat. Res.* 33: 563–573.

Allred D.R. (1973) Effects of a nuclear detonation on arthropods at the Nevada Test Site. *Brigham Young Univ. Sci. Bull., Biol. Ser.* 18: 1–20.

Bacq, Z.M. and Alexander, P. (1961) *Fundamentals of Radiobiology.* Pergamon Press, London.

Behrens, B. and Karber, G. (1935) Wie sind Reihenversuche fur biologische Auswertungen am Zweckmassigsten anzvorden. *Arch. Exp. Pathol. Pharmacol.* 177: 379–388.

Bocquet, P.L. and Fromageot, P. (1967) Sur l'origine de la taurine urinaire excrétée par le rat soumis à une irradiation par les rayons gamma du ^{60}Co. *Intl. J. Radiat. Biol.* 13: 343–353.

Cabanac, M. and Le Guelte, L. (1980) Temperature regulation and prostaglandin E1 fever in scorpion. *J. Physiol.* 303: 365–370.

Cociancich, S., Goyffon, M., Bontems, F., Bulet, P., Bouet, F., Menez, A. and Hoffmann J. (1993) Purification and characterization of a scorpion defensin, a 4kDa antibacterial peptide presenting structural similarities with insect defensins and scorpion toxins. *Biochem. Biophys. Res. Comm.* 194: 17–22.

Deloince, R., Poter, H., Blanc, P. and Niaussat, P. (1969) A propos des hémocytes d'un scorpion saharien *Androctonus australis* (L.). *Bull. Mus. Nat. Hist.* 39: 36–39.

Dresco-Derouet, L. (1973) Le métabolisme respiratoire des scorpions - IV - Intensité respiratoire de plusieurs espèces d'Afrique du Nord et ses relations avec l'altitude. *Bull. Soc. Hist. Nat. Toulouse* 109: 235–240.

Ducoff, H.S. (1972) Causes of death in irradiated adult insects. *Biol. Rev. Cambr. Phil. Soc.* 47: 211–240.

Ely M.J. and Jungreis, J. (1977) Effects of X-irradiation on egg hatchability, larval and pupal survival in the tobacco hornworm, *Manduca sexta*. *J. Insect Physiol.* 23: 95–101.
Goyffon, M. (1973) Actions biologiques des radiations ionisantes. *Rev. Quest. Scientif.* 144: 337–372.
Goyffon, M. (1975) *Effets physiopathologiques de l'irradiation par radiations ionisantes chez le Scorpion* (thesis PhD). University of Paris.
Goyffon, M. (1977) Organotropisme du cuivre chez le scorpion. *Rev. Arachnol.* 1: 9–12.
Goyffon, M. (1978) Amines biogènes et activité électrique spontanée du système nerveux prosomien du scorpion. *Comp. Biochem. Physiol.* 59: 65–73.
Goyffon, M. (1983) Panchronisme et résistance aux agressions de l'environnement chez les scorpions. *Bull. Soc. Zool. Fr.* 108: 585–592.
Goyffon, M., Drouet, J. and Francaz, J.M. (1980) Neurotransmitter amino-acids and spontaneous electrical activity of the prosomian nervous system of the scorpion. *Comp. Biochem. Physiol.* C 66: 59–64.
Goyffon, M. and Dupont, J. (1973) Les phospholipides de l'hépatopancréas du scorpion *Androctonus mauretanicus* (Poc). *Comp. Biochem. Physiol.* B 44: 1243–1247.
Goyffon, M., Luyckx, J. and Vachon, M. (1975) Sur l'existence d'une activité électrique rythmique spontanée du système nerveux céphalique du scorpion. *C.R. Acad. Sci.* (Paris) ser. D 280: 873–876.
Goyffon, M. and Martoja, R. (1983) Cytophysiological aspects of digestion and storage in the liver of a scorpion *Androctonus australis* (Arachnida). *Cell Tiss. Res.* 228: 661 -675.
Goyffon, M. and Ridet, J.M. (1969) Fractionnement électrophorétique et dosage des protéines de l'hémolymphe du scorpion *Androctonus australis* (L.). *Bull. Mus. Nat. Hist.* 41: 30–35.
Goyffon, M., Schantz, R. and Niaussat, P. (1969) Evolution de la glycémie du scorpion *Androctonus australis* (L.) soumis à une irradiation par les rayons gamma du ^{60}Co. *C.R. Soc. Biol.* 163: 601–605.
Gysin, J., Le Coroller, Y. and Irunberry, J. (1969) Cytologie sanguine et hémopoïèse chez le scorpion. *Arch. Inst. Past. Algérie* 47: 1–10
Kiefer, J. (1990) *Biological Radiation Effects.* Springer Verlag, New York.
Lamy, J. and Goyffon, M. (1969) Etude de la catalase de l'hépatopancréas du scorpion *Androctonus australis* (L.). Cinétique de la décomposition de l'eau oxygénée. Action de l'aminotriazole. *C.R. Acad. Sci.* (Paris) ser. D 269: 2449–2452.
Lane, N.J., Harrison, B.J. and Bowerman, R.F. (1981) A vertebrate-like blood-brain barrier, with intraganglionic blood channels and occluding junctions, in the scorpion. *Tissue Cell* 13: 557–576.
Mazzotti, L., Rhode, R.H., Lopez, F. and Telich, J. (1961) Radiacion con rayos gamma de escorpiones de la especie *Centruroides limpidus. Rev. Inst. Salubr. Enferm. Trop.* 21: 125–127.
Morel, G. (1974) Histopathologie du scorpion *Buthus occitanus* (Am.) au cours d'infections expérimentales par *Bacillus thuringiensis* Berliner. *Bull. Soc. Zool. Fr.* 99: 479–487.
Niaussat, P., Vachon, M., Grenot, C. and Pierre, F. (1963) Comportement du scorpion saharien *Androctonus amoreuxi* (Aud. and Sav.) après exposition à des doses importantes de rayonnement gamma et troubles consécutifs à cette irradiation. *C.R. Acad. Sci.* (Paris) ser. D 257: 1177–1180.
Queinnec, E., Gardes-Albert, M., Goyffon, M., Ferradini, C. and Vuillaume, M. (1990) Anti-oxidant activity of hemocyanin: a pulse radiolysis study. *Biochim. Biophys. Acta.* 1041: 153–159.
Queinnec, E., Gardes-Albert, M., Ducancel, F., Goyffon, M., Ferradini, C. and

Vuillaume, M. (1991) Etude cinétique des propriétés catalytiques de l'hémocyanine du scorpion. *J. Chim. Phys.* 88: 927–935.

Ravindranath, M.H. (1974) The hemocytes of a scorpion *Palamnaeus swammerdammi. J. Morphol.* 114: 1–9.

Schantz, R., Goyffon, M. and Niaussat, P. (1969) La glycémie du scorpion *Androctonus australis* (L.). *Bull. Soc. Zool. Fr.* 94: 159–166.

Schantz, R. and Goyffon, M. (1970) Les acides aminés libres de l'hémolymphe du scorpion *Androctonus australis* (L.). Effets d'une irradiation gamma. *C.R. Soc. Biol.* 164: 1225–1227.

Serfaty, A. (1942) Influence du jeûne précédant le repas sur l'activité trypsique, au cours de la digestion, chez les scorpions. *Bull. Soc. Entom. Fr.* 47: 89–90.

Vachon, M., Aeberhardt, A., Grenot, C., Niaussat, P. and Pierre, F. (1963) Sur la radiosensibilité du scorpion saharien *Androctonus amoreuxi* (Aud. and Sav.). *C.R. Acad. Sci.* (Paris) ser. D 256: 4290–4293.

Vachon, M., Niaussat, P., Ebersole, J.H. and Grenot, C. (1964) Sur la radiosensibilité comparée vis-à-vis du rayonnement gamma de quelques espèces de scorpions. *C.R. Acad. Sci.* (Paris) ser. D 259: 3389–3391.

Van Aardt, W.J. (1991) Oxygen consumption and haemocyanin oxygen affinity in the scorpion *Opisthophtalmus latimanus* Koch. *J. Entomol. Soc. S. Afr.* 54: 129–139

Venkatachari, S.A.T. (1971) Analysis of spontaneity in the nervous system of the scorpion *Heterometrus fulvipes. Ind. J. Exp. Biol.* 9: 338–344.

18

Scorpion Diversity in Tropical South America

Implication for Conservation Programs

WILSON LOURENÇO

Tropical South America is one of the most biologically diverse regions on earth (Wilson, 1988). To understand such diversity, the biogeographic patterns for several groups of plants and animals have been analyzed (Prance, 1982a; Gentry, 1990). Although these patterns vary somewhat, certain regions exhibit especially high endemism, or a large number of species native to a localized region. Many of these endemic centers are found in Amazonia, Guayana, and in the tropical Andes, the latter region being considered by some to be the epicenter of global biodiversity (Gentry, 1992). Biologists now seek a rational basis for selecting which areas should receive the highest priority for conservation (Prance, 1990). In this chapter, I illustrate how studies of neotropical scorpion populations have contributed to this emerging consensus for preservation of South American tropical biomes.

Selection of areas for protection requires, a priori, a basic knowledge of several diverse taxa within it. In particular, knowledge of biosystematics and historical and ecological factors in biogeography must be considered. For this reason, most research on tropical evolutionary biogeography is based on a limited number of taxa, primarily vertebrates and woody plants. It is significant, therefore, that the data collected for neotropical scorpions describing their patterns of geographical distribution, differentiation, and life-history strategies are now sufficient to incorporate this invertebrate taxon into consideration of areas to be preserved (Lourenço, 1991a,b, 1995). A demonstration of this point is attempted in the following sections.

THE MORPHOCLIMATICAL REGIONS OF TROPICAL SOUTH AMERICA

An accurate biogeographical view of any region, especially at the continental scale, requires a precise knowledge of its morphoclimatical subdivisions—that is, the subregions within which specific conditions of climate and soil produce a particular type of vegetation. The heterogeneity of tropical South American vegetation (fig. 18.1) can be described by major type-formations based mainly on physiognomic and ecological conditions of climate and soil (Eiten, 1974; Ab'Saber, 1977a). For the studies described here, several major vegetation formations and transition zones are recognized: (1) rainforests or wet forest, where annual rainfall varies from 1500 mm to 4000 mm, as in Amazonia and Guayana, to 9000 mm in the Pacific Colombian-Ecuadorian forest; (2) open vegetation

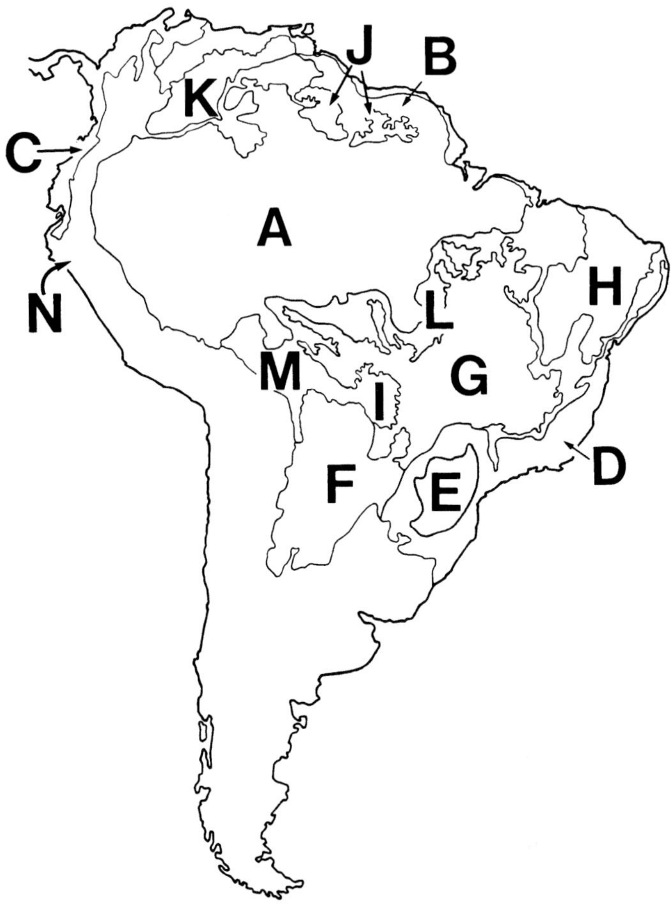

Figure 18.1 The morphoclimatic regions of tropical South America. A, Amazon; B, Guayana; C, Choco; D, Atlantic forest; E, Araucaria; F, Chaco; G, Cerrados; H, Caatingas; I, Pantanal; J, Gran Sabana; K, Llanos; L, Cerrados Amazon transition region; M, Chaco Amazon transition region; N, Pacific coastal desert. After Eiten (1974) and Ab'Saber (1977a).

(e.g., dry formations as in Chaco and Brazilian northeastern Caatingas, where the annual average rainfall is 300–1000 mm with 7–10 months of strong dry season; (3) nonxerophitic formations as in Central Brazil Cerrados, Gran Sabana and Llanos, where average annual rainfall is 750–2000 mm with a moderately strong dry season of 4–5 months and average temperatures ranging from 20°C to 26°C; (4) transition regions (e.g., Chaco-Amazon and Cerrados-Amazon), where intermediate climate condition can be observed (see Eiten, 1974, for details). Vegetation is not uniform within these major subregions, and gradients can occur within them. The species diversity observed in tropical South America can be explained largely by this heterogeneity of vegetational zones.

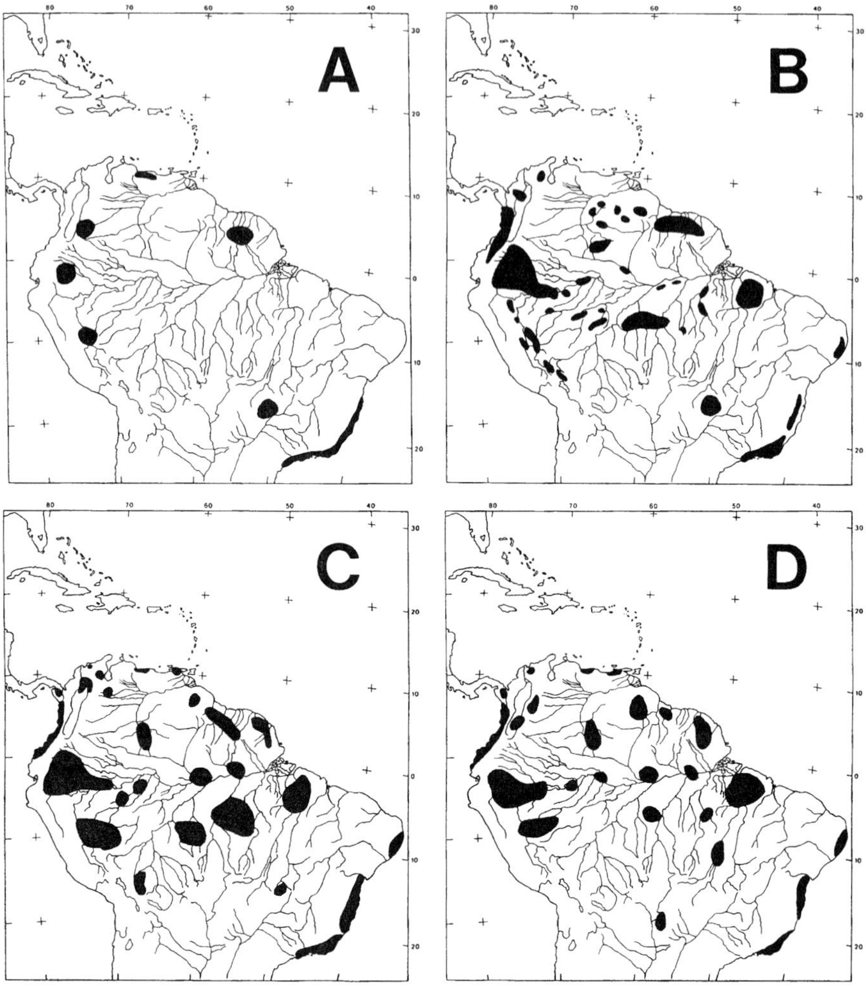

Figure 18.2 Major refugia of scorpion biodiversity in comparison to other taxa. (A) After Vanzolini (1970) based on lizards. (B) After Haffer (1969, 1974) based on birds. (C) After Prance (1982) based on woody plants. (D) After Lourenço (1986) based on scorpions.

HISTORICAL FACTORS IN BIOGEOGRAPHY

Some authors believed that biological diversity in rainforests is high because these environments have been relatively stable for many millions of years (Federov, 1966; Richards, 1969). However, data from ecological, paleoclimatic, and palynologic studies (Prance, 1982a, 1985) indicate that this apparent stability was interrupted by periods of climatic change during several dry/wet/dry episodes of the Cenozoic Period, especially during the Pleistocene and Holocene Epochs. During the earlier Quaternary Period, temperate regions were glaciated; cooler, drier conditions prevailed in the tropical zones of today and reduced the rainforest to savannas or dry forests, except in localized regions where conditions of temperature and humidity allowed the forests to persist. This historical reduction of rainforests to patch refugia is supported by existing biogeographical patterns of distribution and differentiation of several taxa (Haffer, 1969, 1974; Vanzolini and Williams, 1970; Prance, 1973, 1982b; Brown, 1979, 1982) and by palynological and geomorphological evidence (Van der Hammen, 1974, 1982; Ab'Saber, 1977b, 1982). Prance (1982b, 1985) suggests that the reduction of forest to small patches stimulated many changes in plant and animal populations in the refugia, ranging from extinction to an increased rate of speciation.

More recent evidence for the refuge theory comes from biogeographical studies of scorpions in tropical South America (Lourenço, 1986, 1991a,b, 1994; Lourenço and Florez, 1990a,b). These investigations led to the designation of about 25 refugia, or areas of high endemism (fig. 18.2.). The number, size, and location of these refugia correlate well with those described by Prance (1973, 1982b) from his studies of woody plant biogeography and studies by Haffer (1969, 1974) and Brown (1979, 1982) on birds and butterflies, respectively, to identify regions of high endemism. Together these studies provide a basis to identify and size the most important centers of species diversity in South American tropical rainforests.

ECOLOGICAL FACTORS IN BIOGEOGRAPHY

The Cenozoic refuge hypothesis accounts for much of the modern patterns of scorpion biogeography in tropical South America, but ecological factors must also affect their distribution, diversity, and endemism (Endler, 1982; Gentry, 1988, 1990; Lourenço, 1991b, 1992). Although the life history of several scorpion species are described (Francke and Sissom, 1984; Polis and Sissom, 1990), the most complete studies generally focus on temperate and desert species (e.g., genus *Paruroctonus* [Vaejovidae] by Polis and Farley [1979, 1980]; and genus *Urodacus* [Urodacidae] by Smith [1966] and Shorthouse and Marples [1982]). The population dynamics and life-history strategies of tropical South American scorpions exhibit characteristics of equilibrium species: relatively large body size for adults, small numbers of better endowed offspring with a higher probability to survive (Pianka, 1970; Polis and Farley, 1980). Some buthids exhibit life histories of opportunistic species (smaller size, larger numbers of offspring). Thus, scorpions generally appear to be unlike most terrestrial invertebrates (most of which have opportunistic life histories) and should be considered among possible exceptions to Pianka's rule (1970).

As equilibrium species, scorpions often exhibit stable and predictable biogeographic patterns, as in groups found exclusively in open vegetation (e.g., *Rhopalurus* species) or foraging in a specific microhabitat (e.g., *Tityus fasciolatus* Pessôa and *Opisthacanthus cayaporum* Vellard are both termite specialists; Lourenço, 1994). Among buthid scorpions some exceptions (mostly species of *Centruroides, Isometrus,* and *Tityus*) can be found where greater vagility and ability to adapt to new environments favor dispersal. For example, *Tityus serrulatus* Lutz and Mello and *T. bahiensis* (Perty) have invaded many disturbed environments in southern and central Brazil (Lourenço, 1988; Lourenço and

Cuellar, 1995), and *Centruroides*, which originated from Mexico, now occurs (accidentally) in Chile, Argentina, and in the Canary Islands; *Isometrus maculatus* (DeGeer), a buthid originally from the Indo-Malayan region, is currently found in all tropical coastal regions of South America and along rivers (e.g., the Amazon upstream to Iquitos in Peru). These are exceptions, however; many scorpions in tropical South America show equilibrium species characteristics, with strict habitat and microhabitat dependence. These species characteristically produce a single clutch after each insemination, and their intrinsic rates of population growth (r_{max}) and mobility are comparatively low. These characteristics favor endemic distributions, and it is common to find some species of tropical scorpions limited to a single site (Lourenço, 1991a,b, 1994, 1996). Thus, specificity of habitat and microhabitat conditions and the need for relatively stable and predictable environments make scorpions good indicators of highly endemic regions and relatively undisturbed environments.

SCORPION BIOGEOGRAPHIC PATTERNS: BIODIVERSITY AND CONSERVATION

Destruction of Tropical Rainforests in South America

For many years tropical botanists have alerted the scientific community and conservation planners about the alarming rate of rainforest destruction (e.g., Prance, 1977; Mori et al., 1981; Prance and Campbell, 1988). The Brazilian Atlantic coastal forest, for example, once occupied about 1 million km^2, extending from Rio Grande do Norte to Rio Grande do Sul State in a strip ranging from 10 to 160 km in width. Intense human exploitation reduced this forest to a few remnant patches, with 66–93% of the original forest destroyed (Mori et al., 1981). Naturally, these are areas of high endemism, with 53% of all regional tree species (Mori et al., 1981) and about 90% of the scorpion species (Lourenço, 1991a). Moreover, even the remnant patches of rainforest associated with three major endemic centers (e.g., Pernambuco, Bahia, and Espirito Santo/Rio) remain threatened (Prance, 1982b). Forest destruction and loss of endemism is not limited to these coastal areas. In eastern Amazonia in the states of Tocantins, Maranhão, and Pará, and in Rondônia in southwestern Amazonia, economic development projects and human colonization are the major processes threatening further destruction of forests (Daly and Prance, 1989). The justification for preserving these and other centers of endemism will come from further documentation of the great diversity of plants and animals they contain (Prance, 1982b), especially invertebrate taxa that are not easily surveyed.

Scorpion Biodiversity and Conservation Programs

Although studies of scorpion biogeography began only a decade ago, the indications they make of endemic regions is remarkably consistent with the general pattern observed for other taxa studied intensively over a much longer period. South American tropical scorpions exhibit a high degree of endemism in major rainforest formations and localized centers of endemism (figs. 18.3, and 18.4). In both the Amazon and Atlantic forests or the Amazon and Choco forests, for

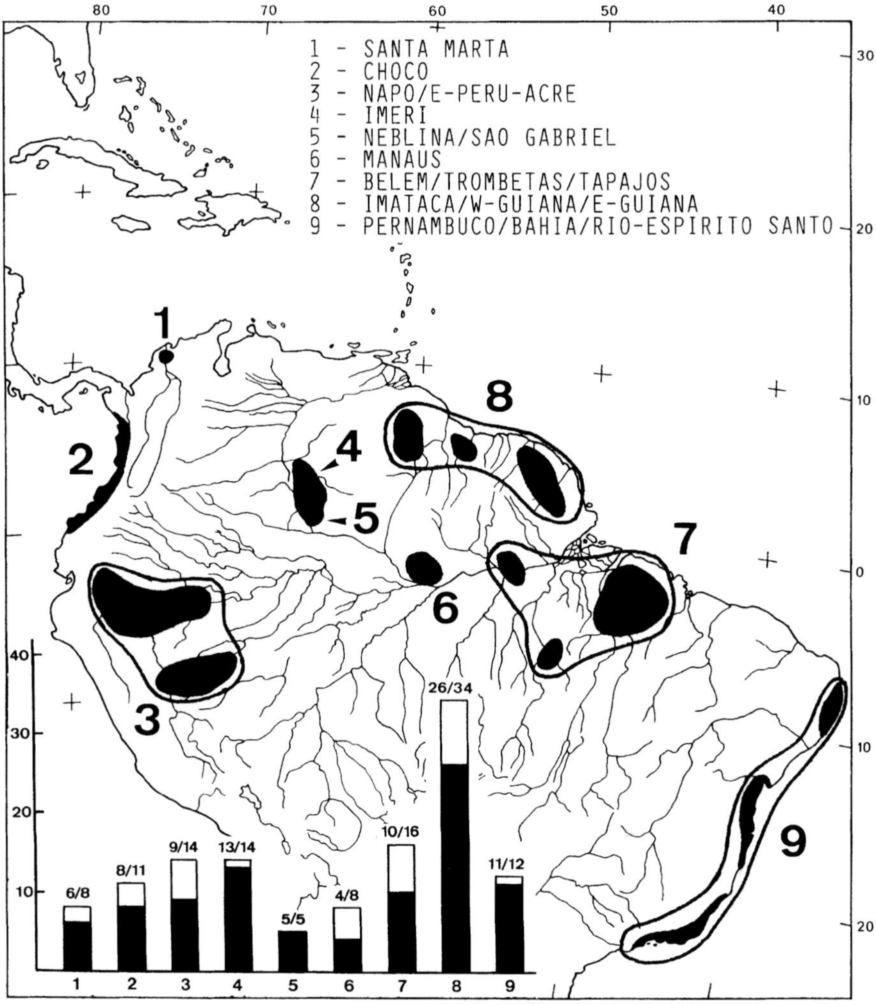

Figure 18.3 Nine centers of scorpion endemism in tropical South America. Total numbers of scorpion species in each center are indicated by bar height; the number of endemic species is represented by the black portion of each bar.

example, each species of scorpion is found only within a relatively localized region. Moreover, only four species are known to be common to the Amazon and Guayana regions (Lourenço, 1991a), with the Atlantic forest having 91.6% (11 of 12 species) endemic species and Guyana 76.4% (26 of 34 species) endemic to the region. More localized areas exhibit the following percentages of endemics: Belém/ Trombetas/ Tapajos, 62.5% (10 of 16 species), Manaus, 50% (4 of 8 species), Imeri, 92.8% (13 of 14 species), Neblina-São Gabriel, 100% (5 species), Napo/E-Peru-Acre, 71.4% (9 of 14 species), Choco, 72.7% (8 of 11 species), and Santa Marta, 75.5% (6 of 8 species) (figs. 18.3, 18.4). Altogether,

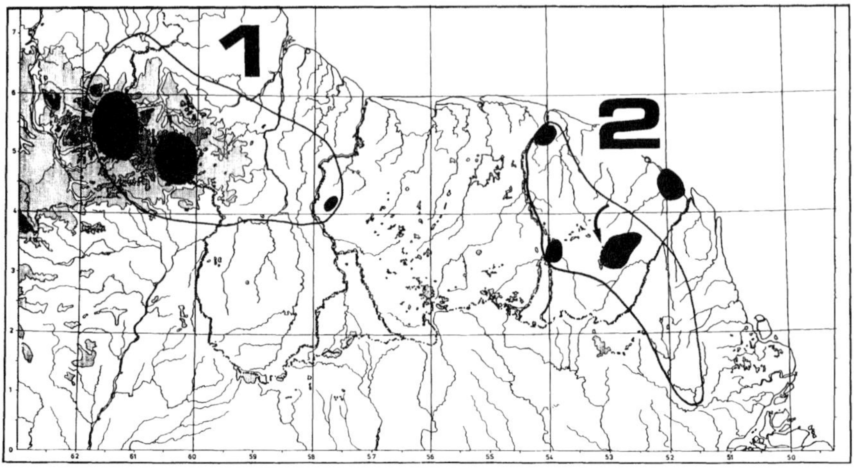

Figure 18.4 Two major endemic areas with priority for preservation in Guayana floristic province. 1, Imataca and West Guiana; 2, East Guiana. The areas defined as having the highest priority for conservation based on scorpion biodiversity are represented in black. The arrow indicates the area of Saül with highest priority for conservation in French Guiana.

with representatives from 7 families, 12 genera, and 68 species (70.6% endemic), this core area of the tropical Andes appears to be the epicenter for diversity of scorpions in the world (Lourenço, 1994).

These findings suggest that scorpions will be particularly useful for defining priority areas for conservation. Indeed, based on the diversity and biogeography of scorpions and other indicative groups of plants and animals, conservation recommendations have already been made in the Guayana region, French Guianan rainforest (fig. 18.4), and a core area of the tropical Andes (Lourenço, 1994, 1996).

CONCLUSIONS

Rainforests are the most extensive biome of tropical South America and they contain the highest diversity of plants and animals in the world. Since 1970, studies of species diversity in tropical South America have developed in parallel with efforts to apply this knowledge to conservation programs. A basic understanding of the biodiversity in different locales is now available. Because it is highly likely that only parcels of the tropics will be effectively protected, those areas containing the highest diversity should be quickly identified and set aside for protection. These studies of plant and animal endemism provide the most reliable data for determining location, size, and number of conservation parks and reserves needed to preserve the most significant areas of biodiversity. In recent years biogeographic studies of neotropical scorpions have aided in the discovery of several centers of endemism. Thus, by their unique biology, behavior, and natural history, scorpions are likely to become useful models in multidisciplinary programs to design conservation units in tropical South America.

Acknowledgments I thank Gary Polis, Jurgen Haffer, David Sissom, Philip Brownell, and Ch. P. Blanc for reviewing this chapter.

References

Ab'Saber, A.N. (1977a) Os domínios morfoclimáticos na América do Sul. Primeira aproximacão. *Geomorfologia* IGEOG-USP. 52: 1–23.

Ab'Saber, A.N. (1977b) Espacos ocupados pela expansão dos climas sêcos na América do Sul, por ocasião dos períodos glaciais quaternários. *Paleoclimas* IGEOG-USP. 3: 1–19.

Ab'Saber, A.N. (1982) The paleoclimate and paleoecology of Brazilian Amazonia. *In* Prance, G.T. (ed.) *Biological Diversification in the Tropics.* Columbia University Press, New York.

Brown, K.S. (1979) Ecologia geográfica e evolução nas florestas tropicais. Ph.D. Thesis, *Universidad Estadual de Campinas,* Sao Paulo. 265p.

Brown, K.S. (1982) Paleoecology and regional patterns of evolution in neotropical forest butterflies. *In* Prance, G.T. (ed.) *Biological Diversification in the Tropics,* pp. 255–308. Columbia University Press, New York.

Daly, D.C. and Prance, G.T. (1989) Brazilian Amazon. *In* Campbell, D.G. and Hammond, H.D. (eds.) *Floristic Inventory of Tropical Countries.* New York Botanical Garden, New York.

Eiten, G. (1974) An outline of the vegetation of South America. *Symp. 5th Congr. Intl. Primatol. Soc.* 529–545.

Endler, J.A. (1982) Pleistocene forest refuges: fact or fancy? *In* Prance, G.T. (ed.) *Biological Diversification in the Tropics,* pp. 644–657. Columbia University Press, New York.

Federov, A.A. (1966) The structure of the tropical rain forest and speciation in the humid Tropics. *J. Ecol.* 54: 1–11.

Francke, O.F. and Sissom, W.D. (1984) The life history of *Vaejovis coahuilae* Williams (Scorpiones, Vaejovidae), with comparaison of the methods used to determine the number of molts to maturity. *J. Arachnol.* 12: 1–20.

Gentry, A.H. (1988) Changes in plant community diversity and floristic composition on geographical and environmental gradients. *Ann. Mo. Botan. Garden 75:* 1–34.

Gentry, A.H. (1990) *Four Neotropical Rainforests.* Yale University Press, New Haven, CT.

Gentry, A.H. (1992) Tropical forest biodiversity: distributional patterns and their conservational significance. *Oikos* 63: 19–28.

Haffer, J. (1969) Speciation in Amazonian forest birds. *Science* 165: 131–137.

Haffer, J. (1974) Avian speciation in Tropical South America. *Nuttall Ornithol. Club,* no. 14 Cambridge, MA.

Lourenço, W.R. (1986) Diversité de la faune scorpionique de la région amazonienne; centres d'endémisme; nouvel appui à la théorie des refuges forestiers du Pléistocene. *Amazoniana* 9: 559–580.

Lourenço, W.R. (1988) Peut-on parler d'une biogéographie du scorpionisme? *C.R. Soc. Biogéogr.* 64: 139–145.

Lourenço, W.R. (1991a) La "Province" biogéographique guyanaise; étude de la biodiversité et des centres d'endémisme en vue de la conservation des patrimoines génétiques. *C.R. Soc. Biogéogr.* 67: 113–131.

Lourenço, W.R. (1991b) Biogéographie évolutive, écologie et les stratégies biodémographiques des Scorpions néotropicaux *C.R. Soc. Biogéogr.* 67: 171–190.

Lourenço, W.R. (1992) Les peuplements des Scorpions des Antilles; facteurs historiques et écologiques en association avec les stratégies biodémographiques. *Stud. Neotrop. Fauna Environ.* 27: 43–62.

Lourenço, W.R. (1994) Diversity and endemism in tropical versus temperate scorpion communities. *Biogeographica* 70(3): 155–160.

Lourenço, W.R. (1995) *Tityus fasciolatus* Pessoa, Scorpion Buthidae a traits caracteristiques d'une espece non-opportuniste. *Biogeographica* 71(2): 69–74

Lourenço, W.R. (1996) The biogeography of scorpions. *Rev. Suisse Zool.*, vol. hors serie II: 437–448.

Lourenço, W.R. and Cuellar, O. (1995) Scorpions, scorpionism, life history strategies and parthenogenesis. *J. Venom. Anim. Toxins* 1(2): 51–62.

Lourenço, W.R. and Florez, E. (1990a) Scorpions (Chelicerata) from Colombia. III. The scorpiofauna of pacific region (Choco), with some biogeographic considerations. *Amazoniana*. 11: 119–134.

Lourenço, W.R. and Florez, E. (1990b) Scorpions (Chelicerata) de Colombia. IV. Biogéographie et diversité biologique des Scorpions de Colombia, avec des commentaires sur les refuges quaternaires. *C.R. Soc. Biogéogr.* 66: 65–74.

Mori, S.A., Boom, B.M. and Prance, G. (1981) Distribution patterns and conservation of eastern Brazilian Coastal forest tree species. *Brittonia* 33: 233–245.

Noss, R.F. (1990) Indicators for monitoring biodiversity: a hierarchical approch. *Conserv. Biol.* 4: 355–364.

Pianka, E.R. (1970) On r- and K-selection. *Am. Nat.* 104: 592–597.

Polis, G.A. (1990) Ecology. *In* Polis, G.A. (ed.) *The Biology of Scorpions*, pp. 247–293. Stanford University Press, Palo Alto, CA.

Polis, G.A. and Farley, R.D. (1979) Characteristics and environmental determinants of natality, growth and maturity in a natural population of the desert scorpion, *Paruroctonus mesaensis* (Scorpionida: Vaejovidae). *J. Zool.* 187: 517–542.

Polis, G.A. and Farley, R.D. (1980) Population biology of a desert scorpion: survivorship, microhabitat and the evolution of a life history strategy. *Ecology* 61: 620–629.

Polis, G.A. and Sissom, W.D. (1990) Life history. *In* Polis, G.A. (ed.) *The Biology of Scorpions*, pp. 161–223. Stanford University Press, Palo Alto, CA.

Prance, G.T. (1973) Phytogeographic support for the theory of Pleistocene forest refuges in the Amazon Basin, based on evidence from distribution patterns in Caryocaraceae, Chrysobalanaceae, Dichapetalaceae and Lecythidaceae. *Acta Amazonica* 3: 5–28.

Prance, G.T. (1977) The phytogeographic subdivisions of Amazonia and their influence on the selection of biological reserves. *In* Prance, G.T. and Elias, T.S. (eds.) *Extinction is Forever*, pp. 195–213. New York Botanical Garden, New York.

Prance, G.T. (ed.) (1982a) *Biological Diversification in the Tropics*. Columbia University Press, New York. 714 pp.

Prance, G.T. (1982b) Forest refuges: evidence from woody angiosperms. *In* Prance, G.T. (ed.) *Biological Diversification in the Tropics*, pp. 137–158. Columbia University Press, New York.

Prance, G.T. (1985) The changing forests. *In* Prance, G.T. and Lovejoy, T.E. (eds.) *Amazonia*, pp. 146–165. Pergamon Press, Oxford.

Prance, G.T. (1990) Consensus for conservation. *Nature* 345: 384.

Prance, G.T. and Campbell, D.G. (1988) The present state of tropical floristics. *Taxon* 37: 519–548.

Richards, P.W. (1969) Speciation in the tropical rain forest and the concept of the niche. *Biol. J. Linn. Soc.* 1: 149–153.

Shorthouse, D.J. and Marples, T. (1982) The life stages and population dynamics of an arid zone scorpion *Urodacus yaschenkoi* (Birula, 1903). *Austral. J. Ecol.* 7: 109–118.

Smith, G.T. (1966) Observations on the life history of the scorpion *Urodacus abruptus*. Pocock (Scorpionida) and an analysis of its home sites. *Austral. J. Zool.* 14: 383–398.

Van der Hammen, T. (1974) The Pleistocene changes of vegetation and climate in tropical South America. *J. Biogeogr.* 1: 3–26.

Van der Hammen, T. (1982) Paleoecology of tropical South America. *In* Prance, G.T. (ed.) *Biological Diversification in the Tropics*, pp. 60–66. Columbia University Press, New York.

Vanzolini, P.E. and Williams, E.E. (1970) South American anoles: geographic differentiation and the evolution of the *Anolis chrysolepis* species group (Sauria, Iguanidae). *Arq. Zool.* 19: 1–298.

Wilson, E.O. (ed.) (1988) *Biodiversity.* National Academy Press, Washington, DC. 521 pp.

Taxonomic Index

Subject Index